THE ESSENTIAL GUIDE TO SERIAL ATA AND SATA EXPRESS

THE ESSENTIAL GUIDE TO SERIAL ATA AND SATA EXPRESS

David A. Deming

CRC Press is an imprint of the
Taylor & Francis Group, an **informa** business
AN AUERBACH BOOK

CRC Press
Taylor & Francis Group
6000 Broken Sound Parkway NW, Suite 300
Boca Raton, FL 33487-2742

First issued in paperback 2019

ISBN-13: 978-1-4822-4331-4 (hbk)
ISBN-13: 978-0-367-37831-8 (pbk)

Visit the Taylor & Francis Web site at
http://www.taylorandfrancis.com

and the CRC Press Web site at
http://www.crcpress.com

Contents

Foreword

The ATA interface is thirty years old, first introduced with the IBM PC AT in 1984. Since then, the capacity of hard disk drives has grown nearly a millionfold. This guide describes the technical innovations that have kept the ATA interface useful over all these years. The latest interface developments, SATA and SATA Express, are just the most recent improvements. There have been many things added to ATA and many things made obsolete over its long history. That is why this guide to SATA and SATA Express (by David Deming) is essential.

An example of the evolution of ATA can be seen in Chapter Three. There are seven commands to read data from a drive listed in the tables. As technology improved, new read commands were introduced. Read Sectors was the first, using the PIO data transfer method. Then came Read Multiple to cut down on system interrupts. Read DMA improved performance again, while Read Stream was an attempt to provide timely data for audio-video applications. The EXT versions of Read were developed to extend the address space, allowing larger drives. The latest new read command, Read First-Party DMA Queued, is the high-performance SATA read command, which allows multiple commands to be queued in a drive.

If Read FPDMA Queued is the latest and best Read command, then why does SATA support the older commands? A large part of ATA's success over thirty years has been the practice of backward compatibility. New features are added while the older features are maintained. The SATA interface was a necessary break in the backward compatibility of the ATA electrical interface, but it maintained compatibility for software. SATA drives can successfully process a thirty-year-old Read Sectors command, even though the command is transferred over a wholly new communication method. This book contains all the commands back to the beginning because knowledge of the history of ATA is essential.

Backward compatibility maintains the popularity of ATA but at the cost of complexity. As new features are added to ATA, the impact of the new feature on all of the existing features and functions must be assessed and documented. The layered model of SATA, described in this book, helps keep changes from propagating throughout the ATA design. For example, the SATA link layer power management system, with Partial and Slumber low power states, are used only in the SATA link layer and below. They do not affect and are not affected by the application layer power management Idle and Standby low power states.

The latest evolution in SATA is SATA Express, a system that allows the choice of either a SATA or a PCIe interfaced device to be attached to a single system connector. A laptop computer designer does not have to decide between SATA, which supports hard drives, optical drives, and solid state drives, and PCIe, with a potentially higher throughput for solid state drives. SATA Express follows the proven example of maintaining backward compatibility by supporting SATA drives, while at the same time providing the choice to use the PCIe interface.

The growth in features and capabilities of the ATA, SATA, and SATA Express interfaces matched pace with advancing computer technology, but at the cost of increasing complexity. This guidebook describing the complexities of the ever-adapting ATA interface is truly essential.

Dan Colegrove
Director of Technology (Industry Standards)
Toshiba America Electronic Components
Chairman of INCITS Technical Committee T13

Preface

Welcome to *The Essential Guide to Serial ATA and SATA Express*. This guide is part of The Data Center Technology Professional Series—a set of technology-focused implementation guides for data center hardware and software interfaces and applications.

This book focuses on the SATA storage interface. SATA is the pervasive disk storage technology in use today that is found in almost every computing application. In today's desktops and servers, SATA is integrated within the computer motherboard to reduce cost and increase functionality. Integrated SATA connections provide a simple disk storage solution where multiple disk drives can be turned into RAID storage.

In the Enterprise, SATA has been and is still being used as commodity, disk-to-disk backup, or even local replication storage. SATA disks also have sufficient performance and availability capabilities to serve applications that access files in a sequential manner.

Enterprise SATA devices have also emerged as a mid-tier of disk storage. Enterprise SATA provides high-capacity storage with mid-level performance characteristics (better than Client SATA). This storage tier is meant to fill the gap between low-cost, low-performance, commodity SATA storage and high-performance Enterprise class storage, such as Serial Attached SCSI.

The Essential Guide to Serial ATA and SATA Express is written for any technically adept person supporting, designing, developing, or deploying Enterprise or Client storage solutions. This is not a book for novices. It was written by an engineer specifically for people unafraid of digging into technical details. This book contains the necessary information and references required to architect, analyze, and troubleshoot any data center application utilizing SATA or SATA Express (SATAe) technology.

If you work in a data center or manage your company's storage resources, the chances are you will encounter storage solutions that require SATA/SATAe software or hardware. This book will assist you in any tasks associated with the installation, configuration, and troubleshooting of SATA/SATAe-based storage applications. You will learn how SATA/SATAe powers data center applications and how it influences and interacts with all protocol layers and system components, from host software drivers to the physical interface to the storage array.

If you are a test, quality, field service, failure analysis, performance, or tech support engineer, this book will assist you in your efforts. If you are a design engineer, system architect, device driver coder, or firmware writer, this is a book that will assist you with all technical details associated with programming, configuring, and architecting SATA designs and solutions. If you are a technically skilled gamer and you like to build your own systems, this book will assist you with your technical questions concerning SATA/SATAe.

When authoring books on computer technology, every technical detail in the book must be as accurate as possible. This means the content must reflect current reference materials as well as industry standards and specifications. Technical content and references were found in the following industry standards and specifications:

- Serial ATA Specification 3.2
- ATA/ATAPI-7 Standards Volume 1 and 3
- ATA Command Sets Standards (ACS)
- Advanced Host Controller Interface Specification (AHCI)
- ATA Serial Transport Standards (AST)

I think you will find this publication has most everything you will need to implement a SATA or SATA Express storage solution.

Please note: Answers to the Review Questions sections found in Chapters 2 through 6 appear in Appendix A: Chapter Review Answers, which can be found at the back of this book.

David A. Deming

Acknowledgments

As with any undertaking of this magnitude, a number of people were involved in the success of this project.

First, I would like to acknowledge the following groups and their contributions to the technical content:

- www.t13.org – ATA/SATA Standards community
- www.SATA-IO.org – Serial ATA Specification community
- www.t10.org – SCSI Standards community

These groups of individuals have the thankless task of defining the industry standards and specifications that allow all computers on the planet earth to properly interact. Without these groups, technical chaos would reign and every computer system and application would be vendor unique.

Next, I would like to thank my publishing team at DerryField Publishing Services—Theron Shreve and Marje Pollack—for their support, encouragement, and patience in completing this project. It has been a tremendous learning experience for me.

I especially want to acknowledge my wife, Jennifer, for her boundless support during my long and late hours while I worked on this project.

I would like to extend a special acknowledgment to Robert Kembel, my trusted advisor, mentor, and friend. Bob and I have been working together educating the worldwide engineering community for over 25 years. Bob, I hope you enjoy your bonus years with the grandkids. It has been an honor working with you, and I appreciate everything you contributed to our organization.

Finally, I would like to acknowledge the students I have met and with whom I have shared wisdom and knowledge. I thank you for trusting my ability to give you the most technically accurate and useful documentation of its kind.

About the Author

David Deming is the President and Chief Technologist of Solution Technology. With a degree in Electronics Engineering and more than two decades of industry experience, David has designed many courses covering a wide range of storage and networking technologies. David's organization is the industry's leading supplier of storage technology education, and David has personally trained over 50,000 engineers worldwide. David has authored over a dozen storage technology–related publications and is an expert instructional designer. He has assisted many companies with detailed job and task analysis, resulting in high-stakes certification programs and curriculum.

David is an active member of numerous industry associations and committees. In the early stages of Fibre Channel technology, David organized and coordinated the efforts of the Interoperability Testing Program for the Fibre Channel Industry Association (FCIA). As Interoperability Program Chair, David played a major role in organizing competing switch vendor efforts to create the first ever FC switch interoperability demonstration. This technology accomplishment was the centerpiece of the Storage Networking Industry Association (SNIA) Technology Center grand opening in Colorado Springs, Colorado. David also authorized the SANmark program, which became the cornerstone test cases developed and administered by the University of New Hampshire Interoperability Lab (UNH-IOL).

David architected and designed the first ever multi-vendor (heterogeneous) Fibre Channel SAN featured in Computer Technology Review, and he coordinated the industry's first and largest Interoperability Lab demonstration for Storage Networking World.

David has been a voting member of the NCITS T11 Fibre Channel standards committee and has served on T10 (SCSI & SAS) and T10.1 (SSA) standards committees. He is a founder and active member of the SNIA Education Committee and Certification Task Force for which, in 2014, he received recognition for his efforts and contributions. David has also received recognition for his volunteer efforts as President and board member of the FCIA.

Acronyms and Glossary

Bits, Bytes, Words, and Dword Format

The figure below demonstrates the relationship between bytes, words, and dwords.

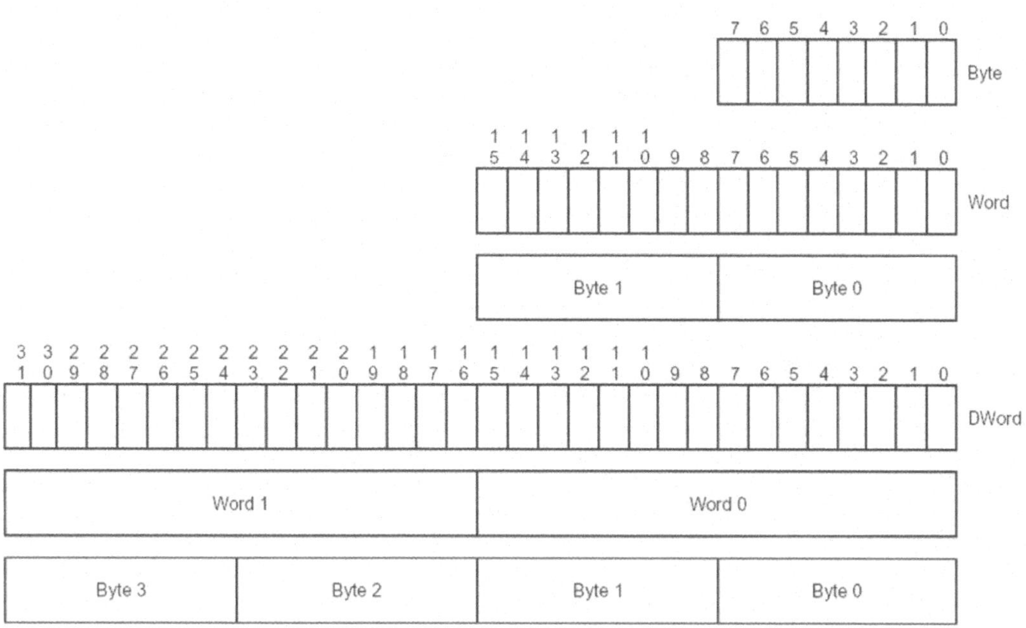

Acronyms

/	division
μA	microampere (10^{-6} amperes)
μs	microsecond (10^{-6} seconds)
A.C.	alternating current
AC	Alternating Current
ACK	acknowledge primitive

ACS	ATA Command Set
AFE	Analog Front End
AnyDword	Any Dword
ASCII	American Standard Code for Information Interchange
ATA	Advanced Technology Attachment
ATA/ATAPI-7	AT Attachment with Packet Interface - 7 standard
ATA8-ACS	AT Attachment 8 - ATA Command Set
ATAPI	Advanced Technology Attachment Packet Interface
AWG	American wire gauge
b	bit
B	byte
BBU	Battery Back-up Unit
BER	bit error rate
BERT	Bit Error Rate Tester
BIOS	Basic Input/Output System
BIST	Built-In Self-Test
CBDS	Continuous Background Defect Scanning
CDB	command descriptor block
CEM	Card Electromechanical
CFA	The CompactFlash™ Association
CIC	Compliance Interconnect Channel
CJTPAT	compliant jitter test pattern
CLTF	Closed Loop Transfer Function
CM	Common Mode
COMP	Composite Pattern
CRC	Cyclic Redundancy Check
D.C.	direct current
DAS	Device Activity Signal
dB	Decibel
DC	Direct Current
DCB	Direct Current Block
DETO	DevSleep Exit Timeout
DevSleep	Device Sleep interface power state
DEVSLP	Device Sleep Signal
DHU	Direct Head Unload
DIPM	Device Initiated Power Management
DJ	Deterministic Jitter
DMA	Direct Memory Access
DMDT	DEVSLP Minimum Detection Time
DNU	Do Not Use
DP	Device Present
DSS	Disable Staggered Spinup
DW	Dword
DWord, Dword, or dword	Double Word • 32 bits of pre-encoded bytes • 40 bits of 8b/10b encoded bytes
Dxx.y	data character
e	2.71828..., the base of the natural (i.e., hyperbolic) system of logarithms
EIA	Electronic Industries Alliance

EMI	Electromagnetic Interference
EOF	end of frame primitive
eSATA	External SATA usage model
ESD	Electrostatic Discharge
Fbaud	Frequency baud
FCOMP	Frame Composite Pattern
FER	Frame Error Rate
FIFO	First In First Out
FIS	Frame Information Structure
FPC	Flexible Printed Circuit
FS	Feature Specific
FUA	force unit access
Gbps (Gb/s)	Gigabits per second (i.e., 109 bits per second)
Gen1	generation 1 physical link rate (1.5 Gbps)
Gen2	generation 2 physical link rate (3.0 Gbps)
Gen3	generation 3 physical link rate (6.0 Gbps)
GHz	gigaherz (109 transitions per second)
GND	Ground
GPL	General Purpose Logging (see ACS-3)
GT/s	Giga Transfers per second (i.e., 109 transfers per second)
HBA	Host Bus Adapter
HBWS	High Bandwidth Scope
HDD	Hard Disk Drive
HFTP	High Frequency Test Pattern
HHD	Hybrid Hard Disk – HHD with Flash Memory
HTDP	High Transition Density Pattern
IOps (IO/s)	I/O's per second
ISI	Inter-Symbol Interference
IU	information unit
JBOD	Just a Bunch of Disks
JMD	Jitter Measurement Device
JTF	Jitter Transfer Function
JTPAT	jitter test pattern
kHz	kilohertz (103 bits per second)
Kxx.y	control character
LBA	Logical Block Address
LBP	Lone Bit Pattern
LED	Light Emitting Diode
LFSCP	Low Frequency Spectral Content Pattern
LFSR	Linear Feedback Shift Register
LFTP	Low Frequency Test Pattern
LL	Lab-Load
LSB	Least Significant Bit
LSS	Lab-Sourced Signal
LTDP	Low Transition Density Pattern
M.2	M.2 usage model
Mbaud	megabaud (106 transitions per second)
MBps (MB/s)	megabytes per second (106 bytes per second)
MDAT	Minimum DEVSLP Assertion Time

MFTP	Mid Frequency Test Pattern)
MHz	megahertz (106 bits per second)
Micro SATA	Micro SATA connector
MicroSSD	MicroSSD usage model
ms	millisecond (10^{-3} seconds)
mSATA	mSATA usage model
MSB	Most Significant Bit
mV	millivolt (10^{-3} volts)
N/A	not applicable
NAK	negative acknowledge primitive
NCQ	Native Command Queuing
nF	nanofarad (10^{-9} Farads)
ns	nanosecond (10^{-9} seconds)
ODD	Optical Disk Drive
OOB	Out-of-Band Signaling
PATA	Parallel ATA
PCB	Printed Circuit Board
PCIe	Peripheral Component Interconnect Express
Phy (PHY)	Physical
PIO	Programmed Input/Output
PLL	Phase Lock Loop
P-P	peak-to-peak
ppm	parts per million (10^{-6})
PRD	Physical Region Descriptor
ps	picosecond (10^{-12} seconds)
Qword	Quad Word
RA	Right Angle
RAID	Redundant Array of Independent Disks
RCDT	rate change delay time
RD–	Running Disparity negative
RD+	Running Disparity positive
RJ	Random Jitter
RRDY	receiver ready primitive
RSN	Reset Speed Negotiation
Rx	Receiver
s	second (unit of time)
SAS	Serial Attached SCSI
SATA	Serial ATA
SATA Express	SATA Express usage model
SCSI	Small Computer System Interface family of standards
SDB	Set Device Bits
SEMB	Serial ATA Enclosure Management Bridge
SEP	Storage Enclosure Processor
SES	SCSI Enclosure Services
SFF	Small Form Factor
SMA	SubMiniature version A
SMART	Self-Monitoring, Analysis, and Reporting Technology
SMT	Surface Mount Technology
SOF	start of frame primitive

SRIS	Separate RefClk Independent SSC
SRST	Software reset
SSC	Spread Spectrum Clocking
SSD	Solid State Drive
SSHD	Solid State Hybrid Device
SSOP	Simultaneous Switching Outputs Pattern
SSP	Software Settings Preservation
TCTF	Transmitter Compliance Test Pattern
TDR	Time Domain Reflectometry
TIA	Time Interval Analyzer
TJ	Total Jitter
Tx	Transmitter
UHost	Universal Host
UI	Unit Interval
μs	microsecond (i.e., 10^{-6} seconds)
USM	Universal Storage Module
UUT	Unit Under Test
V	volt
VNA	Vector Network Analyzer
x	multiplication
XOR	Exclusive Logical OR

Glossary

28-bit command:	A command that uses Feature (7:0), Count (7:0), LBA (27:0), Device (15:8), and Command (7:0) to specify its arguments.
48-bit command:	A command that uses Feature (15:0), Count (15:0), LBA (47:0), Device (15:8), and Command (7:0) to specify its arguments.
acoustics	Measurement of airborne noise emitted by information technology and telecommunications equipment [ISO 7779:1999(E)]
Active mode	The power condition specified by the PM0: Active state.
Active Port	The active port is the currently selected host port on a Port Selector.
application	Software that is dependent on the services of an operating system.
application client	An object in the host that is the source of commands and device management functions (see ATA8-AAM).
ASCII Character	Designates 8-bit value that is encoded using the ASCII Character set.
ATA (AT Attachment)	ATA defines the physical, electrical, transport, and command protocols for the internal attachment of storage devices to host systems.
ATA (AT Attachment) device	A device implementing the General feature set.
ATA string	A set of ASCII characters.
ATA/ATAPI-4 device	A device that complies with ANSI INCITS 317-1998, the AT Attachment Interface with Packet Interface Extensions-4.
ATA/ATAPI-5 device	A device that complies with ANSI INCITS 340-2000, the AT Attachment with Packet Interface-5.
ATA/ATAPI-6 device	A device that complies with ANSI INCITS 361-2002, the AT Attachment with Packet Interface-6.
ATA/ATAPI-7 device	A device that complies with this standard.
ATA-1 device	A device that complied with ANSI X3.221-1994, the AT Attachment Interface for Disk Drives. ANSI X3.221-1994 has been withdrawn.
ATA-2 device	A device that complied with ANSI X3.279-1996, the AT Attachment Interface with Extensions. ANSI X3.279-1996 has been withdrawn.
ATA-3 device	A device that complies with ANSI X3.298-1997, the AT Attachment-3 Interface. ANSI X3.298-1997 has been withdrawn.
ATAPI	A device implementing the PACKET feature set (AT Attachment Packet Interface).
ATAPI device	A device implementing the Packet Command feature set.
AU (Allocation Unit)	The minimum number of logically contiguous sectors on the media as used in the Streaming feature set. An Allocation Unit may be accessed with one or more requests.
AV (Audio-Video)	Audio-Video applications use data that is related to video images and/or audio. The distinguishing characteristic of this type of data is that accuracy is of lower priority than timely transfer of the data.
backchannel	When transmitting a FIS, the backchannel is the receive channel.
Background Activities	Activities initiated by a command that occur after command completion has been reported.
BER (bit error rate)	The statistical probability of a transmitted encoded bit being erroneously received in a communication system.
BIOS (Basic Input/ Output System)	An initial application client that is run when power is applied. The primary function of BIOS is to initialize various components (e.g., storage devices).

bit synchronization	The state in which a receiver has synchronized the internal receiver clock to the external transmitter and is delivering retimed serial data.
block erase	An internal media operation supported by some devices that sets a block of data to a vendor specific value (i.e., replacing previous data) and may precondition the media for write operations.
bus release	For devices implementing overlap, the term bus release is the act of clearing both DRQ and BSY to zero before the action requested by the command is completed. This allows the host to select the other device or deliver another queued command.
byte	A sequence of eight contiguous bits considered as a unit.
byte count	The value placed in the Byte Count register by the device to indicate the number of bytes to be transferred during this DRQ assertion when executing a PACKET PIO data transfer command.
byte count limit	The value placed in the Byte Count register by the host as input to a PACKET PIO data transfer command to specify the maximum byte count that may be transferred during a single DRQ assertion.
cache	A data storage area outside the area accessible by application clients that may contain a subset of the data stored in the non-volatile data storage area.
CFA (CompactFlash™ Association)	The CompactFlash™ Association which created the specification for compact flash memory that uses the ATA interface.
CFA device	A device implementing the CFA feature set. CFA Devices may implement the ATA8-APT transport or the ATA8-AST transport
CFast™ Device	A CF form factor device that conforms to the SATA device requirements in this standard, implements the ATA8-AST transport and does not implement the ATA8-APT transport. CFast devices may support the CompactFlash™ feature set.
check condition	For devices implementing the PACKET Command feature set, this indicates that an error or exception condition has occurred.
CHS (cylinder-head-sector)	An obsolete method of addressing the data on the device by cylinder number, head number, and sector number.
circular buffer	A buffer that is filled starting at the first byte, continuing to the last byte, and then wrapping to store data in the first byte of the buffer again.
code violation	In a serial interface implementation, a code violation is an error that occurs in the decoding of an encoded character. (See Volume 3, Clause 15.)
command aborted	Command completion with ABRT set to one in the Error register and ERR set to one in the Status register.
command acceptance	A command is considered accepted whenever the currently selected device has the BSY bit cleared to zero in the Status register and the host writes to the Command register. An exception exists for the DEVICE RESET command (see Clause 6) In a serial implementation; command acceptance is a positive acknowledgment of a host to device register FIS.
Command Block registers	Interface registers used for delivering commands to the device or posting status from the device. In a serial implementation, the command block registers are FIS payload fields.
command completion	Command completion is the completion by the device of the action requested by the command or the termination of the command with an error, the placing of the appropriate error bits in the Error register, the placing of the appropriate status bits in the Status register, the clearing of both BSY and DRQ to zero, and Interrupt pending.
command packet	A data structure transmitted to the device during the execution of a PACKET command that includes the command and command parameters.

command released	When a device supports overlap or queuing, a command is considered released when a bus release occurs before command completion.
COMRESET	A commanded hardware reset in the Serial ATA transport (see SATA 2.6).
Control Block registers	In a parallel implementation, interface registers used for device control and to post alternate status. In a serial interface implementation, the logical field of a FIS corresponding to the Device Register bits of a parallel implementation.
control character	In a serial interface implementation, an encoded character that represents a non-data byte. (See Serial ATA Volume 3, Clause 15.)
CRC (Cyclical Redundancy Check)	A means used to check the validity of certain data transfers.
Cylinder High register	The name used for the LBA High register in previous ATA/ATAPI standards.
Cylinder Low register	The name used for the LBA Mid register in previous ATA/ATAPI standards.
data character	In a serial interface implementation, an encoded character that represents a data byte. (See Serial ATA Volume 3, Clause 15.)
Data Set	A set of LBA ranges used by the device as a single group.
data-in	The protocol that moves data from the device to the host. Such transfers are initiated by READ commands.
data-out	The protocol that moves data from the host to the device. Such transfers are initiated by WRITE commands.
Delayed LBA	Any sector for which the performance specified by the Streaming Performance Parameters log is not valid.
device	A storage peripheral. Traditionally, a device on the interface has been a hard disk drive, but any form of storage device may be placed on the interface provided the device adheres to this standard.
device selection	In a parallel implementation, a device is selected when the DEV bit of the Device register is equal to the device number assigned to the device by means of a Device 0/Device 1 jumper or switch, or use of the CSEL signal. In a serial implementation, the device ignores the DEV bit; the host adapter may use this bit to emulate device selection.
disparity	The difference between the number of ones and the number of zeros in an encoded character. (See Serial ATA Volume 3, Clause 15.)
DMA (direct memory access) data transfer	A means of data transfer between device and host memory without host processor intervention.
don't care	A term to indicate that a value is irrelevant for the particular function described.
driver	The active circuit inside a device or host that sources or sinks current to assert or negate a signal on the bus.
DRQ (Data Request) data block	A number of logical sectors with available status when using either the PIO Data-In command protocol or the PIO Data-Out command protocol.
Dword or dword or DWord	A sequence of four contiguous bytes considered as a unit.
elasticity buffer	In a serial interface implementation, a portion of the receiver where character slipping and/or character alignment is performed.
encoded character	In a serial interface implementation, the output of the 8b10b encoder. (See Serial ATA Volume 3, Clause 15.)
First-party DMA access	A method by which a device accesses host memory. First-party DMA differs from DMA in that the device sends a DMA Setup FIS to select host memory regions, whereas for DMA, the host configures the DMA controller.

FIS (Frame Information Structure)	A data structure and is the payload of a frame and does not include the SOF primitive, CRC, and EOF primitive.
frame	A unit of information exchanged between the host adapter and a device. A frame consists of an SOF primitive, a Frame Information Structure, a CRC calculated over the contents of the FIS, and an EOF primitive.
free fall	A vendor-specific condition of acceleration.
FUA (Forced Unit Access)	Forced Unit Access requires that user data be transferred to or from the device media before command completion even if caching is enabled.
Gen1 DWORD Time	The time it takes to transmit a 40-bit encoded value at 1.5 Gb/Sec.
hardware reset	The routine performed by a device after a hardware reset event, as defined in ATA8-AAM.
host	The computer system executing the software BIOS and/or operating system device driver controlling the device and the adapter hardware for the ATA interface to the device.
host adapter	The implementation of the host transport, link, and physical layers.
host interface	The service delivery subsystem (see ATA8-AAM).
Idle mode	The power condition specified by the PM1 Idle state.
Interrupt Pending	In a parallel implementation, an internal state of a device. In this state, the device asserts INTRQ if nIEN is cleared to zero and the device is selected. In a serial implementation, the Interrupt Pending state is an internal state of the host adapter. This state is entered by reception of a FIS with the I field set to one. (See Volume 3, Clause 16.)
Invalid LBA	An LBA that is greater than or equal to the largest value reported in IDENTIFY DEVICE data words 60..61, IDENTIFY DEVICE data words 100..103, or IDENTIFY DEVICE data words 230..233.
LBA (logical block address)	The virtual addressing of data on the device that masks the devices physical addressing (e.g., cylinders, heads, and sectors).
LFSR	Linear Feedback Shift Register.
link	The link layer manages the Phy layer to achieve the delivery and reception of frames.
log	A named sequence of one or more log pages.
log address	A numeric value that a log command uses to identify a specific log.
log command	A SMART READ LOG command, SMART WRITE LOG command, or GPL feature set command.
log page	A 512-byte block of data associated with a log.
logical block	See logical sector—usually 512, 520, 528, or 4096 bytes.
logical sector	A uniquely addressable set of 256 words (512 bytes).
LSB (least significant bit)	In a binary code, the bit or bit position with the smallest numerical weighting in a group of bits that, when taken as a whole, represent a numerical value (e.g., in the number 0001b, the bit that is set to one).
Master Password Capability	Indicates if the Master password may be used to unlock the device.
Media	The material on which user data is stored.
Media Access Command	Any command that causes the device to access non-volatile media.
MSB (most significant bit)	In a binary code, the bit or bit position with the largest numerical weighting in a group of bits that, when taken as a whole, represents a numerical value (e.g., in the number 1000b, the bit that is set to one).

native max address	The highest LBA that a device accepts as reported by DEVICE CONFIGURATION IDENTIFY data or as reduced by the DEVICE CONFIGURATION SET command (i.e., the highest LBA that is accepted by a device using the SET MAX ADDRESS command or, if the 48-bit Address feature set is supported, then the highest value accepted by a device using the SET MAX ADDRESS EXT command).
Non-Volatile cache	Cache that retains data through all reset events (e.g., power-on reset). Non-volatile cache is a subset of the non-volatile media.
Non-Volatile Write cache	Write cache that retains data through all power and reset events. NOTE: Industry practice could result in conversion of a vendor-specific bit, byte, field, or code value into a defined standard value in a future standard.
organizationally unique identifier (OUI)	A numeric identifier that is assigned by the IEEE such that no assigned identifiers are identical. OUI is equivalent to company_id or IEEE company_id. The numeric identifier is called an OUI when it is assigned by the IEEE. The IEEE maintains a tutorial describing the OUI at http://standards.ieee.org/regauth/oui/.
overlap	A protocol that allows devices that require extended command time to perform a bus release so that commands may be executed by the other device (if present) on the bus.
packet delivered command	A command that is delivered to the device using the PACKET command via a command packet that contains the command and the command parameters. See also register delivered command.
partition	A range of LBAs specified by an application client.
Password Attempt Counter Exceeded	IDENTIFY DEVICE, word 128, bit 4.
PATA (Parallel ATA) device	A device implementing the parallel ATA transport (see ATA8-APT).
phy	Physical layer electronics. (See Serial ATA Volume 3, Clause 14.)
physical sector	A group of contiguous logical sectors that are read from or written to the device media in a single operation.
PIO (programmed input/output) data transfer	PIO data transfers are performed by the host processor utilizing accesses to the Data register.
power condition	One of the following power management substates: Idle_a, Idle_b, Idle_c, Standby_y, or Standby_z
power cycle	The period from when power is removed from a host or device until the subsequent power-on event (see ATA8-AAM).
power-on reset	The host-specific routine performed by the host or the routine performed by a device after detecting a power-on event (see ATA8-AAM).
primitive	In a serial interface implementation, a single DWORD of information that consists of a control character in byte 0 followed by three additional data characters in byte 1 through 3.
queued	Command queuing allows the host to issue concurrent commands to the same device. Only commands included in the Overlapped feature set may be queued. In this standard, the queue contains all commands for which command acceptance has occurred but command completion has not occurred.
Queued Command	A NCQ command that has reported command acceptance but not command completion.
QWord	A sequence of eight contiguous bytes considered as a unit.
read command	A command that causes the device to transfer data from the device to the host. The following commands are read commands: READ DMA, READ DMA EXT, READ DMA QUEUED, READ FPDMA QUEUED, READ MULTIPLE, READ MULTIPLE EXT, READ SECTOR(S), READ SECTOR(S) EXT, READ STREAM EXT, READ STREAM DMA EXT, READ VERIFY SECTOR(S), or READ VERIFY SECTOR(S) EXT.

read stream command	A command that causes the device to transfer data from the device to the host. The following commands are read stream commands: READ STREAM EXT and READ STREAM DMA EXT.
register	A register may be a physical hardware register or a logical field.
register delivered command	A command that is delivered to the device by placing the command and all of the parameters for the command in the device Command Block registers. See also packet delivered command.
register transfers	The host reading and writing any device register except the Data register. Register transfers are 8 bits wide.
released	In a parallel interface implementation, indicates that a signal is not being driven. For drivers capable of assuming a high-impedance state, this means that the driver is in the high-impedance state. For open-collector drivers, the driver is not asserted.
SATA (Serial ATA) device	A device implementing the serial ATA transport (see ATA8-AST).
SCT Command	A command that writes to the SCT command/status log.
SCT Status	A command that reads from the SCT command/status log.
sector	A uniquely addressable set of 256 words (512 bytes).
Sector Number register	The LBA Low register in previous ATA/ATAPI standards. Sector Number register: The LBA Low register in previous ATA/ATAPI standards.
Security Level	See Master Password Capability.
Serial ATAPI device	A device implementing the serial ATA transport (see ATA8-AST) and the PACKET feature set.
Shadow Command Block	In a serial interface implementation, a set of virtual fields in the host adapter that map the **Command** Block registers defined at the command layer to the fields within the FIS content.
Shadow Control Block	In a serial interface implementation, a set of virtual fields in the host adapter that map the **Control** Block registers defined at the command layer to the fields within the FIS content.
signature	A unique set of values placed in the Command Block registers by the device to allow the host to distinguish devices implementing the PACKET Command feature set from those devices not implementing the PACKET Command feature set.
signed	A value that is encoded using two's complement.
Sleep mode	The power condition specified by the PM3: Sleep state.
SMART	Self-Monitoring, Analysis, and Reporting Technology for prediction of device degradation and/or faults. For prediction of device degradation and/or faults.
software reset	The routine performed by a device after a software reset event, as defined in ATA8-AAM. The software reset routine includes the actions defined in ATA8-AAM, this standard, and the applicable transport standards.
spin-down	The process of bringing a rotating media device's media to a stop.
spin-up	The process of bringing a rotating media device's media to operational speed.
Standby mode	The power condition specified by the PM2: Standby state.
Stream	A set of operating parameters specified by a host using the CONFIGURE STREAM command to be used for subsequent READ STREAM commands and WRITE STREAM commands.
transport	The transport layer manages the lower layers (link and Phy) as well as constructing and parsing FISes.
Ultra DMA burst	An Ultra DMA burst is defined as the period from an assertion of DMACK—to the subsequent negation of DMACK—when an Ultra DMA transfer mode has been enabled by the host.

unaligned write	A write command that does not start at the first logical sector of a physical sector or does not end at the last logical sector of a physical sector.
unit attention condition	A state that a device implementing the PACKET Command feature set maintains while the device has asynchronous status information to report to the host.
unrecoverable error	When the device sets either the ERR bit or the DF bit to one in the Status register at command completion.
user data	Data that is transferred between the application client and the device using read commands and write commands.
user data area	Any area of the device's media that stores user data and is addressable by the host from LBA 0 to DEVICE CONFIGURATION IDENTIFY data words 3..6.
vendor specific	Bits, bytes, fields, and code values that are reserved for vendor-specific purposes. These bits, bytes, fields, and code values are not described in this standard, and implementations may vary among vendors. This term is also applied to levels of functionality whose definition is left to the vendor.
Volatile Cache	Cache that does not retain data through power cycles.
word	A sequence of two contiguous bytes considered as a unit.
write command	A command that causes the device to transfer data from the host to the device. The following commands are write commands: SCT Write Same, WRITE DMA, WRITE DMA EXT, WRITE DMA FUA EXT, WRITE FPDMA QUEUED, WRITE MULTIPLE, WRITE MULTIPLE EXT, WRITE MULTIPLE FUA EXT, WRITE SECTOR(S), WRITE SECTOR(S) EXT, WRITE STREAM DMA EXT, or WRITE STREAM EXT.
write stream command	A command that causes the device to transfer data from the host to the device. The following commands are write stream commands: WRITE STREAM DMA EXT and WRITE STREAM EXT.
WWN (worldwide name)	A 64-bit worldwide unique name based upon a company's IEEE identifier. See IDENTIFY DEVICE Words (108:111).

Chapter 1

Introduction

Objectives

This chapter provides an introduction to the Serial ATA and SATA Express storage technologies. You will learn about the following topics:

1. Where it all started, including a brief introduction to the Advanced Technology Architecture (ATA) and how it led to SATA technology.
2. The key advantages of serial storage interfaces and how they are deployed in today's servers and data centers.
3. The key technological advantages of SATA and the goals and objectives of the original governing body.
4. An introduction to SATA Express (SATAe) and how all Peripheral Component Interconnect (PCI)-based storage alternatives may impact today's SATA architecture.
5. The eventual migration path from ATA-based software drivers to Non-Volatile Memory Express-based software and how PCIe-based storage impacts all storage devices.
6. Basic storage interface evolution, including all SATA alternatives as they relate to capacity, cost, performance, and reliability.
7. SATA markets and applications and SATA position within storage tiers.
8. SATA connectivity, including port multipliers and port selectors and how they can be used to construct high-performance storage arrays.
9. New SATA form factors and connection alternatives, including Micro SATA, mSATA, M.2, SATA Universal Storage Module, microSSD, and SATA Express.

1.1 Introduction to SATA

In the 1990s, parallel interfaces such as IBM's PC-AT, IDE, and the Small Computer System Interface (SCSI) bus provided host computers with storage attachment capabilities for over a decade. Although today these technologies mostly serve for historical context, the ATA and SCSI architectures and

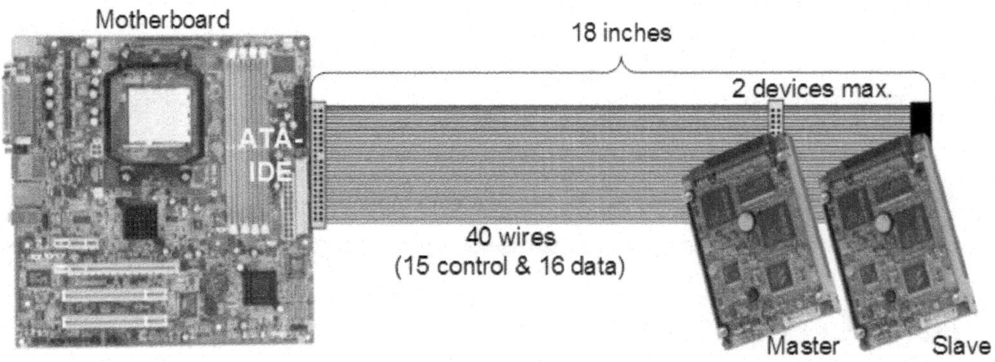

Figure 1.1 ATA configuration—where it all started.

software interfaces for storage devices have been preserved, enhanced, and mapped to serial transport alternatives, including SATA, Serial Attached SCSI (SAS), Fibre Channel (FC), InfiniBand (IB), PCI Express (PCIe), and Ethernet (iSCSI). SATA technology is based upon the original parallel ATA architecture, which allowed for two disk devices on an 18-inch-long 40-pin daisy-chained cable (Figure 1.1).

At the turn of the century, a group of computer industry leaders came together to develop the next generation disk storage interface based upon serial interface technology and the current ATA command set. The original SATA architecture mainly consisted of a serial hard disk drive (HDD) interface intended to provide a scalable high-capacity/high-performance client storage architecture for PC-Desktop-Server applications. During the spring 2000 Intel Developer Forum (IDF), the formation of a Serial ATA Working Group was announced to begin development of a 1.5-Gb/s serial interface compatible with existing ATA drivers and software. The marketing efforts of five major participating companies eventually formed the SATA International Organization (SATA-IO) and delivered the first Serial ATA 1.0 Specification on August 29, 2001. Today, the Serial ATA Specification stands at Revision 3.2 (released: August 2013) and outlines SATA implementations with link speeds up to 6 Gb/s, along with a new connectivity alternative that allows direct attachment to the PCI Express (PCIe) bus (www.sata-io.org).

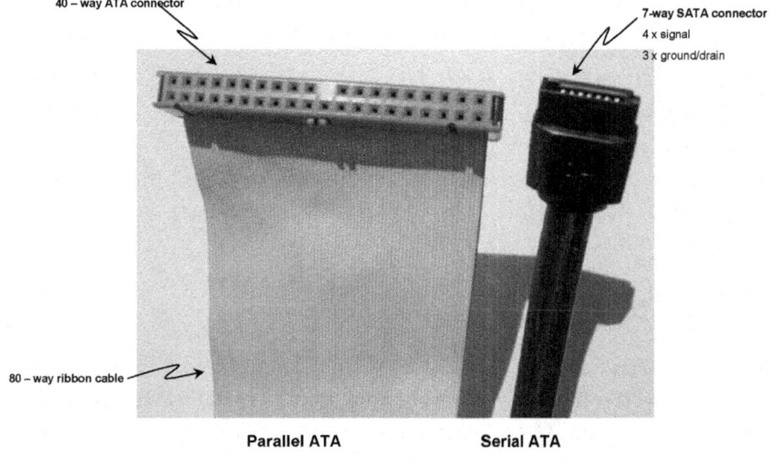

Figure 1.2 Parallel ATA versus SATA cabling.

1.1.1 Serial Interface Advantage

The SATA architecture, as the name denotes, is a serial technology. Serial interfaces dramatically reduce the number or wires needed to carry information between a host and a storage device. SATA technology replaced the old parallel physical cable plant with cabling and connector alternatives that made design, installation, and maintenance extremely easy. The parallel ATA 18-inch cable length limitation and two-device maximum was a serious issue that complicated peripheral expansion designs, making some internal drive configurations impossible to implement.

Figure 1.2 shows the significantly reduced form factor of the SATA connector and cable.

SATA Cabling

The SATA architecture replaces the parallel ribbon cable with a thin, flexible cabling that can be up to 1 meter in length. The serial cable is smaller and easier to route inside the chassis. The small-diameter cable can help improve air flow inside the PC system chassis and will facilitate the designs of smaller PC or hand-held systems.

The low pin count SATA connector eliminates the need for large cumbersome failure-prone connectors that consume valuable printed circuit board real estate.

SATA configurations are all point-to-point serial links that allow users to attach as many SATA devices as SATA connectors on the host motherboard (or SATA host adapter) (see Figure 1.3).

SATA characteristics:

- The original focus was to be mainly implemented in server and host systems for high-capacity storage.
  - Initial interfaces ran at 1.5 Gb/s (150MB/s) with 3 Gb/s upgrades.
  - Today's interfaces run at 6 Gb/s (600MB/s) and are compatible with all slower interface speeds.
- SATA was originally designed to be 100% software compatible with legacy ATA drivers.
  - Legacy ATA driver compatibility will sacrifice SATA queuing capabilities and other SATA features/enhancements (e.g., power management, security, etc.).

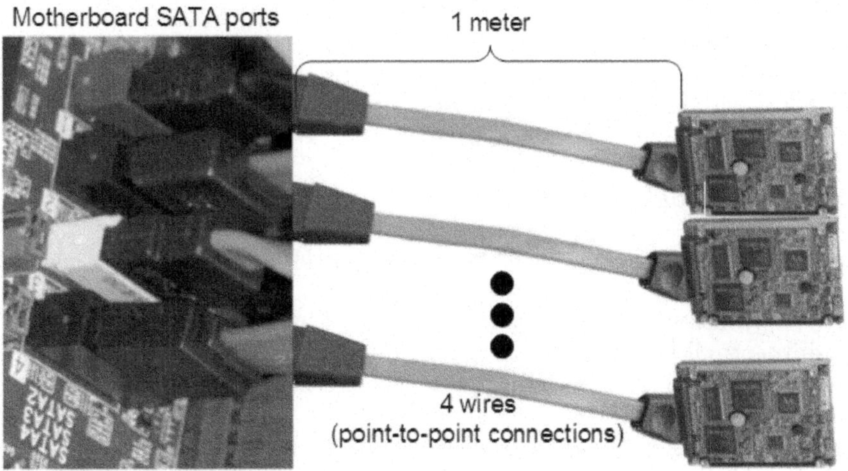

Figure 1.3 SATA configuration is point-to-point.

- SATA will continue to play a fundamental role in mobile storage for the near future.
- The SATA physical interface will most likely achieve obsolescence around 2020.
    - The SATA physical interface will eventually be replaced by PCI Express connectivity.
    - However, SATA will continue to be used in low-cost computing applications requiring internal storage with reliable external storage alternatives.

1.2 The Move to Serial Technology

The trend towards serial interfaces in all areas of data communications is a reality that is here to stay for a very long time. USB and Firewire provide serial links to connect external peripheral devices to PCs and mobile computing platforms. Networks have mostly used serial links, Ethernet being the most commonly deployed with speeds ranging from 10 up to 10,000 Megabit/s. The Fibre Channel interface provided a high-performance 4-Gb/s serial disk drive attachment in conjunction with a switched fabric with speeds up to 16 Gb/s. The PCI bus even moved to serial technology (i.e., PCI Express) and is now starting to impact the storage device industry.

In general, serial interfaces provide a number of compelling benefits over parallel busses:

- Fewer conductors are required, making smaller cables and connectors feasible.
- Greater distances can be achieved using serial links.
- It is possible to build complex topologies using switches, hubs, and loops.
- Converting a serial bit stream from a copper medium to optical and back to copper again is viable with serial links.
- Serial links can easily provide full-duplex communications, whereas their parallel counterparts are half duplex or unidirectional.
- Serial interfaces enable high availability designs, fault isolation, and cable management.

Serial interfaces are common in all applications, from mobile to Enterprise . . .

- USB 1, 2, and 3
- 1394 (Firewire, iLink)
- Ethernet: 10, 100, and 1000 BASE-T, and 10Gb Ethernet
- Fibre Channel and Fibre Channel over Ethernet (FCoE)
- Serial Attached SCSI (SAS)
- PCI Express (PCIe)
- InfiniBand and RDMA over Converged Ethernet (RoCE)

1.2.1 Serial ATA Goals and Objectives

Primary objectives for the new serial interface, as defined by the SATA-IO (www.sata-io.org):

- ✓ Provide a technology whose performance can be scaled over several generations to accommodate disk drive and other peripheral attachment for the next 10 years
- ✓ Require the absolute minimum change to existing PC implementations in order to support the new technology
- ✓ Provide a greater degree of reliability and integrity than that delivered by parallel ATA
- ✓ Define a publicly accessible standard interface that promotes wide adoption and interoperability
- ✓ Address low power and power management requirements for mobile devices

- ✓ Eliminate the requirement for the device "jumper" used by parallel ATA
- ✓ Provide features and characteristics that enable the construction of drive arrays and thus help drive SATA into new markets (Enterprise)

1.2.2 SATA Benefits

The original Serial ATA 1.0 Specification defined a direct replacement for parallel ATA in desktop and laptop applications (www.sata-io.org). The many benefits of SATA include the following:

- A performance roadmap specifying three link speed generations providing 150, 300, and 600 Megabyte/s (MB/s) data transfer capability.

Name	Speed	Date	Units	Description	Data Rate
G1	1.5	2000	Gb/s	1st Generation 10b bit rate	150 MB/s
G2	3.0	2008	Gb/s	2nd Generation 10b bit rate	300 MB/s
G3	6.0	2012	Gb/s	3rd Generation 10b bit rate	600 MB/s

- SATA originally had to have cost parity with existing parallel ATA.
    - Today SATA disks are the only alternative for PC/Laptop/Desktop storage.
    - Conversely, SAS will mainly be used in high-performance servers and Enterprise class storage arrays.
- Cables that are thin and flexible, permitting much easier handling and routing.
    - Air flow obstruction problems are also significantly reduced.
    - The maximum cable length has been extended to 1 meter.
- Reduced voltage levels on the serial link.
- Only four signal wires are required for a Serial ATA link.
    - This results in considerably smaller connectors, less traces to be routed across PCBs, and fewer pins on the protocol chips that implement the interface.
    - A considerable saving in board space can be achieved; Serial ATA requires about 25% of the space required by Parallel ATA.
- Defines a connector suitable for blind mating and hot plugging.
    - This is required for hot-plug applications in drive arrays.
- SATA deployment can be totally transparent to application and operating systems.
    - To take full advantage of all the additional features offered by Serial ATA, enhancements to BIOSs and device drivers will need to be made.
- Two low power modes are available.
    - These are in addition to other power-saving modes, such as spinning down the disk, which are implemented in ATA drives.
    - This is of special significance in mobile/laptop applications to help prolong battery life.

1.3 Almost Exactly 10 Years Later . . .

Almost exactly 10 years after the initial SATA devices started to materialize (2002), the SATA-IO devised a method to attach a SATA protocol-based nonvolatile storage device directly to the PCI Express system bus, hence the name SATA Express (SATAe). This new alternative leverages two proven technologies: the PCIe bus for connectivity and the SATA protocol, which provides the ATA command set and software interface. SATAe and flash-based storage provides a whole new level of creativity for future designs of most any application, from mobile devices to Enterprise class storage arrays.

1.3.1 Introduction to SATA Express

SATA Express is purely PCIe when it comes to the physical connections and transport protocols, as demonstrated in Figure 1.4.

- There is no SATA link or transport layer, so there's no translation overhead.
- SATA Express is the standardization of PCIe as an interface for a client storage device.
  - Standards provide interoperability and interchangeability, meaning that devices from different manufacturers work together and can be interchanged with no loss of functionality.
  - To achieve these goals, SATA Express needs standard connectors and common operating system drivers.

The SATA protocol stack is made up of SATA Specification-defined Transport, Link, and Phy layers. These layers work together to facilitate information movement between the Advanced Host Channel Interface (ACHI) driver and the physical device ports. SATA Express simply replaces the SATA layers and physical components with PCI Express Transaction, Link, and Phy layers and PCIe physical components. A major difference is the number of available ports (or PCIe Lanes), in that SATA will generally utilize only one port, and SATAe devices could have up to four ports (Lanes). From a market perspective, legacy SATA–connected devices will provide connectivity to both HDD and SSDs, and SATAe devices will most likely only be used on SSDs.

Both SATA and SATAe utilize the ATA command set, and the Application, operating system, and file system components are mostly ignorant as to which hardware is being used. Although the application and AHCI drivers are the same, the hardware will required a PCIe driver for SATAe devices.

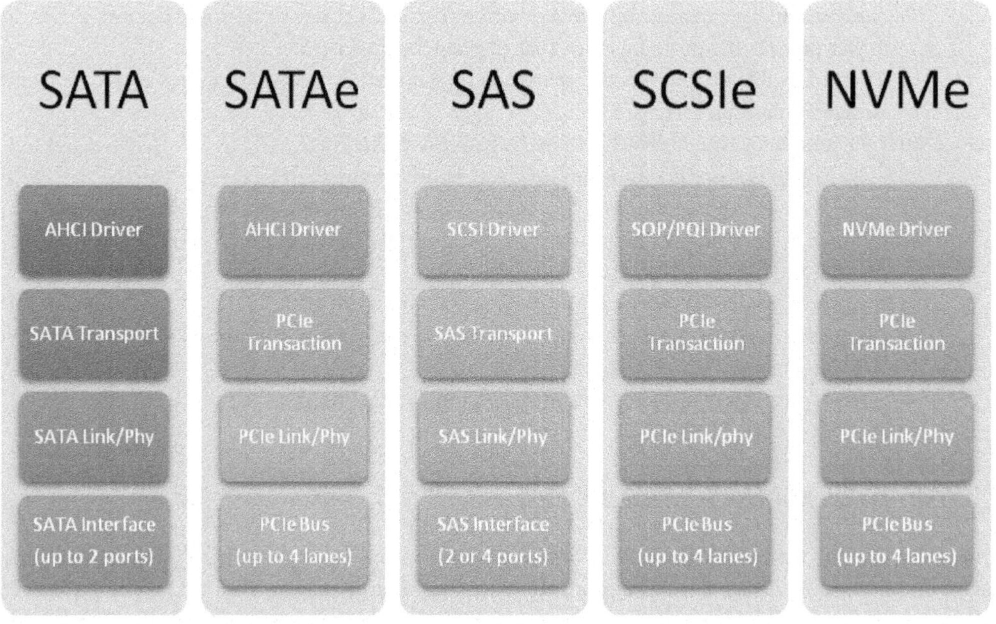

Figure 1.4 SATA, SATA Express, SAS, and NVM Express stacks.

Just as the SATAe alternative replaced SATA hardware with PCIe hardware, the same can be true of the SAS. One difference between SAS and SATA is found in the device driver, which reflects the use of the SCSI commands versus ATA. SCSI Express will be the Enterprise class PCIe-based storage device alternative, whereas SATAe will satisfy Client class storage requirements.

As operating systems vendor NVM Express adoption accelerates, legacy SATA and even newer SATA Express SSDs will fade away and be replaced by PCIe-based storage technology. The NMVe architecture provides comparable command functionality as a SATA command set—the bandwidth of PCIe multi-lane connections—and was purposely built for SSD and flash technology. This combination provides extremely low latency I/O, which satisfies many I/O-intensive applications requirements.

The SATA-IO defined SATAe host and device connectors.

- Both connectors are slightly modified standard SATA connectors and are mechanically compatible with today's SATA connector.
- This plug compatibility is important, as it enables SATA and PCIe to co-exist (see Figure 1.5).
- The new host connector supports
  - up to two SATA ports; and
  - up to two PCIe lanes.
- There is a separate signal driven by a SATAe drive that tells the host the device is PCIe.
  - This signal determines which drivers the host is utilizing when communicating with the device.
  - The capability allows a motherboard to have a single connector that supports a legacy SATA drive or a PCIe-based SATAe drive.

The SCSI Trade Association (STA) is the governing body responsible for both SAS and SCSIe technology.

- The legacy SAS connector is compatible with the legacy SATA connector.
- Eventually, all backplanes will use the SFF-8639 backplane connector, which will accept SATA, SATAe, SAS dual-port, SAS multi-port, and any device with up to four lanes of PCIe connectivity.
- SCSIe will require an SCSI over PCIe (SOP) and PCIe Queuing Interface (PQI) device driver in the host.

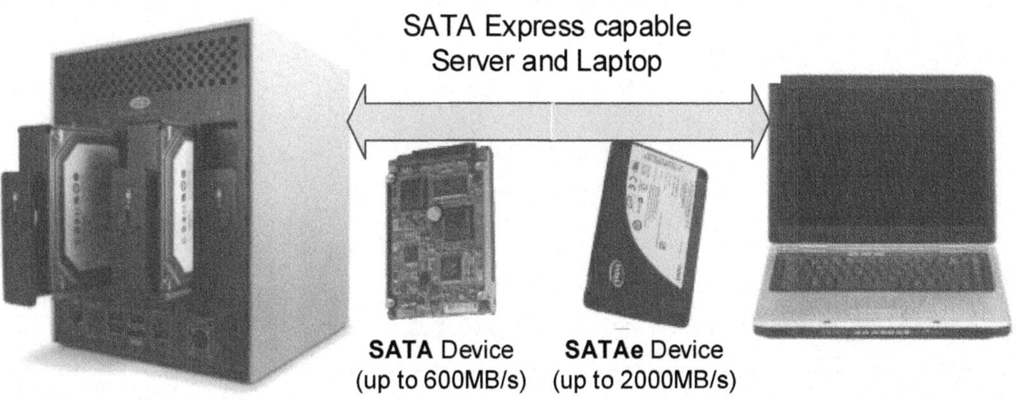

Figure 1.5 SATAe-capable hosts accept SATA or SATA Express devices.

The NVM Promotional Group is the governing body responsible for NVMe technology.

- NVMe is a PCIe-based storage alternative that can use up to four global PCIe lanes.
- NVMe will require an NVMe device driver in the host.

Migration from ATA to NVMe

As SATA has evolved, so will SATAe. All storage stacks like SATA require many layers of drivers as well as protocol translations from technology to technology. Translations add unnecessary overhead to each process and can double to triple the need for expensive compute cycles (see Figure 1.6). The goal of the next technological phase is to streamline the command overhead associated with processing I/O requests and to optimize file transfers via direct memory access (DMA) data transfer engines. This work has been going on for some time under the auspices of the "Enterprise Non-Volatile Controller Host Channel Interface" (Enterprise NVMHCI), which is now shortened to Non-Volatile Memory Express (NVMe).

NVMe is the combination of a new set of Administrative and I/O commands on top of PCIe transport, link, and physical components. Other than brief mentions like this, NVMe is beyond the scope of this book.

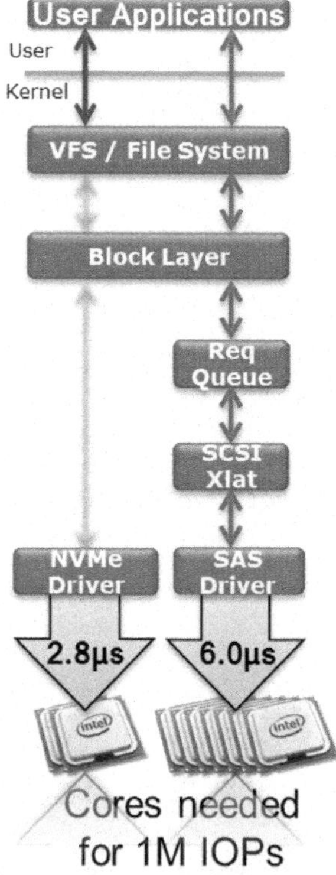

Figure 1.6 NVM versus SCSI storage stack overhead example. (*Source*: Flash Memory Summit 2013; www.nvmexpress.org/wp-content/uploads/2013-FMS-NVMe-Track.pdf; Amber Huffman, Sr. Principal Engineer, Intel)

ATA command protocol will eventually be replaced with lightweight protocols such as NVMe.

- Software driver transition will take up to five years for migration (complete by 2018).
  - This is mainly due to the software drivers, management services, security, and hot plugging capabilities already built into current operating systems drivers and management tools.
  - Some applications are tightly coupled to storage access methods (e.g., databases use RAW access).
- Operating system vendors must support and build a storage stack based upon new I/O commands and data transfer semantics.
- New administrative functions and architectural concepts need to be dramatically changed over previous implementations.
  - This will be a learning curve for an engineering ecosystem of architects, designers, developers, and programmers.

Although there are numerous advantages associated with NVMe, legacy software applications, storage stacks, and management tools will impede the adaption process.

As demonstrated in Figure 1.6, NVMe reduces latency overhead by more than 50%:

- SCSI/SAS: 6.0 µs – 19,500 cycles
- NVMe: 2.8 µs – 9,100 cycles

NVMe is defined to scale for future NVM:

- NVMe supports future memory developments that will drive latency overhead below 1 ms.
- Measurements were taken on Intel® Core™ i5-2500K 3.3GHz 6MB L3 Cache Quad-Core Desktop Processor using Linux RedHat EL6.0 2.6.32-71 Kernel.

1.4 Storage Evolution

When Serial ATA was first designed, most storage interfaces were parallel (with the exception of Fibre Channel Arbitrated Loop). ATA/IDE devices provided low-cost, high-capacity client storage, and SCSI was mainly used for high-performance Enterprise class storage or high-end server applications. Table 1.1 outlines a high level comparison of the numerous device alternatives as they relate to capacity, cost, performance, and reliability.

Table 1.1 HDD Storage Applications Overview

Application	Capacity	Cost	Performance	Reliability
Client SATA	High	Low	Low	Low
Enterprise SATA	High	Medium	Medium	Medium
Nearline SAS	Medium	Medium	Medium	High
Serial Attached SCSI (SAS)	Low	High	High	High
NVM Express (NVMe)	Low	Very high	Very high	Very high
Fibre Channel Arbitrated Loop (FC-AL)	Obsolete			
Parallel SCSI	Obsolete			
Parallel ATA	Obsolete			

Table 1.2 Disk Interface Details

Interface	SATA	SAS		PCIe		
Device Type	SATA	SAS	MultiLink SAS	SCSI Express	NVM Express	SATA Express
Form Factor	1.8, 2.5, 3.5, mSATA	2.5, 3.5	2.5	2.5	2.5, M.2	2.5, M.2
Driver	AHCI	SCSI/SAS	SCSI/SAS	SCSI & SOP/ PQI	NVMe	AHCI
Command Set	ATA	SCSI	SCSI	SCSI	NVMe	ATA
Queues	1	16	16	64k	64k	1
Commands	32	64k	64k	64k	64k	32
Drive connector	SFF-xxxx	SFF-8680	SFF-8630 SFF-8639	SFF-8639	SFF-8639, CEM	SFF-xxxx
Ports/Lanes	1	1, 2	1, 2, 4	1, 2, 4	1, 2, 4	1, 2
Link Speed (Gb/s)	6	12	12	8	8	8
Max Bandwidth (MB/s)	600	4800 (x2)	9600 (x4)	8000 (x4)	8000 (x4)	2000 (x2)
Express Bay Compatible	Yes	Yes	Yes	Yes	Yes	Yes
Drive Power	9W	9W	Up to 25W	Up to 25W	Up to 25W	Up to 25W
Surprise Removal	Yes	Yes	Yes	Future	Future	Yes

Life used to be simple until PCIe-based storage started to muddy the disk interface market. For a number of years there were only either SATA or SAS interface disks and no SSDs. SATA disks, of course, addressed the high-capacity, low-cost needs of scalable capacity applications, and SAS fulfilled the high-performance, high-availability, Enterprise class storage functions of data center storage arrays. Table 1.2 summarizes all current and future PCIe-based storage alternatives.

The good news is that most PCIe-based storage are flash-based; therefore, HDDs will still mainly come with either SATA or SAS interfaces (i.e., there are no commercially available PCIe-based HDDs, as of this printing). However, there are hybrid PCIe-based storage cards that have both flash and rotational media on a standard PCIe adapter card.

SATA Storage Today . . .

SATA hard disk drive (HDD) storage has been used in client and, yes, even Enterprise applications for over 10 years.

Since 2010, SATA-interfaced solid state disks (SSD) have been used in client and mobile applications that require non-volatile mass storage and lower power consumption.

- SATA-interfaced SSDs have also penetrated many Enterprise applications due to device latencies 100 to 1000 times lower than rotational media devices.
- Typical SSDs can execute over 10,000 random operations per second.

Hybrid hard disks (HHDs) are a combination of flash and hard disk storage that can be used to keep hot data in high-speed memory while ultimately protecting the data by copying it to the hard disk component. These devices are popular in laptop and desktop systems that constantly deal with hot data.

SATA Applications

SATA forever altered the HDD landscape and penetrated market areas for which it was never intended, again due to the low cost per gigabyte that it offers. SATA applications might include local low-latency warm storage, staging of back-up data prior to committing to tape, hierarchical storage management (HSM), and utility storage for non-mission critical data.

Figure 1.7 demonstrates basic access patterns relative to the amount of storage capacity required for typical applications. For example, hot storage will make up less than 10% of the total storage in the data center. Hot storage is made up of files and data that have been accessed over the past month. Industry research analysts dissected thousands of application I/O statistics and found that once a file is older than 30 days, it is rarely used again. These data access patterns have been researched and categorized into information classes that are used to manage every file, from creation to destruction. The process and data classification is referred to as Information Lifecycle Management. Different data access frequency patterns create varying I/O workloads, which leads to a "not-all-data-is-equal" mind-set of storage tiers. Tiered storage allows a data center architect to ensure that all corporate data is stored on the appropriate storage media, thus reducing the overall total cost of storage. For example, storing video files on corporate SSDs would be an extreme waste of the most expensive storage in the data center.

The ability to track data use patterns and workloads is extremely important in data centers that can perform chargeback to cost centers. This trend is making its way into every corporate data center to pay for the storage services provided to the organization. Now, when a department requests a RAID-5 array capable of a million IO/s, IT management has a method of tracking and charging for a specific level of service to a storage array that meets their requirements. Everything is in writing, with exact expectations from response times to capacity to availability, and so forth.

For at least the immediate future, SATA will continue to enjoy the client/server market space it has occupied for the last decade. Because SATA technology is 100% plug compatible with Serial Attached SCSI (SAS), it enables multi-tiered storage arrays that require both high-capacity, low-cost storage and Enterprise class high-performance solutions.

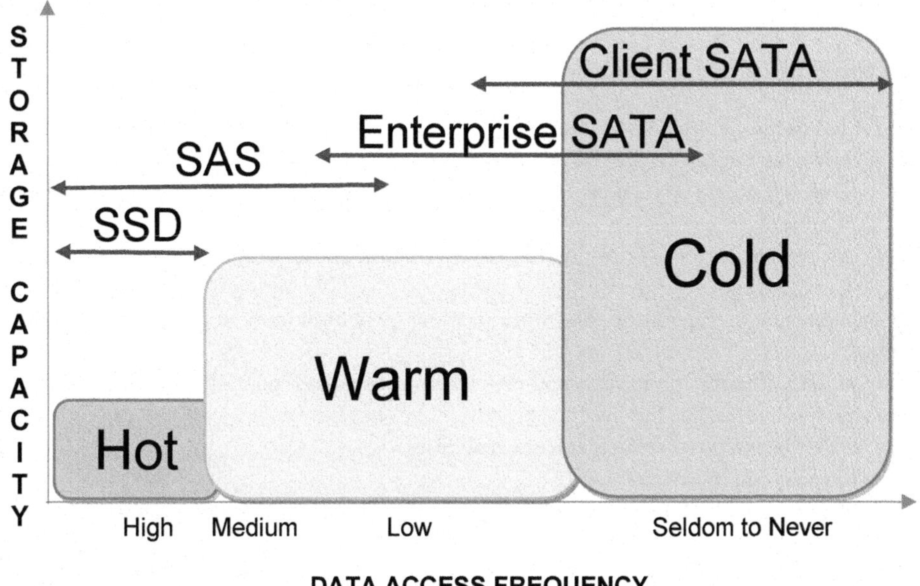

Figure 1.7 Basic SATA-SAS-SSD use model.

- Client SATA devices
    - SATA has dominated the client/server market for over a decade
    - Made up of high-capacity (up to 4TB – 2013, 6TB in 2015) low-cost storage
    - Connects through the original standard SATA interface connector
    - Data transfer rates from 300 to 600 MB/s (3 and 6 Gb/s, respectively)
    - Lower performance and reliability, with rotational speeds of 7200 rpm and 600,000 MTBF
    - Mobile devices at 5400 rpm and 1.8-inch form factors
- Enterprise SATA
    - HDD made up of mid-capacity storage; today: <1TB
    - Higher rotation speeds at 10,000 rpm and MTBF at 1M hours
    - Data transfer rates at 600 MB/s
    - Usually 3.5-inch form factor and standard SATA disk connector
    - Local (aka: near line) warm or cold storage
    - Tape back-up staging (D2D2T)
    - Hierarchical Storage Management (HSM)
- SATA Express (SATAe)
    - Will eventually populate both client and enterprise applications
    - PCI Express (PCIe) connectivity
    - Form factor freedom to choose from many sizes and profiles
    - Up to 2000 MB/s data transfer rates
    - SATAe most likely be completely replaced with NVMe by 2018, once NVMe drive stacks have become mainstream for all operating systems

Technology Note: Any capacity statements made in this book do not take into consideration the latest developments in shingled recording technology. Shingled technology is used on disk drives (and eventually tapes and CDs, too) to increase the bit density of the disk drive by overlapping tracks like roof shingles. By overlapping tracks, the device decreases the size of track guard-bands, thus creating denser tracks, which, of course, leads to higher bit densities.

SATA in the Enterprise

- SATA has permeated the data centers
    - Low-end and mid-range storage arrays
    - Due to low cost, some users feel it is easier and cheaper to use appliance storage (throwaway, if it breaks)
    - Due to the extreme price differential, SATA disk arrays are a fraction of the cost of SAS-based arrays
- Incorporates many capabilities and optimizations previously only available in Enterprise class storage
- Compatible with SAS architecture and uses the same backplane contained within all current storage arrays
- Hot swapping and power-management capabilities
- Data integrity improvements—CRC
- Disk array designs must compensate for high availability and connectivity issues

SATA Discs Compared to SAS and SSD

SATA was born into a Desktop inheritance, and its incorporation into disc drives is consequently motivated by an inherent need to meet the overriding market expectation of low-cost storage. But that doesn't mean the drive is a slouch or that its quality is in any way jeopardized by its production cost targets. However, there's no free lunch, and compromises have to be made if cost goals are to be realized. So let's digress for a moment and spend a little time talking about the disc drive at the end of the interface cable.

Table 1.3 summarizes the major differences between a Client SATA, Enterprise SATA, SAS, and SSDs as they apply to an industry standard tiered storage model.

Rotational Vibration (RV) is typically caused by cabinet excitation due to random actuator movements in neighbouring drives. This type of vibration generates a force on the data heads that tends to move them off track—like trying to stay in your lane while driving in a high crosswind. Extensive tests conducted in Seagate labs have shown that the RV occurring in commercial cabinets is typically between 10 and 20 radians/sec/sec. This is well outside the RV tolerance of SATA drives and results in a significant reduction in performance due to excessive retries while writing and reading to and from the medium.

Duty Cycle on Desktop drives is 8 hours per day, 5 days per week. On Enterprise drives, it's all day, every day.

Interactive Error Management allows the host to define the type and depth of the error recovery procedures to be instigated and managed by the drive in the event of a failure.

Internal Data Integrity Checks ensure that the data written to the drive is not only accurately recorded, but also stored at the right location on the medium. During subsequent reads, the data is verified for location and for content prior to its return to the initiator. There's little point in jumping through hoops to guarantee the integrity of the data if it's not the data you wanted.

This is doubly devastating if you don't know that you got the wrong data!

Dual Port provides access to the drive from two independent sources. This feature is intended to provide an alternate path to the data in the event of a failure at the primary port. In other words, it circumvents the "single point of failure" scenario at the interface level. In line with the adage "you've paid for it, so you might as well use it," it also provides a substantial performance increase in those systems with the power to manage it.

Factors That Influence Device Performance

The bit-rate or speed of the interface is only one component that determines device performance. For example, if the interface is capable of transferring bits at 6 Gb/s then the maximum data transfer rate is 600 MB/s (half-duplex). If the interface was only capable of 3 Gb/s bit-rates, then, of course, the device could only transmit data at half the data rate of 6 Gb/s interfaces. Following is a list of the most influential components that increase or decrease device performance:

- Interface speeds
  - SATA up to 6.0 Gb/s = 600 MB/s
  - SAS up to 12 Gb/s = 1200 MB/s
  - SATAe 8 Gb/s per lane = 1000 MB/s/lane = 2000 MB/s
- Media type
  - Type of hard disk assembly (HDA): Client versus Enterprise
  - Rotational speed: 5400, 7200, 10k, and 15k rpm
  - Flash Type: SLC, eMLC, MLC, TLC, and so forth.
- Capacity
  - Size of disk platters and bit density
  - Seek times

Table 1.3 Disk Type Comparison

Interface	SSD	SAS	Enterprise SATA	Client SATA
Applications	Tier 0	Tier 1	Tier 2 or 3	Tier 3
Interface Speeds (Gb/s)	6 or 12	12 and 6	6 and 3	6 and 3
Capacities (GB)	128 to 1,000	37 to 500	500 to 2,000	1 to 4,000
Rotational Speeds	na	15K	10K & 7.2K	7.2K
Seek Times (typical)	na	3.5 to 4.7ms	8 to 11ms	11ms
Access Times (typical)	<1ms	5 to 8ms	8 to 15ms	13 to 15ms
MTBF (hours)	1.2 million	1.4 million	1 million	600,000
Duty Cycle (hours × days)	24 × 7	24 × 7	24 × 7	8 × 5
Form Factors	3.5″ and 2.5″	3.5″ and 2.5″	3.5″	3.5″ and Mobile
Sustained Data Rates	200MB/s per channel	75 MB/s to 120 MB/s	65 to 90 MB/s	less than 60 MB/s
Dual Ports	Yes	Yes	Yes	No
Rotational Vibration	UV sensitive	21 rad/sec/sec	5 to 12 rad/sec/sec	
Typical I/Os per sec/drive (no RV)	na	319 (Queue depth = 16)	77	
Typical I/Os per sec/drive (10 rad/sec^2)	na	319 (Queue depth = 16)	35	
Typical I/Os per sec/drive (20 rad/sec^2)	na	310 (Queue depth = 16)	< 7	
Interactive Error Management	Yes	Yes	No	

    - Inner versus outer track transfer rates (80 versus 125 MB/s)
        - Short stroking disks only use outermost tracks
- I/O profile: How the device is accessed
    - Random or sequential
    - Read or write
    - Transfer size (how much data is transferred per operation)
    - Mixed workloads: random & sequential or read and write or varied transfer sizes
- Software
    - AHCI (advanced host controller interface) and ATA command set
    - SCSI (small computer system interface) architecture and commands
    - NMVe (nonvolatile memory express) command set

1.5 Serial ATA System Connection Model

The introduction of the SATA adapter is wholly transparent to the driver and the operating system. The upper layers of the legacy ATA adapter are retained, and the driver is presented with an interface identical to the previous parallel ATA model.

Although legacy drivers can be supported in this way without modification, and with no cognizance of the underlying serial infrastructure, such a system will be unable to take advantage of some of the new features offered by SATA. In a desktop PC situation this may be quite satisfactory.

In the configuration shown here (see Figure 1.8), in which two SATA drives are attached to the SATA adapter, the adapter is required to either perform Parallel ATA Master/Slave drive emulation or alternatively emulate the behaviour of two completely separate parallel ATA busses, each with a single drive attached.

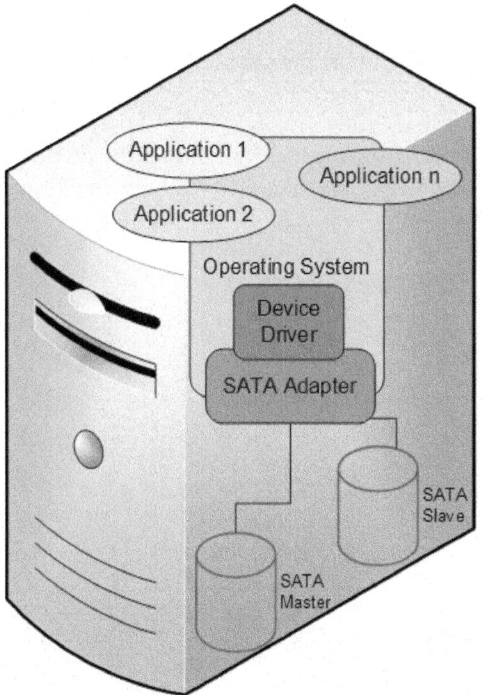

Figure 1.8 SATA connectivity.

1.5.1 SATA Connectivity

The SATA environment is a point-to-point connection scheme with very limited connectivity. SATA standards do not define switches or other hardware mechanisms that allow the architecture to go beyond a single device to host connection. In fact, as mentioned above, the original ATA architecture components are emulated in a SATA environment so that a two device to host connection scheme can be realized. SATA-IO realized these connectivity issues early on, and in the Serial ATA Specifications, they developed special hardware to provide both connectivity and availability.

To overcome this restriction, the SATA community has defined a hardware and software scheme that will allow more than one (or two for ATA emulation) device to be attached to a host system. This is accomplished through a mechanism known as a Port Multiplier.

Port Multipliers

A Port Multiplier (PM) is a mechanism in which one active host connection can communicate with multiple devices. A Port Multiplier can be thought of as a simple multiplexer in which one active host connection is multiplexed to multiple device connections, as shown in Figure 1.9.

A Port Multiplier enables a SATA host to communicate with multiple devices:

- Effectively multiplexes a single host connection to multiple device connections
- Implemented as a discreet chip
- Uses a simple addressing mechanism
- Can connect a single host with up to 15 devices
- A port multiplier is ***not*** a switch
- Port multipliers ***cannot*** be cascaded
- Host must be "Port Multiplier aware"
- Device is unaware of Port Multiplier presence

Note: To support PMs, the host must have the ability to set the proper control mechanisms to access the devices attached to the PM. These mechanisms will be covered later, in detail. For now, it must

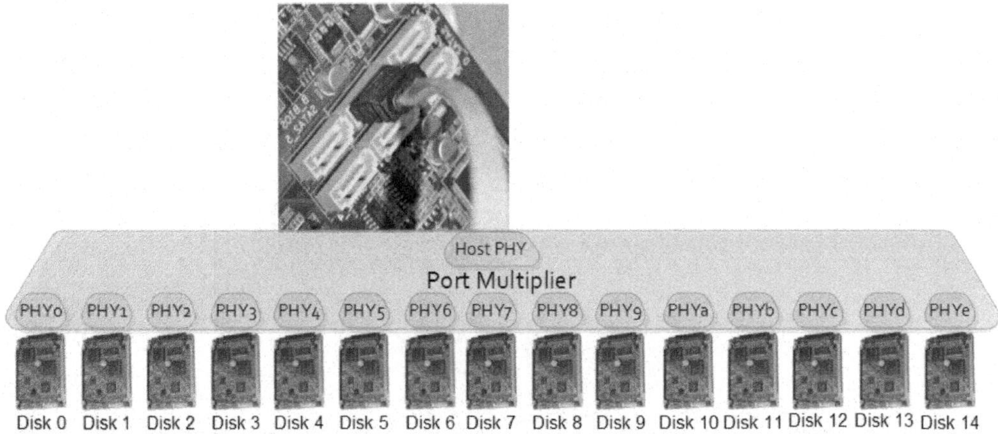

Figure 1.9 Port multiplier diagram.

be understood that the host must be "PM aware," which requires the SATA device driver to set and examine bits in frames transmitted to and received from devices, as seen in the diagram below. The PM Port # will be transmitted in the first word of all frames.

First Dword of all FIS Types

As Specified In SATA 1.0a																				PM Port				FIS Type							
7	6	5	4	3	2	1	0	7	6	5	4	3	2	1	0	7	6	5	4	3	2	1	0	7	6	5	4	3	2	1	0

Example Application SATA Host

One possible application of the Port Multiplier is to increase the number of Serial ATA connections in an enclosure that does not have a sufficient number of Serial ATA connections for all of the drives in the enclosure. An example of this is shown in Figure 1.10. A multilane cable (×4) with four Serial ATA connections is attached to the enclosure. The enclosure contains 16 Serial ATA drives. To create the appropriate number of Serial ATA connections, four 1-to-4 Port Multipliers are used to create 16 Serial ATA connections.

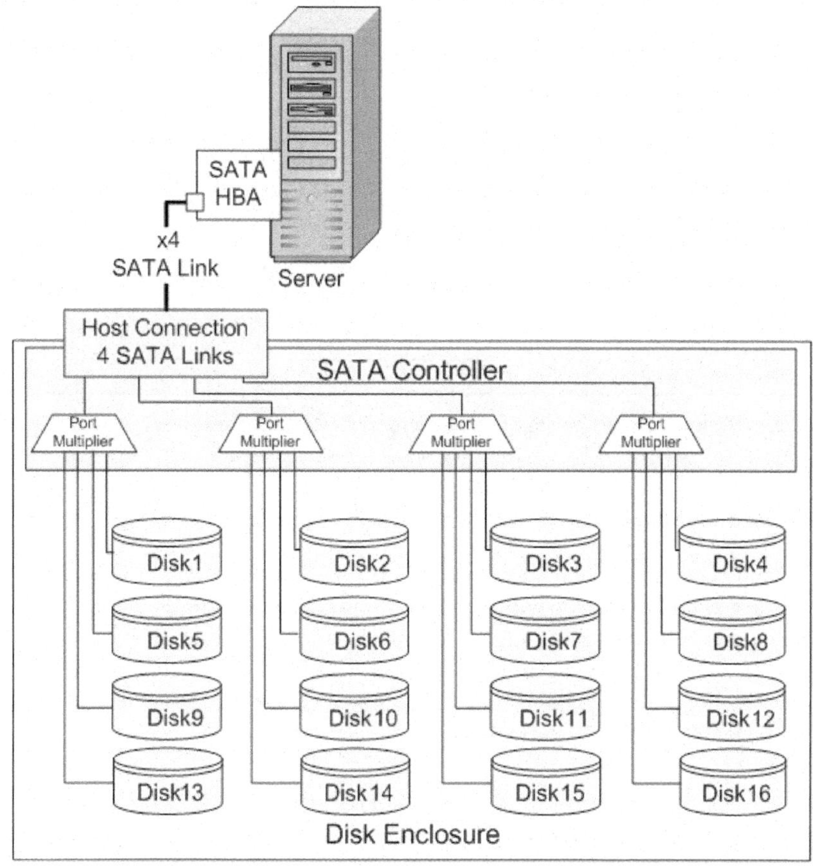

Figure 1.10 SATA-based enclosure example using Port Multipliers.

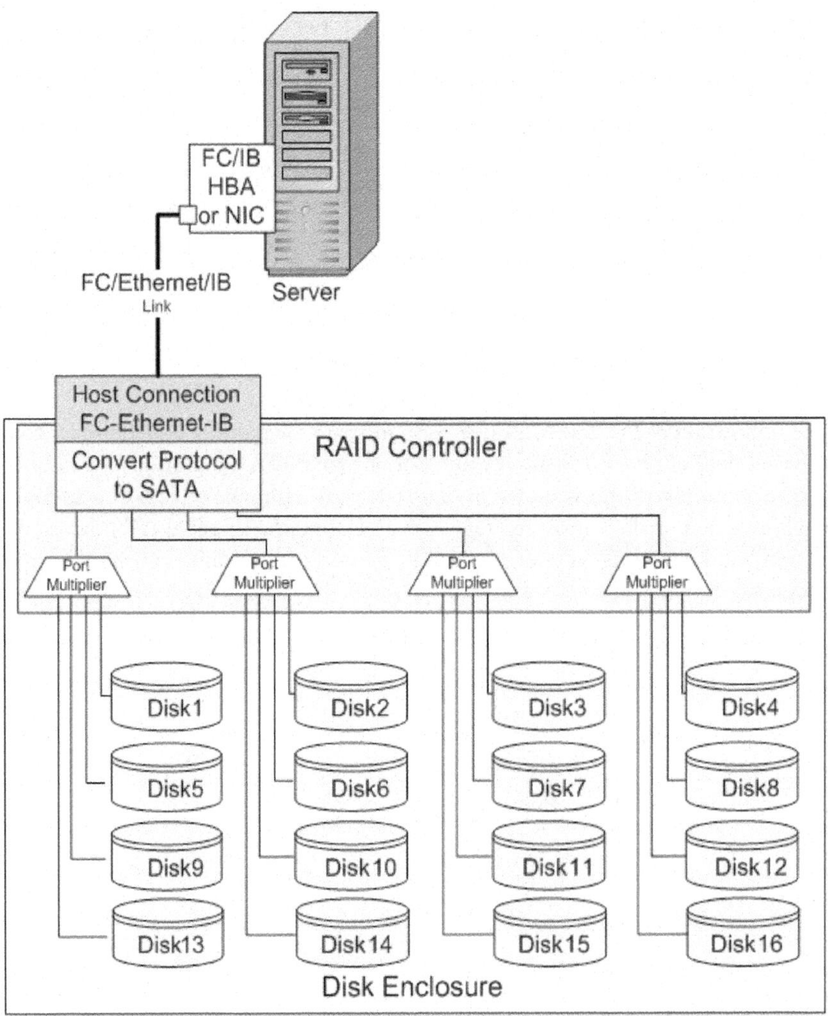

Figure 1.11 Example of External host SAN connectivity (FC-Ethernet-IB).

Example of Application of FC-IB-Ethernet Host

Figure 1.11 demonstrates how a Fibre Channel (FC), InfiniBand (IB), SAS, or Gigabit Ethernet external interface is used as the connection within the rack to the enclosure. Inside the enclosure, a host controller creates four Serial ATA connections from the host connection. The enclosure contains 16 Serial ATA drives. To create the appropriate number of Serial ATA connections, four 1-to-4 Port Multipliers are used to create eight Serial ATA connections. This design will require the use of a protocol conversion device—for example, FC to SATA controller ASIC.

1.5.2 External SATA Storage

Another example application is using a Port Multiplier to increase the number of Serial ATA connections in a mobile docking station. The example shown in Figure 1.12 has a proprietary interface between

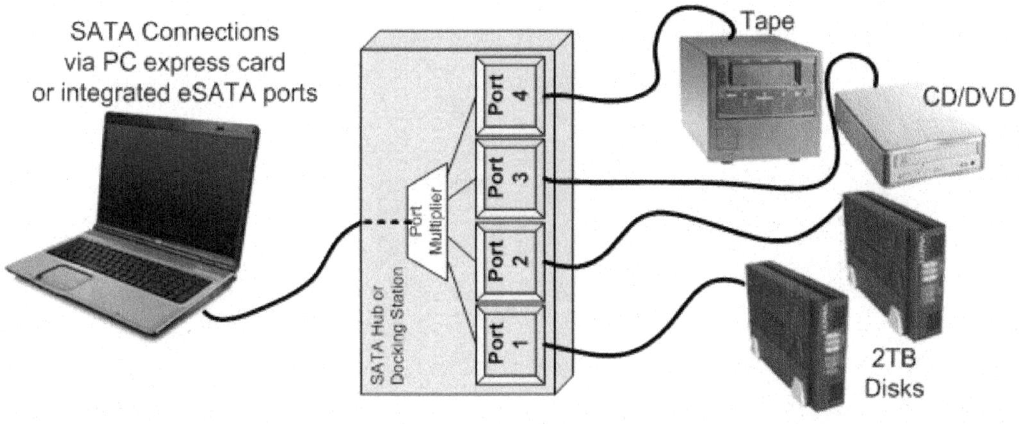

Figure 1.12 Mobile docking station example using a Port Multiplier.

the laptop and the docking station. The proprietary interface may route a Serial ATA connection from the laptop to the docking station, or the docking station may create a Serial ATA connection itself. The docking station routes the Serial ATA connection to a Port Multiplier to create an appropriate number of Serial ATA connections for the number of devices to be attached. This is similar to a USB hub.

1.5.3 SATA High Availability

SATA standards do not provide for high-availability designs. In order to accomplish this feat, the SATA-IO developed and maintains a document that details the use of a device known as a Port Selector (www.sata-io.org).

A Port Selector is a mechanism that allows two different host ports to connect to the same device in order to create a redundant path to that device. In combination with RAID, the Port Selector allows system providers to build fully redundant solutions. The upstream ports of a Port Selector can also be attached to a Port Multiplier or a Serial ATA Switch to provide redundancy in a more complex topology. A Port Selector can be thought of as a simple multiplexer, as shown in Figure 1.13.

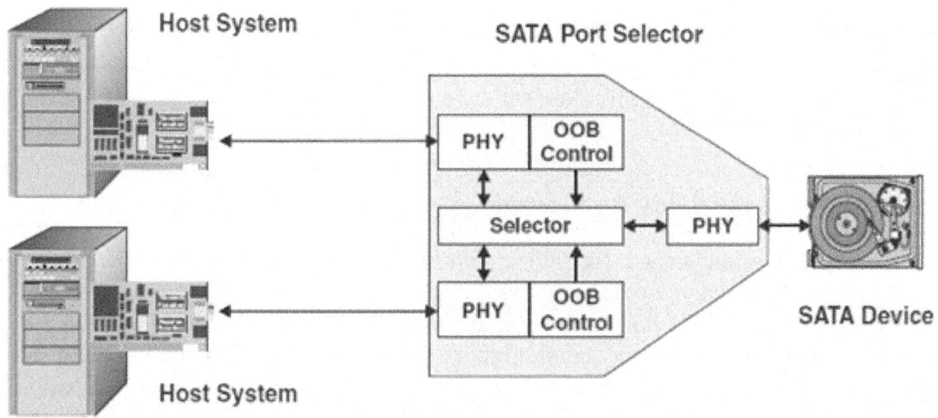

Figure 1.13 Port Selector diagram.

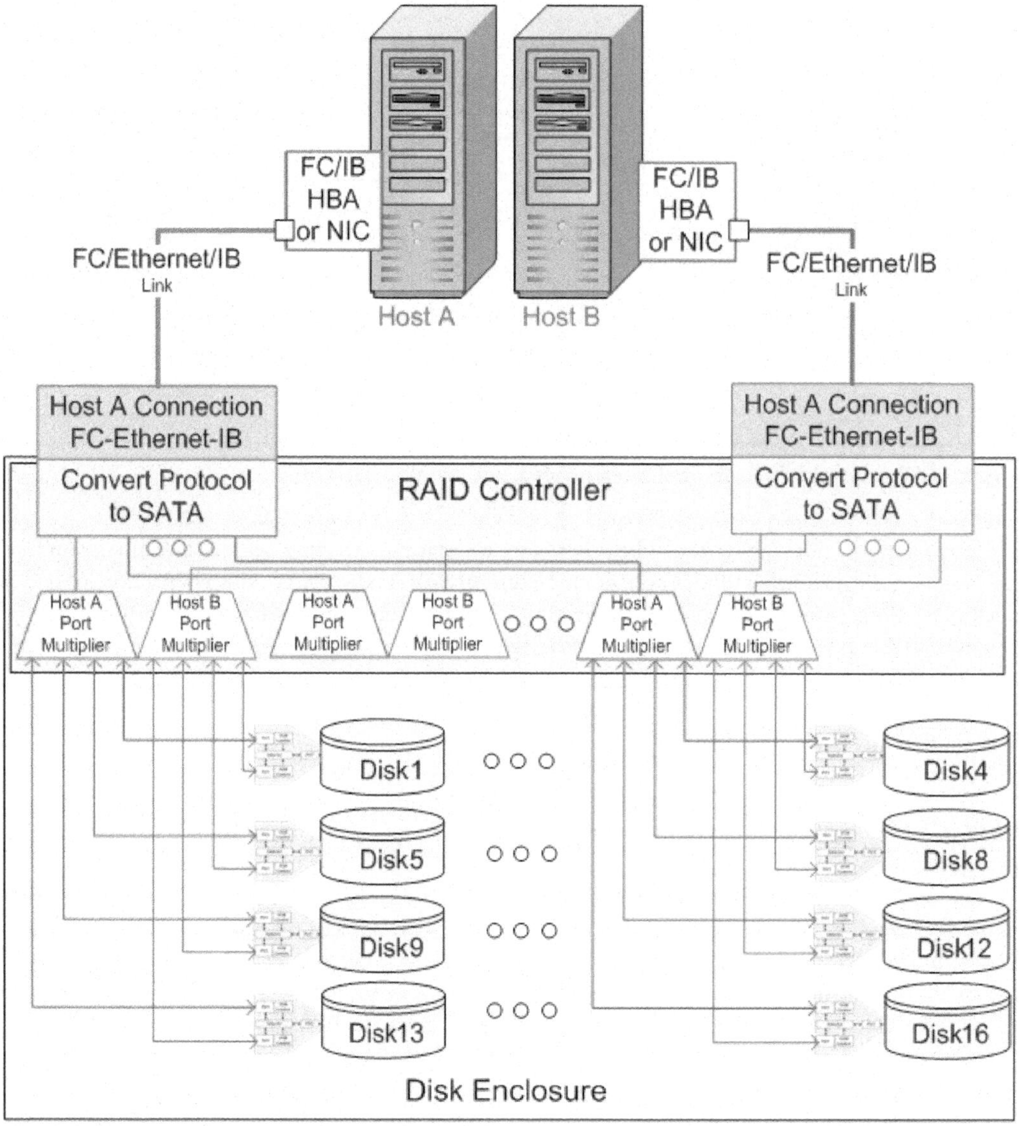

Figure 1.14 Example failover application with two hosts.

Example of High-Availability Application

A typical example of an application that utilizes Port Selectors is shown in Figure 1.14. Port Selectors provide a means for redundant access to a disk device. This ingredient, along with RAID, allows a system with no single point of failure to be built. Typically, the Port Selector would be packaged in the hard drive carrier to create a single serviceable unit in case the hard drive failed. The total system would consist of two hosts, each connected to a RAID array where each drive in the system had a Port Selector attached that was connected to each host. One host could be considered the live host and the other host may be the spare. In this configuration, the live host would maintain access to all of the devices, and the spare host would only take over access to the devices if the live host had a failure.

1.6 SATA Applications—More than Disk Storage

SATA has become a family of various form factors and connector alternatives that serve many different application requirements. Although SATA mainly started as a disk interface, it has grown to encompass many applications that require some type of storage, whether it be solid state or disk based (see Figure 1.15). Most of these additional connectivity alternatives were specifically designed to accommodate flash and SSD technology.

1.6.1 Client and Enterprise SATA Devices

The SATA specifications have added numerous connectors that all serve one application requirement or another. The original legacy SATA and SAS connectors are plug compatible in a SAS backplane. This allowed for multitier storage enclosures that could be populated with both high-performance SAS HDDs and low-cost, high-capacity SATA HDDs.

The easiest method to determine if the disk interface is SAS or SATA is to look for the gap (mind the gap) on the device-side connector. In Figure 1.16, the top disk drive has a gap between the power pins (right of gap) and signal pins (left of gap); therefore, the device has a SATA interface and inherits the operational characteristics of SATA HDDs (i.e., high capacity, low cost, low performance, single point of failure).

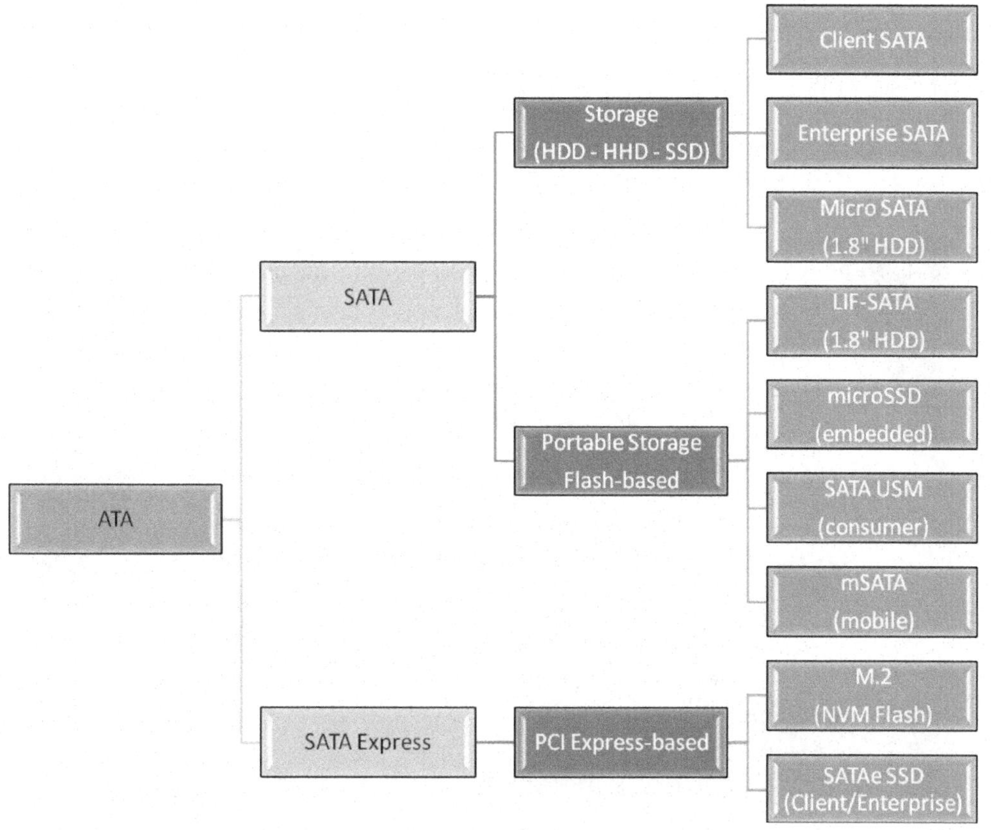

Figure 1.15 SATA technology.

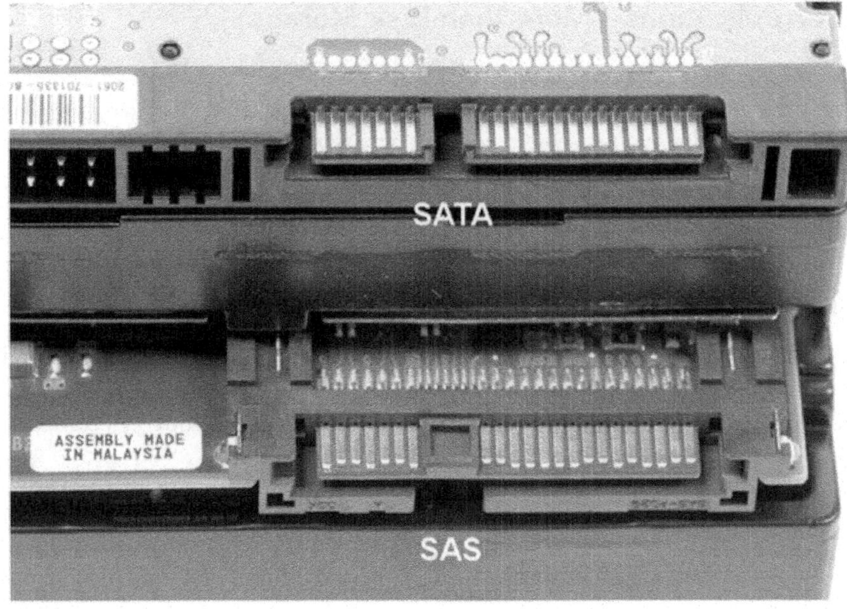

Figure 1.16 SATA disk has a gap—SAS disk gap is filled in with pins on underside.

If the gap is filled in as seen in the lower disk drive, then the device has a SAS interface, is usually considered to be Enterprise class, and is a high-performance, high-availability storage device that is more reliable than SATA-interfaced devices.

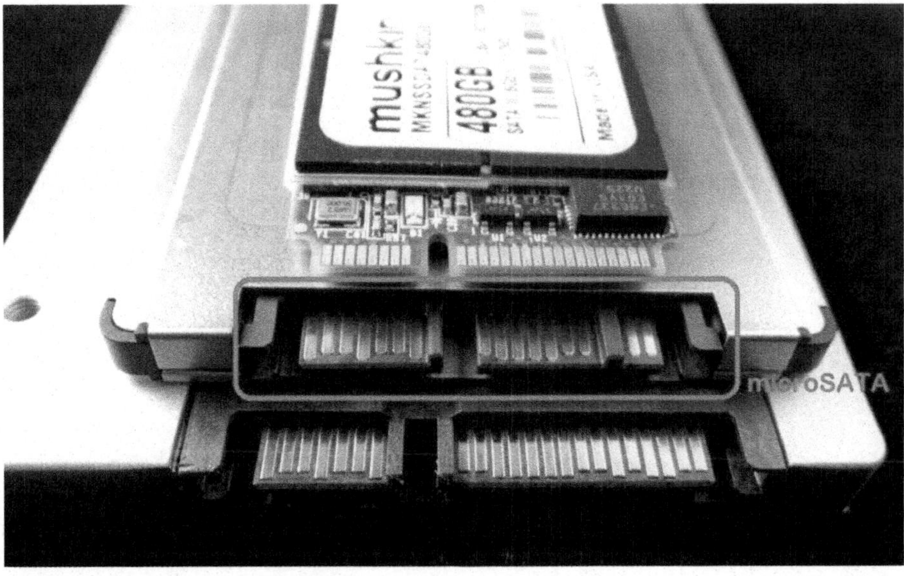

Figure 1.17 mSATA, microSATA, and SATA interfaces. (*Source*: www.thessdreview.com/our-reviews/mushkin-chronos-go-deluxe-1-8-sata-3-ssd-review/)

Micro SATA

The internal Micro SATA connector (see Figure 1.17) is specifically designed to enable connection of a slim 1.8-inch form factor HDD to the Serial ATA interface.

- The internal Micro SATA connector uses the 1.27-mm pitch configuration for both the signal and power segments.
- The signal segment has the same configuration as the internal standard SATA connector.
- The power segment provides the present voltage requirement support of 3.3 V and includes a provision for a future voltage requirement of 5 V.
- In addition, there is a reserved pin, P7.
- Finally, there are two optional pins, P8 and P9, for vendor-specific use.

The internal Micro SATA connector is designed with staggered pins for hot plug backplane (non-cabled) applications.

- A special power segment key is located between pins P7 and P8.
- This feature prevents insertion of other SATA cables.

Care should be taken in the application of this device, so that excessive stress is not exerted on the device or connector.

- Backplane configurations should pay particular attention so that the device and connector are not damaged due to excessive misalignment.

1.6.2 mSATA

The mSATA connector is designed to enable connection of a new family of small form factor devices (see Figure 1.18) to the Serial ATA interface.

- All mSATA physical dimensions are under the control of JEDEC and provided in SATA as informative.
- See JEDEC MO-300 for all physical requirements (www.jedec.org).

Figure 1.18 mSATA SSD. (*Source*: www.anandtech.com/show/6710/intel-ssd-525-review-240gb)

Figure 1.19 mSATA application example. (*Source*: www.thessdreview.com/daily-news/latest-buzz/asus-p8z77-v-premium-motherboard-equipped-with-fully-functioning-liter-on-msata-ssd/)

mSATA devices save a tremendous amount of space, operate at extremely low temperatures, and draw minimal power. These devices can be used as a system cache or to serve hot data, resident applications, or as boot devices. Figure 1.19 demonstrates how an mSATA SSD can be directly inserted into the motherboard. This device can be used to store the boot image for the server therefore providing fast-boot capable recovery from operating system or application failures.

1.6.3 M.2 SATA

M.2 connector with support for SATA as well as PCI Express signaling.

This board format is specifically designed to match commonly used SSD memory components to ensure maximum use of circuit board area, as seen in Figure 1.20.

The SATA M.2 definition supports the following capabilities:

- SATA transfer rates:
  - Gen1 (i.e., 1.5 Gbps)
  - Gen2 (i.e., 3.0 Gbps)
  - Gen3 (i.e., 6.0 Gbps)
- PCI Express generations:
  - V1 (i.e., 2.5 GT/s per lane)
  - V2 (i.e., 5 GT/s per lane)
  - V3 (i.e., 8 GT/s per lane)

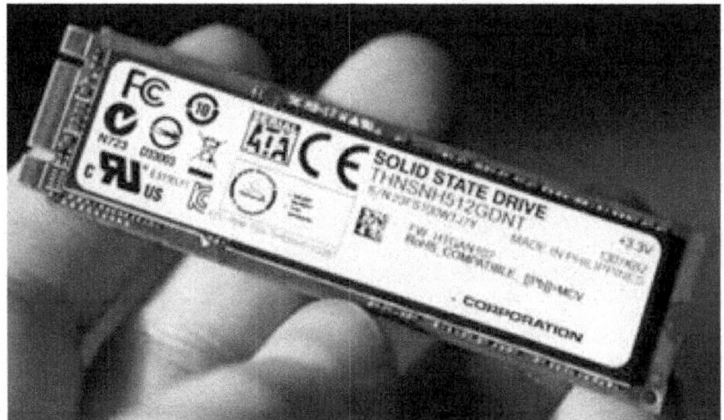

Figure 1.20 ATA M.2 SSD example. (*Source:* www.thessdreview.com/our-reviews/toshiba-hg5d-series-sata-m-2-ssd-review-512gb/)

The purpose of the mated connector impedance requirement is to optimize signal integrity by minimizing reflections.

- The host may support the use of PCIe or SATA signaling over the same interconnect.
- The nominal characteristic differential impedance of PCIe is 85 ohm, whereas the nominal characteristic differential impedance of SATA is 100 ohm.

As seen in Figure 1.21, the M.2 specification describes in detail a set of module sizes (e.g., 22 mm × 42 mm, 22 mm × 80 mm), connector heights (e.g., 2.25 mm, 2.75 mm, and 4.2 mm), and keying options for use in M.2 SSD modules.

Figure 1.22 shows two M.2 connector examples that accommodate different connection characteristics:

- 1 notch PCIe x4 = 2000MB/s
- 2 notch PCIe x2/SATA = max 1000MB/s or 550MB/s

To gain the perspective of form factors and connector differences, Figure 1.23 shows an M.2 SSD (top), an mSATA SSD (middle), and a legacy 2.5-inch enclosure mounted SATA SSD (background).

LIF-SATA

This internal LIF-SATA connector is designed to enable connection of a new family of slim 1.8-inch form factor HDDs to the Serial ATA interface (see Figure 1.24).

The internal low-insertion force (LIF) Serial ATA connector uses the 0.5-mm pitch configuration for both the signal and power segments.

- The signal segment has the same configuration as the internal standard Serial ATA connector, but the power segment provides the present voltage requirement support of 3.3 V, as well as provision for a future voltage requirement of 5 V.
- In addition, there is P8 (defined as Device Activity Signal/Disable Staggered Spin up).
- Finally, there are six vendor pins: P18, 19, 20, and 21 for HDD customer usage and P22 and P23 for HDD manufacturing usage.

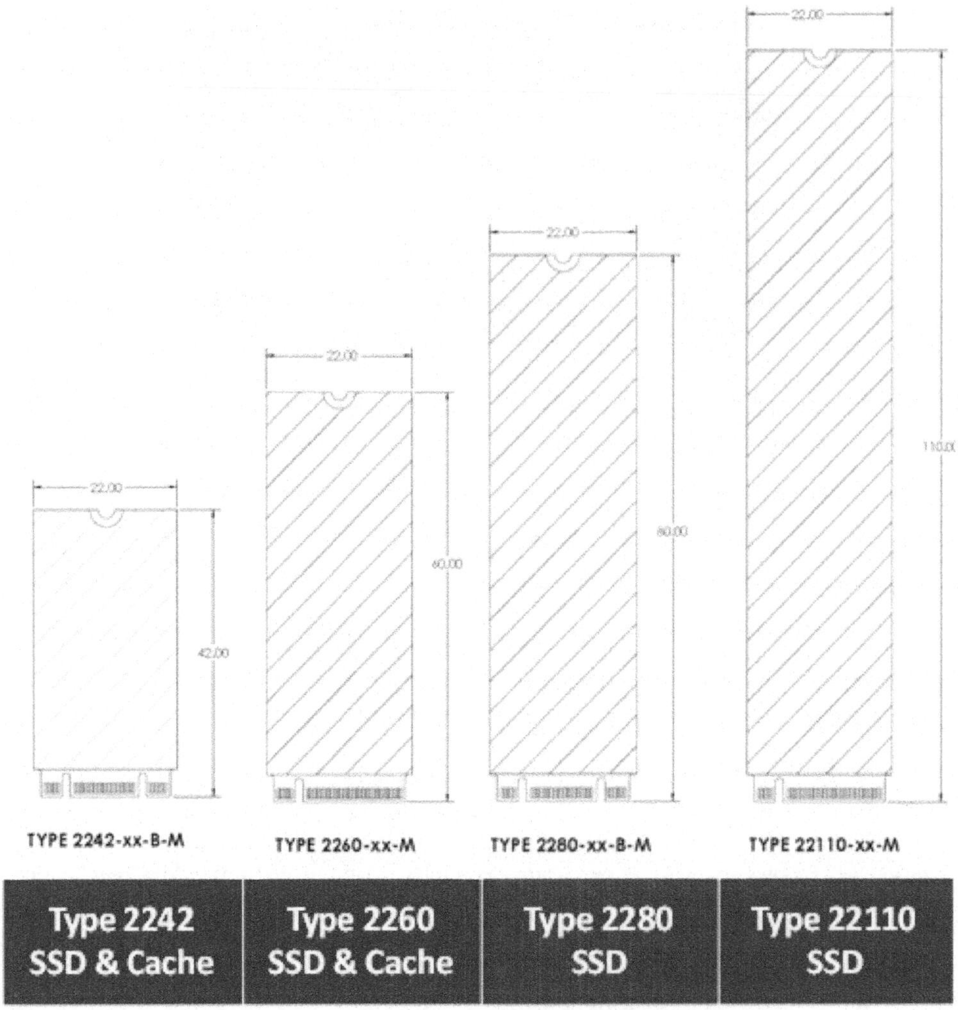
22.00
42.00
60.00
80.00
110.00
TYPE 2242-xx-B-M
TYPE 2260-xx-M
TYPE 2280-xx-B-M
TYPE 22110-xx-M
Type 2242
SSD & Cache
Type 2260
SSD & Cache
Type 2280
SSD
Type 22110
SSD

Figure 1.21 M.2 form factors and applications. (*Source*: www.PCI-SIG.com)

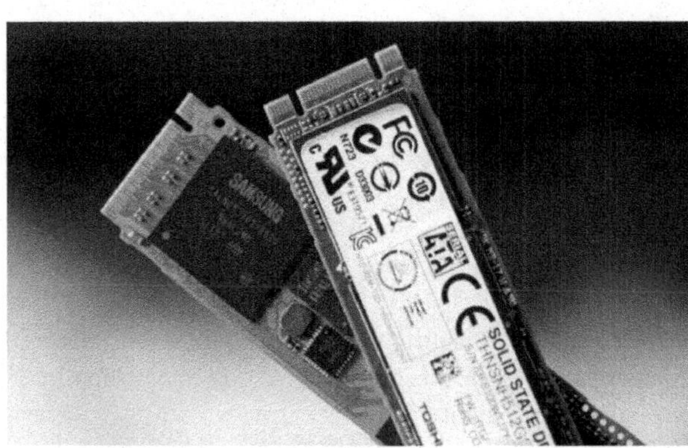

Figure 1.22 M.2 SSD examples. (*Source*: www.thessdreview.com/our-reviews/toshiba-hg5d-series-sata-m-2-ssd-review-512gb/)

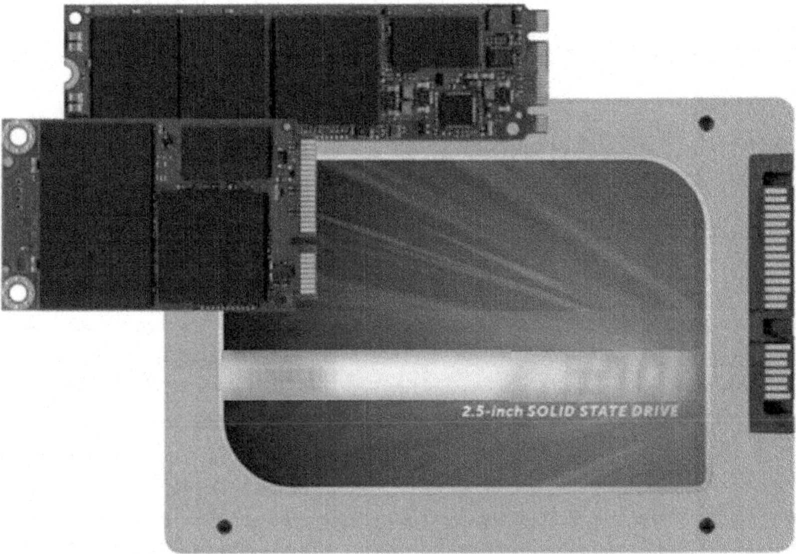

Figure 1.23 M.2 versus mSATA application example. (*Source*: http://www.dailytech.com/Micron+Announces+True+20+nm+SSD+Trio+Late+2013+DDR4+Launch/article29633.htm)

1.6.4 microSSD™

The SATA microSSD™ standard for embedded solid state drives (SSDs) offers a high-performance, low-cost, embedded storage solution for mobile computing platforms, such as ultra-thin laptops. The microSSD specification eliminates the module connector from the traditional SATA interface, enabling developers to produce a single-chip SATA implementation for embedded storage applications.

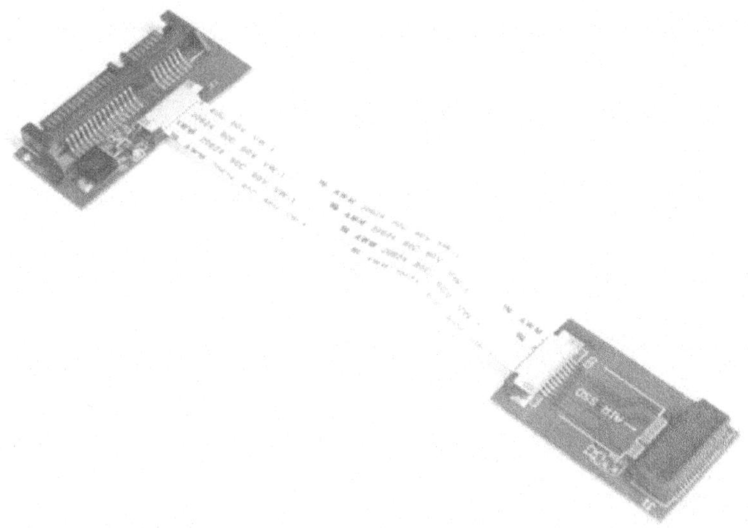

Figure 1.24 Low insertion force (LIF) example. (*Source*: www.microsatacables.com)

Figure 1.25 Example of a microSSD BGA. (*Source*: http://americas.micross.com/products-services/standard-plastic-packages/microssd.stml)

The specification defines a new electrical pin-out that allows SATA to be delivered using a single ball grid array (BGA) package as seen in Figure 1.25. The BGA package sits directly on the motherboard, supporting the SATA interface without a connecting module. By eliminating the connector, the microSSD standard enables the smallest physical SATA implementation to date, making it an ideal solution for embedding storage in small form factor devices.

1.6.5 SATA Universal Storage Module

Revision 3.1 of the SATA specification introduced a variety of performance improvements, reliability enhancements, and new features to expand the functionality and convenience of SATA-based storage devices. In addition to providing new power-management capabilities and mechanisms to maximize device efficiency, Revision 3.1 also includes the SATA Universal Storage Module (USM) specification for implementing removable and expandable portable storage applications (www.sata-io.org).

SATA is more than just a disk interface. SATA-IO also supports and fosters the efforts of the gaming and portable media industry (www.sata-io.org):

- Introduced: Jan. 4, 2011
- USM defines slot connectors for data modules with an integrated SATA interface and defined small form factor(s) standards (SFF)
- Modules plug into TVs, DVRs, game consoles, and computers for portable storage applications
- First powered 6Gb/s SATA connector for consumer applications
- First industry standard for external SATA removable storage device
- Targeted for completion in 1H 2011

USM was designed to meet the storage needs of a variety of digital media applications, including the following:

- **PCs and Laptops:** USM cartridges simplify adding storage to PCs and laptops (see Figure 1.26). They also enable the use of smaller density internal storage (e.g., SSD), allowing laptops to be lighter and thinner, with the larger capacity storage to reside on USM cartridges.
- **Digital Video Recorders (DVR)/Set-Top Boxes (STB):** Consumers can add additional capacity for storing more digital media content without service providers having to provide new equipment or pay for a truck roll.
- **TVs:** USM enables TVs to store content directly, eliminating the need for a separate DVR unit. In Japan, for example, one third of all TVs already have storage inside them. With the industry

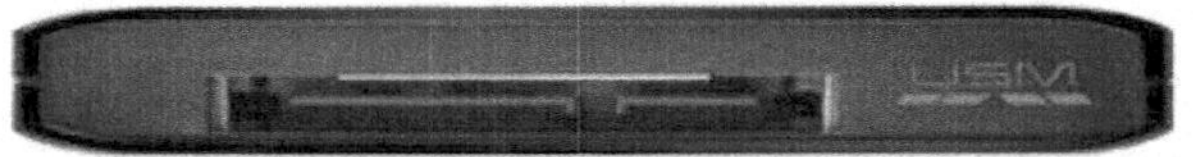

Figure 1.26 SATA USM drive. (*Source*: www.seagate.com)

adopting "Smart TV" technology and DNLA compliancy, USM allows users to play their downloaded and user-created content directly on the TV.

- **Game Consoles:** Game consoles employ a variety of storage configurations, including embedded HDDs, drives that can be upgradable by opening the system, and those that utilize a proprietary interface. USM enables console OEMs to lower initial system cost while giving users access to a wider variety of storage options.
- **Video Surveillance:** Video surveillance systems today use either internal hard drives or removable storage with limited capacity, such as tape cartridges. USM provides for unlimited capacity and simple storage exchange to maximize recording time and flexibility.

San Francisco-based Cloud Engines Inc. launched the fourth generation of its "build your own cloud" hardware platform, the Pogoplug Series 4. This new revision arrived after the company launched its Pogoplug Cloud service, which provides 5 GB of free cloud-based storage. This Series 4 device essentially allows the user to add additional storage to the virtual locker by plugging in an SD card, USB drives, and/or a 2.5-inch hard drive (see Figure 1.27).

To do this, the Pogoplug Series 4 drive hub connects directly to a local network via a Gigabit Ethernet port. Once connected, users can then add their own storage devices via ports for two USB 3.0 drives on the back, an SD card slot on the side, and one USB 2.0 port and a SATA/USM port on top, residing underneath a removable hood.

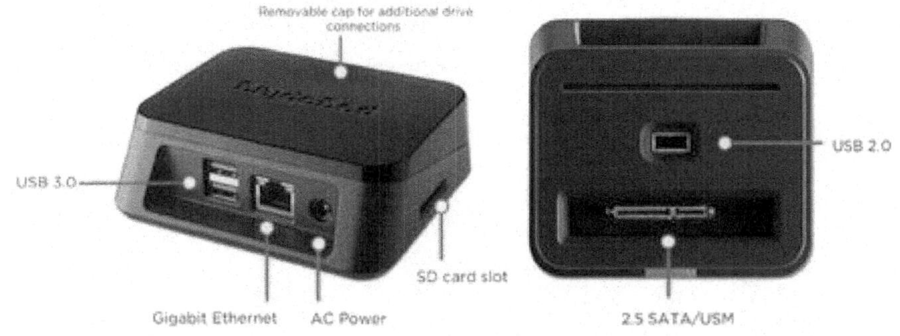

Figure 1.27 USM application of the Pogoplug Series 4 drive hub. (*Source*: http://www.tomsguide.com/us/Pogoplug-Series-4-Cloud-Storage-preview-review,news-13531.html)

USM Slim

The USM specification defines a 9mm form factor, making it an ideal storage solution for notebooks, tablets, and other portable devices (see Figure 1.28). The specification allows manufacturers to develop external storage offerings that seamlessly pair with these thin and light mobile devices so that consumers can still have instant access to their music, movies, photos and other content at any time or place.

Figure 1.28 USM provides ultra-slim profile.

1.7 Summary

Key points covered in Chapter 1 include the following:

1. SATA is based upon the parallel ATA interface of the 1990s.
2. SATA still uses the ATA command set to communicate with disk devices.
3. The SATA-IO is the governing body for all SATA specifications, and the current specification is at revision 3.2.
4. SATA is a point-to-point serial interface that supports up to 1m cable lengths.
5. Every SATA device requires a dedicated connection to the host.
6. Serial interfaces reduce cable and connector cost and make PCB trace routing easier.
7. SATA's smaller cables makes cable routing a breeze and reduces airflow restrictions.
8. SATA links can operated at 1.5, 3, and 6 Gb/s.
9. SATA devices have been used in client applications for over 10 years.
10. SATA has also evolved to serve Enterprise applications.
11. SATA and SAS are plug compatible (i.e., SATA devices can plug into SAS-based array backplanes).
12. SATA Express replaces the SATA Transport, Link, and Phy layers with PCIe Transaction, Link, and Phy layers.
13. NVM Express replaces the ATA command set with a lightweight set of I/O and Administrative commands.
14. SATAe-capable hosts and enclosures will accept both SATA and SATAe devices.
15. Client SATA devices have up to 4TB of storage capacity.
16. Enterprise SATA devices demonstrate increased reliability and performance over their client counterparts.
17. SATA requires port multipliers (PM) to provide connectivity. Up to 15 devices can be attached to a single PM.
18. PM-aware device drivers are required to use PMs in the storage array.
19. SATA requires port selectors to provide high availability. Port selectors can switch between primary and secondary host connections in case of host failure.
20. SATA has evolved over the past decade to include mSATA, SATA USM, M.2, Micro SATA, and microSSD connectivity and form factor alternatives.

Chapter 2

SATA Technical Overview

Objectives

In order to understand SATA, you must first understand ATA. This includes understanding the ATA Architecture, the Protocol, the Task Register Interface, and the Command Sets used to store and retrieve information. This chapter provides an introduction to Serial ATA technology. You will learn about the following topics:

1. What ATA is and how it impacts SATA
2. SATA Standards and Specifications
    a. Standards work done by INCITS T-13 Working Group
        i. ATA/ATAPI 1-6 for parallel ATA
        ii. ATA/ATAPI 7 for Serial ATA
        iii. ATA-8 for SCSI, such as protocol layer segmentation (e.g., Command, Transport, and Physical layer Standards)
        iv. Approved ANSI references for all ATA-related Standards
    b. Specification work done by the SATA International Organization (SATA-IO)
        i. Chronological list of Specifications
        ii. Key technology entry points (e.g., port selectors, port multipliers, etc.)
3. SATA protocol layers
    a. Application
    b. Transport
    c. Link
    d. Physical
4. I/O register model and how it relates to SATA
5. Important registers, including COMMAND, STATUS, ERROR, and DEVICE
6. CHS LBA addressing, translation, and 48-bit addressing
7. ATA protocols, including PIO, DMA, and PACKET
8. Serial link characteristics, including differential signaling, encoding, and primitives
9. Idle links
10. Frame transmission
11. Link layer protocol

12. Transport layer protocol, including basic Frame Information Structures (FIS) and numerous frame payloads
13. Application layer basics, including register input and outputs

2.1 What Is ATA?

The acronym "ATA" stands for "AT Attachment." The "AT" stands for "Advanced Technology," its usage derived from the IBM PC/AT personal computer that was produced in the 1980s. ATA referred to the technology that was used to connect both floppy and hard disk drives in those early PC systems. Bear in mind that a 40-MB capacity 5.25" hard drive was considered leading edge technology at this time and that CD-ROMs had yet to make an appearance. For over two decades, ATA has become a ubiquitous technology in the PC market place, and the ATA legacy command set may survive for the next ten years. Around 2002, the parallel ATA interface achieved obsolescence and was replaced with SATA in all client systems.

So what, exactly, is ATA?

- ATA is a physical and logical interface for connecting peripheral devices, such as hard disk drives, to host systems.
- These host systems, in most cases, are the processor motherboards used in personal computers.

2.1.1 Physical Interface

There are ATA Standards that define the physical interface components, including the wires, connectors, and type of electrical signals that enable hard drives, tape drives, CD-ROMs, and DVDs to be connected to host systems in an interoperable fashion.

- The Standards also describe a protocol using signaling and messaging schemes that allow a host to both store and retrieve data from the attached storage device or devices.
- Put another way, ATA is a physical layer and protocol layer interface definition for storage devices.

Many different acronyms have been associated with this interface over the years, as seen in Table 2.1.

Table 2.1 PC Bus Acronyms

Acronym	Definition	Description
PC/AT	Personal computer advanced technology	In the mid-1980s, IBM produced the PC/AT, which used a combined hard/floppy disk controller: • Controller card separate from drives • Controller occupied an ISA slot • Made for IBM by Western Digital using WD chipset
IDE	Integrated Device Electronics	This term refers to the integration of a hard drive mechanical assembly and its associated controller electronics.
EIDE	Enhanced IDE	This term was used by Western Digital Corporation to differentiate the "added value feature set" from its drive products.
UDMA	UltraDMA	This is one of the more recent data transfer modes defined by the ATA Specification.
Ultra ATA		See UDMA.
Fast ATA		This was a term used by Seagate to differentiate drives with higher performance interfaces.

Note: All of these terms refer to some implementation of the ATA interface.

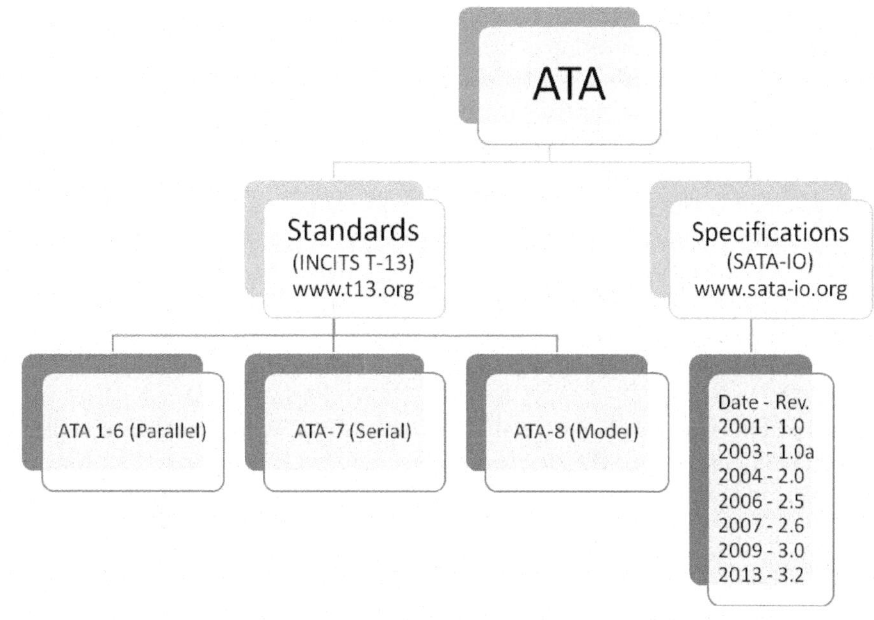

Figure 2.1 ATA Standards and Specifications.

2.2 Specifications and Standards

There are a number of organizations that are involved in the generation of industry Standards and Specifications. SATA has two development paths that almost run in parallel. There is a "Standards" path and a "Specifications" path (see Figure 2.1).

2.2.1 Industry Standards

Industry Standards are developed in an open committee format where any organization can participate. Standards meetings are held every other month and are composed of both voting and non-voting members.

The ATA Standards are generated and maintained by the International Committee for Information Technology Standards (INCITS), which is associated with the American National Standards Institute (ANSI). The group specifically concerned with ATA is known as T13. Other such groups within INCITS deal with different interface technologies:

- T10 for SCSI and SAS
- T11 for Fibre Channel
- T1 for SONET

Sometimes, the standards process can be slow and cumbersome, and an alternative method for creating an open systems specification is to form a group of companies with a mutual interest to produce one. These groups are known as industry or trade associations, special interest groups, forums, and so on. The original SATA Specification was created by the Serial ATA Trade Association (now the Serial ATA International Organization [SATA-IO]). This body was formed in February 2000 and set to work on creating a serial replacement for parallel ATA. Each of the founding members had a power of veto over any proposals for new material to be placed in the SATA Specifications. Other members are known as contributors. In 2003, the membership of the SATA-IO stood at around 130 contributor companies. SATA-IO will continue to maintain specifications, promote and market the benefits of the technology, foster quality and interoperability in products, and define new technology and future interface speeds that carry the storage industry into the next decade.

One other industry group that engages in activities concerning ATA implementers is the Small Form Factor Committee (SFF). The SFF produces specifications for drive form factors and connector positions, along with protocol specifications for ATAPI CDs and DVDs. The new Serial ATA Specification does not completely replace the ATA/ATAPI Standards. So that maximum legacy software support can be maintained, certain mechanisms defined in the parallel ATA Standards must be emulated in a serial ATA environment. In particular, the ATA "Task File registers" and their associated protocols must be emulated. The Serial ATA Specification provides a new transport mechanism for the ATA and ATAPI command set defined in the ATA/ATAPI Standards but completely replaces the parallel bus defined in those Standards. Serial ATA expects the ATA command layer implementation to be at ATA/ATAPI-5 or higher. ATAPI support over serial ATA is also required for communicating with CDs, DVDs, and tape devices.

ATA Standards

Table 2.2 shows the three main generations of ATA Standards as they relate to the protocol layers.

- ATA/ATAPI 1-6 Standards only covered parallel ATA technology.
  - Parallel ATA (PATA) Standards may still be important to legacy implementations.
  - SATA had to emulate PATA mechanisms to provide legacy support.
  - SATA provided a new transport mechanism for ATA commands.
- Serial ATA was added in the ATA/ATAPI-7 Volume 3 Standard.
  - ATA/ATAPI7 Standard received a major overhaul and was divided into three volumes.
  - The original SATA Specifications completed by the SATA-IO was forwarded to INCITS T-13 for formal standardization (InterNational Computer Information Technology Standards).
- ATA-8
  - This set of Standards follows the SCSI Standards model:
    - Architectural model – ATA Architecture Model (AAM)
    - Command Sets – ATA Command Set (ACS)
    - Transport/Physical Interface – ATA Serial Transport (AST)

Table 2.2 ATA Standards Overview

Layer	ATA	Serial ATA ATA7	SATA Now ATA8
Architecture	ATA 1-6	Volume 1	ATA Architecture Model (AAM)
Application	ATA 1-6	Volume 1	ATA8 Command Set (ACS)
Transport	ATA 1-6	Volume 1	ATA8 Serial Transport (AST)
Physical	ATA 1-6 Parallel only	Serial – Volume 3 Parallel – Volume 2	ATA8 Serial Transport (AST) ATA8 Parallel Transport (APT)

Note: ATA = ATA/ATAPI.

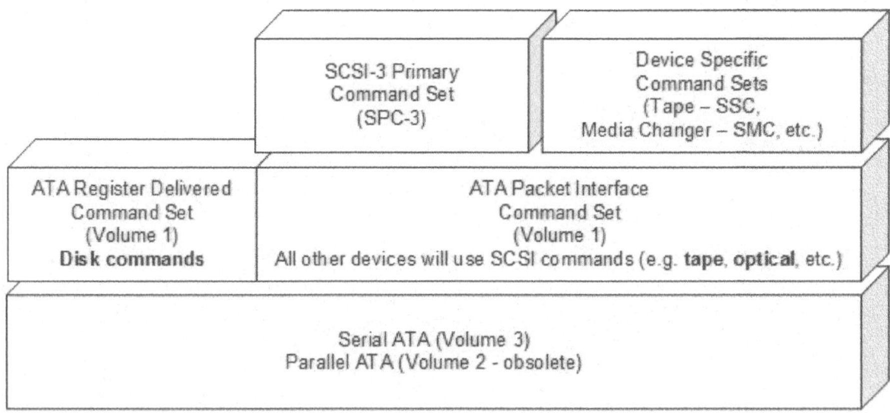

Figure 2.2 ATA/ATAPI-7 Standards hierarchy. (*Source:* www.t13.org)

ATA-7 SATA Standards

Because of the addition of Serial ATA to the Standards, a slightly different approach needed to be taken to document ATA technology. In other words, we still had register and packet delivered command sets, but we now had both parallel and serial transport, as well as physical layers, to document.

To accommodate both parallel and serial technology, the ATA/ATAPI-7 Standards were developed in three distinctly different volumes.

- Volume 1 defines the register delivered commands used by devices implementing the Standard. Consists of two methods to deliver commands:
  - Register delivered, which is used only for disk drives.
    - This is the only method needed to communicate with disk storage.

Packet delivered, which is used for all other device types.

- Tape, Optical, Enclosure, etc.
- This is the packetizing of SCSI commands into standard register delivered packets.
- Maps SCSI command protocol to SATA.

- Volume 2 defines the ***parallel*** interface connectors and cables for physical interconnection between host and storage device; the electrical and logical characteristics of the interconnecting signals; and the protocols for the transporting commands, data, and status over the PATA interface.
- Volume 3 defines the ***serial*** interface connectors and cables for physical interconnection between host and storage device; the electrical and logical characteristics of the interconnecting signals; and the protocols for the transporting commands, data, and status over the SATA interface.

Figure 2.2 shows the hierarchical relationship of these documents.

ATA-8 Standards

ATA-8 is the most recent set of Standards that define the SATA architecture. As seen in Figure 2.3, the ATA-8 Standards (www.t13.org) comprise the following:

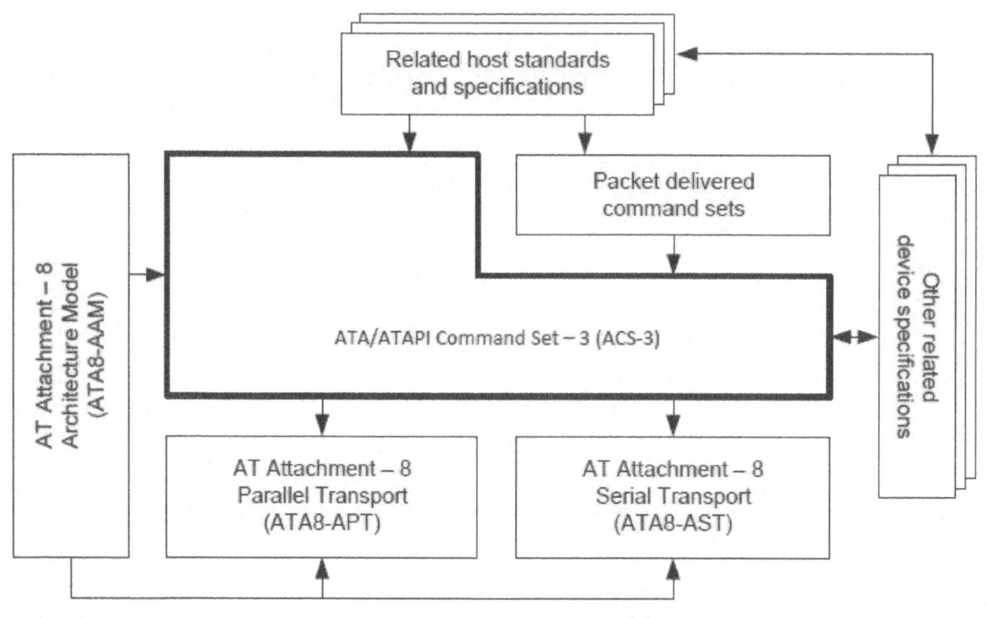

Figure 2.3 ATA-8 document relationships. (*Source*: www.t13.org)

- Architectural model – ATA Architecture Model (AAM) defines host and device responsibilities and provides a foundation for compatibility across all Standards.
- Command Sets – ATA Command Set (ACS) defines the register delivered commands used by devices implementing the Standard.
- Transport/Physical Interface – ATA Serial Transport (AST) defines the connectors and cables for physical interconnection between host and storage device; the electrical and logical characteristics of the interconnecting signals; and the protocols for the transporting commands, data, and status over the interface for the serial interface.

Table 2.3 Approved ANSI References

Name	Reference
AT Attachment with Packet Interface Extension - 5 (ATA/ATAPI-5)	ANSI INCITS 340-2000
AT Attachment with Packet Interface Extension - 6 (ATA/ATAPI-6)	ANSI INCITS 361-2002
AT Attachment with Packet Interface Extension - 7 (ATA/ATAPI-7)	ANSI INCITS 397-2005 ISO/IEC 14776-971
ATA/ATAPI-7 Amendment 1	ANSI INCITS 397-2005/AM 1-2006
AT Attachment 8 - ATA/ATAPI Command Set (ATA8-ACS)	ANSI INCITS 452-2008
AT Attachment 8 - ATA/ATAPI Architecture Model (ATA8-AAM)	ANSI INCITS 451-2008
SMART Command Transport (SCT)	ANSI INCITS TR38-2005
Acoustics—Measurement of airborne noise emitted by information technology and telecommunications equipment	ISO/IEC 7779:1999(E)

Table 2.4 References under Development

Name	Project Number
AT Attachment 8 - Parallel Transport (ATA8-APT)	INCITS 1698D ISO/IEC 14776-881
AT Attachment 8 - ATA Serial Transport (ATA8-AST)	INCITS 1697D
Host Bus Adapter - 2 (HBA-2)	INCITS 2014D
SCSI Block Commands - 3 (SBC-3)	INCITS 1799D
SCSI Primary Commands - 4 (SPC-4)	INCITS 1731D

Keep in mind the following points:

- SATA is not a command layer protocol.
  - SATA is a physical interface that delivers ATA commands.
  - AST-8 defines the transport layer and physical connectors and cables.
- ATA is the command protocol.
  - ACS-8 defines the ATA command sets for register and packet delivered commands.

The current ANSI-approved versions of Standards can be seen in Table 2.3. One design methodology is to always design according to the latest version of Standards while preparing for the next release. Currently, the ANSI-approved command set Standard has a date of 2008 (Table 2.4), while work is being done on a 2014 version (as of publication: Standard rev. 4q). Major releases should always be a target objective, considering users will most likely purchase products based upon approved Standards.

ATA Standards Evolution (see Table 2.5)

This information is provided for historical purposes. The ATA-1 Standard was formally ratified by ANSI in 1994. However, by 1990 the Specification was sufficiently rigorous and technically stable enough to lead to multi-vendor interoperable implementations. The original Standard only defined a mechanism for hard disk attachment. Connectors, cable, and signaling characteristics were specified, along with register definitions and command descriptions.

2.2.2 SATA Specifications

SATA Specifications have been developed by the Serial ATA International Organization.

- The original Specification was handed to INCITS T13 for formal ratification as an industry Standard.
- The base SATA 1.0 Specification was included in the ATA/ATAPI-7 Standard, which now consists of three separate volumes.
- The Serial ATA 1.0 Specification was completed in August 2001.
  - The first SATA drives were available in late 2002.
- SATA is a Specification-driven technology. Due to the lengthy process to develop and approve Standards, SATA implementations have been mainly driven by the Specifications developed by the SATA-IO.

Table 2.5 ATA Standards

Version	Date	Detailed Descriptions
ATA-1	1994	• Programmed I/O (PIO) modes 0, 1, and 2, a host CPU intensive data transfer method with three different transfer speed options. • Single Word Direct Memory Access (DMA) modes 0,1, and 2, a data transfer method that slightly reduced the load on the host CPU. • Multi-Word DMA Mode 0, a data transfer method that significantly reduced host CPU involvement. • This Standard was finally withdrawn in 1999. At that point, there had been many subsequent iterations of the ATA Standard.
ATA-2	1996	At this point, considerations for use of ATA in a laptop environment were also considered. Some of the key new features it introduced are as follows: • Faster data transfer modes • Logical Block Addressing (LBA) mode • Block transfers that permitted single sector (512-byte) reads or writes to be aggregated into a single multi-sector read or write, thus improving interface performance • Power management features
ATA-3	1997	• Self-Monitoring Analysis and Reporting Technology (SMART), a mechanism intended to provide a warning of degraded device quality before device failure • Drive Security Feature, enabling devices to be protected by a password
ATA/ATAPI-4	1998	• UltraDMA Modes 0, 1, 2 giving data transfer speeds of up to 33.3 MB/s • High-performance 80-way cable for improved data transmission reliability • CRC protection on data transfers – but not on commands • Limited command queuing capability • The Packet Interface for tape and CD-ROM drives
ATA/ATAPI-5	2000	• UltraDMA Modes 3 and 4 provide data transfer rates of 44.4 and 66.7 MB/s, respectively. • The enhanced 80-way cable defined in ATA/ATAPI-4 is mandated for UltraDMA Modes 3 and 4.
ATA/ATAPI-6	2002	• UltraDMA mode 5 giving a data transfer speed of 100MB/s. • A 48-bit LBA addressing method is defined that overcomes the maximum addressable device size of 137GB imposed by the 28-bit LBA. • CHS addressing is obsoleted. • Automatic Acoustic Management feature set – allows acoustic emanation to be reduced, normally with corresponding reduction in performance.
ATA/ATAPI-7	2008	Introduction of the Serial Interface Division of Standards into three volumes: • Volume 1 describes the ATA command protocol. • Volume 2 describes the ATA parallel interface. • Volume 3 describes the Serial ATA interface.
ATA/ATAPI-8	2014	Consists of three different Standards: • ASC: ATA Command Set – Rev 4q – October 23, 2013 • AAM: ATA Architectural Model currently at Rev 3 – 09/16/08 • AST: ATA Serial Transport currently at Rev 3 – 02/26/10

SATA-IO Specifications

The Specifications described in Table 2.6 are the defining documents for implementing SATA components.

Table 2.6 SATA-IO Specifications

Specifications	Description
Serial ATA 1.0	Date: August 29, 2001 Original Serial ATA Specification Major contributors: APT, Dell, IBM, Intel, Maxtor, Seagate
Serial ATA 1.0a	Date: January 7, 2003
Extension to Serial ATA 1.0a aka: Serial ATA II	Revision 1.2 Date: August 27, 2004
Serial ATA Integrated Specification	Revision 2.5 Date: Reviews due by October 22, 2005
SATA 2.5	Ratification Date: October 27, 2005
SATA 2.6	Ratification Date: February 15, 2007
SATA 3.0	Ratification Date: June 2, 2009
SATA 3.2	Ratification Date: August 2013
Other Specifications	
Port Multiplier	Revision 1.2 Released: June 27, 2005
Port Selector	Revision 1.0 Date: August 28, 2003
Serial ATA II Cables and Connectors Volume 1	Revision 1.0 Date: February 28, 2003 Errata Released: July 7, 2004
Serial ATA II Cables and Connectors Volume 2	Revision 1.0 Date: May 20, 2004

Source: www.sata-io.org.

SATA-IO Specifications and Naming Conventions (see Table 2.7)

The term SATA II has grown in popularity as the moniker for the SATA 3Gb/s data transfer rate, causing great confusion with customers because, quite simply, it's a misnomer. The following information was taken directly from the SATA-IO website (www.sata-io.org/sata-naming-guidelines).

- The first step toward a better understanding of SATA is to know that SATA II is not the brand name for SATA's 3Gb/s data transfer rate, but the name of the organization formed to author the SATA Specifications. The group has since changed names, to the Serial ATA International Organization, or SATA-IO.
- The 3Gb/s capability is just one of many defined by the former SATA II committee, but because it is among the most prominent features, 3Gb/s has become synonymous with SATA II.

For a description of SATA-IO Specifications and the official guidelines to SATA product naming, please see Table 2.7, which comprises common confusing terms and misconceptions . . .

- Gen 3 is not = 3 Gb/s.
- Third Generation SATA = 6 Gb/s.
- Gen 2 is = 3 Gb/s ó often confused with SATA II.
- Do not use the terms "SATA II" or "SATA III."

Table 2.7 SATA-IO Specifications and Naming Conventions

Revision/Generation	Interface Name	Data Transfer Rate	Naming Convention
SATA Revision 3.x (Not "SATA III")	Third generation (not "Gen 3")	Up to 6Gb/s	SATA 6Gb/s + [product name]
SATA Revision 2.x (Not "SATA II")	Second generation (not "Gen 2")	Up to 3Gb/s	SATA 3Gb/s + [product name]
SATA Revision 1.x (Not "SATA 1")	First generation (not "Gen 1")	Up to 1.5Gb/s	SATA 1.5Gb/s + [product name]
Mechanisms/Devices	**Terms to Avoid**	**Interface Speeds**	**Additional Capabilities**
Cable Connections Disk Drives Host Devices Port Multiplier Devices Port Selector Devices	Gen 3 SATA III SATA 3.0	1.5Gb/s, 3Gb/s, 6Gb/s SATA 6Gb/s is backward compatible with SATA 3Gb/s, and SATA 3Gb/s is backward compatible with SATA 1.5Gb/s	Asynchronous Notification ClickConnect eSATA HotPlug Link Power Management Native Command Queuing (NCQ) Staggered Spin-Up xSATA

Source: www.sata-io.org/sata-naming-guidelines.

Examples of Serial ATA capabilities and the official SATA product naming conventions can be seen in Table 2.8.

Table 2.8 SATA Naming Conventions—Examples

Device Name +	SATA Speed +	Additional Capabilities
ACME Disk Drive	SATA 6Gb/s or SATA 3Gb/s	NCQ, Hot Plug, Staggered Spin-Up
ACME Port Multiplier	SATA 6Gb/s or SATA 3Gb/s	NCQ, Staggered Spin-Up
ACME Port Selector	SATA 1.5 Gb/s	eSATA, NCQ
ACME Cable	Up to 6Gb/s	std, eSATA, xSATA
ACME Connectors	Up to 6Gb/s	

Source: www.sata-io.org/sata-naming-guidelines.

Sources of Information

- Serial ATA Specifications and design guides: www.sata-io.org
- ATA/ATAPI Standards: www.t13.org
- Other ATAPI Specifications: ftp.seagate.com/sff
- SCSI Command Sets (Primary and Multimedia): www.t10.org

2.3 SATA Layered Architecture

Serial ATA is a layered architecture (see Figure 2.4). Layers 1, 2, and 3 are SATA layers, whereas Layer 4, shown as the Application layer, represents the ATA Command layer. The basic responsibilities of the layers are as follows:

- Layer 1 (Physical Layer) provides cables, connectors, signaling, initialization.
- Layer 2 (Link Layer) provides framing, flow control, error checking, transmission word generation.

- Layer 3 (Transport Layer) provides formats for conveying ATA commands for processing by the link layer.
- Layer 4 (Application Layer) provides the commands to communicate between a host and device.

Although both host and device implement peer architectural layers, operations within them can differ between host and device. The detailed protocol that must be implemented at each layer is defined in a series of state diagrams. There is a set of state diagrams for each architectural layer.

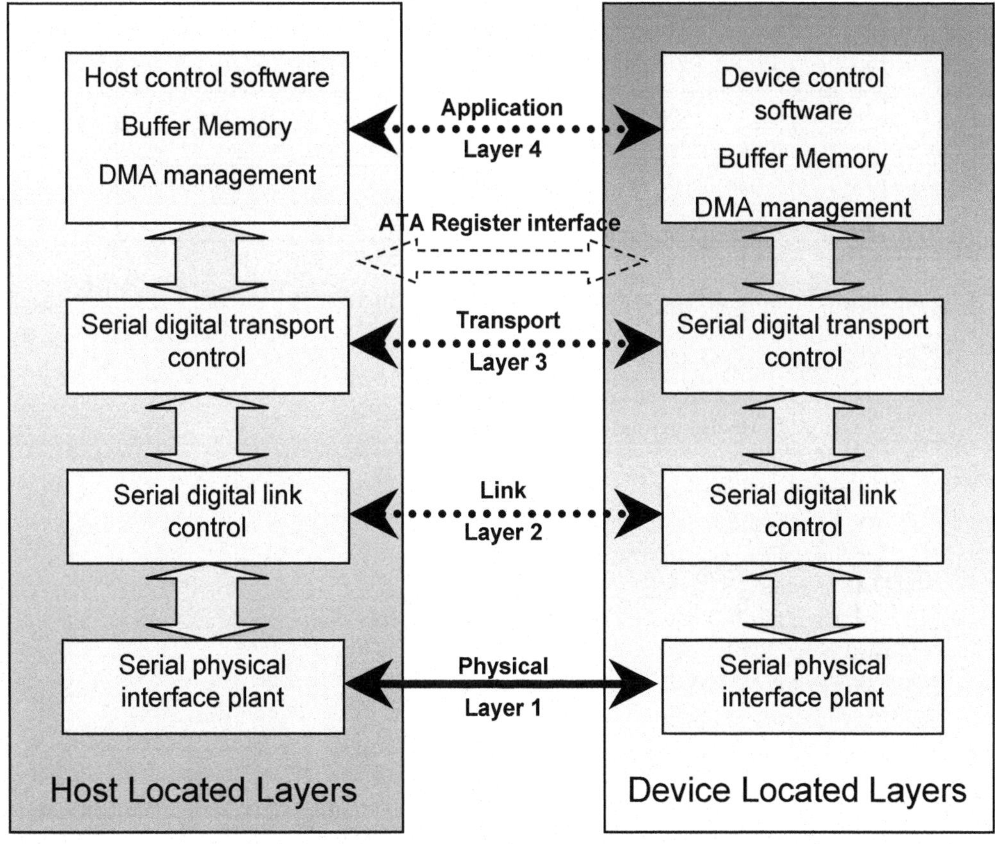

Figure 2.4 SATA protocol layers.

2.4 The I/O Register Model

SATA follows the ATA model and uses a register driven interface. It is important to have at least a basic understanding of the ATA registers because SATA will use frames to transmit register information in Standard-defined structures (see Table 2.9).

- In the parallel ATA days
  a. the host/device would set the address signals (DA0, DA1, and DA2) on the control portion of the parallel bus; and
  b. the host/device would load each 8-bit register one at a time.

- In SATA
  a. the host/device programs all registers at one time; and
  b. the host/device transmits them in Frame Information Structures (FISes).

Table 2.9 I/O Registers

<table>
<tr><th>DA2</th><th>DA1</th><th>DA0</th><th>Host -> Device (Write)</th><th>Device -> Host (Read)</th></tr>
<tr><td>0</td><td>0</td><td>0</td><td>Data Out</td><td>Data In</td></tr>
<tr><td>0</td><td>0</td><td>1</td><td>Features</td><td>Error</td></tr>
<tr><td>0</td><td>1</td><td>0</td><td colspan="2">Sector Count</td></tr>
<tr><td>0</td><td>1</td><td>1</td><td colspan="2">Sector Number</td></tr>
<tr><td>1</td><td>0</td><td>0</td><td colspan="2">Cylinder High</td></tr>
<tr><td>1</td><td>0</td><td>1</td><td colspan="2">Cylinder Low</td></tr>
<tr><td>1</td><td>1</td><td>0</td><td colspan="2">Device/Head</td></tr>
<tr><td>1</td><td>1</td><td>1</td><td>Command</td><td>Status</td></tr>
</table>

Communications between host and device use a set of registers:

- Command block registers
  - Used by host to send commands to device
  - Used by device to send status to the host
  - Also known as "Task File Registers"
- Control block registers
  - Used by host for device control
  - Used by device to send alternate status
- Data registers
  - Used for PIO data transfers
- Data ports
  - Used for DMA data transfers

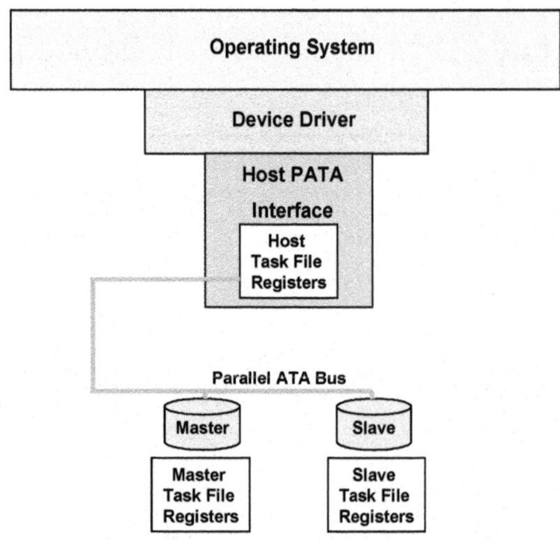

Figure 2.5 Task File Register model.

2.4.1 Host and Device Register

The ATA Command and Control Registers are often referred to as Task File Registers (see Figure 2.5). A copy of the Task File Registers resides on the host as well as on both the master and slave devices. When the host loads a Task File Register, then all registers are simultaneously written.

2.4.2 Parallel Register Operations

The following series of diagrams (see Figures 2.6 through 2.9) provide a visual representation of the register loading by host and/or device to accomplish an operation. As such, they are considerably simplified and do not demonstrate all of the control signal protocol and state changes that occur in a real implementation.

Technical Note: This section is not of major importance and could be skipped. It is provided to demonstrate how parallel ATA worked as far as loading registers were concerned. In SATA, all these registers are basically virtualized on the device and are loaded with a single frame versus one byte or word at a time (as in parallel).

Figure 2.6 shows the organization of the Task File Registers. When the device or host loads their registers, the corresponding registers are loaded. When the host loads the Device/Head register, then the devices—Device/Head register is loaded (shadow registers).

Parallel ATA requires the registers to be loaded in a particular sequence (see Figure 2.10). The first register that is loaded is the Device/Head register. Once the Device/Head register is loaded the host loads the Cylinder High, Cylinder Low, Sector Number, Sector Count, and Features registers, as seen in Figure 2.7.

To initiate the operation, the host loads the Command Register with the operation that we want the device to perform (see Figure 2.8). After the Command Register is loaded, the device accesses the command code to determine if it supports the command.

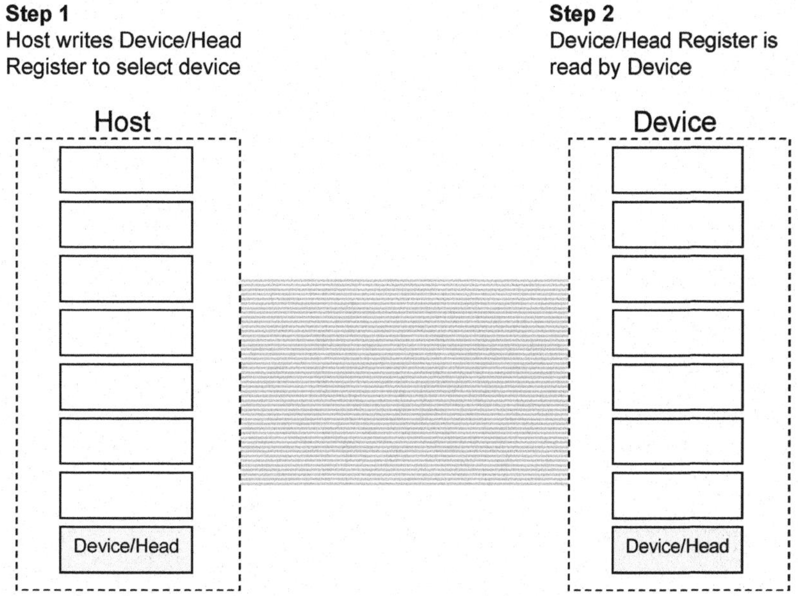

Figure 2.6 Host loads Device/Head Register.

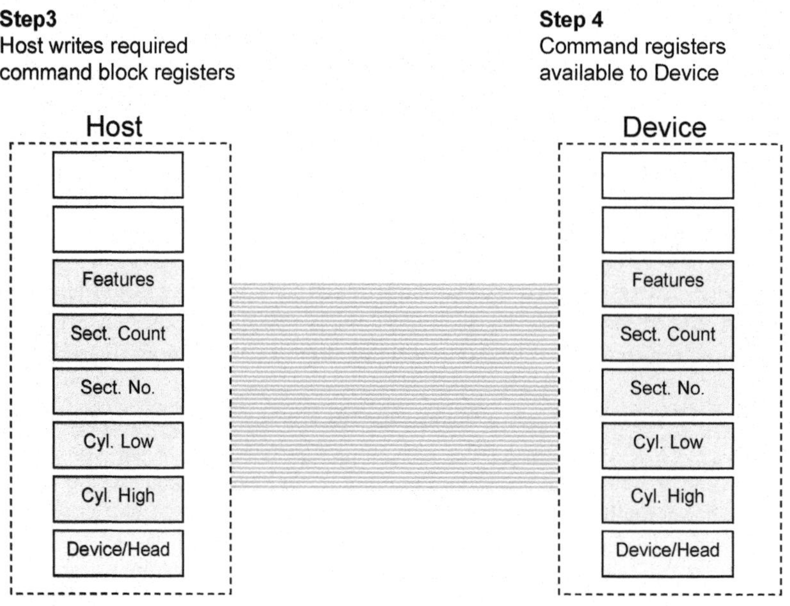

Figure 2.7 Host loads cylinder high, low, sector, count, and features registers.

Once the Command Task file is loaded, the Device sets the BSY bit and executes the command (see Figure 2.9).

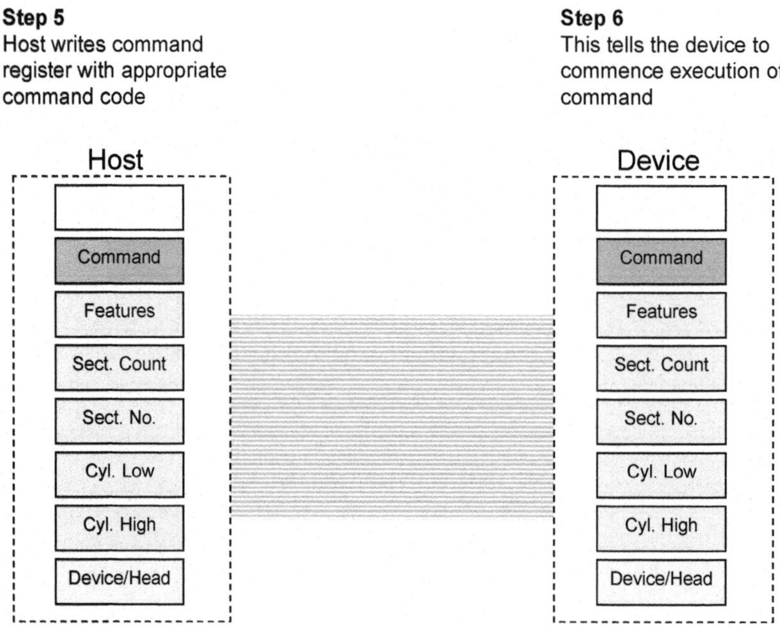

Figure 2.8 Host loads Command Register.

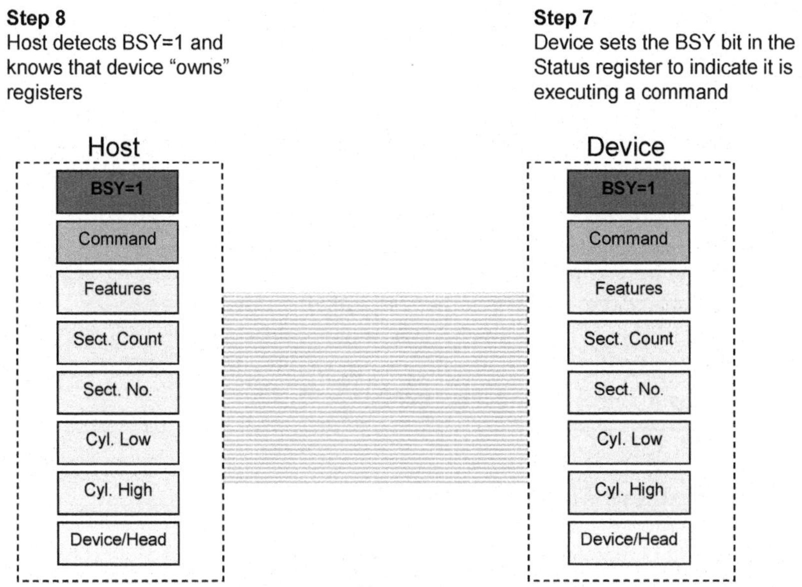

Step 8
Host detects BSY=1 and knows that device "owns" registers

Step 7
Device sets the BSY bit in the Status register to indicate it is executing a command

Host: BSY=1, Command, Features, Sect. Count, Sect. No., Cyl. Low, Cyl. High, Device/Head

Device: BSY=1, Command, Features, Sect. Count, Sect. No., Cyl. Low, Cyl. High, Device/Head

Figure 2.9 Device sets BSY and executes operation.

PATA Protocol Specifics

(1) In traditional controller operation, only the selected controller receives commands from the host following selection.
- In ATA, the register contents go to both drives (and their embedded controllers).
- The host discriminates between the two by using the DEV bit in the Drive/Head Register.
  - DEV bit 0 selects device 0
  - DEV bit 1 selects device 1

(2) Devices using this interface are programmed by the host computer to perform commands and return status to the host at command completion.
- When two devices are daisy chained on the same interface
  - commands are written in parallel to both devices; and
  - except for the EXECUTE DEVICE DIAGNOSTICS Command, only the selected device executes the command.

(3) Data is transferred in parallel (8 or 16 bits) either to or from host memory to the device's buffer under the direction of commands previously transferred from the host.
- The device performs all of the operations necessary to properly write data to, or read data from, the disk media.
- Data read from the media is stored in the device's buffer pending transfer to the host memory, and data is transferred from the host memory to the drive's buffer to be written to the media.

(4) Protocol is determined by the class of command.
- ○ PIO data-in
- ○ PIO data-out commands
- ○ Non-data commands
- ○ DMA transfer commands

(5) There are four architectural concepts that control protocol:
- BSY (Busy) bit in Status or Alternate Status Registers. This bit informs the host that the device is busy. If this bit is set (BSY=1), the host cannot access the Command Block

Registers, and the host will wait until this bit is cleared (BSY=0) before it proceeds with the I/O process.

- DRQ (Data Request) bit in Status or Alternate Status Registers. This bit informs the host that the device is ready to transfer a word or byte of data between the host and the device.
- INTRQ (Device Interrupt) is a control signal on the ATA bus. This signal is set by the device when it needs to interrupt the host system. Except in a few cases, this signal is generated by the device whenever the device changes the BSY bit from 1 to 0. INTRQ is "advice" to the host telling the host that it should fetch the device status. The ATA interface does **not require** the host to use the INTRQ signal. A host may choose to ignore or disable the INTRQ signal and use polling of the Status register instead.
- Command execution begins when the host writes the Command register. Once command execution begins, the device must have either the BSY bit set (BSY=1) or the DRQ bit set (DRQ=1) until the command is completed.

2.4.3 PATA I/O Process Flow

The sequence diagram in Figure 2.10 demonstrates the basic flow of an I/O process between a host and a device. It only addresses a higher level of understanding and doesn't include detailed descriptions of the two main protocol bits—BSY (Busy) and DRQ (Data Request)—and the control signal INTRQ (Device Interrupt).

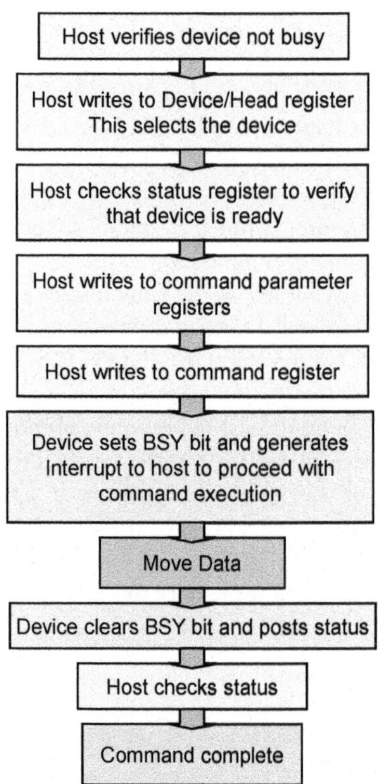

Figure 2.10 I/O process sequence diagram.

- Command parameter registers include Cylinder High, Cylinder Low, Sector Number, Sector Count, and Features registers.
- Once the host writes to the Command register, the device executes the command.

Again, this information is provided to get a general understanding of how parallel ATA protocol works. In SATA, these protocol mechanisms will be emulated inside SATA frames. SATA frames will be covered later.

2.5 Register Formats

Some registers have fixed formats. This section provides details for the major registers used in SATA, including the following:

- Command (see Table 2.10)
- Status (see Table 2.11)
- Error (see Table 2.13)
- Device (see Table 2.17)

2.5.1 Command Register

The COMMAND field contains the command code (see Table 2.10). The heading for each command starts with the name of the command.

- The name is followed by "-" and then the command code, subcommand code if applicable, and protocol used to process the command.
- An example heading reads:
  - READ SECTOR(S) – 20h, PIO Data-In
- In this example, heading the name of the command is READ SECTOR(S).
  - The command code is 20h.
  - The protocol used to transfer the data is PIO Data-In.

Examples:

A **Read Sectors** command has a command code of **0x20.**
A **Write Sectors** command has a command code of **0x30.**

Table 2.10 Command Code Register

Register	7	6	5	4	3	2	1	0
Error								
Sector Count								
LBA Low								
LBA Mid								
LBA High								
Device								
Command	Command Code							

2.5.2 Status Register

The Status Register (see Table 2.11) may have variable or command-specific bits. Most commands result in the Status Register being transferred from the device to the host. Table 2.12 provides detailed descriptions of all the Status Register bits.

Table 2.11 Status Register Fields

Register	7	6	5	4	3	2	1	0
Error								
Sector Count								
LBA Low								
LBA Mid								
LBA High								
Device								
Status	BSY	DRDY	DF/SE	DWE	DRQ	AE	SNS	CC/ERR

Table 2.12 Status Register Bit Descriptions

Bit	Name	Description
7	BUSY	BSY set to one when device is busy performing operations.
6	DEVICE READY	DDRY set to one to indicate that the device is ready.
5	DEVICE FAULTDF	DF the device is in a condition where continued operation may affect the integrity of user data on the device (e.g., failure to spin-up without error, or no spares remaining for reallocation), then the device shall: • return command aborted with the DEVICE FAULT bit set to one in response to all commands (e.g., IDENTIFY DEVICE commands, IDENTIFY PACKET DEVICE commands), • except REQUEST SENSE DATA EXT commands.
	STREAM ERROR	SE set to one if an error occurred during the processing of a command in the Streaming feature set.
4	DEFERRED WRITE ERROR	DWE set to one if an error was detected in a deferred write to the media for a previous WRITE STREAM DMA EXT command or WRITE STREAM EXT command.
3	DATA REQUEST	DRW set to one to indicate that the requested data is ready to be transmitted.
2	ALIGNMENT ERROR	AE denotes the first byte of data transfer does not begin at the first byte of a physical sector or the last byte of data transfer does not end at the last byte of a physical sector.
1	SENSE DATA AVAILABLE	SNS bit shall be set to one if: (a) the SENSE DATA SUPPORTED bit is set to one; (b) the SENSE DATA ENABLED bit is set to one; and (c) the device has sense data to report after processing any command.
0	CHECK CONDITION	CC is set to one if the ATAPI device sets the CHECK CONDITION bit: (a) value in the SENSE KEY field is greater than zero; (b) ABORT bit is set to one; (c) END OF MEDIA bit is set to one; or (d) ILLEGAL LENGTH INDICATOR bit is set to one.
	ERROR	ERR is set to one if an ATA device sets any bit in the ERROR Register field to one.

2.5.3 Error Register

The Error Register format is dependent upon the type of device that contains the register. Table 2.13 shows the format of the Error Register for register delivered devices, and Table 2.15 shows the format of the Error Register for packet delivered devices.

At command completion of any command except EXECUTE DEVICE DIAGNOSTIC or DEVICE RESET, the contents of this register are valid when the ERR bit is set to one in the Status Register. Error bit descriptions for register delivered devices can be found in Table 2.14, and packet delivered devices can be found in Table 2.16.

Table 2.13 Error Register—Register Delivered

Register	7	6	5	4	3	2	1	0
Error	ICRC	UNC	obsolete	INDF	obsolete	ABRT	EOM	ILI CMDTO
Sector Count								
LBA Low								
LBA Mid								
LBA High								
Device								
Status	BSY	DRDY	DF/SE	DWE	DRQ	AE	0	CC/ERR

Table 2.14 Error Register Bit Descriptions

Bit	Name	Description
7	INTERFACE CRC	ICRC bit shall be set to one if an interface CRC error occurred during an Ultra DMA data transfer. • The value of the INTERFACE CRC bit may be applicable to Multiword DMA transfers and PIO data transfers. • If the INTERFACE CRC bit is set to one, the ABORT bit is set to one.
6	UNCORRECTABLE ERROR	UNC bit shall be set to one if the data contains an uncorrectable error.
5	OBSOLETE	NA
4	ID NOT FOUND	The IDNF bit shall be set to one if: • a user-addressable address was not found; or • an address outside of the range of user-addressable addresses is requested and the ABORT bit is cleared to zero.
3	OBSOLETE	NA
2	ABORT	ABRT set to one if the device aborts the command.
1	END OF MEDIA	EOM bit set to one indicates that the end of the medium has been reached by an ATAPI device (see SFF 8020i).
0	COMMAND COMPLETION TIME OUT	CMDTO set to one if the STREAMING SUPPORTED bit is set to one (i.e., the Streaming feature set is supported) and a command completion time-out has occurred in response to a Streaming feature set command.
	ILLEGAL LENGTH INDICATOR	ILI bit is specific to the SCSI command set implemented by ATAPI devices (e.g., tape).

Table 2.15 Status Register—Packet Device—ATAPI/SCSI

Register	7	6	5	4	3	2	1	0
Error	SENSE KEY (0h-Fh)				obsolete	ABRT	EOM	ILI CMDTO
Sector Count								
LBA Low								
LBA Mid								
LBA High								
Device								
Status	BSY	DRDY	DF/SE	DWE	DRQ	AE	1	CC/ERR

Table 2.16 Sense Key Descriptions

Sense Key	Description
0h	NO SENSE. Indicates that there is no specific sense key information to be reported. This may occur for a successful command or a command that received CHECK CONDITION status because one of the filemark, EOM, or ILI bits is set to one.
1h	RECOVERED ERROR. Indicates that the last command completed successfully with some recovery action performed by the device server. Details may be determinable by examining the additional sense bytes and the information field. When multiple recovered errors occur during one command, the choice of which error to report (first, last, most severe, etc.) is vendor specific.
2h	NOT READY. Indicates that the logical unit addressed cannot be accessed. Operator intervention may be required to correct this condition.
3h	MEDIUM ERROR. Indicates that the command terminated with a non-recovered error condition that was probably caused by a flaw in the medium or an error in the recorded data. This sense key may also be returned if the device server is unable to distinguish between a flaw in the medium and a specific hardware failure (sense key 4h).
4h	HARDWARE ERROR. Indicates that the device server detected a non-recoverable hardware failure (e.g., controller failure, device failure, parity error, etc.) while performing the command or during a self-test.
5h	ILLEGAL REQUEST. Indicates that there was an illegal parameter in the command descriptor block (CDB) or in the additional parameters supplied as data for some commands (FORMAT UNIT, SEARCH DATA, etc.). If the device server detects an invalid parameter in the CDB, then it shall terminate the command without altering the medium. If the device server detects an invalid parameter in the additional parameters supplied as data, then the device server may have already altered the medium.
6h	UNIT ATTENTION. Indicates that the removable medium may have been changed or the target has been reset.
7h	DATA PROTECT. Indicates that a command that reads or writes the medium was attempted on a block that is protected from this operation. The read or write operation is not performed.
8h	BLANK CHECK. Indicates that a write-once device or a sequential-access device encountered blank medium or format-defined end-of-data indication while reading or a write-once device encountered a non-blank medium while writing.
9h	Vendor Specific. This sense key is available for reporting vendor-specific conditions.
Ah	COPY ABORTED. Indicates a COPY, COMPARE, or COPY AND VERIFY command was aborted due to an error condition on the source device, the destination device, or both.
Bh	ABORTED COMMAND. Indicates that the device server aborted the command. The application client may be able to recover by trying the command again.

2.5.4 *Device Register*

The Device Register (previously Device/Head Register) (see Tables 2.17 and 2.18) has two bits that may be used. You will have to look at each command usage model to determine if these bits need to be set.

Table 2.17 Device Register (ATA-8)

Register	7	6	5	4	3	2	1	0
Error	SENSE KEY (0h-Fh)				obsolete	ABRT	EOM	ILI CMDTO
Sector Count								
LBA Low								
LBA Mid								
LBA High								
Device	obsolete	LBA	obsolete	DEV	Reserved			
Status	BSY	DRDY	DF/SE	DWE	DRQ	AE	1	CC/ERR

Note: LBA—Used to select LBA or CHS addressing – always 1 in ATA-8.
DEV—Selects master or slave device – ATA emulation mode only.

Table 2.18 Device Register pre-ATA-8 (Device/Head Register)—Historical

Bit	7	6	5	4	3	2	1	0
	obsolete	LBA	obsolete	DEV	Head number or LBA			

2.6 CHS and LBA Addressing

2.6.1 *CHS Addressing*

Addressing data on a spinning disk device requires the positioning of mechanical components. Prior to logical block addressing, file systems had to organize the platters into geometric cylinders, tracks and sectors (see Figure 2.11). Original ATA command sets used this addressing scheme.

- Early drives used physical sector addressing
- A data block is identified by its cylinder, head, and sector (CHS) position
- ATA Registers allow CHS to be defined by the host
- CHS schemes are actually a logical mapping
- Permits large capacity drives to be supported
- Allows the drive to perform its own defective sector management

2.6.2 *LBA Addressing*

Logical Block Addressing (LBA) hides (or virtualizes) the physical geometry of the device and provides a simple method to access data that is organized into fixed sized blocks (e.g., 512 bytes, 4k bytes, etc.).

- All data blocks are addressed as a contiguous LBA range—from 0 to the highest block
  - Easier for the host to manage

- Only the drive itself knows the physical CHS location of an LBA
  - Performs a translation from LBA to physical CHS
- Permits larger capacity addressing ranges
- Allows the drive to perform defect management

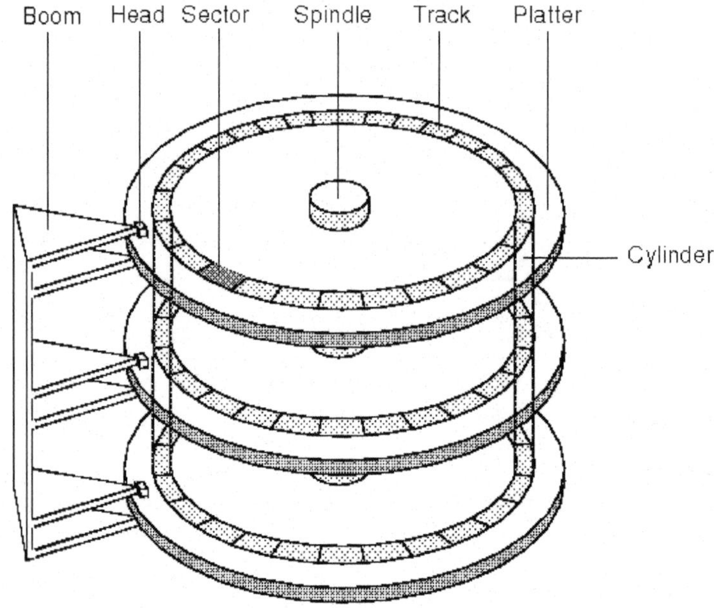

Figure 2.11 Cylinder, head, sector addressing. (*Source*: © 2003 Caldera International, Inc.)

Figure 2.12 demonstrates LBA to CHS relationship and the typical characteristics of file systems:

- File systems organize a devices logical blocks (CHS) into "clusters" (or inodes).
- Clusters are groups of contiguous logical blocks that have a direct relationship to how much data is transferred during each operation.
  - For example, when a file system has organized its clusters into 64k chunks, then every read or write operation moves at least 64k of data.
  - If a file is 64k+1byte in size, then the host would either send two 64k operations, or a single 128k operation.
  - Clusters/inodes are used to simplify file management and provide specific data set formats (or even metadata constructs)
- File Allocation Table – organizes clusters, tracks the clusters' status (Free/Next/Final) and provides an index to the appropriate LBA.

When the LBA arrives at the device it must translate it to ATA Cylinder, Head, and Sector registers. In the original ATA implementations there was only translation to a 28-bit LBA:

- 28-bit addressing only allowed devices with capacities up to 128GB.
- ATA/ATAPI-6 Standard has defined an enhanced 48-bit LBA scheme to allow even higher capacities.

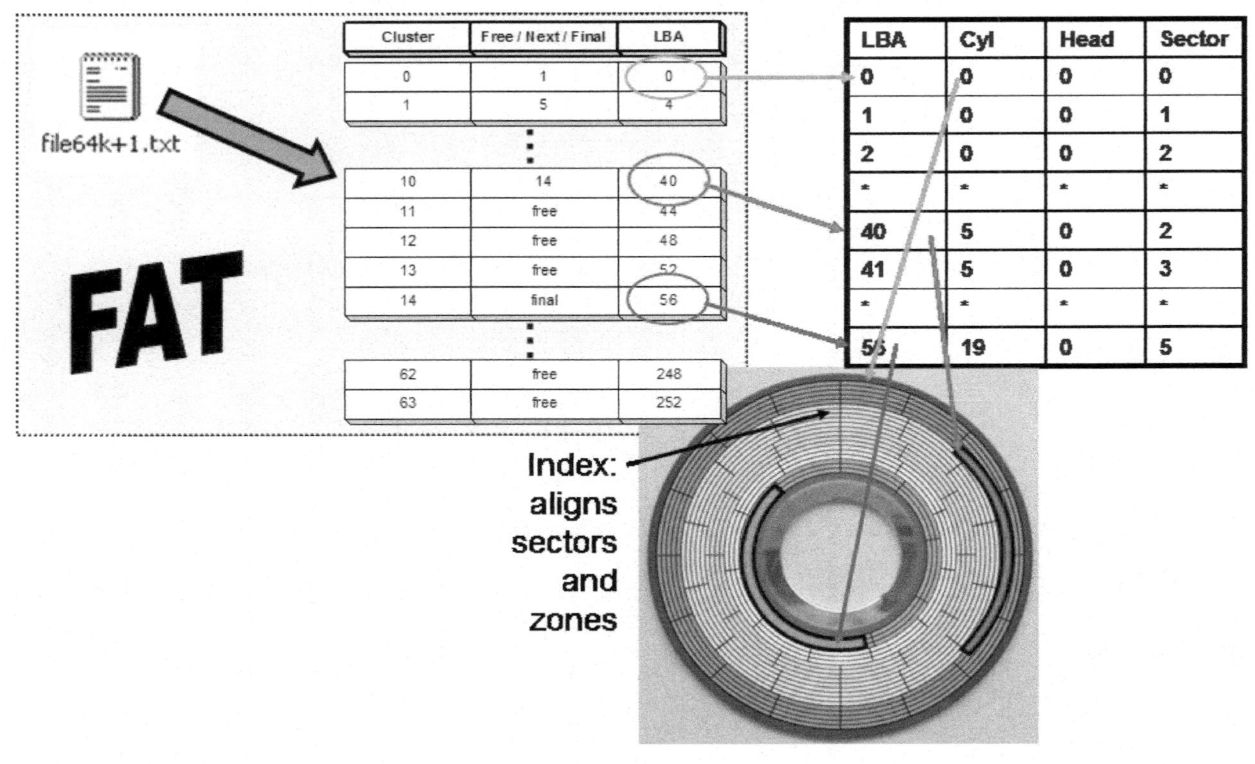

Figure 2.12 Example of CHS to LBA translation.

48-Bit Addressing

When ATAPI-6 introduced the extended 48-bit sector addressing mechanism, it also had to create a set of commands to support the extended addressing feature (EXT).

- This provided support for larger capacity devices.
    - Up to approximately 144 petabytes.
- In the first implementations the LBA and Sector Count Registers are each a 2-byte deep FIFO.
- Each time a register is written
    - new content is placed in "most recently written" location; and
    - old content is moved to "previous content" location.
- Legacy Register appearance is maintained.
- There is a special set of 48-bit ATA commands.
    - READ SECTORS EXT, WRITE DMA EXT, etc.
    - Device support for 48-bit addressing is indicated via the Identify Device command.
- 48-bit addressing will add extension registers, as shown in Table 2.19.

Table 2.19 Extended 48-Bit Address Registers

Register	"most recently written"	"previous content"
Features	Reserved	Reserved
Sector Count	Sector count (7:0)	Sector Count Ext. (15:8)
LBA Low	LBA (7:0)	LBA Ext. (31:24)
LBA Mid	LBA (15:8)	LBA Ext. (39:32)
LBA High	LBA (23:16)	LBA Ext. (47:40)
Device Register	Bits 7 and 5 are obsolete, the LBA bit shall be set to one, the DEV bit shall indicate the selected device, bits (3:0) are reserved	Reserved

2.7 ATA Protocols

ATA Protocols are described in the Transport Standards (e.g., ATA8-APT and ATA8-AST). The protocols listed below are implemented by all transports that use ATA commands defined in the Standard. The following list of protocols is described in ATA8-AAM, and the implementation of each protocol is described in the Transport Standards:

a. Non-Data Command Protocol
b. PIO Data-In Command Protocol
c. PIO Data-Out Command Protocol
d. DMA Command Protocol
e. PACKET Command Protocol
f. DMA Queued Command Protocol
g. Execute Device Diagnostic Command Protocol
h. Device Reset Command Protocol

2.7.1 Data Transfer Modes – PIO

Programmed I/O (PIO) was the first data transfer method defined for ATA.

- Host CPU directly controls data transfer to/from the device, one word at a time
  - Considerable host processing overhead
  - Multiple PIO modes are defined
    - Mode 0 3.3 MB/s
    - Mode 1 5.2 MB/s
    - Mode 2 8.3 MB/s
    - Mode 3 11.1 MB/s
    - Mode 4 16.7 MB/s

2.7.2 Data Transfer Modes—DMA

DMA modes enable data transfer between host memory and device without intervention from host CPU. There are two types of DMA Mode defined:

- Single word DMA
  - Requires DMA setup for each word to be transferred
  - Single word DMA is not much more efficient than PIO
  - No longer used
- Multiword DMA
  - Allows a "burst" of data words in a single DMA operation
  - "First-party DMA" used today:
    - First-Party DMA refers to the device that controls the DMA operation.
    - Third-Party DMA refers to host-implemented DMA control, which is no longer used.

Data Transfer Modes—UltraDMA

Increasing performance from the Multiword DMA Mode 2 data rate of 16.7 MB/s by simply increasing the clock frequency of the ATA bus did not really present a viable option for reasons of signal integrity across the 40-way ATA bus. Poor signal integrity would result in poor data integrity, and the ATA interface, at this point, did not provide mechanisms to check that data integrity had been preserved after transmission from host to device or vice versa. The changes and additions made to the ATA/ATAPI Standard (see Table 2.20) to provide ongoing performance enhancement took several forms:

- The introduction of "double transition clocking" on the ATA bus
- The definition of a new 80-conductor ribbon cable
- The introduction of CRC checking on data transfers

Table 2.20 DMA Transfer Modes, Transfer Rates, and ATA Version

UltraDMA Mode	Cycle Time (ns)	Max Transfer Rate (MB/s)	ATA/ATAPI
Mode 0	240	16.7	4
Mode 1	160	25.0	4
Mode 2	120	33.3	4
Mode 3	90	44.4	5
Mode 4	60	66.7	5
Mode 5	40	100.0	6
Mode 6	30	133.3	7

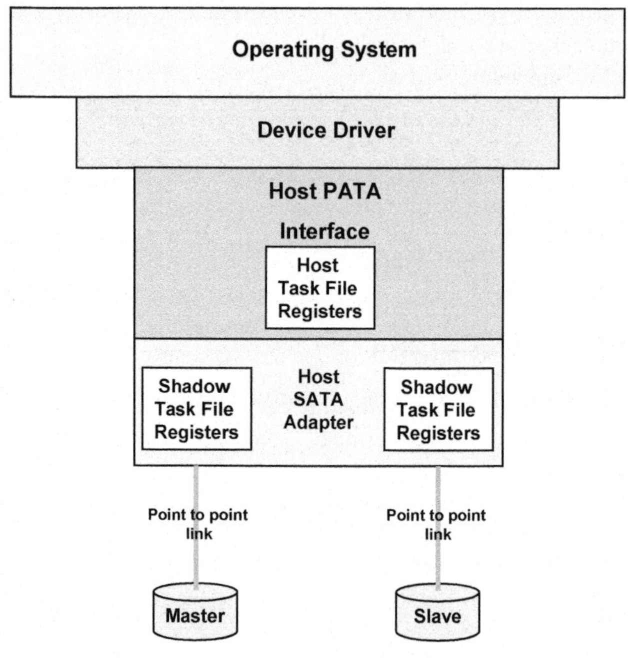

Figure 2.13 Shadow Task File registers.

2.7.3 Parallel ATA Emulation

Emulation of parallel implementations of ATA device behavior as perceived by the host BIOS or software driver is a cooperative effort between the device and serial interface host adapter hardware. The behavior of Command and Control Block registers, PIO and DMA data transfers, resets, and interrupts are all emulated.

The host adapter contains a set of registers that ***Shadow*** the contents of the traditional device registers (see Figure 2.13), referred to as the Shadow Command Block and Shadow Control Block.

- The Command Block registers are used for sending commands to the device or posting status from the device.
    - These registers include the LBA High, LBA Mid, LBA Low, Device, Sector Count, Command, Status, Features, Error, and Data registers.
- The Control Block registers are used for device control and to post alternate status. These registers include the Device Control and Alternate Status registers.

All serial implementations of ATA devices behave like Device 0 devices. Devices shall ignore the DEV bit in the Device field of received register FISes, and it is the responsibility of the host adapter to gate transmission of register FISes to devices, as appropriate, based on the value of the DEV bit.

After a reset or power-on, the host bus adapter emulates the behavior of a traditional ATA system during device discovery.

- Immediately after reset, the host adapter shall place the value 0x7Fh in its Shadow Status Register and Shadow Alternate Status Register and shall place the value 0xFFh in all the other Shadow Command Block registers (0xFFFFh in the Data Register).

- In this state, the host bus adapter shall not accept writes to the Shadow Command Block and Shadow Control Block.
- When the host Phy detects the presence of an attached device, the host bus adapter shall set bit 7 in the Shadow Status Register, yielding the value 0xFFh or 0x80h, and the host bus adapter shall allow writes to the Shadow Command Block and Shadow Control Block.
- If a device is present, the Phy shall take no longer than 10 ms to indicate that it has detected the presence of a device and has set bit 7 in the Shadow Status Register.
- The Serial implementation of ATA time limit of 10 ms is different from the parallel implementation of ATA.
- When the attached device establishes communication with the host bus adapter, it shall send a Register FIS to the host, resulting in the Shadow Command Block and Shadow Control Block being updated with values appropriate for the attached device.
    - This is known as a "Signature."

The host adapter may present a Device 0-only emulation to host software, that is, each device is a Device 0, and each Device 0 is accessed at a different set of host bus addresses. The host adapter may optionally present a Device 0/Device 1 emulation to host software, that is, two devices on two separate serial ports are represented to host software as a Device 0 and a Device 1 accessed at the same set of host bus addresses.

2.8 SATA Technical Overview

This section will provide the basic concepts of the SATA architecture. Whereas the details for each topic are covered extensively in the following chapters, this section provides a technical overview of the architecture.

2.8.1 Serial Links

SATA physical links use high-speed serial electrical signaling. Like most other high-speed serial interfaces, SATA uses low-voltage differential signaling with separate transmit and receive lines. This results in a four-wire interface (plus shielding) as shown in Figure 2.14.

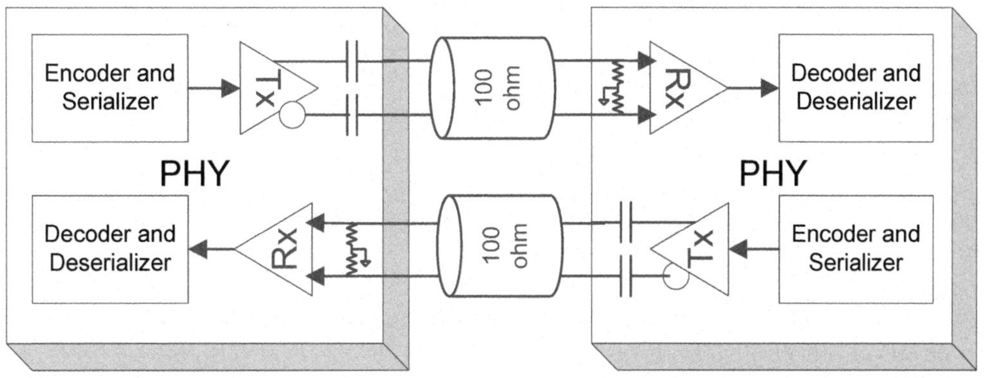

Figure 2.14 SATA link.

A physical copper link is a set of four wires used as two differential signal pairs.

- One differential signal transmits in one direction, whereas the other differential signal transmits in the opposite direction.
- Although the bi-directional links indicate that data could be transmitted in both directions simultaneously, SATA is a half-duplex data transfer architecture. This occurs because the receiver is always communicating its protocol state to the transmitting device/host.

A PHY is a transceiver; it is the object in a device that electrically interfaces to a physical link.

2.8.2 Differential Signaling

SATA physical links use differential signaling (as do most high-speed serial links such as Fibre Channel and InfiniBand). In differential signaling, one signal line carries a positive or non-inverted output, and the other signal line carries a negative or inverted output, as shown in Figure 2.15.

A logical one is indicated when the voltage level of the positive line is greater than that of the negative line. A logical zero is indicated when the voltage level of the negative line is greater than that of the positive line.

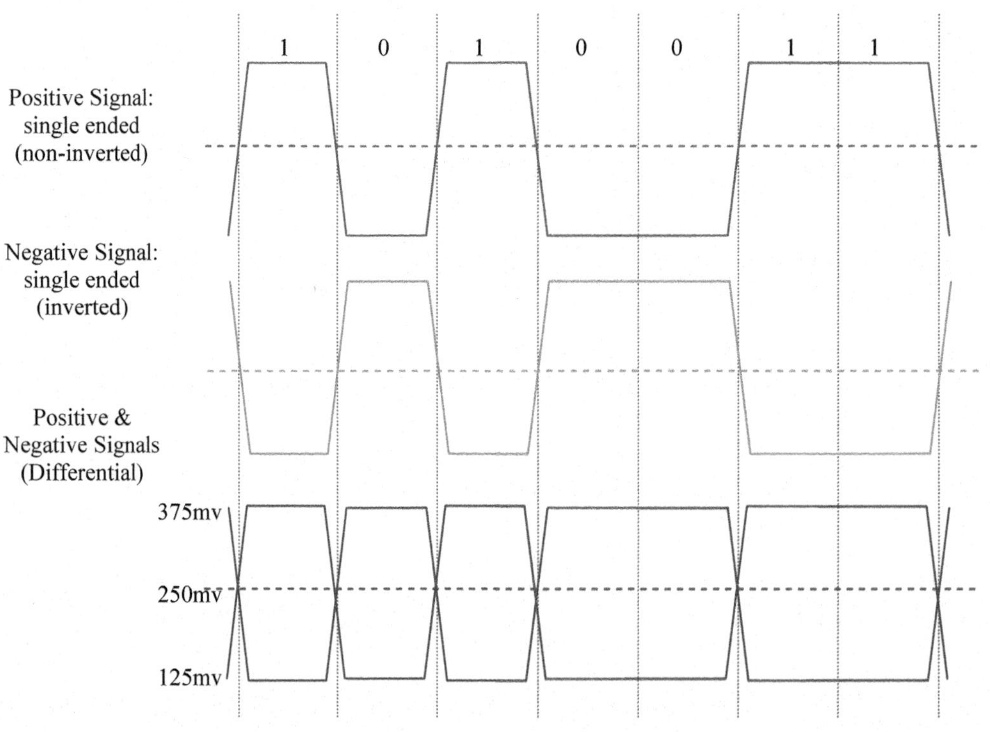

Figure 2.15 Differential signaling levels.

2.9 Encoding

All data bytes transferred in SATA are encoded into 10-bit transmission characters prior to transmission and decoded back into eight-bit bytes upon reception. The encoding scheme used is the 8b10b coding scheme (8b10b encoding is used by a number of other data interfaces, including Fibre Channel, Gigabit Ethernet optical links, SAS, PCI Express, and InfiniBand among others).

- Encoding provides the mechanism to allow for transmitting information across a serial link.
- Encoding turns arbitrary data into patterns (signals) that are suitable for transmission across serial connections.
- Clock recovery and error detection are built into the encoding scheme.

2.9.1 8b10b Encoding

The encoding scheme used by SATA encodes 8-bit data bytes into 10-bit transmission characters to improve the transmission characteristics of the serial data stream. This scheme is called "8b10b" encoding, referring to the number of data bits input to the encoder and the number of bits output from the encoder.

- 8b10b coding converts 8-bit bytes into 10-bit data characters for transmission on the wire.
- Reasons for encoding:
    - Clock recovery
    - DC balance
    - Special characters
    - Error detection
    - Provides a constant bit rate
- Invented by IBM in 1983.
- Used by Fibre Channel, Gigabit Ethernet, 1394b, and many other Standards.
- All high-speed serial interfaces require a signal always present, even when the link is "idle"—that is, no data is being transmitted.
    - In this case, there will always be some type of signal present.

Figure 2.16 shows how 8 bits of data is encoded to 10 bits before being transmitted across the link and how after 10 bits are received, they are decoded on the receiver. Other blocks demonstrate how clocks are generated and recovered.

2.9.2 Encoding Characteristics

To facilitate the processing of information, data fields and structures are aligned on double-word (dword) boundaries, when possible (see Table 2.21). A dword consists of four 8-bit bytes, or a total of 32 bits. However, when each of the 8-bit bytes is encoded prior to transmission, the resulting transmission dword consists of 40 bits, representing the four 10-bit transmission characters.

- Byte = 8 bits (xxh)
- Character = 10 bits (symbolized by Dxx.y for Data Characters or Kxx.y for Control Characters)
- dword = 4 characters (or 4 bytes, depending on context)
- Double word (a computer word = 16 bits)
    - Represents 32 bits of data
    - 40 bits on the wire

- A dword is either
  - a **data dword**;
  - a **primitive**;
  - or an invalid dword.

Bits shown on the first (top) row are numbered from left to right, 31–00. Each group of 8 bits becomes a character (second row), which is the encoded representation of a hexadecimal byte (third row). Two bytes are equal to a word (fourth row), and two words make up a dword (fifth row).

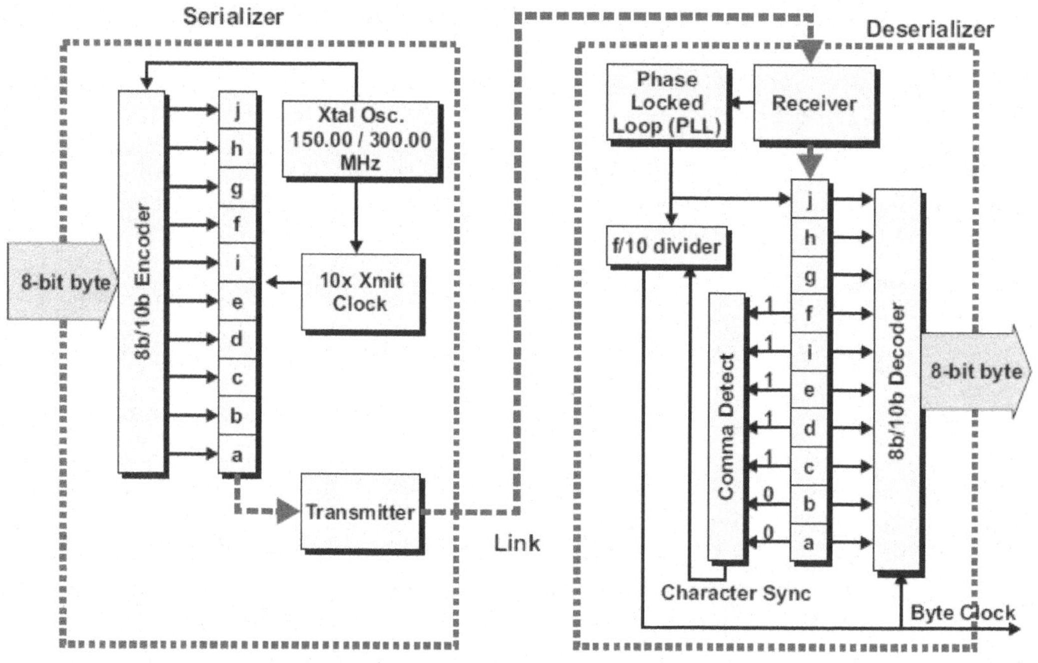

Figure 2.16 Transmitter and receiver logic.

Table 2.21 SATA Character Format

31	30	29	28	27	26	25	24	23	22	21	20	19	18	17	16	15	14	13	12	11	10	09	08	07	06	05	04	03	02	01	00
4th character								3rd character								2nd character								1st character							
byte 3								byte 2								byte 1								byte 0							
word 1																word 0															
dword																															

Dword Notation

Byte orientation depends upon interface and, to some extent, how an analyzer may present primitives. In the olden days of PC-AT, all words were read from right to left, which was referred to as "little-endian" notation (Table 2.22). This notation positions byte 0 of the dword in the right most position, and byte 3 is the left most byte, similar to reading from right to left.

Table 2.22 SATA Primitives Notation

Primitive Name	Byte 3	Byte 2	Byte 1	Byte 0
ALIGN	D27.3	D10.2	D10.2	K28.5
CONT	D25.4	D25.4	D10.5	K28.3
- - -	- - -	- - -	- - -	K28.3

SAS primitives use "big-endian" (Table 2.23) or first byte/last byte notation. This notation positions byte 0 (1st byte) of the dword in left most position and byte 3 (4th byte) is the right most byte, similar to reading from left to right.

Table 2.23 SAS Primitives Notation

Primitive Name	1st Byte 0	2nd Byte 1	3rd Byte 2	4th Byte 3
ALIGN (0)	K28.5	D10.2	D10.2	D27.3
SATA_CONT	K28.3	D10.5	D25.4	D25.4
- - -	Kxx.y	- - -	- - -	- - -

2.10 SATA Primitives and Data Words

Primitives consist of 1 control character and 3 data characters (Table 2.24).

- The first character is K28.3 for most of the SATA primitives.
- The last three characters are data characters.

The "xx.y" value is determined by the 8b/10b encoding scheme outlined in the link layer.

- Example: C0h = D05.6

Table 2.24 SATA Primitive Dword Format and SOF Example

<table>
<tr><td>31</td><td>30</td><td>29</td><td>28</td><td>27</td><td>26</td><td>25</td><td>24</td><td>23</td><td>22</td><td>21</td><td>20</td><td>19</td><td>18</td><td>17</td><td>16</td><td>15</td><td>14</td><td>13</td><td>12</td><td>11</td><td>10</td><td>09</td><td>08</td><td>07</td><td>06</td><td>05</td><td>04</td><td>03</td><td>02</td><td>01</td><td>00</td></tr>
<tr><td colspan="8">byte 3</td><td colspan="8">byte 2</td><td colspan="8">byte 1</td><td colspan="8">byte 0</td></tr>
<tr><td colspan="8">Dxx.y</td><td colspan="8">Dxx.y</td><td colspan="8">Dxx.y</td><td colspan="8">K28.3</td></tr>
<tr><td colspan="8">D23.1</td><td colspan="8">D23.1</td><td colspan="8">D21.5</td><td colspan="8">K28.3</td></tr>
</table>

Data words are made up of just Data Characters, as seen in Table 2.25.

Table 2.25 SATA Data Dword Format with Data Dword—Example

<table>
<tr><td>31</td><td>30</td><td>29</td><td>28</td><td>27</td><td>26</td><td>25</td><td>24</td><td>23</td><td>22</td><td>21</td><td>20</td><td>19</td><td>18</td><td>17</td><td>16</td><td>15</td><td>14</td><td>13</td><td>12</td><td>11</td><td>10</td><td>09</td><td>08</td><td>07</td><td>06</td><td>05</td><td>04</td><td>03</td><td>02</td><td>01</td><td>00</td></tr>
<tr><td colspan="8">byte 3</td><td colspan="8">byte 2</td><td colspan="8">byte 1</td><td colspan="8">byte 0</td></tr>
<tr><td colspan="8">Dxx.y</td><td colspan="8">Dxx.y</td><td colspan="8">Dxx.y</td><td colspan="8">Dxx.y</td></tr>
<tr><td colspan="8">D23.1</td><td colspan="8">D23.1</td><td colspan="8">D21.5</td><td colspan="8">D21.5</td></tr>
</table>

2.10.1 SATA Primitives

Tables 2.26 and 2.27 show all SATA primitives, their descriptions, and their decoded characters.

Table 2.26 lists most of the primitives, with brief descriptions. Primitives are used to determine the Link Layer Protocol. They are used to request the transfer of frames and acknowledge the receipt of frames. They are even used to determine the start and end of frames.

Table 2.27 shows the dword representation of some primitives.

Table 2.26 SATA Primitives

Primitive	Name	Description
ALIGN	Physical layer control	Upon receipt of an ALIGN, the physical layer readjusts internal operations, as necessary, to perform its functions correctly. This primitive is always sent in pairs; there is no condition in which an odd number of ALIGN primitives shall be sent (except as noted for retimed loopback).
EOF	End of frame	EOF marks the end of a frame. The previous non-primitive DWORD is the CRC for the frame.
R_ERR	Reception error	Current node (host or device) detected error in received payload.
R_IP	Reception in progress	Current node (host or device) is receiving payload.
R_OK	Reception with no error	Current node (host or device) detected no error in received payload.
R_RDY	Receiver ready	Current node (host or device) is ready to receive payload.
SOF	Start of frame	Start of a frame. Payload and CRC follow to EOF.
SYNC	Synchronization	Synchronizing primitive—always idle.
WTRM	Wait for frame termination	After transmission of any of the EOFs, the transmitter will transmit WTRM while waiting for reception status from receiver.
X_RDY	Transmission data ready	Current node (host or device) has payload ready for transmission.

Table 2.27 SATA Primitives Decodes

Primitive name	Byte 3 contents	Byte 2 contents	Byte 1 contents	Byte 0 contents
ALIGN	D27.3	D10.2	D10.2	K28.5
EOF	D21.6	D21.6	D21.5	K28.3
R_ERR	D22.2	D22.2	D21.5	K28.3
R_IP	D21.2	D21.2	D21.5	K28.3
R_OK	D21.1	D21.1	D21.5	K28.3
R_RDY	D10.2	D10.2	D21.4	K28.3
SOF	D23.1	D23.1	D21.5	K28.3
SYNC	D21.5	D21.5	D21.4	K28.3
WTRM	D24.2	D24.2	D21.5	K28.3
X_RDY	D23.2	D23.2	D21.5	K28.3

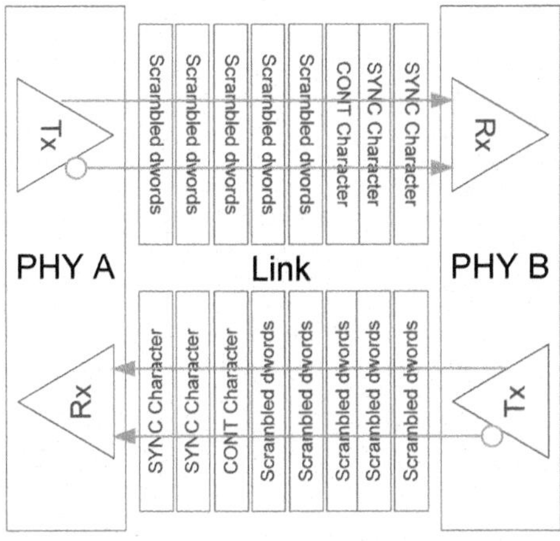

Figure 2.17 Idle link.

2.11 Idle Serial Links

In SATA, either SYNC primitives or scrambled dwords (see Figure 2.17) are transmitted when there are no frames or other primitives to transmit (all serial interfaces require this). This is required in all high-speed serial interfaces and provides two main functions: (1) It keeps the PHYs synchronized; and (2) it provides a mechanism whereby link failures can be immediately detected.

The CONT primitive is used for all repetitive primitives. This primitive sequence is used to minimize the effect of electromagnetic interference (EMI) caused by continually transmitting the same information over and over. An analyzer trace can be seen in Figure 2.18, which demonstrates how an idle link would look from an analytical perspective.

Time Stamp	I1	T1
720.880 (us)	XXXX (SATA_SYNC)	XXXX (SATA_SYNC)
720.906 (us)	XXXX (SATA_SYNC)	XXXX (SATA_SYNC)
720.933 (us)	XXXX (SATA_SYNC)	XXXX (SATA_SYNC)
720.960 (us)	XXXX (SATA_SYNC)	XXXX (SATA_SYNC)
720.986 (us)	XXXX (SATA_SYNC)	XXXX (SATA_SYNC)
721.013 (us)	XXXX (SATA_SYNC)	XXXX (SATA_SYNC)
721.040 (us)	XXXX (SATA_SYNC)	XXXX (SATA_SYNC)
721.066 (us)	XXXX (SATA_SYNC)	XXXX (SATA_SYNC)
721.093 (us)	XXXX (SATA_SYNC)	XXXX (SATA_SYNC)
721.120 (us)	XXXX (SATA_SYNC)	XXXX (SATA_SYNC)
721.146 (us)	XXXX (SATA_SYNC)	XXXX (SATA_SYNC)
721.173 (us)	XXXX (SATA_SYNC)	XXXX (SATA_SYNC)
721.200 (us)	XXXX (SATA_SYNC)	XXXX (SATA_SYNC)
721.226 (us)	XXXX (SATA_SYNC)	XXXX (SATA_SYNC)
721.253 (us)	XXXX (SATA_SYNC)	XXXX (SATA_SYNC)
721.280 (us)	XXXX (SATA_SYNC)	XXXX (SATA_SYNC)
721.306 (us)	XXXX (SATA_SYNC)	XXXX (SATA_SYNC)
721.333 (us)	XXXX (SATA_SYNC)	XXXX (SATA_SYNC)
721.360 (us)	XXXX (SATA_SYNC)	XXXX (SATA_SYNC)
721.386 (us)	XXXX (SATA_SYNC)	XXXX (SATA_SYNC)
721.413 (us)	XXXX (SATA_SYNC)	XXXX (SATA_SYNC)
721.440 (us)	XXXX (SATA_SYNC)	XXXX (SATA_SYNC)
721.466 (us)	XXXX (SATA_SYNC)	XXXX (SATA_SYNC)
721.493 (us)	XXXX (SATA_SYNC)	XXXX (SATA_SYNC)
721.520 (us)	XXXX (SATA_SYNC)	XXXX (SATA_SYNC)
721.546 (us)	XXXX (SATA_SYNC)	XXXX (SATA_SYNC)

Figure 2.18 Analyzer trace display of idle link. (Trace display captured using a Teledyne/LeCroy Analyzer.)

2.12 Frame Transmission

All information is transmitted across the link in frames.

- All frames (Figure 2.19) include
  - a start-of-frame (SOF) primitive;
  - one or more data words, which contains the information;
  - CRC (the last data word before the EOF);
  - an end-of-frame primitive (EOF).
- Frames require a permission primitive before frame transmission.
- Frames are verified once transmitted.

	Frame delimiter	Payload (2048 dwords)				Frame check	Frame delimiter	
Idle	SOF	DW 0	DW 1	DW n	DW last	CRC	EOF	Idle

Figure 2.19 Frame format.

Figure 2.20 shows the logical display of dwords and primitives when a SATA frame is transmitted. When the link is idle, SYNC (idle) primitives or scrambled dwords are continuously transmitted. When a host or device originates a frame, it encapsulates the frame contents between a SOF and EOF primitive. Data dwords make up the contents of a frame, and all frames must align on 4-byte boundaries.

Description	Byte n+3	Byte n+2	Byte n+1	Byte n
Scrambled dwords	D21.6	D11.5	D21.7	D25.4
Scrambled dwords	D20.0	D27.3	D11.4	D15.5
SOF	D23.1	D23.1	D21.5	K28.3
Data dword	Features	Command	various bits	FIS Type (27h)
Data dword	Dev/Head	Cyl High	Cyl Low	Sector Number
Data dword	Features (exp)	Cyl High (exp)	Cyl Low (exp)	Sector # (exp)
Data dword	Control	Reserved	Sector Cnt (exp)	Sector Count
Data dword	Reserved	Reserved	Reserved	Reserved
CRC	D23.4	D13.7	D25.4	D13.3
EOF	D21.6	D21.6	D21.5	K28.3
WTRM	D24.2	D24.2	D21.5	K28.3
WTRM	D24.2	D24.2	D21.5	K28.3

Figure 2.20 Frame transmission example shows host to device FIS).

Time Stamp	I1	T1
619.146 (us)	XXXX (SATA_X_RDY)	XXXX (SATA_R_RDY)
619.173 (us)	XXXX (SATA_X_RDY)	XXXX (SATA_R_RDY)
619.200 (us)	XXXX (SATA_X_RDY)	XXXX (SATA_R_RDY)
619.226 (us)	XXXX (SATA_X_RDY)	XXXX (SATA_R_RDY)
619.253 (us)	XXXX (SATA_X_RDY)	XXXX (SATA_R_RDY)
619.280 (us)	XXXX (SATA_X_RDY)	XXXX (SATA_R_RDY)
619.306 (us)	XXXX (SATA_X_RDY)	XXXX (SATA_R_RDY)
619.333 (us)	XXXX (SATA_X_RDY)	XXXX (SATA_R_RDY)
619.360 (us)	XXXX (SATA_X_RDY)	XXXX (SATA_R_RDY)
619.386 (us)	XXXX (SATA_X_RDY)	XXXX (SATA_R_RDY)
619.413 (us)	XXXX (SATA_X_RDY)	XXXX (SATA_R_RDY)
619.440 (us)	XXXX (SATA_X_RDY)	XXXX (SATA_R_RDY)
619.466 (us)	XXXX (SATA_X_RDY)	XXXX (SATA_R_RDY)
619.493 (us)	SATA_SOF	XXXX (SATA_R_RDY)
619.520 (us)	2780A000	XXXX (SATA_R_RDY)
619.546 (us)	00000200	XXXX (SATA_R_RDY)
619.573 (us)	00000000	XXXX (SATA_R_RDY)
619.600 (us)	00000000	XXXX (SATA_R_RDY)
619.626 (us)	00000000	XXXX (SATA_R_RDY)
619.653 (us)	CRC: 2DADF342	XXXX (SATA_R_RDY)
619.680 (us)	SATA_EOF	XXXX (SATA_R_RDY)
619.706 (us)	SATA_WTRM	XXXX (SATA_R_RDY)
619.733 (us)	SATA_WTRM	XXXX (SATA_R_RDY)
619.760 (us)	SATA_CONT	XXXX (SATA_R_RDY)
619.786 (us)	XXXX (SATA_WTRM)	XXXX (SATA_R_RDY)
619.813 (us)	XXXX (SATA_WTRM)	XXXX (SATA_R_RDY)

Figure 2.21 Trace example of frame transmission. (Trace example captured using a Teledyne/LeCroy Analyzer.)

An example analyzer trace display of the Host-to-Device FIS can be seen in Figure 2.21. The trace demonstrates:

- For each line: Time stamp Host (I1) Device (T1)
  - Each column shows what the corresponding device is transmitting.
- The FIS begins with SATA_SOF.
- The FIS contents are displayed in big-endian format.
  - Notice 27h appears on left (verses right).
  - Programmers read from left to right (see Figure 2.22), therefore analyzers allow information to be presented in either format (hardware engineers typically prefer little-endian).
- When a transmitter sends the EOF, it is immediately followed by the WTRM primitive (wait for termination) sequence.

Index	Hex				B0	B1	B2	B3
000000	27	80	A0	00	STP Frame Type (0x27) Register Host to Device	PM Port 0x00 Res... Res... Res... C 0x00 0x00 0x00 0x01	Command (0xA0) Packet	Features 0x00
000001	00	00	02	00	LBA Low 0x00	LBA Mid 0x00	LBA High 0x02	Device 0x00
000002	00	00	00	00	LBA Low (exp) 0x00	LBA Mid (exp) 0x00	LBA High (exp) 0x00	Features (exp) 0x00
000003	00	00	00	00	Sector Count 0x00	Sector Count (exp) 0x00	Reserved 0x00	Control 0x00
000004	00	00	00	00	Reserved 0x00	Reserved 0x00	Reserved 0x00	Reserved 0x00
000005	2D	AD	F3	42	CRC 0x2DADF342			

Figure 2.22 SATA frame view display (shown in big-endian format). (Captured using a Teledyne/LeCroy Analyzer.)

Frames are preceded by primitives or scrambled Dwords. Scrambled Dwords are used to minimize the impact of EMI created by repeating bit patterns on high-speed serial interfaces. By scrambling the data, the encoded transmitted bit streams continuously change, therefore eliminating EMI issues. The SATA protocol to enable this mode can be seen in Figure 2.21 by looking for the SATA_WTRM right after the SATA_EOF.

- The primitive is transmitted twice SATA_WRTM – SATA_WTRM.
- They are immediately followed by SATA_CONT.
- The transmitter transmits scrambled data Dwords using an industry standard algorithm. Each scrambled Dword is shown as "XXXX" in the trace, and, in this example, they represent the SATA_WTRM primitive.

2.13 Link Layer Protocol

The SATA interface uses a half-duplex protocol, wherein while one device is transmitting, the receiving device sends an indication in the opposite direction as to what is being received. For example, when a device transmits a frame, the receiving device sends a primitive in the opposite direction indicating that receipt of frame is in progress (RIP). This section covers the Link layer protocol that is used by the SATA architecture.

The series of diagrams that follow demonstrate the Link Layer protocol of SATA. Again, due to the characteristics of serial links, there always must be a signal present on the wire.

Step 1 – Host and device are both in IDLE state transmitting SYNC primitives.

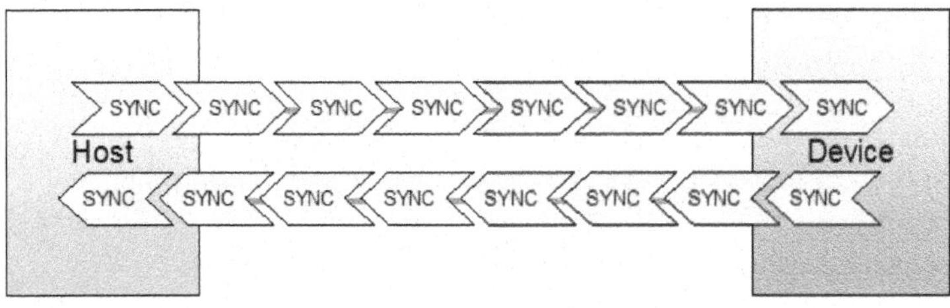

Step 2 – When the Host has data ready to transmit, the Host sends X_RDY primitives and waits for a response from the device.

Step 3 – When the device receives X_RDY, it responds by sending R_RDY primitives indicating that it is ready to receive data.

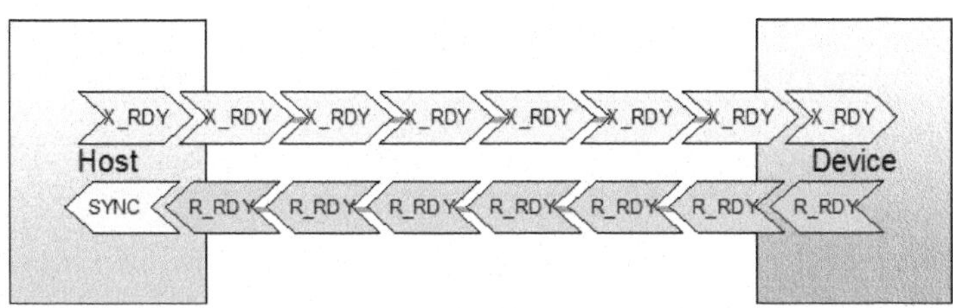

Step 4 – When the host receives R_RDY from the device, it starts transmitting a frame by sending an SOF primitive followed by data Dwords.

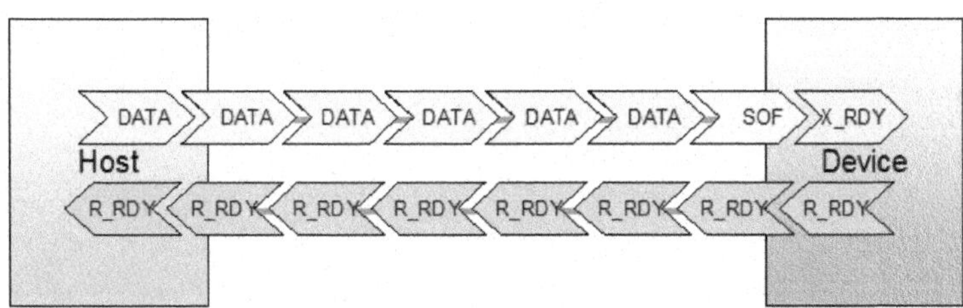

Step 5 – When the device receives the SOF, it responds by sending RIP primitives indicating frame reception in progress.

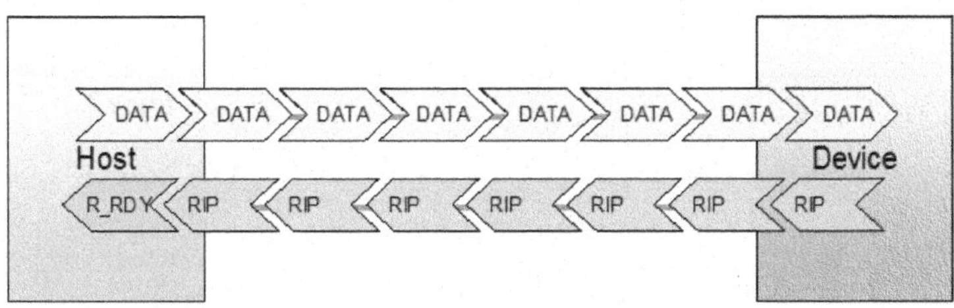

Step 6 – At the end of the data, the host sends the calculated CRC followed by an EOF primitive. The host then sends WTRM primitives and awaits a response from the device.

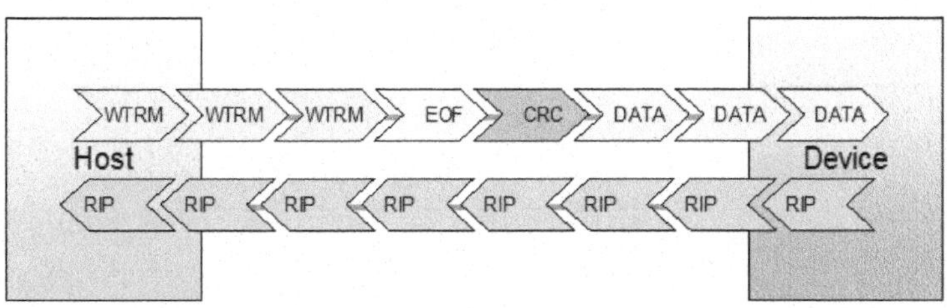

Step 7 – When the device receives the EOF, it knows the previous Dword was CRC; it then compares the received CRC with its own computed CRC value.

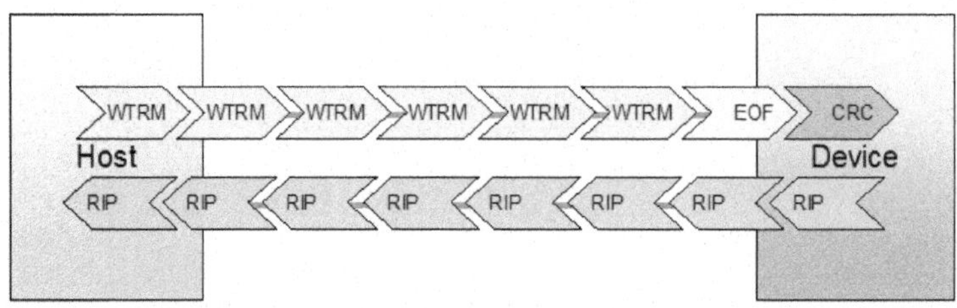

Step 8 – If the CRC indicates that the frame was received without error, the device transmits R_OK primitives.

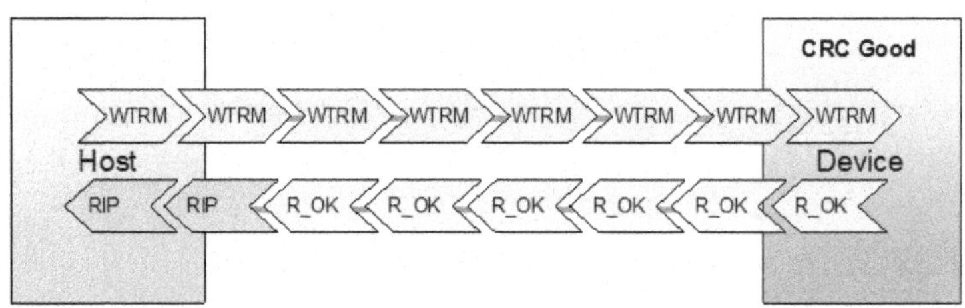

Step 9 – When the host receives R_OK, it knows that the frame was received without error and returns to the IDLE state and transmits SYNC primitives.

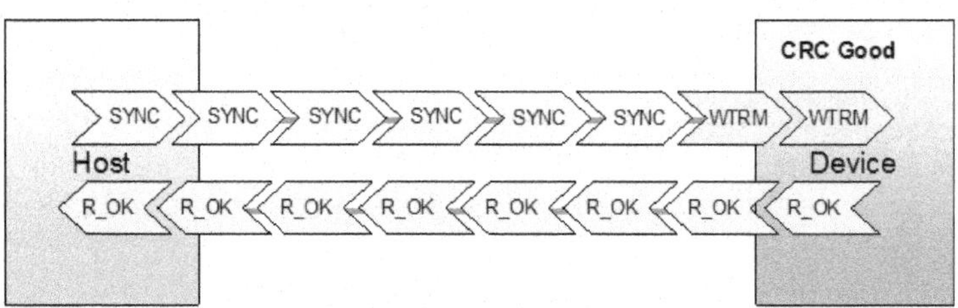

Step 10 – When the device receives SYNC, it stops sending R_OK, returns to the IDLE state, and sends SYNC primitive. At this point, the frame transfer is complete.

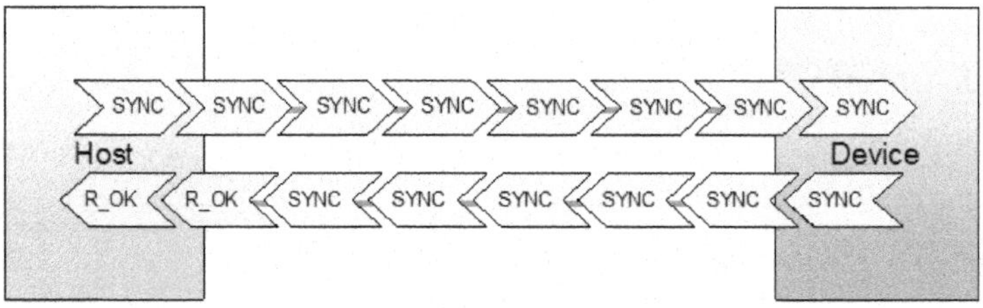

2.13.1 Link Layer Protocol Summary

In SATA, primitives are continuously transmitted across the links to indicate the state of the device transmitting the primitive. In order to make Figure 2.23 more readable, only one instance of each primitive is shown.

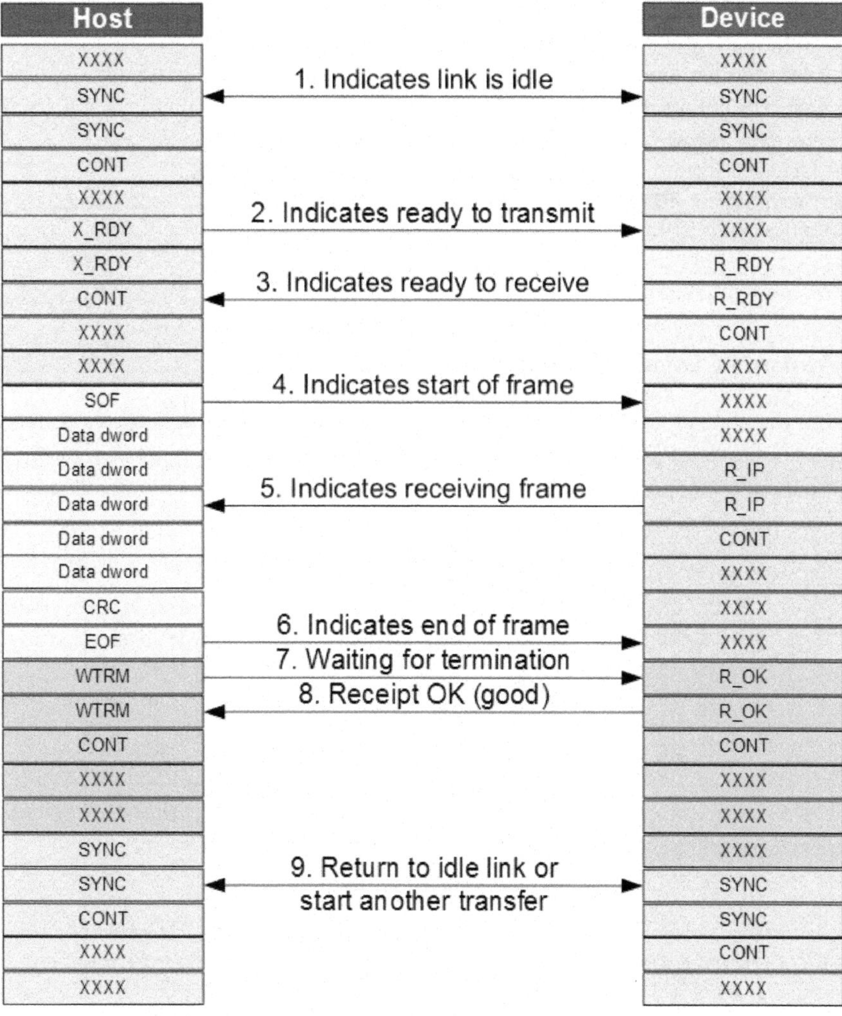

Figure 2.23 Link layer protocol of frame transmission.

Link Layer Protocol Steps

1. Link idle
    - SYNC transferred by both devices
2. Request permission to transfer
    - Host signals request to transfer frame by sending an X_RDY

3. Permission granted
   - Device signals permission to transmit frame by sending R_RDY
4. Transmit SATA frame
   - Host signals beginning of frame with SOF primitive
   - Follows with Data Words
   - Ends with CRC and EOF primitive
5. Receipt in progress
   - Device informs host of receipt of frame by sending R_IP
6. End of frame
7. Wait for termination
   - After EOF host transmits WTRM
8. Acknowledge frame
   - Device informs host of successful receipt of frame with R_OK
   - Or SATA_R_ERROR if an error occurred
9. Link idle
   - Device and host return to idle link by transmitting SYNCs

2.13.2 Transport Layer Protocol

SATA utilizes the same protocol as its parallel ATA ancestor but replaces the register loads with Frame Information Structures or FISs. Table 2.28 provides an overview of the types of frames used by the SATA transport layer.

Table 2.28 FIS Codes and Descriptions

FIS Type (Code)	Description	Direction
27h	Register host to device Used to transmit Command information	H => D
34h	Register device to host Used to transmit Status information	D => H
39h	DMA Activate Used during DMA Write operations	D => H
41h	DMA Setup Used during first Party DMA transfers	D => H
5Fh	PIO Setup Used during Programed IO transfers	D => H
46h	Data Carries customer Data – 8k maximum	D => H

Tables 2.29 to 2.31 show the three FISes:

- Host to device—used to transmit command information
- Device to host—used to transmit status and error information
- Data—used to move data between devices

The protocol used is not new to storage technology in that the host sends a command to the device, which in turn decodes and executes the command. Some commands require the transfer of data, and

Table 2.29 Host to Device FIS—Example

	byte n+3	byte n+2	byte n+1	byte n
SOF	D23.1	D23.1	D21.5	K28.3
Data dword 0	Features 7:0	Command	**c** | r | r | r | PM Port	FIS Type (**27h**)
Data dword 1	Device [Dev/Head]	LBA 23:16 [Cyl High]	LBA 15:8 [Cyl Low]	LBA 7:0 [Sector Number]
Data dword 2	Features 15:8	LBA 47:40 [Cyl High (exp)]	LBA 39:32 [Cyl Low (exp)]	LBA 31:24 [Sector # (exp)]
Data dword 3	Control	***ICC***	Count 15:8 [Sector Cnt (exp)]	Count 7:0 [Sector Count]
Data dword 4	Reserved	Reserved	Reserved	Reserved
CRC	Dxx.y	Dxx.y	Dxx.y	Dxx.y
EOF	D21.6	D21.6	D21.5	K28.3

Note: c = Command bit; r = reserved; ICC = Isochronous Command Completion.

Table 2.30 Device to Host FIS—Example

	byte n+3	byte n+2	byte n+1	byte n
SOF	D23.1	D23.1	D21.5	K28.3
Data dword 0	Error	Status	r | **i** | r | r | PM Port	FIS Type (**34h**)
Data dword 1	Device [Dev/Head]	LBA 23:16 [Cyl High]	LBA 15:8 [Cyl Low]	LBA 7:0 [Sector Number]
Data dword 2	Reserved	LBA 47:40 [Cyl High (exp)]	LBA 39:32 [Cyl Low (exp)]	LBA 31:24 [Sector # (exp)]
Data dword 3	Reserved	Reserved	Count 15:8 [Sector Cnt (exp)]	Count 7:0 [Sector Count]
Data dword 4	Reserved	Reserved	Reserved	Reserved
CRC	Dxx.y	Dxx.y	Dxx.y	Dxx.y
EOF	D21.6	D21.6	D21.5	K28.3

Table 2.31 Data FIS—Example

	byte n+3	byte n+2	byte n+1	byte n
SOF	D23.1	D23.1	D21.5	K28.3
Data dword 0	Reserved			FIS Type (46h)
Data dword 1	Data dword			
Data dword 2	(minimum 1 dword – maximum 2048 dwords)			
Data dword 3	Data dword			
Data dword 4	Data dword			
Data dword -	Data dword			
Data dword n	Data dword			
CRC	Dxx.y	Dxx.y	Dxx.y	Dxx.y
EOF	D21.6	D21.6	D21.5	K28.3

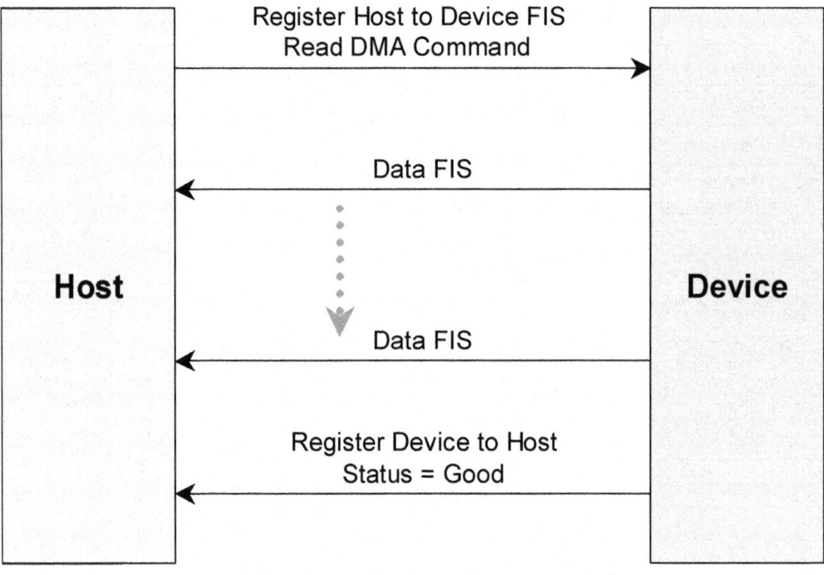

Figure 2.24 Read operation example.

every commands ends with some sort of status indication detailing the success or failure of the command. Figure 2.24 shows the frame-level protocol for a Read operation.

2.14 Application Layer

In order to make the transition to SATA a smooth one, SATA utilizes the same Application layer mechanisms to communicate with storage devices. This allows the system integrator to essentially use the same device drivers to communicate with the newer SATA devices with very little code modification. Of course, if the system integrator wants to take advantage of the enhancements provided by SATA, then modification would have to be made. However, the essential components of the device driver, including error handling schemes, will remain mostly unchanged.

In SATA (just as in PATA), two application layer options exist to communicate with storage devices. Most disc drives today will use the proven register Interface command sets, as previously discussed. Most of the other devices, including tapes, CDs, and DVDs will use the ATAPI interface to access and control the device. These interfaces will be covered in great detail in the Application Layer chapter.

Table 2.32 CHS to LBA Equivalents

ATA Registers	Host to device	Device to host
Features/Error	Features	Error
Sector Count	Sector Count Low	Sector Count Low
Sector Number	LBA Low	LBA Low
Cylinder Low	LBA Mid	LBA Mid
Cylinder High	LBA High	LBA High
Device/Head	Device	Device
Command/Status	Command	Status

2.14.1 CHS to LBA Translation

SATA uses the following addressing equivalents (see Table 2.32). The change from CHS addressing to LBA addressing was made in ATA/ATAPI-6 and above.

Inputs

The inputs describes the Command Block Register data that the host (driver) supplies. Basic descriptions for read and write type operations include the following:

- Command Code determines operation.
- Sector Count provides details as to how much data will follow the command.
- LBA Low-Mid-High provides the location of the information in logical block form.
- Features allows for different variations of a command.
- Device is mainly obsolete and provides no useful information beyond ATA-8.

Register	7	6	5	4	3	2	1	0
Features								
Sector Count								
LBA Low								
LBA Mid								
LBA High								
Device								
Command	Command Code							

Outputs

The outputs describes the Command Block registers returned by the device at the end of a command. The two main pieces of information are the Status and Error registers.

Register	7	6	5	4	3	2	1	0
Error								
Sector Count								
LBA Low								
LBA Mid								
LBA High								
Device								
Status								

2.14.2 General Feature Set Commands

The following general feature set commands are mandatory for all devices that are capable of ***both reading and writing*** their media and do not implement the PACKET command feature set:

- EXECUTE DEVICE DIAGNOSTIC
- FLUSH CACHE
- IDENTIFY DEVICE
- READ DMA
- READ MULTIPLE
- READ SECTOR(S)
- READ VERIFY SECTOR(S)
- SET FEATURES
- SET MULTIPLE MODE
- WRITE DMA
- WRITE MULTIPLE
- WRITE SECTOR(S)

2.15 Summary

Figure 2.25 demonstrates what happens at each layer of the architecture. Each layer will be discussed in detail in subsequent chapters.

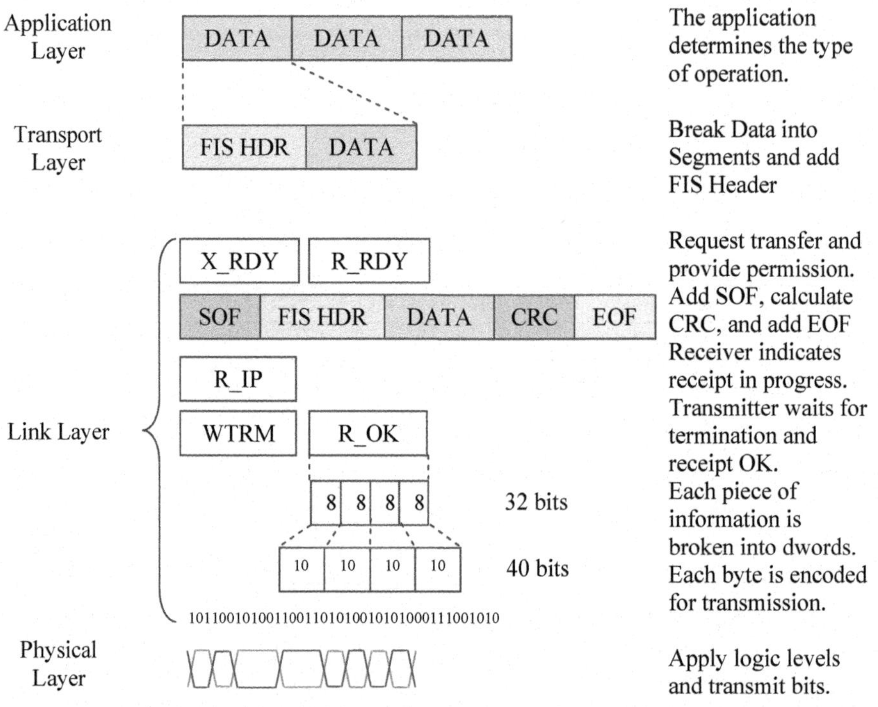

Figure 2.25 SATA protocol layer functions.

2.16 Review Questions

1. Which industry Standard first defined serial transport?

   a. ATA-6
   b. ATA-7 Volume 3
   c. ATA-7 Volume 1
   d. ATA-8 Serial Transport

2. What is the name of the Standards organization responsible for SATA?

   a. INCITS T-10
   b. SATA-IO
   c. INCITS T-13
   d. INCITS T-11

3. Which organization and website provides the SATA Specifications?

   a. www.t10.org
   b. www.sata-io.org
   c. www.t13.org
   d. www.t11.org

4. Which protocol layer provides framing formats for conveying ATA commands?

   a. PHY
   b. Link
   c. Transport
   d. Application

5. Which protocol layer specifies cables, connectors, and signaling levels?

   a. PHY
   b. Link
   c. Transport
   d. Application

6. Which protocol layer defines the command sets?

   a. PHY
   b. Link
   c. Transport
   d. Application

7. Which protocol layer provides encoding, primitives, flow control, and framing definition?

   a. PHY
   b. Link
   c. Transport
   d. Application

8. Which register tells the device what operation to perform?

   a. ERROR
   b. SECTOR COUNT
   c. LBA LOW
   d. LBA MID
   e. LBA HIGH
   f. DEVICE
   g. STATUS
   h. COMMAND

9. Which register tells the device how much data to transfer?

   a. ERROR
   b. SECTOR COUNT
   c. LBA LOW
   d. LBA MID
   e. LBA HIGH
   f. DEVICE
   g. STATUS
   h. COMMAND

10. Which register is filled by the device at the end of the operation?

   a. ERROR
   b. SECTOR COUNT
   c. LBA LOW
   d. LBA MID
   e. LBA HIGH
   f. DEVICE
   g. STATUS
   h. COMMAND

11. Which register is filled by the device if the operation completed unsuccessfully?

   a. ERROR
   b. SECTOR COUNT
   c. LBA LOW
   d. LBA MID
   e. LBA HIGH
   f. DEVICE
   g. STATUS
   h. COMMAND

12. What are the Standard protocols supported by ATA?

   a. Non-Data Command Protocol
   b. PIO Data-In Command Protocol
   c. PIO Data-Out Command Protocol
   d. DMA Command Protocol

e. PACKET command Protocol
f. DMA Queued Command Protocol
g. All of the above

13. What is the PACKET command protocol used for?

    a. To communicate with other host computers
    b. To pass SCSI commands to devices
    c. To pass Disk commands to device
    d. To allow device to device copies

14. SATA links use what type of cabling?

    a. TWINAX
    b. CAT-6
    c. Differential signal pairs
    d. Fibre optic

15. What type of encoding scheme is used in SATA?

    a. 64b/66b
    b. 128b/130b
    c. 8b/10b
    d. Spread spectrum clocking

16. What is a dword?

    a. 8-bits
    b. 16-bits
    c. 32-bits
    d. 64-bits

17. What is a SATA primitive?

    a. Begins with K28.7 followed by three data dwords
    b. Begins with K28.3 followed by three data dwords
    c. Begins with K28.1 followed by three data dwords

18. What is transmitted on serial links when no information is being transmitted?

    a. Nothing
    b. Specific repeated pattern
    c. Idle primitives
    d. Scrambled dwords

19. What primitive is transmitted when a device wants to transmit a frame?

    a. EOF
    b. R_ERR
    c. R_IP

d. R_OK
e. R_RDY
f. SOF
g. SYNC
h. WTRM
i. X_RDY

20. What primitive is transmitted when a device is receiving a frame?

a. EOF
b. R_ERR
c. R_IP
d. R_OK
e. R_RDY
f. SOF
g. SYNC
h. WTRM
i. X_RDY

21. What primitive is transmitted when a device completes frame transmission?

a. EOF
b. R_ERR
c. R_IP
d. R_OK
e. R_RDY
f. SOF
g. SYNC
h. WTRM
i. X_RDY

22. What primitive is transmitted when a device wants to allow another device to start frame transmission?

a. EOF
b. R_ERR
c. R_IP
d. R_OK
e. R_RDY
f. SOF
g. SYNC
h. WTRM
i. X_RDY

23. What primitive is transmitted to inform the other device that frame transmission was successful?

a. EOF
b. R_ERR
c. R_IP
d. R_OK

e. R_RDY
f. SOF
g. SYNC
h. WTRM
i. X_RDY

24. What does the acronym FIS mean?

a. Framing Instrument Structure
b. Frame Information Station
c. Framing Information Structure
d. Frame Information Structure

25. Which FIS is used to transmit the COMMAND information?

a. Register transfer host to device
b. Register transfer device to host
c. DMA Setup
d. DMA Activate
e. PIO Setup

26. Which FIS is used to transmit the STATUS information?

a. Register transfer host to device
b. Register transfer device to host
c. DMA Setup
d. DMA Activate
e. PIO Setup

Chapter 3

SATA Application Layer

Objectives

- Basic overview of ATA Standards
- ATA addressing characteristics concerning 24- and 48-bit addressing
- Numerical list of all current ATA command codes
- Brief command descriptions with grouping into Feature Sets (categories):
  - General Feature Set
  - Optional
  - Packet
  - Power Management
  - Security
  - SMART (Self-Monitoring Analysis Reporting Technology)
  - Removable Media
  - Streaming
  - Logging
  - Overlapped (Queuing)
  - Compact Flash
- Detailed descriptions of almost every ATA command, including
  - Command Code
  - Feature Set
  - Protocol: non-data, PIO, DMA
  - Register Inputs
  - Normal and Error Outputs
  - Prerequisites
  - Detailed Command Description

3.1 SATA Application/Command Layer

This chapter provides details for almost every ATA command. It provides the same command-related information and tables you would find in either the SATA-IO Specification or T-13 Standards. This chapter is not meant to be a replacement for any standard or specification, and you should always keep the latest electronic versions of those documents on hand.

3.1.1 Command Set

ATA-8 is the most recent set of standards that defines the SATA architecture (see Figure 3.1):

- Architectural model – ATA Architecture Model (AAM) defines host and device responsibilities and provides a foundation for compatibility across all standards.
- Command Sets – ATA Command Set (ACS) defines the register delivered commands used by devices implementing the standard.
- Transport/Physical Interface – ATA Serial Transport (AST) defines the connectors and cables for physical interconnection between the host and storage devices, the electrical and logical characteristics of the interconnecting signals, and the protocols for the transporting commands, data, and status over the interface for the serial interface.

Keep in mind the following points:

- SATA is not a command layer protocol.
  - SATA is a physical interface that delivers ATA commands.
  - AST-8 defines the transport layer and physical connectors and cables.
- ATA is the command protocol.
  - ACS-8 defines the ATA command sets for register delivered and packet delivered commands.

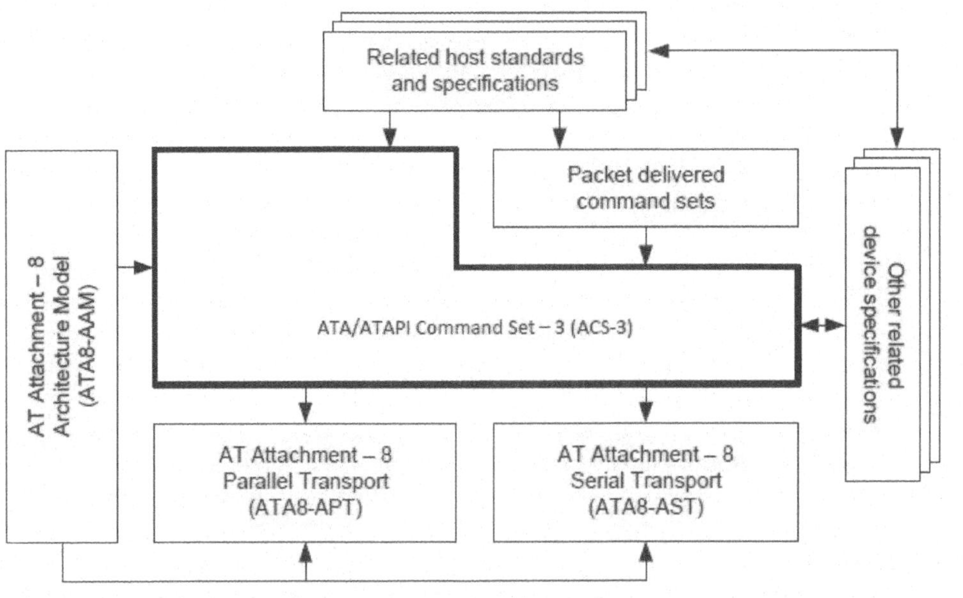

Figure 3.1 ATA-8 document relationships. (*Source:* www.t13.org)

3.1.2 Register Delivered Command Sector Addressing

For register delivered data transfer commands, all addressing of data sectors recorded on the device's media is by a logical sector address. There is no implied relationship between logical sector addresses and the actual physical location of the data sector on the media. All devices will support LBA translation.

In Standards ATA/ATAPI-5 and earlier, a CHS translation was defined (see Table 3.1). This translation is obsolete, but if implemented, it will be implemented as defined in ATA/ATAPI-5.

SATA uses the following equivalents (see Tables 3.2 and 3.3). These changes were made in ATA/ATAPI-6 and above. CHS addressing references were replaced with LBA addressing in all standards at ATA-8.

Table 3.1 – CHS Addressing—Now Obsolete

Register	7	6	5	4	3	2	1	0
Features								
Sector Count								
Sector Number								
Cylinder Low								
Cylinder High								
Device/Head								
Command	Command Code							
NOTE – No entry indicates register or bit not used by the device. If the register is written by the host, bits with no entry will be written to zero.								

Table 3.2 – 24-Bit Address Commands

ATA Registers	ATA/ATAPI-6 and Above Write	ATA/ATAPI-6 and Above Read
Features/Error	Features	Error
Sector Count	Sector Count Low	Sector Count Low
Sector Number	LBA Low	LBA Low
Cylinder Low	LBA Mid	LBA Mid
Cylinder High	LBA High	LBA High
Device/Head	Device	Device
Command/Status	Command	Status

Table 3.3 – 48-Bit Addressing Adds Extension Registers

Register	"most recently written"	"previous content"
Features	Reserved	Reserved
Sector Count	Sector count (7:0)	Sector Count Ext. (15:8)
LBA Low	LBA (7:0)	LBA Ext. (31:24)
LBA Mid	LBA (15:8)	LBA Ext. (39:32)
LBA High	LBA (23:16)	LBA Ext. (47:40)
Device register	Bits 7 and 5 are obsolete, the LBA bit will be set to one, the DEV bit will indicate the selected device, bits (3:0) are reserved.	Reserved

3.2 Command Codes from ATA/ATAPI Command Set R4I

Command Codes (sorted by command name)

Command	Code	ATA device	ATAPI	Protocol	Argument
ACCESSIBLE MAX ADDRESS CONFIGURATION	78h	O	P	ND	48-bit
CHECK POWER MODE	E5h	M	M	ND	28-bit
CONFIGURE STREAM	51h	O	O	ND	48-bit
DATA SET MANAGEMENT	06h	O	P	DM	48-Bit
DEVICE RESET	08h	N	M	DR	28-bit
DOWNLOAD MICROCODE	92h	O	N	PO	28-bit
DOWNLOAD MICROCODE DMA	93h	O	N	DM	28-bit
EXECUTE DEVICE DIAGNOSTIC	90h	M	M	DD	28-bit
FLUSH CACHE	E7h	O	O	ND	28-bit
FLUSH CACHE EXT	EAh	O	N	ND	28-bit
IDENTIFY DEVICE	ECh	M	M	PI	28-bit
IDENTIFY PACKET DEVICE	A1h	N	M	PI	28-bit
IDLE	E3h	M	O	ND	28-bit
IDLE IMMEDIATE	E1h	M	M	ND	28-bit
NCQ QUEUE MANAGEMENT	63h	O	N	ND	48-bit
NOP	00h	O	M	ND	28-bit
PACKET	A0h	N	M	P	
READ BUFFER	E4h	O	N	PI	28-bit
READ BUFFER DMA	E9h	O	N	DM	28-bit
READ DMA	C8h	O	N	DM	28-bit
READ DMA EXT	25h	O	N	DM	48-bit
READ FPDMA QUEUED	60h	O	N	DMQ	48-bit
READ LOG DMA EXT	47h	O	O	DM	48-bit
READ LOG EXT	2Fh	O	O	PI	48-bit
READ MULTIPLE	C4h	O	N	PI	28-bit
READ MULTIPLE EXT	29h	O	N	PI	48-bit
READ SECTOR(S)	20h	O	M	PI	28-bit
READ SECTOR(S) EXT	24h	O	N	PI	48-bit
READ STREAM DMA EXT	2Ah	O	N	DM	48-bit
READ STREAM EXT	2Bh	O	N	PI	48-bit
READ VERIFY SECTOR(S)	40h	O	N	ND	28-bit
READ VERIFY SECTOR(S) EXT	42h	O	N	ND	48-bit
RECEIVE FPDMA QUEUED	65h	O	N	DMQ	48-bit
REQUEST SENSE DATA EXT	0Bh	O	P	ND	48-bit

(Continued on following page)

Command Codes (sorted by command name) (*Continued*)

Command	Code	ATA device	ATAPI	Protocol	Argument
Sanitize Device	B4h	O	N	ND	48-bit
SECURITY DISABLE PASSWORD	F6h	O	O	PO	28-bit
SECURITY ERASE PREPARE	F3h	O	O	ND	28-bit
SECURITY ERASE UNIT	F4h	O	O	PO	28-bit
SECURITY FREEZE LOCK	F5h	O	O	ND	28-bit
SECURITY SET PASSWORD	F1h	O	O	PO	28-bit
SECURITY UNLOCK	F2h	O	O	PO	28-bit
SEND FPDMA QUEUED	64h	O	N	DMQ	48-bit
SET DATE & TIME EXT	77h	O	N	ND	48-bit
SET FEATURES	EFh	M	M	ND	28-bit
SET MULTIPLE MODE	C6h	O	N	ND	28-bit
SLEEP	E6h	M	M	ND	28-bit
SMART	B0h	O	N	ND	
STANDBY	E2h	M	O	ND	28-bit
STANDBY IMMEDIATE	E0h	M	M	ND	28-bit
TRUSTED NON-DATA	5Bh	O	P	ND	28-bit
TRUSTED RECEIVE	5Ch	O	P	PI	28-bit
TRUSTED RECEIVE DMA	5Dh	O	P	DM	28-bit
TRUSTED SEND	5Eh	O	P	PO	28-bit
TRUSTED SEND DMA	5Fh	O	P	DM	28-bit
WRITE BUFFER	E8h	O	N	PO	28-bit
WRITE BUFFER DMA	EBh	O	N	DM	28-bit
WRITE DMA	CAh	O	N	DM	28-bit
WRITE DMA EXT	35h	O	N	DM	48-bit
WRITE DMA FUA EXT	3Dh	O	N	DM	48-bit
WRITE FPDMA QUEUED	61h	O	N	DMQ	48-bit
WRITE LOG DMA EXT	57h	O	O	DM	48-bit
WRITE LOG EXT	3Fh	O	O	PO	48-bit
WRITE MULTIPLE	C5h	O	N	PO	28-bit
WRITE MULTIPLE EXT	39h	O	N	PO	48-bit
WRITE MULTIPLE FUA EXT	CEh	O	N	PO	48-bit
WRITE SECTOR(S)	30h	O	N	PO	28-bit
WRITE SECTOR(S) EXT	34h	O	N	PO	48-bit
WRITE STREAM DMA EXT	3Ah	O	N	DM	48-bit
WRITE STREAM EXT	3Bh	O	N	PO	48-bit
WRITE UNCORRECTABLE EXT	45h	O	N	ND	48-bit

(*Continued on following page*)

Command Codes (sorted by command name) (*Continued*)

Command	List of codes . . .
Reserved	01h..02h, 04h..05h, 07h, 09h..0Ah, 0Ch..0Fh, 28h, 2Ch..2Fh, 43h..44h, 46h, 48h..4Fh, 52h..56h, 58h..5Ah, 62h, 66h..6Fh, 9Bh..9Fh, A3h..AFh, B2h..B3h, B5h, BCh..BFh, CFh, D0h, D2h..D9h
Reserved for Compact Flash Association	03h, 38h, 87h, B7h..BBh, C0h, and CDh
Retired	11h..1Fh, 71h..76h, 79h..7Fh, 94h..99h, DBh..DDh
Obsolete	10h, 21h..23h, 26h, 27h, 31h..33h, 36h, 37h, 3Ch, 3Eh, 41h, 50h, 70h, 91h, A2h, B1h, B6h, C7h, C9h, CBh..CCh, D1h, DAh, DEh, DFh, EDh..EEh, F8h, F9h
Vendor Specific	80h..86h, 88h..8Fh, 9Ah, C1h..C3h, F0h, F7h, FAh..FFh

ND	Non-Data	DMQ	DMA QUEUED	M	Mandatory
PI	PIO Data-In	O	Optional	PO	PIO Data-Out
N	Use prohibited	DM	DMA	R	Reserved
DR	Device Reset	DD	Execute Device Diagnosti		

Brief Command Descriptions

Command	Description
Accessible Max Address Configuration	The Accessible Max Address Configuration feature set provides a method for an application client to discover the native max address and control the accessible max address. The following commands are mandatory for devices that support the Accessible Max Address Configuration feature set: – GET NATIVE MAX ADDRESS EXT – SET ACCESSIBLE MAX ADDRESS EXT – FREEZE ACCESSIBLE MAX ADDRESS EXT
Acknowledge media change	This command is reserved for use by removable media devices. The implementation of the command is vendor specific.
Check power mode	If the device is in, going to, or recovering from the Standby Mode.
Configure Stream	The Streaming feature set allows a host to request delivery of data within an allotted time, placing a priority on the time to transfer the data rather than the integrity of the data. While processing commands in the Streaming feature set, devices may process background tasks if the specified command processing time limits for the commands are met.
Data Set Management	The DATA SET MANAGEMENT command provides information for device optimization (e.g., file system information).
Door lock	This command either locks the device media or provides the status of the media change request button.
Door unlock	This command will unlock the device media, if it is locked, and will allow the device to respond to the media change request button.
Download Microcode	This command enables the host to alter the device's Microcode. The data transferred using the Download Microcode is vendor specific.
Execute drive diagnostic	This command will perform the internal diagnostic tests implemented by the device.

(*Continued on following page*)

Brief Command Descriptions (*Continued*)

Command	Description
Flush Cache	The FLUSH CACHE command requests the device to flush the volatile write cache. If there is data in the volatile write cache, that data will be written to the nonvolatile media. This command will not indicate completion until the data is flushed to the nonvolatile media or an error occurs.
Identify device	This command enables the host to receive parameter information from the device. Some devices may have to read the media in order to complete this command.
Identify device DMA	This command enables the host to receive parameter information from the device. This command transfers the same 256 words of device identification data as transferred by the Identify Device command.
Idle	This command causes the device to set the BSY bit, enter the Idle Mode, clear the BSY bit, and assert INTRQ.
Idle immediate	This command causes the device to set the BSY bit, enter the Idle Mode, clear the BSY bit, and assert INTRQ. INTRQ is asserted even though the device may not have fully transitioned to Idle Mode.
Initialize device parameters	This command allows the host to set the number of logical sectors per track and the number of logical heads minus 1 per logical cylinder for the current CHS translation mode.
Media Eject	This command completes any pending operations, spins down the device if needed, unlocks the door or media if locked, and initiates a media eject, if required.
NOP	This command enables a host that can only perform 16-bit register access to check device status.
Packet	The PACKET command transfers a SCSI CDB (see SPC-4) via a command packet. If the native form of the encapsulated command is shorter than the packet size reported in IDENTIFY PACKET DEVICE data word 0 bits 1:0, then the encapsulated command will begin at byte 0 of the packet. Packet bytes beyond the end of the encapsulated command are reserved.
Read buffer	This command enables the host to read the current contents of the device's sector buffer.
Read DMA (w/retry)	This command executes in a similar manner to the Read Sectors command, except for the following: – The host initializes the DMA channel prior to issuing the command. – Data transfers are qualified by DMARQ and are performed by the DMA channel. – The device issues only one interrupt per command to indicate that data transfer has terminated and status is available.
Read DMA (w/o retry)	Same as Read DMA except no retires.
Read Log	The READ LOG command is for devices that implement the General Purpose Logging feature set. The READ LOG command returns the specified log to the host.
Read long (w/retry)	This command performs similarly to the Read Sectors command except that it returns the data and a number of vendor-specific bytes appended to the data field of the desired sector.

(*Continued on following page*)

Brief Command Descriptions (*Continued*)

Command	Description
Read long (w/o retry)	Same as Read Long except no retires.
Read multiple	This command performs similarly to the Read Sectors command. Interrupts are not generated on every sector but on the transfer of a block that contains the number of sectors defined by a Set Multiple Mode command.
Read sector(s) (w/retry)	This command reads from 1 to 256 sectors, as specified in the Sector Count register. A sector count of 0 requests 256 sectors. The transfer begins at the sector specified in the Sector Number register.
Read sector(s) (w/o retry)	Same as Read Sectors but without retires enabled.
Read Stream	The READ STREAM DMA command provides a method for a host to read data within an allotted time. This command allows the host to specify that additional actions are to be performed by the device prior to the completion of the command.
Read verify sector(s) (w/retry)	This command is identical to the Read Sectors command, except that the DRQ bit is never set, and no data is transferred to the host.
Read verify sector(s) (w/o retry)	Same as Read Verify Sectors but no retry.
Recalibrate	The function performed by this command is vendor specific.
Request Sense Data	This command is for devices that implement the Sense Data Reporting feature set. When sense data is available, the sense key, additional sense code, and additional sense code qualifier fields will be set to values that are defined in the SPC-4 standard.
Sanitize Device	The Sanitize Device feature set allows hosts to request that devices modify the content of all user data areas in the device in a way that results in previously existing data in these areas becoming unreadable. Sanitize operations are initiated using one of the sanitize operation commands. If the Sanitize Device feature set is supported, at least one of the following commands will be supported: – CRYPTO SCRAMBLE EXT – BLOCK ERASE EXT or – OVERWRITE EXT
Secure Disable	When the device is in Secure Mode RO or RW, unlocked, with an existing set of valid passwords, this command will remove the device from Secure Mode.
Secure Enable RO	This command will set the device into Secure Mode Read Only and define the valid set of passwords.
Secure Enable RW	This command will set the device into Secure Mode Read/Write and define the valid set of passwords.
Secure Enable WP	This command will set the device into Secure Mode Write Protect. In this mode, the entire device can be read, but all write command will be rejected.
Secure Lock	This command will lock the device anytime the device is in secure mode xx, unlocked.
Secure State	This command will return the Secure Mode state of a device that implements the Secure Mode Function Set.

(*Continued on following page*)

Brief Command Descriptions (*Continued*)

Command	Description
Secure Unlock	This command will lock the device anytime the device is in secure mode xx, unlocked.
Seek	The function performed by this command is vendor specific.
Set Data and Time	This command sets the Date and Time TimeStamp device statistic in milliseconds using January 1, 1970 UT 12:00 am as the baseline. If the device processes a Power on reset, then the TimeStamp will be set to the Power On Hours device statistic.
Set features	This command is used by the host to establish all sorts of optional features that affect the execution of certain device features.
Set multiple mode	This command enables the device to perform Read and Write Multiple operations and establishes the block count for these commands.
Security set password	The SECURITY SET PASSWORD command transfers 512 bytes of data from the host. If the SECURITY SET PASSWORD commands return command completion without error, the command sets only one of the following: – the User password or – the Master password
Sleep	This command is the only way to cause the device to enter Sleep Mode.
Standby	This command causes the device to set the BSY bit, enter the Standby Mode, clear the BSY bit, and assert INTRQ.
Standby immediate	Same as Standby, except it doesn't use the Standby Timer.
SMART	Provides SMART capable host with command set to send Self Monitoring Analysis and Reporting Technology commands to disk devices (e.g., Read/Write Logs, Enable/Disable Operations, etc.).
Trusted Send/Receive	The Trusted Computing feature set provides an interface between a security component embedded in a device and an application client (host).
Write buffer	This command enables the host to overwrite the contents of one sector in the device's buffer.
Write DMA (w/retry)	This command executes in a similar manner to the Write Sectors command except for the following: • The host initializes the DMA channel prior to issuing the command. • Data transfers are qualified by DMARQ and are performed by the DMA channel. • The device issues only one interrupt per command to indicate that data transfer has terminated and status is available.
Write DMA (w/o retry)	Same as Write DMA but without retries.
Write Log	The WRITE LOG command is for devices that implement the General Purpose Logging feature set. The WRITE LOG command writes a specified number of 512-byte blocks of data to the specified log.
Write long (w/retry)	This command performs similarly to the Write Sectors command except that it writes the data and a number of vendor-specific bytes: The device does not generate vendor-specific bytes itself.
Write long (w/o retry)	Same as Write Long except no retries.

(*Continued on following page*)

Brief Command Descriptions (*Continued*)

Command	Description
Write multiple	This command performs similarly to the Write Sectors command except that it writes the data and a number of vendor-specific bytes: The device does not generate vendor-specific bytes itself.
Write same	This command is similar to the Write Sector(s) except that only one sector of data is transferred. The contents of the sector are written to the media one or more times.
Write sector(s) (w/retry)	This command performs similarly to the Write Sectors command. Interrupts are not generated on every sector but on the transfer of a block that contains the number of sectors defined by a Set Multiple Mode command.
Write sector(s) (w/o retry)	Same as Write Sectors but no retry.
Write stream	The WRITE STREAM command writes data within an allotted time. This command specifies that additional actions are to be performed by the device prior to the completion of the command.
Write uncorrectable	The WRITE UNCORRECTABLE command causes the device to report an uncorrectable error when the specified logical sectors are subsequently read.
Write verify	This command is similar to the Write Sector(s) except that each sector is verified from the media after being written and before the command is completed.

3.2.1 General Feature Set Commands

The following General feature set commands are mandatory for all devices that are capable of both reading and writing their media and do not implement the PACKET command feature set:

- EXECUTE DEVICE DIAGNOSTIC
- FLUSH CACHE
- IDENTIFY DEVICE
- READ DMA
- READ MULTIPLE
- READ SECTOR(S)
- READ VERIFY SECTOR(S)
- SET FEATURES
- SET MULTIPLE MODE
- WRITE DMA
- WRITE MULTIPLE
- WRITE SECTOR(S)

The following General feature set commands are mandatory for all devices that are capable of ***only reading*** their media and do not implement the PACKET command feature set:

- EXECUTE DEVICE DIAGNOSTIC
- IDENTIFY DEVICE
- READ DMA
- READ MULTIPLE

- READ SECTOR(S)
- READ VERIFY SECTOR(S)
- SET FEATURES
- SET MULTIPLE MODE

3.2.2 Optional General Feature Set Commands

The following General feature set commands are optional for devices not implementing the PACKET command feature set:

- DOWNLOAD MICROCODE
- NOP
- READ BUFFER
- WRITE BUFFER

The following General feature set command is prohibited for use by devices not implementing the PACKET command feature set:

- DEVICE RESET

3.2.3 PACKET General Feature Set Commands

The following General feature set commands are mandatory for all devices implementing the PACKET command feature set:

- EXECUTE DEVICE DIAGNOSTIC
- IDENTIFY DEVICE
- NOP
- READ SECTOR(S)
- SET FEATURES

The following General feature set commands are ***optional*** for all devices implementing the PACKET command feature set:

- FLUSH CACHE

The following General command set commands are ***prohibited*** for use by devices implementing the PACKET command feature set because ***these functions are supported by Packet commands:***

- DOWNLOAD MICROCODE
- READ BUFFER
- READ DMA
- READ MULTIPLE
- READ VERIFY
- SET MULTIPLE MODE
- WRITE BUFFER
- WRITE DMA
- WRITE MULTIPLE
- WRITE SECTOR(S)

3.2.4 Unique PACKET Command Feature Set

- The optional PACKET Command feature set provides for devices that require command
- parameters that are too extensive to be expressed in the Command Block registers.
- Devices implementing the PACKET Command feature set exhibit responses different from those exhibited by devices not implementing this feature set.
- The command ***unique*** to the PACKET Command feature set is
  - PACKET
- The PACKET command allows a host to send a command to the device via a command packet. The command packet contains the command and command parameters that the device is to execute.
  - DEVICE RESET
- The DEVICE RESET command is provided to allow the device to be reset without affecting the other device on the bus.
  - IDENTIFY PACKET DEVICE
- The IDENTIFY PACKET DEVICE command is used by the host to get identifying parameter information for a device implementing the PACKET Command feature set.

Upon receipt of the PACKET command, the device sets BSY to one and prepares to receive the command packet. When ready, the device sets DRQ to one and clears BSY to zero. The command packet is then transferred to the device by PIO transfer. When the last word of the command packet is transferred, the device sets BSY to one and clears DRQ to zero.

3.2.5 Power Management Feature Set

- The Power Management feature set permits a host to modify the behavior of a device in a manner that reduces the power required to operate.
- The Power Management feature set provides a set of commands and a timer that enable a device to implement low power consumption modes.
- A register delivered command device that implements the Power Management feature set will implement the following minimum set of functions:
  - A Standby timer
  - CHECK POWER MODE command
  - IDLE command
  - IDLE IMMEDIATE command
  - SLEEP command
  - STANDBY command
  - STANDBY IMMEDIATE command

The CHECK POWER MODE command allows a host to determine if a device is currently in, going to or leaving Standby or Idle mode. The CHECK POWER MODE command will not change the power mode or affect the operation of the Standby timer.

The IDLE and IDLE IMMEDIATE commands move a device to Idle mode immediately from the Active or Standby modes. The IDLE command also sets the Standby timer count and enables or disables the Standby timer.

The STANDBY and STANDBY IMMEDIATE commands move a device to Standby mode immediately from the Active or Idle modes. The STANDBY command also sets the Standby timer count and enables or disables the Standby timer.

The SLEEP command moves a device to Sleep mode. The device›s interface becomes inactive at command completion of the SLEEP command. A hardware or software reset or DEVICE RESET command is required to move a device out of Sleep mode.

The Standby timer provides a method for the device to automatically enter Standby mode from either Active or Idle mode following a host programmed period of inactivity. If the Standby timer is enabled and if the device is in the Active or Idle mode, the device waits for the specified time period. If no command is received, the device automatically enters the Standby mode.

3.2.6 PACKET Power Management Feature Set

A device that implements the PACKET Command feature set and implements the Power Management feature set will implement the following minimum set of functions:

- CHECK POWER MODE command
- IDLE IMMEDIATE command
- SLEEP command
- STANDBY IMMEDIATE command

3.2.7 Security Mode Feature Set

The optional Security Mode feature set is a password system that restricts access to user data stored on a device. The system has two passwords, User and Master, and two security levels, High and Maximum. The security system is enabled by sending a user password to the device with the SECURITY SET PASSWORD command. When the security system is enabled, access to user data on the device is denied after a power cycle until the User password is sent to the device with the SECURITY UNLOCK command.

A device that implements the Security Mode feature set ***will implement the following minimum set*** of commands:

- SECURITY SET PASSWORD
- SECURITY UNLOCK
- SECURITY ERASE PREPARE
- SECURITY ERASE UNIT
- SECURITY FREEZE LOCK
- SECURITY DISABLE PASSWORD

3.2.8 SMART Feature Set Commands

These commands use a single command code and are differentiated from one another by the value placed in the Features register.

If the SMART feature set is implemented, the following commands will be implemented:

- SMART DISABLE OPERATIONS
- SMART ENABLE/DISABLE AUTOSAVE
- SMART ENABLE OPERATIONS
- SMART RETURN STATUS

If the SMART feature set is implemented, the following commands may be implemented:

- SMART EXECUTE OFF-LINE IMMEDIATE
- SMART READ DATA
- SMART READ LOG
- SMART WRITE LOG
- READ LOG EXT
- WRITE LOG EXT

The intent of self-monitoring, analysis, and reporting technology (the SMART feature set) is to protect user data and minimize the likelihood of unscheduled system downtime that may be caused by predictable degradation and/or fault of the device. By monitoring and storing critical performance and calibration parameters, SMART feature set devices attempt to predict the likelihood of near-term degradation or fault condition. Providing the host system with the knowledge of a negative reliability condition allows the host system to warn the user of the impending risk of data loss and advise the user of appropriate action. Support of this feature set is indicated in the IDENTIFY DEVICE data.

Devices that implement the PACKET Command feature set will not implement the SMART feature set as described in this standard. Devices that implement the PACKET Command feature set and SMART will implement SMART as defined by the command packet set implemented by the device. This feature set is optional if the PACKET Command feature set is not supported.

3.2.9 Removable Media Status Notification Feature Set

The following commands are defined to implement the Removable Media Status Notification feature set.

- GET MEDIA STATUS
- MEDIA EJECT
- SET FEATURES (Enable media status notification)
- SET FEATURES (Disable media status notification)

Note: Devices implementing the PACKET Command feature set control the media eject mechanism via the START/STOP UNIT command packet.

The Removable Media Status Notification feature set is the preferred feature set for securing the media in removable media storage devices. This feature set uses the SET FEATURES command to enable Removable Media Status Notification. Removable Media Status Notification gives the host system maximum control of the media. The host system determines media status by issuing the GET MEDIA STATUS command and controls the device eject mechanism via the MEDIA EJECT command (for devices not implementing the PACKET Command feature set) or the START/STOP UNIT command (for devices implementing the PACKET Command feature set). While Removable Media Status Notification is enabled, devices not implementing the PACKET Command feature set execute MEDIA LOCK and MEDIA UNLOCK commands without changing the media lock state (no-operation). While Removable Media Status Notification is enabled, the eject button does not eject the media.

3.2.10 Removable Media Feature Set

The following commands are defined to implement the Removable Media feature set:

- MEDIA EJECT
- MEDIA LOCK
- MEDIA UNLOCK

The preferred sequence of events to use the Removable Media feature set is as follows:

- Host system checks whether the device implements the PACKET Command feature set via the device signature in the Command Block registers.
- Host system issues the IDENTIFY DEVICE command and checks that the device is a removable media device and that the Removable Media feature set is supported.
- Host system periodically issues MEDIA LOCK commands to determine if:
  - no media is present in the device (NM)—media is locked if present;
  - a media change request has occurred (MCR).

The Removable Media feature set is intended only for devices not implementing the PACKET Command feature set. This feature set operates with Media Status Notification disabled. The MEDIA LOCK and MEDIA UNLOCK commands are used to secure the media, and the MEDIA EJECT command is used to remove the media. While the media is locked, the eject button does not eject the media. Media status is determined by checking the media status bits returned by the MEDIA LOCK and MEDIA UNLOCK commands.

Power-on reset, hardware reset, and the EXECUTE DEVICE DIAGNOSTIC command clear the Media Lock (LOCK) state and the Media Change Request (MCR) state. Software reset clears the Media Lock (LOCK) state, clears the Media Change Request (MCR) state, and preserves the Media Change (MC) state.

3.2.11 CompactFlash™ Association (CFA) Feature Set

The optional CompactFlash™ Association (CFA) feature set provides support for solid-state memory devices.

A device that implements the CFA feature set ***will implement*** the following minimum set of commands:

- CFA REQUEST EXTENDED ERROR CODE
- CFA WRITE SECTORS WITHOUT ERASE
- CFA ERASE SECTORS
- CFA WRITE MULTIPLE WITHOUT ERASE
- CFA TRANSLATE SECTOR
- SET FEATURES Enable/Disable 8-bit transfer

- Devices reporting the value 848Ah in IDENTIFY DEVICE data word 0 or devices having bit 2 of IDENTIFY DEVICE data word 83 set to one will support the CFA feature set. If the CFA feature set is implemented, all five commands will be implemented.

Support of DMA commands is optional for devices that support the CFA feature set.

The CFA ERASE SECTORS command preconditions the sector for a subsequent CFA WRITE SECTORS WITHOUT ERASE or CFA WRITE MULTIPLE WITHOUT ERASE command to achieve higher performance during the write operation. The CFA TRANSLATE SECTOR command provides information about a sector, such as the number of write cycles performed on that sector, and an indication of the sector's erased precondition. The CFA REQUEST EXTENDED ERROR CODE command provides more detailed error information.

3.2.12 48-Bit Address Feature Set

The optional 48-bit Address feature set allows devices with capacities up to 281,474,976,710,655 sectors. This allows device capacity up to 144,115,188,075,855,360 bytes. In addition, the number of sectors that may be transferred by a single command are increased by increasing the allowable sector count to 16 bits.

The commands in the 48-bit Address feature set are prohibited from use for devices implementing the PACKET Command feature set. Commands unique to the 48-bit Address feature set are as follows:

- FLUSH CACHE EXT
- READ DMA EXT
- READ DMA QUEUED EXT
- READ MULTIPLE EXT
- READ NATIVE MAX ADDRESS EXT
- READ SECTOR(S) EXT
- READ VERIFY SECTOR(S) EXT
- SET MAX ADDRESS EXT
- WRITE DMA EXT
- WRITE DMA FUA EXT
- WRITE DMA QUEUED EXT
- WRITE DMA QUEUED FUA EXT
- WRITE MULTIPLE EXT
- WRITE MULTIPLE FUA EXT
- WRITE SECTOR(S) EXT

3.2.13 Streaming Feature Set

The Streaming feature set is an optional feature set that allows a host to request delivery of data from a contiguous logical block address range within an allotted time. This places a priority on time to access the data rather than the integrity of the data. Streaming feature set commands only support 48-bit addressing.

A device that implements the Streaming feature set ***will implement the following minimum set*** of commands:

- CONFIGURE STREAM
- READ STREAM EXT
- WRITE STREAM EXT
- READ STREAM DMA EXT
- WRITE STREAM DMA EXT
- READ LOG EXT
- WRITE LOG EXT

- Support of the Streaming feature set is indicated in IDENTIFY DEVICE data word 84 bit 4.

Note: PIO versions of these commands limit the transfer rate (16.6 MB/s), provide no CRC protection, and limit status reporting as compared to a DMA implementation.

3.2.14 General Purpose Logging Feature Set

The General Purpose Logging feature set provides a mechanism for accessing logs in a device.

- These logs are associated with specific feature sets, such as SMART.
- Support of the individual logs is determined by support of the associated feature set.
- If the device supports a particular feature set, support for any associated log(s) is mandatory.

Support for the General Purpose Logging feature set will not be disabled.

- If the feature set associated with a requested log is disabled, the device will return command abort.

If the General Purpose Logging feature set is implemented, the following commands will be supported:

- READ LOG EXT
- WRITE LOG EXT

3.2.15 Overlapped Feature Set

The Queued feature set allows the host to issue concurrent commands to the same device. The Queued feature set is optional if the Overlap feature set is supported. Only commands included in the Overlapped feature set may be queued. The queue contains all commands for which command acceptance has occurred but command completion has not occurred. If a queue exists when a non-queued command is received, the non-queued command will be command aborted, and the commands in the queue will be discarded. The ending status will be command aborted and the results are indeterminate.

The optional Overlap feature set allows devices that require extended command time to perform a bus release so that the other device on the bus may be used.

- To perform a bus release, the device will clear both DRQ and BSY to zero.
- When selecting the other device during overlapped operations, the host will disable assertion of INTRQ via the nIEN bit on the currently selected device before writing the Device register to select the other device and then may re-enable interrupts.

The only commands that may be overlapped are as follows:

- NOP (with a subcommand code other than 00h)
- PACKET
- READ DMA QUEUED
- READ DMA QUEUED EXT

- SERVICE
- WRITE DMA QUEUED
- WRITE DMA QUEUED EXT
- WRITE DMA QUEUED FUA EXT

3.3 Command Descriptions

3.3.1 Command Overview

Commands are issued to the device by loading the required registers in the command block with the needed parameters and then writing the command code to the Command register. Required registers are those indicated by a specific content in the Inputs table for the command, that is, not noted as na or obs.

Any references to parallel implementation bus signals (e.g., DMACK, DMARQ, etc.) apply only to parallel implementations. See Volume 3 of the ATA Standards for additional information on serial protocol. Some register bits (e.g., nIEN, SRST, etc.) have different requirements in the serial implementation (see Volume 3).

Each command description in the following sections contains the following subsections:

Command Code

Specifies the command code for this command.

Feature Set

Specifies feature set and if the command is mandatory or optional.

Protocol

Specifies which protocol is used by the command.

Inputs

Describes the Command Block register data that the host will supply.

Register	7	6	5	4	3	2	1	0
Features								
Sector Count								
LBA Low								
LBA Mid								
LBA High								
Device								
Command	Command Code							
NOTE – na specifies the content of a bit or field is not applicable to the particular command. Obs specifies that the use of this bit is obsolete.								

Normal Outputs

Describes the Command Block register data returned by the device at the end of a command. Where command completes without an error.

Register	7	6	5	4	3	2	1	0
Error	na							
Sector Count								
LBA Low								
LBA Mid								
LBA High								
Device								
Status	BSY	DRDY	DF/SE	X	DRQ	Obs	Obs	ERR
NOTE – na indicates the content of a bit or field is not applicable to the particular command. Obs indicates that the use of this bit is obsolete.								

Error Outputs

Describes the Command Block register data that will be returned by the device at command completion with an unrecoverable error.

Register	7	6	5	4	3	2	1	0
Error	ICRC	WP UNC	MC	IDNF	MCR	ABRT	NM	MED
Sector Count								
LBA Low								
LBA Mid								
LBA High								
Device								
Status	BSY	DRDY	DF SE	X	DRQ	Obs	Obs	ERR
NOTE – na specifies the content of a bit or field is not applicable to the particular command. Obs specifies that the use of this bit is obsolete.								

Prerequisites

Any prerequisite commands or conditions that will be met before the command is issued.

Description

The description of the command function(s).

3.3.2 Device Configuration

Individual Device Configuration Overlay feature set commands are identified by the value placed in the Features register. The table below shows these Features register values.

Device Configuration Overlay Features Register Values

Value	Command
C0h	DEVICE CONFIGURATION RESTORE
C1h	DEVICE CONFIGURATION FREEZE LOCK
C2h	DEVICE CONFIGURATION IDENTIFY
C3h	DEVICE CONFIGURATION SET
00h-BFh, C4h-FFh	Reserved

DEVICE CONFIGURATION RESTORE

Command code

B1h with a Features register value of C0h.

Feature set

Device Configuration Overlay feature set.

- Mandatory when the Device Configuration Overlay feature set is implemented.

Protocol

Non-data

Inputs

The Features register will be set to C0h.

Register	7	6	5	4	3	2	1	0
Features	C0h							
Sector Count	na							
LBA Low	na							
LBA Mid	na							
LBA High	na							
Device	na			DEV	na			
Command	B1h							

Device –

DEV will specify the selected device.

Normal outputs

Register	7	6	5	4	3	2	1	0
Error	na							
Sector Count	na							
LBA Low	na							
LBA Mid	na							
LBA High	na							
Device	obs	na	obs	DEV	na			
Status	BSY	DRDY	DF	na	DRQ	na	na	ERR

Device register –
DEV will indicate the selected device.
Status register –
BSY will be cleared to zero, indicating command completion.
DRDY will be set to one.
DF (Device Fault) will be cleared to zero.
DRQ will be cleared to zero.
ERR will be cleared to zero.

Error outputs

Register	7	6	5	4	3	2	1	0
Error	na	na	na	na	na	ABRT	na	na
Sector Count	na							
LBA Low	na							
LBA Mid	na							
LBA High	na							
Device	obs	na	obs	DEV	na			
Status	BSY	DRDY	DF	na	DRQ	na	na	ERR

Error register –
ABRT will be set to one if the device does not support this command, if a Host Protected Area has been set by a SET MAX ADDRESS or SET MAX ADDRESS EXT command, or if DEVICE CONFIGURATION FREEZE LOCK is set.
Device register –
DEV will indicate the selected device.
Status register –
BSY will be cleared to zero, indicating command completion.
DRDY will be set to one.
DF (Device Fault) will be set to one if a device fault has occurred.
DRQ will be cleared to zero.
ERR will be set to one if an Error register bit is set to one.

Prerequisites

DRDY set to one.

Description

The DEVICE CONFIGURATION RESTORE command disables any setting previously made by a DEVICE CONFIGURATION SET command and returns the content of the IDENTIFY DEVICE or IDENTIFY PACKET DEVICE command data to the original settings, as indicated by the data returned from the execution of a DEVICE CONFIGURATION IDENTIFY command.

DEVICE CONFIGURATION FREEZE LOCK

Command code

B1h with a Features register value of C1h.

Feature set

Device Configuration Overlay feature set.
- Mandatory when the Device Configuration Overlay feature set is implemented.

Protocol

Non-data

Inputs

The Features register will be set to C1h.

Register	7	6	5	4	3	2	1	0
Features	C1h							
Sector Count	na							
LBA Low	na							
LBA Mid	na							
LBA High	na							
Device	na			DEV	na			
Command	B1h							

Device –
DEV will specify the selected device.

Normal outputs

Register	7	6	5	4	3	2	1	0
Error	na							
Sector Count	na							
LBA Low	na							
LBA Mid	na							
LBA High	na							
Device	obs	na	obs	DEV	na			
Status	BSY	DRDY	DF	na	DRQ	na	na	ERR

Device register –
DEV will indicate the selected device.
Status register –
BSY will be cleared to zero, indicating command completion.
DRDY will be set to one.
DF (Device Fault) will be cleared to zero.
DRQ will be cleared to zero.
ERR will be cleared to zero.

Error outputs

Register	7	6	5	4	3	2	1	0
Error	na	na	na	na	na	ABRT	na	na
Sector Count	na							
LBA Low	na							
LBA Mid	na							
LBA High	na							
Device	obs	na	obs	DEV	na			
Status	BSY	DRDY	DF	na	DRQ	na	na	ERR

Error register –
 ABRT will be set to one if the device does not support this command or the device has executed a previous DEVICE CONFIGURATION FREEZE LOCK command since power-up.
Device register –
 DEV will indicate the selected device.
Status register –
 BSY will be cleared to zero, indicating command completion.
 DRDY will be set to one.
 DF (Device Fault) will be set to one if a device fault has occurred.
 DRQ will be cleared to zero.
 ERR will be set to one if an Error register bit is set to one.

Prerequisites

DRDY set to one.

Description

The DEVICE CONFIGURATION FREEZE LOCK command prevents accidental modification of the Device Configuration Overlay settings. After successful execution of a DEVICE CONFIGURATION FREEZE LOCK command, all DEVICE CONFIGURATION SET, DEVICE CONFIGURATION FREEZE LOCK, DEVICE CONFIGURATION IDENTIFY, and DEVICE CONFIGURATION RESTORE commands will be aborted by the device. The DEVICE CONFIGURATION FREEZE LOCK condition will be cleared by a power-down. The DEVICE CONFIGURATION FREEZE LOCK condition will not be cleared by hardware or software reset.

DEVICE CONFIGURATION IDENTIFY

Command code

B1h with a Features register value of C2h.

Feature set

Device Configuration Overlay feature set.

- Mandatory when the Device Configuration Overlay feature set is implemented.

Protocol

PIO data-in

Inputs

The Features register will be set to C2h.

Register	7	6	5	4	3	2	1	0
Features	C2h							
Sector Count	na							
LBA Low	na							
LBA Mid	na							
LBA High	na							
Device	na			DEV	na			
Command	B1h							

Device –
DEV will specify the selected device.

Normal outputs

Register	7	6	5	4	3	2	1	0
Error	na							
Sector Count	na							
LBA Low	na							
LBA Mid	na							
LBA High	na							
Device	obs	na	obs	DEV	na			
Status	BSY	DRDY	DF	na	DRQ	na	na	ERR

Device register –
DEV will indicate the selected device.
Status register –
BSY will be cleared to zero, indicating command completion.
DRDY will be set to one.
DF (Device Fault) will be cleared to zero.
DRQ will be cleared to zero.
ERR will be cleared to zero.

Error outputs

Register	7	6	5	4	3	2	1	0
Error	na	na	na	na	na	ABRT	na	na
Sector Count	na							
LBA Low	na							
LBA Mid	na							
LBA High	na							
Device	obs	na	obs	DEV	na			
Status	BSY	DRDY	DF	na	DRQ	na	na	ERR

Error register –

ABRT will be set to one if the device does not support this command or the device has executed a previous DEVICE CONFIGURATION FREEZE LOCK command since power-up.

Device register –

DEV will indicate the selected device.

Status register –

BSY will be cleared to zero, indicating command completion.

DRDY will be set to one.

DF (Device Fault) will be set to one if a device fault has occurred.

DRQ will be cleared to zero.

ERR will be set to one if an Error register bit is set to one.

Prerequisites

DRDY set to one.

Description

The DEVICE CONFIGURATION IDENTIFY command returns a 512-byte data structure via PIO data-in transfer. The content of this data structure indicates the selectable commands, modes, and feature sets that the device is capable of supporting. If a DEVICE CONFIGURATION SET command has been issued reducing the capabilities, the response to an IDENTIFY DEVICE or IDENTIFY PACKET DEVICE command will reflect the reduced set of capabilities, and the DEVICE CONFIGURATION IDENTIFY command will reflect the entire set of selectable capabilities.

The term "is allowed" indicates that the device may report that a feature is supported and/or enabled.

If the device is not "allowed" to report support, then the device will not support and will report that the selected feature is both "not supported" and, if appropriate, "not enabled."

The format of the Device Configuration Overlay data structure is shown in Table 3.4.

Word 0: Data structure revision

Word 0 will contain the value 0002h.

Word 1: Multiword DMA modes supported

Word 1 bits (2:0) contain the same information as contained in word 63 of the IDENTIFY DEVICE or IDENTIFY PACKET DEVICE command data. Bits (15:3) of word 1 are reserved.

Word 2: Ultra DMA modes supported

Word 2 bits (6:0) contain the same information as contained in word 88 of the IDENTIFY DEVICE or IDENTIFY PACKET DEVICE command data. Bits (15:7) of word 2 are reserved.

Words (6:3): Maximum LBA

Words (6:3) define the maximum LBA. This is the highest address accepted by the device in the factory default condition. If no DEVICE CONFIGURATION SET command has been executed modifying the factory default condition, this is the same value as that returned by a READ NATIVE MAX ADDRESS or READ NATIVE MAX ADDRESS EXT command.

Word 7: Command/features set supported

Word 7 bit 0 if set to one indicates that the device is allowed to report support for the SMART feature set.

Table 3.4 – Device Configuration Identify Data Structure

Word	Content
0	Data structure revision
1	Multiword DMA modes supported 15–3 Reserved 2 1 = Reporting support for Multiword DMA mode 2 and below is allowed 1 1 = Reporting support for Multiword DMA mode 1 and below is allowed 0 1 = Reporting support for Multiword DMA mode 0 is allowed
2	Ultra DMA modes supported 15–7 Reserved 6 1 = Reporting support for Ultra DMA mode 6 and below is allowed 5 1 = Reporting support for Ultra DMA mode 5 and below is allowed 4 1 = Reporting support for Ultra DMA mode 4 and below is allowed 3 1 = Reporting support for Ultra DMA mode 3 and below is allowed 2 1 = Reporting support for Ultra DMA mode 2 and below is allowed 1 1 = Reporting support for Ultra DMA mode 1 and below is allowed 0 1 = Reporting support for Ultra DMA mode 0 is allowed
3–6	Maximum LBA
7	Command set/feature set supported 15–14 Reserved 13 1 = Reporting support for SMART Conveyance self-test is allowed 12 1 = Reporting support for SMART Selective self-test is allowed 11 1 = Reporting support for Forced Unit Access is allowed 10 Reserved 9 1 = Reporting support for Streaming feature set is allowed 8 1 = Reporting support for 48-bit Addressing feature set is allowed 7 1 = Reporting support for Host Protected Area feature set is allowed 6 1 = Reporting support for Automatic acoustic management is allowed 5 1 = Reporting support for READ/WRITE DMA QUEUED commands is allowed 4 1 = Reporting support for Power-up in Standby feature set is allowed 3 1 = Reporting support for Security feature set is allowed 2 1 = Reporting support for SMART error log is allowed 1 1 = Reporting support for SMART self-test is allowed 0 1 = Reporting support for SMART feature set is allowed
8–9	Reserved for serial ATA
10–254	Reserved
255	Integrity word 15–8 Checksum 7–0 Signature

Word 7 bit 1 if set to one indicates that the device allowed to report support for SMART self-test including the self-test log.

Word 7 bit 2 if set to one indicates that the device is allowed to report support for SMART error logging.

Word 7 bit 3 if set to one indicates that the device is allowed to report support for the Security feature set.

Word 7 bit 4 if set to one indicates that the device is allowed to report support for the Power-up in Standby feature set.

Word 7 bit 5 if set to one indicates that the device is allowed to report support for the READ DMA QUEUED and WRITE DMA QUEUED commands.

Word 7 bit 6 if set to one indicates that the device is allowed to report support for the Automatic Acoustic Management feature set.

Word 7 bit 7 if set to one indicates that the device is allowed to report support for the Host Protected Area feature set.

Word 7 bit 8 if set to one indicates that the device is allowed to report support for the 48-bit Addressing feature set.

Word 7 bit 9 if set to one indicates that the device is allowed to report support for Streaming feature set.

Word 7 bit 10 – Reserved

Word 7 bit 11 if set to one indicates that the device is allowed to report support for Force Unit Access commands.

Word 7 bit 12 if set to one indicates that the device is allowed to report support for SMART Selective self-test.

Word 7 bit 13 if set to one indicates that the device is allowed to report support for SMART Conveyance self-test.

Word 8–9: Reserved for serial ATA

These words are reserved for future serial ATA use.

Words (254:10): Reserved

Word 255: Integrity word

Bits (7:0) of this word will contain the value A5h. Bits (15:8) of this word will contain the data structure checksum. The data structure checksum will be the two's complement of the sum of all bytes in words (154:0) and the byte consisting of bits (7:0) of word 255. Each byte will be added with unsigned arithmetic, and overflow will be ignored. The sum of all bytes is zero when the checksum is correct.

DEVICE CONFIGURATION SET

Command code

B1h with a Features register value of C3h.

Feature set

Device Configuration Overlay feature set.

- Mandatory when the Device Configuration Overlay feature set is implemented.

Protocol

PIO data-out

Inputs

The Features register will be set to C3h.

Register	7	6	5	4	3	2	1	0
Features	C3h							
Sector Count	na							
LBA Low	na							
LBA Mid	na							
LBA High	na							
Device	na			DEV	na			
Command	B1h							

Device –
DEV will specify the selected device.

Normal outputs

Register	7	6	5	4	3	2	1	0
Error	na							
Sector Count	na							
LBA Low	na							
LBA Mid	na							
LBA High	na							
Device	obs	na	obs	DEV	na			
Status	BSY	DRDY	DF	na	DRQ	na	na	ERR

Device register –
DEV will indicate the selected device.
Status register –
BSY will be cleared to zero, indicating command completion.
DRDY will be set to one.
DF (Device Fault) will be cleared to zero.
DRQ will be cleared to zero.
ERR will be cleared to zero.

Error outputs

Register	7	6	5	4	3	2	1	0
Error	na	na	na	na	na	ABRT	na	na
Sector Count	Vendor specific							
LBA Low	Bit location low							
LBA Mid	Bit location high							
LBA High	Word location							
Device	obs	na	obs	DEV	na			
Status	BSY	DRDY	DF	na	DRQ	na	na	ERR

Error register –

ABRT will be set to one if the device does not support this command, if a DEVICE CONFIGURATION SET command has already modified the original settings as reported by a DEVICE CONFIGURATION IDENTIFY command, if DEVICE CONFIGURATION FREEZE LOCK is set, if any of the bit modification restrictions described in 0 are violated, or if a Host Protected Area has been established by the execution of a SET MAX ADDRESS or SET MAX ADDRESS EXT command, or if an attempt was made to modify a mode or feature that cannot be modified with the device in its current state.

Sector Count –

This register may contain a vendor-specific value.

LBA Low –

If the command was aborted because an attempt was made to modify a mode or feature that cannot be modified with the device in its current state, this register will contain bits (7:0) set in the bit positions that correspond to the bits in the device configuration overlay data structure words 1, 2, or 7 for each mode or feature that cannot be changed. If not, the value will be 00h.

LBA Mid –

If the command was aborted because an attempt was made to modify a mode or feature that cannot be modified with the device in its current state, this register will contain bits (15:8) set in the bit positions that correspond to the bits in the device configuration overlay data structure words 1, 2, or 7 for each mode or feature that cannot be changed. If not, the value will be 00h.

LBA High –

If the command was aborted because an attempt was made to modify a bit that cannot be modified with the device in its current state, this register will contain the offset of the first word encountered that cannot be changed. If an illegal maximum LBA is encountered, the offset of word 3 will be entered. If a checksum error occurred, the value FFh will be entered. A value of 00h indicates that the Data Structure Revision was invalid.

Device register –

DEV will indicate the selected device.

Status register –

BSY will be cleared to zero, indicating command completion.

DRDY will be set to one.

DF (Device Fault) will be set to one if a device fault has occurred.

DRQ will be cleared to zero.

ERR will be set to one if an Error register bit is set to one.

Prerequisites

DRDY set to one.

Description

The DEVICE CONFIGURATION SET command allows a device manufacturer or a personal computer system manufacturer to reduce the set of optional commands, modes, or feature sets supported by a device, as indicated by a DEVICE CONFIGURATION IDENTIFY command. The DEVICE CONFIGURATION SET command transfers an overlay that modifies some of the bits set in words 63, 82, 83, 84, and 88 of the IDENTIFY DEVICE or IDENTIFY PACKET DEVICE command data (see Table 3.5). When the bits in these words are cleared, the device will no longer support the indicated command, mode, or feature set. If a bit is set in the overlay transmitted by the device that is not set in the overlay received from a DEVICE CONFIGURATION IDENTIFY command, no action is taken for that bit. Modifying the maximum LBA of the device also modifies the address value returned by a READ NATIVE MAX ADDRESS or READ NATIVE MAX ADDRESS EXT command.

The term "is allowed" indicates that the device may report that a feature is supported and/or enabled.

If the device is not "allowed" to report support, then the device will not support and will report that the selected feature is both "not supported" and, if appropriate, "not enabled."

Table 3.5 – Device Configuration Overlay Data Structure

Word	Content	
0	Data structure revision	
1	Multiword DMA modes supported	
	15–3	Reserved
	2	1 = Reporting support for Multiword DMA mode 2 and below is allowed
	1	1 = Reporting support for Multiword DMA mode 1 and below is allowed
	0	1 = Reporting support for Multiword DMA mode 0 is allowed
2	Ultra DMA modes supported	
	15–7	Reserved
	6	1 = Reporting support for Ultra DMA mode 6 and below is allowed
	5	1 = Reporting support for Ultra DMA mode 5 and below is allowed
	4	1 = Reporting support for Ultra DMA mode 4 and below is allowed
	3	1 = Reporting support for Ultra DMA mode 3 and below is allowed
	2	1 = Reporting support for Ultra DMA mode 2 and below is allowed
	1	1 = Reporting support for Ultra DMA mode 1 and below is allowed
	0	1 = Reporting support for Ultra DMA mode 0 is allowed
3–6	Maximum LBA	
7	Command set/feature set supported	
	15–14	Reserved
	13	1 = Reporting support for SMART Conveyance self-test is allowed
	12	1 = Reporting support for SMART Selective self-test is allowed
	11	1 = Reporting support for Forced Unit Access is allowed
	10	– Reserved for technical report
	9	1 = Reporting support for Streaming feature set is allowed
	8	1 = Reporting support for 48-bit Addressing feature set is allowed
	7	1 = Reporting support for Host Protected Area feature set is allowed
	6	1 = Reporting support for Automatic acoustic management is allowed
	5	1 = Reporting support for READ/WRITE DMA QUEUED commands is allowed
	4	1 = Reporting support for Power-up in Standby feature set is allowed
	3	1 = Reporting support for Security feature set is allowed
	2	1 = Reporting support for SMART error log is allowed
	1	1 = Reporting support for SMART self-test is allowed
	0	1 = Reporting support for SMART feature set is allowed
8–9	Reserved for serial ATA	
10–254	Reserved	
255	Integrity word	
	15–8	Checksum
	7–0	Signature

Word 0: Data structure revision

Word 0 will contain the value 0002h.

Word 1: Multiword DMA modes supported

Word 1 bits (15:3) are reserved.

Word 1 bit 2 is cleared to disable support for Multiword DMA mode 2 and has the effect of clearing bit 2 in word 63 of the IDENTIFY DEVICE or IDENTIFY PACKET DEVICE response. This bit will not be cleared to zero if Multiword DMA mode 2 is currently selected.

Word 1 bit 1 is cleared to disable support for Multiword DMA mode 1 and has the effect of clearing bit 1 to zero in word 63 of the IDENTIFY DEVICE or IDENTIFY PACKET DEVICE response. This bit will not be cleared to zero if Multiword DMA mode 2 is supported or Multiword DMA mode 1 or 2 is selected.

Word 1 bit 0 will not be cleared to zero.

Word 2: Ultra DMA modes supported

Word 2 bits (15:7) are reserved.

Word 2 bit 6 is cleared to zero to disable support for Ultra DMA mode 6 and has the effect of clearing bit 6 to zero in word 88 of the IDENTIFY DEVICE or IDENTIFY PACKET DEVICE response. This bit will not be cleared to zero if Ultra DMA mode 6 is currently selected.

Word 2 bit 5 is cleared to zero to disable support for Ultra DMA mode 5 and has the effect of clearing bit 5 to zero in word 88 of the IDENTIFY DEVICE or IDENTIFY PACKET DEVICE response. This bit will not be cleared to zero if Ultra DMA mode 5 is currently selected.

Word 2 bit 4 is cleared to zero to disable support for Ultra DMA mode 4 and has the effect of clearing bit 4 to zero in word 88 of the IDENTIFY DEVICE or IDENTIFY PACKET DEVICE response. This bit will not be cleared to zero if Ultra DMA mode 5 is supported or if Ultra DMA mode 5 or 4 is selected.

Word 2 bit 3 is cleared to zero to disable support for Ultra DMA mode 3 and has the effect of clearing bit 3 to zero in word 88 of the IDENTIFY DEVICE or IDENTIFY PACKET DEVICE response. This bit will not be cleared to zero if Ultra DMA mode 5 or 4 is supported or if Ultra DMA mode 5, 4, or 3 is selected.

Word 2 bit 2 is cleared to zero to disable support for Ultra DMA mode 2 and has the effect of clearing bit 2 to zero in word 88 of the IDENTIFY DEVICE or IDENTIFY PACKET DEVICE response. This bit will not be cleared to zero if Ultra DMA mode 5, 4, or 3 is supported or if Ultra DMA mode 5, 4, 3, or 2 is selected.

Word 2 bit 1 is cleared to zero to disable support for Ultra DMA mode 1 and has the effect of clearing bit 1 to zero in word 88 of the IDENTIFY DEVICE or IDENTIFY PACKET DEVICE response. This bit will not be cleared to zero if Ultra DMA mode 5, 4, 3, or 2 is supported or if Ultra DMA mode 5, 4, 3, 2, or 1 is selected.

Word 2 bit 0 is cleared to zero to disable support for Ultra DMA mode 0 and has the effect of clearing bit 0 to zero in word 88 of the IDENTIFY DEVICE or IDENTIFY PACKET DEVICE response. This bit will not be cleared to zero if Ultra DMA mode 5, 4, 3, 2, or 1 is supported or if Ultra DMA mode 5, 4, 3, 2, 1, or 0 is selected.

Words (6:3): Maximum LBA

Words (6:3) define the maximum LBA. This will be the highest address accepted by the device after execution of the command. When this value is changed, the content of IDENTIFY DEVICE data words (61:60) and (103:100) will be changed as described in the SET MAX ADDRESS and SET MAX ADDRESS EXT command descriptions to reflect the maximum address set with this command. This value will not be changed and command aborted will be returned if a Host Protected Area has been established by the execution of a SET MAX ADDRESS or SET MAX ADDRESS EXT command with an address value less than that returned by a READ NATIVE MAX ADDRESS or READ NATIVE MAX ADDRESS EXT command. Any data contained in the Host Protected Area is not affected.

Word 7: Command/features set supported

Word 7 bits (15:14) are reserved.

Word 7 bit 13 is cleared to zero to disable support for the SMART Conveyance self-test. Subsequent attempts to start this test via the SMART EXECUTE OFF-LINE IMMEDIATE command will cause that command to abort. In addition, the SMART READ DATA command will clear bit 5 to zero in the "Off-line data collection capabilities" field. If this bit is supported by DEVICE CONFIGURATION SET, then this feature will not be disabled by bit 1 of word 7.

Word 7 bit 12 is cleared to zero to disable support for the SMART Selective self-test. Subsequent attempts to start this test via the SMART EXECUTE OFF-LINE IMMEDIATE command will cause that command to abort. In addition, the SMART READ DATA command will clear bit 6 to zero in the "Off-line data collection capabilities" field. If this bit is supported by DEVICE CONFIGURATION SET, then this feature will not be disabled by bit 1 of word 7.

Word 7 bit 11 is cleared to zero to disable support for the Force Unit Access commands and has the effect of clearing bits 6 and 7 to zero in word 84 and word 87 of the IDENTIFY DEVICE or IDENTIFY PACKET DEVICE response.

Word 7 bit 10 is –Reserved

Word 7 bit 9 is cleared to zero to disable support for the Streaming feature set and has the effect of clearing bits 4, 9 and 10 to zero in word 84 and word 87 and clearing the value in words (99:95) and word 104 of the IDENTIFY DEVICE or IDENTIFY PACKET DEVICE response.

Word 7 bit 8 is cleared to zero to disable support for the 48-bit Addressing feature set and has the effect of clearing bit 10 to zero in word 83 and word 86 and clearing the value in words (103:100) of the IDENTIFY DEVICE or IDENTIFY PACKET DEVICE response.

Word 7 bit 7 is cleared to zero to disable support for the Host Protected Area feature set and has the effect of clearing bit 10 to zero in word 82 and word 85 and clearing bit 8 to zero in word 83 and word 86 of the IDENTIFY DEVICE or IDENTIFY PACKET DEVICE response. If a Host Protected Area has been established by use of the SET MAX ADDRESS or SET MAX ADDRESS EXT command, these bits will not be cleared to zero and the device will return command aborted.

Word 7 bit 6 is cleared to zero to disable for the Automatic Acoustic Management feature set and has the effect of clearing bit 9 to zero in word 83 and word 94 of the IDENTIFY DEVICE or IDENTIFY PACKET DEVICE response.

Word 7 bit 5 is cleared to zero to disable support for the READ DMA QUEUED and WRITE DMA QUEUED commands and has the effect of clearing bit 1 to zero in word 83 and word 86 of the IDENTIFY DEVICE or IDENTIFY PACKET DEVICE response.

Word 7 bit 4 is cleared to zero to disable support for the Power-up in Standby feature set and has the effect of clearing bits (6:5) to zero in word 83 and word 86 and clearing the value in word 94 of the IDENTIFY DEVICE or IDENTIFY PACKET DEVICE response. If Power-up in Standby has been enabled by a jumper, these bits will not be cleared.

Word 7 bit 3 is cleared to zero to disable support for the Security feature set and has the effect of clearing bit 1 to zero in word 82 and word 85 of the IDENTIFY DEVICE or IDENTIFY PACKET DEVICE response. These bits will not be cleared if the Security feature set has been enabled.

Word 7 bit 2 is cleared to zero to disable support for the SMART error logging and has the effect of clearing bit 0 to zero in word 84 and word 87 of the IDENTIFY DEVICE or IDENTIFY PACKET DEVICE response.

Word 7 bit 1 is cleared to zero to disable support for the SMART self-test and has the effect of clearing bit 1 to zero in word 84 and word 87 of the IDENTIFY DEVICE or IDENTIFY PACKET DEVICE response.

Word 7 bit 1 disables support for the offline, short, extended self-tests (off-line and captive modes). If bit 12 or bit 13 of word 7 are not supported, Word 7 bit 1 may also disable support for conveyance self-test and selective self-test.

Word 7 bit 0 is cleared to zero to disable support for the SMART feature set and has the effect of clearing bit 0 to zero in word 82 and word 85 of the IDENTIFY DEVICE or IDENTIFY PACKET DEVICE response. If bits (2:1) of word 7 are not cleared to zero or if the SMART feature set has been enabled by use of the SMART ENABLE OPERATIONS command, these bits will not be cleared and the device will return command aborted.

Words 8–9: Reserved for serial ATA

These words are reserved for future serial ATA use.

Words (254:10): Reserved

Word 255: Integrity word

Bits (7:0) of this word will contain the value A5h. Bits (15:8) of this word will contain the data structure checksum. The data structure checksum will be the two's complement of the sum of all bytes in words (254:0) and the byte consisting of bits (7:0) of word 255. Each byte will be added with unsigned arithmetic, and overflow will be ignored. The sum of all bytes is zero when the checksum is correct.

DEVICE RESET

Command code

08h

Feature set

General feature set

- Use prohibited when the PACKET Command feature set is not implemented.
- Mandatory when the PACKET Command feature set is implemented.

Protocol

Device reset

Inputs

Register	7	6	5	4	3	2	1	0
Features	na							
Sector Count	na							
LBA Low	na							
LBA Mid	na							
LBA High	na							
Device	obs	na	obs	DEV	na	na	na	na
Command	08h							

Device register –
DEV will specify the selected device.

Normal outputs

Register	7	6	5	4	3	2	1	0
Error	Diagnostic results							
Sector Count	Signature							
LBA Low	Signature							
LBA Mid	Signature							
LBA High	Signature							
Device	0	0	0	DEV	0	0	0	0
Status	See Status Codes							

Error register –
The diagnostic code as described in the table below is placed in this register.

Code	Description
When this code is in the Device 0 Error register	
01h	Device 0 passed, Device 1 passed or not present
00h, 02h–7Fh	Device 0 failed, Device 1 passed or not present
81h	Device 0 passed, Device 1 failed
80h, 82h–FFh	Device 0 failed, Device 1 failed
NOTE – Codes other than 01h and 81h may indicate additional information about the failure(s).	

Sector Count, LBA Low, LBA Mid, LBA High –
Signature of device

Signature for a Device That Does Not Support the Packet Command

Register	Value
Sector Count	01h
Sector Number	01h
Cylinder Low	00h
Cylinder High	00h
Device/Head	00h
Error	01h
Status	00h–70h

Signature for a Device That Supports the Packet Command

Register	Value
Sector Count	01h
Sector Number	01h
Cylinder Low	14h
Cylinder High	EBh
Device/Head	00h
Error	01h
Status	00h

Device register –

DEV will indicate the selected device.

Status register – standard status

Error outputs

If supported, this command will not end in an error condition. If this command is not supported and the device has the BSY bit or the DRQ bit set to one when the command is written, the results of this command are indeterminate. If this command is not supported and the device has the BSY bit and the DRQ bit cleared to zero when the command is written, the device will respond with command aborted.

Prerequisites

This command will be accepted when BSY or DRQ is set to one, DRDY is cleared to zero, or DMARQ is asserted. This command will be accepted when in Sleep mode.

Description

The DEVICE RESET command enables the host to reset an individual device without affecting the other device.

DOWNLOAD MICROCODE

Command code

92h

Feature set

General feature set

- Optional for devices not implementing the PACKET Command feature set.
- Use prohibited for devices implementing the PACKET Command feature set.

Protocol

PIO data-out.

Inputs

Bits (3:0) of the Device register will always be cleared to zero. The LBA High and LBA Mid registers will be cleared to zero. The LBA Low and Sector Count registers are used together as a 16-bit sector count value. The Feature register specifies the subcommand code.

Register	7	6	5	4	3	2	1	0
Features	Subcommand code							
Sector Count	Sector count (low order)							
LBA Low	Sector count (high order)							
LBA Mid	00h							
LBA High	00h							
Device	obs	na	obs	DEV	0	0	0	0
Command	92h							

Device register –
DEV will specify the selected device.

Normal outputs

Register	7	6	5	4	3	2	1	0
Error	na							
Sector Count	na							
LBA Low	na							
LBA Mid	na							
LBA High	na							
Device	obs	na	obs	DEV	na	na	na	na
Status	BSY	DRDY	DF	na	DRQ	na	na	ERR

Device register –
DEV will indicate the selected device.
Status register –
BSY will be cleared to zero, indicating command completion.
DRDY will be set to one.
DF (Device Fault) will be cleared to zero.
DRQ will be cleared to zero.
ERR will be cleared to zero.

Error outputs

The device will return command aborted if the device does not support this command or did not accept the microcode data. The device will return command aborted if subcommand code is not a supported value.

Register	7	6	5	4	3	2	1	0
Error	na	na	na	na	na	ABRT	na	na
Sector Count	na							
LBA Low	na							
LBA Mid	na							
LBA High	na							
Device	obs	na	obs	DEV	na	na	na	na
Status	BSY	DRDY	DF	na	DRQ	na	na	ERR

Error register –

ABRT will be set to one if the device does not support this command or did not accept the microcode data. ABRT may be set to one if the device is not able to complete the action requested by the command.

Device register –

DEV will indicate the selected device.

Status register –

BSY will be cleared to zero, indicating command completion.

DRDY will be set to one.

DF (Device Fault) will be set to one if a device fault has occurred.

DRQ will be cleared to zero.

ERR will be set to one if an Error register bit is set to one.

Prerequisites

DRDY set to one.

Description

This command enables the host to alter the device's microcode. The data transferred using the DOWNLOAD MICROCODE command is vendor specific.

All transfers will be an integer multiple of the sector size. The size of the data transfer is determined by the contents of the LBA Low register and the Sector Count register. The LBA Low register will be used to extend the Sector Count register to create a 16-bit sector count value. The LBA Low register will be the most significant eight bits and the Sector Count register will be the least significant eight bits. A value of zero in both the LBA Low register and the Sector Count register will specify no data is to be transferred. This allows transfer sizes from 0 bytes to 33,553,920 bytes, in 512-byte increments.

The Features register will be used to determine the effect of the DOWNLOAD MICROCODE command. The values for the Features register are as follows:

- 01h – download is for immediate, temporary use.
- 07h – save downloaded code is for immediate and future use.

Either or both values may be supported. All other values are reserved.

EXECUTE DEVICE DIAGNOSTIC

Command code

90h

Feature set

General feature set

- Mandatory for all devices.

Protocol

Device diagnostic

Inputs

Only the command code (90h). All other registers will be ignored.

Register	7	6	5	4	3	2	1	0
Features	na							
Sector Count	na							
LBA Low	na							
LBA Mid	na							
LBA High	na							
Device	obs	na	obs	na	na	na	na	na
Command	90h							

Device register –
DEV will be ignored.

Normal outputs

The diagnostic code written into the Error register is an 8-bit code. See Diagnostic Codes table to defines these values. Both Device 0 and Device 1 will provide these register contents.

Register	7	6	5	4	3	2	1	0
Error	Diagnostic code							
Sector Count	Signature							
LBA Low	Signature							
LBA Mid	Signature							
LBA High	Signature							
Device	Signature							
Status	See Diagnostic Codes table for values							

Error register –
Diagnostic code.
Sector Count, LBA Low, LBA Mid, LBA High, Device registers –
Device signatures can be found in the DEVICE RESET section in the current chapter.
Device register –
DEV will be cleared to zero.
Status register –
Remaining device Signature.

Diagnostic Codes

Code [see note (1)]	Description
When this code is in the Device 0 Error register	
01h	Device 0 passed, Device 1 passed or not present
00h, 02h–7Fh	Device 0 failed, Device 1 passed or not present
81h	Device 0 passed, Device 1 failed
80h, 82h–FFh	Device 0 failed, Device 1 failed
When this code is in the Device 1 Error register	
01h	Device 1 passed [see note (2)]
00h, 02h–7Fh	Device 1 failed [see note (2)]
NOTE – (1) Codes other than 01h and 81h may indicate additional information about the failure(s). (2) If Device 1 is not present, the host may see the information from Device 0 even though Device 1 is selected.	

Error outputs

The table above shows the error information that is returned as a diagnostic code in the Error register.

Prerequisites

This command will be accepted, regardless of the state of DRDY.

Description

This command will cause the devices to perform the internal diagnostic tests. Both devices, if present, will execute this command, regardless of which device is selected.

If the host issues an EXECUTE DEVICE DIAGNOSTIC command while a device is in or going to a power management mode except Sleep, then the device will execute the EXECUTE DEVICE DIAGNOSTIC sequence.

FLUSH CACHE

Command code

E7h

Feature set

General feature set

- Mandatory for all devices not implementing the PACKET Command feature set.
- Optional for devices implementing the PACKET Command feature set.

Protocol

Non-data

Inputs

Register	7	6	5	4	3	2	1	0
Features	na							
Sector Count	na							
LBA Low	na							
LBA Mid	na							
LBA High	na							
Device	obs	na	obs	DEV	na	na	na	na
Command	E7h							

Device register –

DEV will specify the selected device.

Normal outputs

Register	7	6	5	4	3	2	1	0
Error	na							
Sector Count	na							
LBA Low	na							
LBA Mid	na							
LBA High	na							
Device	obs	na	obs	DEV	na	na	na	na
Status	BSY	DRDY	DF	na	DRQ	na	na	ERR

Device register –
DEV will indicate the selected device.
Status register –
BSY will be cleared to zero, indicating command completion.
DRDY will be set to one.
DF (Device Fault) will be cleared to zero.
DRQ will be cleared to zero.
ERR will be cleared to zero.

Error outputs

An unrecoverable error encountered during execution of writing data results in the termination of the command and the Command Block registers contain the sector address of the sector where the first unrecoverable error occurred. Subsequent FLUSH CACHE commands continue the process of flushing the cache starting with the first sector after the sector in error.

Register	7	6	5	4	3	2	1	0
Error	na	na	na	na	na	ABRT	na	na
Sector Count	na							
LBA Low	LBA (7:0)							
LBA Mid	LBA (15:8)							
LBA High	LBA (23:16)							
Device	obs	na	obs	DEV	LBA (27:24)			
Status	BSY	DRDY	DF	na	DRQ	na	na	ERR

Error register –
ABRT may be set to one if the device is not able to complete the action requested by the command.
LBA Low, LBA Mid, LBA High, Device –
Will be written with the address of the first unrecoverable error. If the device supports the 48-bit Address feature set and the error occurred in an address greater than FFFFFFFh, the value set in the LBA Low, LBA Mid, and LBA High registers will be FFh and the value set in bits (3:0) of the Device register will be Fh.
Device register –
DEV will indicate the selected device.
Status register –
BSY will be cleared to zero, indicating command completion.
DRDY will be set to one.
DF (Device Fault) will be set to one if a device fault has occurred.
ERR will be set to one if an Error register bit is set to one.-

Prerequisites

DRDY set to one.

Description

This command is used by the host to request the device to flush the write cache. If there is data in the write cache, that data will be written to the media. The BSY bit will remain set to one until all data has been successfully written or an error occurs.

Note: This command may take longer than 30 s to complete.

FLUSH CACHE EXT

Command code

EAh

Feature set

48-bit Address feature set
- Mandatory for all devices implementing the 48-bit Address feature.
- Prohibited for devices implementing the PACKET Command feature set.

Protocol

Non-data

Inputs

Register		7	6	5	4	3	2	1	0
Features	Current	Reserved							
	Previous	Reserved							
Sector Count	Current	Reserved							
	Previous	Reserved							
LBA Low	Current	Reserved							
	Previous	Reserved							
LBA Mid	Current	Reserved							
	Previous	Reserved							
LBA High	Current	Reserved							
	Previous	Reserved							
Device		obs	na	obs	DEV	na			
Command		EAh							
NOTE – The value indicated as Current is the value most recently written to the register. The value indicated as Previous is the value that was in the register before the most recent write to the register.									

Device register –
DEV will specify the selected device.

Normal outputs

Register		7	6	5	4	3	2	1	0
Error		na							
Sector Count	HOB = 0	Reserved							
	HOB = 1	Reserved							
LBA Low	HOB = 0	Reserved							
	HOB = 1	Reserved							
LBA Mid	HOB = 0	Reserved							
	HOB = 1	Reserved							
LBA High	HOB = 0	Reserved							
	HOB = 1	Reserved							
Device		obs	na	obs	DEV	Reserved			
Status		BSY	DRDY	DF	na	DRQ	na	na	ERR
NOTE – HOB = 0 indicates the value read by the host when the HOB bit of the Device Control register is cleared to zero. HOB = 1 indicates the value read by the host when the HOB bit of the Device Control register is set to one.									

Device register –
DEV will indicate the selected device.
Status register –
BSY will be cleared to zero, indicating command completion.
DRDY will be set to one.
DF (Device Fault) will be cleared to zero.
DRQ will be cleared to zero.
ERR will be cleared to zero.

Error outputs

An unrecoverable error encountered while writing data results in the termination of the command and the Command Block registers contain the sector address of the sector where the first unrecoverable error occurred. Subsequent FLUSH CACHE EXT commands continue the process of flushing the cache starting with the first sector after the sector in error.

Register		7	6	5	4	3	2	1	0
Error		na	na	na	na	na	ABRT	na	na
Sector Count	HOB = 0	Reserved							
	HOB = 1	Reserved							
LBA Low	HOB = 0	LBA (7:0)							
	HOB = 1	LBA (31:24)							
LBA Mid	HOB = 0	LBA (15:8)							
	HOB = 1	LBA (39:32)							
LBA High	HOB = 0	LBA (23:16)							
	HOB = 1	LBA (47:40)							
Device		obs	na	obs	DEV	Reserved			
Status		BSY	DRDY	DF	na	DRQ	na	na	ERR
NOTE – HOB = 0 indicates the value read by the host when the HOB bit of the Device Control register is cleared to zero. HOB = 1 indicates the value read by the host when the HOB bit of the Device Control register is set to one.									

Error register –
ABRT will be set to one if the device is not able to complete the action requested by the command.
LBA Low –
LBA (7:0) of the address of the first unrecoverable error when read with Device Control register HOB bit cleared to zero.
LBA (31:24) of the address of the first unrecoverable error when read with Device Control register HOB set to one.
LBA Mid –
LBA (15:8) of the address of the first unrecoverable error when read with Device Control register HOB cleared to zero.
LBA (39:32) of the address of the first unrecoverable error when read with Device Control register HOB set to one.
LBA High –
LBA (23:16) of the address of the first unrecoverable error when read with Device Control register HOB cleared to zero.
LBA (47:40) of the address of the first unrecoverable error when read with Device Control register HOB is set to one.

Device register –
DEV will indicate the selected device.
Status register –
BSY will be cleared to zero, indicating command completion.
DRDY will be set to one.
DF (Device Fault) will be set to one if a device fault has occurred.
DRQ will be cleared to zero.
ERR will be set to one if an Error register bit is set to one; however, if SE is set to one, ERR will be cleared to zero.

Prerequisites

DRDY set to one.

Description

This command is used by the host to request the device to flush the write cache. If there is data in the write cache, that data will be written to the media. The BSY bit will remain set to one until all data has been successfully written or an error occurs.

Note: This command may take longer than 30 s to complete.

GET MEDIA STATUS

Command code

DAh

Feature set

Removable Media Status Notification feature set
– Mandatory for devices implementing the Removable Media Status Notification feature set.

Removable Media feature set
– Optional for devices implementing the Removable Media feature set.

Protocol

Non-data

Inputs

Register	7	6	5	4	3	2	1	0
Features	na							
Sector Count	na							
LBA Low	na							
LBA Mid	na							
LBA High	na							
Device	obs	na	obs	DEV	na	na	na	na
Command	DAh							

Device register –
DEV will specify the selected device.

Normal Outputs

Normal outputs are returned if Media Status Notification is disabled or if no bits are set to one in the Error register.

Register	7	6	5	4	3	2	1	0
Error	na							
Sector Count	na							
LBA Low	na							
LBA Mid	na							
LBA High	na							
Device	obs	na	obs	DEV	na	na	na	na
Status	BSY	DRDY	DF	na	DRQ	na	na	ERR

Device register –
DEV will indicate the selected device.
Status register –
BSY will be cleared to zero, indicating command completion.
DRDY will be set to one.
DF (Device Fault) will be cleared to zero.
DRQ will be cleared to zero.
ERR will be cleared to zero.

Error Outputs

If the device does not support this command, the device will return command aborted.

Register	7	6	5	4	3	2	1	0
Error	na	WP	MC	na	MCR	ABRT	NM	obs
Sector Count	na							
LBA Low	na							
LBA Mid	na							
LBA High	na							
Device	obs	na	obs	DEV	na	na	na	na
Status	BSY	DRDY	DF	na	DRQ	na	na	ERR

Error register –
ABRT will be set to one if device does not support this command. ABRT may be set to one if the device is not able to complete the action requested by the command.
NM (No Media) will be set to one if no media is present in the device. This bit will be set to one for each execution of GET MEDIA STATUS until media is inserted into the device.
MCR (Media Change Request) will be set to one if the eject button is pressed by the user and detected by the device. The device will reset this bit after each execution of the GET MEDIA STATUS command and only set the bit again for subsequent eject button presses.
MC (Media Change) will be set to one when the device detects media has been inserted. The device will reset this bit after each execution of the GET MEDIA STATUS command and only set the bit again for subsequent media insertions.

WP (Write Protect) will be set to one for each execution of GET MEDIA STATUS while the media is write protected.

Device register –

DEV will indicate the selected device.

Status register –

BSY will be cleared to zero, indicating command completion.

DRDY will be set to one.

DF (Device Fault) will be set to one if a device fault has occurred.

DRQ will be cleared to zero.

ERR will be set to one if an Error register bit is set to one.

Prerequisites

DRDY set to one.

Description

This command returns media status bits WP, MC, MCR, and NM, as defined above. When Media Status Notification is disabled, this command returns zeros in the WP, MC, MCR, and NM bits.

IDENTIFY DEVICE

Command code

ECh

Feature set

General feature set

- Mandatory for all devices.
- Devices implementing the PACKET Command feature set (See 0).

Protocol

PIO data-in

Inputs

Register	7	6	5	4	3	2	1	0
Features	na							
Sector Count	na							
LBA Low	na							
LBA Mid	na							
LBA High	na							
Device	obs	na	obs	DEV	na			
Command	ECh							

Device register –

DEV will specify the selected device.

Outputs

Normal outputs

Register	7	6	5	4	3	2	1	0
Error	na							
Sector Count	na							
LBA Low	na							
LBA Mid	na							
LBA High	na							
Device	obs	na	obs	DEV	na	na	na	na
Status	BSY	DRDY	DF	na	DRQ	na	na	ERR

Device register –
DEV will indicate the selected device.
Status register –
BSY will be cleared to zero, indicating command completion.
DRDY will be set to one.
DF (Device Fault) will be cleared to zero.
DRQ will be cleared to zero.
ERR will be cleared to zero.

Outputs for PACKET Command feature set devices

In response to this command, devices that implement the PACKET Command feature set will post command aborted and place the PACKET Command feature set signature in the Command Block registers.

Error outputs

Devices not implementing the PACKET Command feature set will not report an error.

Prerequisites

DRDY set to one.

Description

The IDENTIFY DEVICE command enables the host to receive parameter information from the device.

Some devices may have to read the media in order to complete this command.

When the command is issued, the device sets the BSY bit to one, prepares to transfer the 256 words of device identification data to the host, sets the DRQ bit to one, clears the BSY bit to zero, and asserts INTRQ if nIEN is cleared to zero. The host may then transfer the data by reading the Data register. Table 3.6 defines the arrangement and meaning of the parameter words in the buffer. All reserved bits or words will be zero.

Some parameters are defined as a 16-bit value. A word that is defined as a 16-bit value places the most significant bit of the value on signal line DD15 and the least significant bit on signal line DD0.

Some parameters are defined as 32-bit values (e.g., words (61:60)). Such fields are transferred using two successive word transfers. The device will first transfer the least significant bits, bits (15:0) of the value, on signal lines DD(15:0), respectively. After the least significant bits have been transferred, the most significant bits, bits (31:16) of the value, will be transferred on DD(15:0), respectively.

Table 3.6 – IDENTIFY DEVICE Data

Word	O/M	F/V	Description
0	M		General configuration bit – significant information:
		F	15 0 = ATA device
		X	14–8 Retired
		F	7 1 = removable media device
		X	6 Obsolete
		X	5–3 Retired
		V	2 Response incomplete
		X	1 Retired
		F	0 Reserved
1		X	Obsolete
2	O	V	Specific configuration
3		X	Obsolete
4–5		X	Retired
6		X	Obsolete
7–8	O	V	Reserved for assignment by the CompactFlash™ Association
9		X	Retired
10–19	M	F	Serial number (20 ASCII characters)
20–21		X	Retired
22		X	Obsolete
23–26	M	F	Firmware revision (8 ASCII characters)
27–46	M	F	Model number (40 ASCII characters)
47	M	F	15–8 80h
		F	7–0 00h = Reserved
		F	01h–FFh = Maximum number of sectors that will be transferred per interrupt on READ/WRITE MULTIPLE commands
48		F	Trusted Computing feature set options
			15 Will be cleared to zero
			14 Will be set to one
			13:1 Reserved for the Trusted Computing Group
			0 Trusted Computing feature set is supported
49	M		Capabilities
		F	15–14 Reserved for the IDENTIFY PACKET DEVICE command
		F	13 1 = Standby timer values as specified in this standard are supported
			0 = Standby timer values will be managed by the device
		F	12 Reserved for the IDENTIFY PACKET DEVICE command
		F	11 1 = IORDY supported
			0 = IORDY may be supported
		F	10 1 = IORDY may be disabled
		F	9 1 = LBA supported
		F	8 1 = DMA supported
		X	7–0 Retired
50	M		Capabilities
		F	15 Will be cleared to zero
		F	14 Will be set to one
		F	13–2 Reserved
		X	1 Obsolete
		F	0 Will be set to one to indicate a device specific Standby timer value minimum
51–52		X	Obsolete
53			15:8 Free-fall Control Sensitivity 7:3 Reserved
			2 The fields reported in word 88 are valid
			1 The fields reported in words 64..70 are valid
			0 Obsolete

(*Continued on following page*)

Table 3.6 – IDENTIFY DEVICE Data (*Continued*)

Word	O/M	F/V	Description
54–58		X	Obsolete
59	M	F	15 The BLOCK ERASE EXT command is supported 14 The OVERWRITE EXT command is supported 13 The CRYPTO SCRAMBLE EXT command is supported 12 The Sanitize feature set is supported 11:9 Reserved 8 Multiple logical sector setting is valid 7:0 Current setting for number of logical sectors that will be transferred per DRQ data block
60–61	M	F	Total number of user addressable sectors
62		X	Obsolete
63	M	F	15–11 Reserved
		V	10 1 = Multiword DMA mode 2 is selected 0 = Multiword DMA mode 2 is not selected
		V	9 1 = Multiword DMA mode 1 is selected 0 = Multiword DMA mode 1 is not selected
		V	8 1 = Multiword DMA mode 0 is selected 0 = Multiword DMA mode 0 is not selected
		F	7–3 Reserved
		F	2 1 = Multiword DMA mode 2 and below are supported
		F	1 1 = Multiword DMA mode 1 and below are supported
		F	0 1 = Multiword DMA mode 0 is supported
64	M	F	15–8 Reserved
		F	7–0 PIO modes supported
65	M		Minimum Multiword DMA transfer cycle time per word
		F	15–0 Cycle time in nanoseconds
66	M		Manufacturer's recommended Multiword DMA transfer cycle time
		F	15–0 Cycle time in nanoseconds
67	M		Minimum PIO transfer cycle time without flow control
		F	15–0 Cycle time in nanoseconds
68	M		Minimum PIO transfer cycle time with IORDY flow control
		F	15–0 Cycle time in nanoseconds
69		F	15 Reserved for CFA 14 Deterministic data in trimmed LBA range(s) is supported 13 Long Physical Sector Alignment Error Reporting Control is supported 12 Obsolete 11 READ BUFFER DMA is supported 10 WRITE BUFFER DMA is supported 9 Obsolete 8 DOWNLOAD MICROCODE DMA is supported 7 Reserved for IEEE 1667 6 0 = Optional ATA device 28-bit commands supported 5 Trimmed LBA range(s) returning zeroed data is supported 4 Device Encrypts All User Data on the device 3 Extended Number of User Addressable Sectors is supported 2 All write cache is non-volatile 1:0 Reserved
70			Reserved
71–74		F	Reserved for the IDENTIFY PACKET DEVICE command.
75	O		Queue depth
		F	15–5 Reserved
		F	4–0 Maximum queue depth – 1

(*Continued on following page*)

Table 3.6 – IDENTIFY DEVICE Data (*Continued*)

Word	O/M	F/V	Description
76		F	Serial ATA Capabilities 15 Supports READ LOG DMA EXT as equivalent to READ LOG EXT 14 Supports Device Automatic Partial to Slumber transitions 13 Supports Host Automatic Partial to Slumber transitions 12 Supports NCQ priority information 11 Supports Unload while NCQ commands are outstanding 10 Supports the SATA Phy Event Counters log 9 Supports receipt of host initiated power management requests 8 Supports the NCQ feature set 7:4 Reserved for Serial ATA 3 Supports SATA Gen3 Signaling Speed (6.0Gb/s) 2 Supports SATA Gen2 Signaling Speed (3.0Gb/s) 1 Supports SATA Gen1 Signaling Speed (1.5Gb/s) 0 Will be cleared to zero
77		F	Serial ATA Additional Capabilities 15:7 Reserved for Serial ATA 6 Supports RECEIVE FPDMA QUEUED and SEND FPDMA QUEUED commands 5 Supports NCQ Queue Management Command 4 Supports NCQ Streaming 3:1 Coded value indicating current negotiated Serial ATA signal speed 0 Will be cleared to zero
78		F	Serial ATA features supported 15:8 Reserved for Serial ATA 7 Device supports NCQ Autosense 6 Device supports Software Settings Preservation 5 Device supports Hardware Feature Control 4 Device supports in-order data delivery 3 Device supports initiating power management 2 Device supports DMA Setup auto-activation 1 Device supports non-zero buffer offsets 0 Will be cleared to zero
79		F	Serial ATA features enabled 15:8 Reserved for Serial ATA 7 Automatic Partial to Slumber transitions enabled 6 Software Settings Preservation enabled 5 Hardware Feature Control is enabled 4 In-order data delivery enabled 3 Device initiated power management enabled 2 DMA Setup auto-activation enabled 1 Non-zero buffer offsets enabled 0 Will be cleared to zero
80	M		Major version number 15:11 Reserved 10 Supports ACS-3 9 Supports ACS-2 8 Supports ATA8-ACS 7 Supports ATA/ATAPI-7 6 Supports ATA/ATAPI-6 5 Supports ATA/ATAPI-5 4:1 Obsolete 0 Reserved

(*Continued on following page*)

Table 3.6 – IDENTIFY DEVICE Data (*Continued*)

Word	O/M	F/V	Description
81	M	F	Minor version number – example codes (see ATA Standards for all values). 001Dh ATA/ATAPI-7 published ANSI INCITS 397-2005 001Eh ATA/ATAPI-7 T13 1532D version 0 001Fh ACS-3 Revision 3b 0021h ATA/ATAPI-7 T13 1532D version 4a 0022h ATA/ATAPI-6 published, ANSI INCITS 361-2002 0027h ATA8-ACS version 3c 0028h ATA8-ACS version 6 0029h ATA8-ACS version 4 0082h ACS-2 published, ANSI INCITS 482-2012 0107h ATA8-ACS version 2d 0110h ACS-2 Revision 3 011Bh ACS-3 Revision 4
82	M		Command sets supported.
		X	15 Obsolete
		F	14 1 = NOP command supported
		F	13 1 = READ BUFFER command supported
		F	12 1 = WRITE BUFFER command supported
		X	11 Obsolete
		F	10 1 = Host Protected Area feature set supported
		F	9 1 = DEVICE RESET command supported
		F	8 1 = SERVICE interrupt supported
		F	7 1 = Release interrupt supported
		F	6 1 = Look-ahead supported
		F	5 1 = Write cache supported
		F	4 Will be cleared to zero to indicate that the PACKET Command feature set is not supported.
		F	3 1 = Mandatory Power Management feature set supported
		F	2 1 = Removable Media feature set supported
		F	1 1 = Security Mode feature set supported
		F	0 1 = SMART feature set supported
83	M		Command sets supported.
		F	15 Will be cleared to zero
		F	14 Will be set to one
		F	13 1 = FLUSH CACHE EXT command supported
		F	12 1 = Mandatory FLUSH CACHE command supported
		F	11 1 = Device Configuration Overlay feature set supported
		F	10 1 = 48-bit Address feature set supported
		F	9 1 = Automatic Acoustic Management feature set supported
		F	8 1 = SET MAX security extension supported
		F	7 See Address Offset Reserved Area Boot, INCITS TR27:2001
		F	6 1 = SET FEATURES subcommand required to spinup after power-up
		F	5 1 = Power-Up In Standby feature set supported
		F	4 1 = Removable Media Status Notification feature set supported
		F	3 1 = Advanced Power Management feature set supported
		F	2 1 = CFA feature set supported
		F	1 1 = READ/WRITE DMA QUEUED supported
		F	0 1 = DOWNLOAD MICROCODE command supported
84	M		Command set/feature supported extension.
		F	15 Will be cleared to zero
		F	14 Will be set to one
		F	13 1 = IDLE IMMEDIATE with UNLOAD FEATURE supported
		F	12 Reserved for technical report

(*Continued on following page*)

Table 3.6 – IDENTIFY DEVICE Data (*Continued*)

Word	O/M	F/V	Description
		F	11 Reserved for technical report
		F	10 1 = URG bit supported for WRITE STREAM DMA EXT and WRITE STREAM EXT
		F	9 1 = URG bit supported for READ STREAM DMA EXT and READ STREAM EXT
		F	8 1 = 64-bit Worldwide name supported
		F	7 1 = WRITE DMA QUEUED FUA EXT command supported
		F	6 1 = WRITE DMA FUA EXT and WRITE MULTIPLE FUA EXT commands supported
		F	5 1 = General Purpose Logging feature set supported
		F	4 1 = Streaming feature set supported
		F	3 1 = Media Card Pass Through Command feature set supported
		F	2 1 = Media serial number supported
		F	1 1 = SMART self-test supported
		F	0 1 = SMART error logging supported
85	M		Command set/feature enabled.
		X	15 Obsolete
		F	14 1 = NOP command enabled
		F	13 1 = READ BUFFER command enabled
		F	12 1 = WRITE BUFFER command enabled
		X	11 Obsolete
		V	10 1 = Host Protected Area feature set enabled
		F	9 1 = DEVICE RESET command enabled
		V	8 1 = SERVICE interrupt enabled
		V	7 1 = release interrupt enabled
		V	6 1 = look-ahead enabled
		V	5 1 = write cache enabled
		F	4 Will be cleared to zero to indicate that the PACKET Command feature set is not supported
		F	3 1 = Power Management feature set enabled
		F	2 1 = Removable Media feature set enabled
		V	1 1 = Security Mode feature set enabled
		V	0 1 = SMART feature set enabled
86	M		Command set/feature enabled.
		F	15–14 Reserved
		F	13 1 = FLUSH CACHE EXT command supported
		F	12 1 = FLUSH CACHE command supported
		F	11 1 = Device Configuration Overlay supported
		F	10 1 = 48-bit Address features set supported
		V	9 1 = Automatic Acoustic Management feature set enabled
		F	8 1 = SET MAX security extension enabled by SET MAX SET PASSWORD
		F	7 See Address Offset Reserved Area Boot, INCITS TR27:2001
		F	6 1 = SET FEATURES subcommand required to spin-up after power-up
		V	5 1 = Power-Up In Standby feature set enabled
		V	4 1 = Removable Media Status Notification feature set enabled
		V	3 1 = Advanced Power Management feature set enabled
		F	2 1 = CFA feature set enabled
		F	1 1 = READ/WRITE DMA QUEUED command supported
		F	0 1 = DOWNLOAD MICROCODE command supported
87	M		Command set/feature default.
		F	15 Will be cleared to zero
		F	14 Will be set to one
		F	13 1 = IDLE IMMEDIATE with UNLOAD FEATURE supported

(*Continued on following page*)

Table 3.6 – IDENTIFY DEVICE Data (*Continued*)

Word	O/M	F/V	Description
		V	12 Reserved for technical report-
		V	11 Reserved for technical report-
		F	10 1 = URG bit supported for WRITE STREAM DMA EXT and WRITE STREAM EXT
		F	9 1 = URG bit supported for READ STREAM DMA EXT and READ STREAM EXT
		F	8 1 = 64 bit Worldwide name supported
		F	7 1 = WRITE DMA QUEUED FUA EXT command supported
		F	6 1 = WRITE DMA FUA EXT and WRITE MULTIPLE FUA EXT commands supported
		F	5 1 = General Purpose Logging feature set supported
		V	4 1 = Valid CONFIGURE STREAM command has been executed
		V	3 1 = Media Card Pass Through Command feature set enabled
		V	2 1 = Media serial number is valid
		F	1 1 = SMART self-test supported
		F	0 1 = SMART error logging supported
88	O	F	15 Reserved
		V	14 1 = Ultra DMA mode 6 is selected 0 = Ultra DMA mode 6 is not selected
		V	13 1 = Ultra DMA mode 5 is selected 0 = Ultra DMA mode 5 is not selected
		V	12 1 = Ultra DMA mode 4 is selected 0 = Ultra DMA mode 4 is not selected
		V	11 1 = Ultra DMA mode 3 is selected 0 = Ultra DMA mode 3 is not selected
		V	10 1 = Ultra DMA mode 2 is selected 0 = Ultra DMA mode 2 is not selected
		V	9 1 = Ultra DMA mode 1 is selected 0 = Ultra DMA mode 1 is not selected
		V	8 1 = Ultra DMA mode 0 is selected 0 = Ultra DMA mode 0 is not selected
		F	7 Reserved
		F	6 1 = Ultra DMA mode 6 and below are supported
		F	5 1 = Ultra DMA mode 5 and below are supported
		F	4 1 = Ultra DMA mode 4 and below are supported
		F	3 1 = Ultra DMA mode 3 and below are supported
		F	2 1 = Ultra DMA mode 2 and below are supported
		F	1 1 = Ultra DMA mode 1 and below are supported
		F	0 1 = Ultra DMA mode 0 is supported
89	O	F	Time required for security erase unit completion
90	O	F	Time required for Enhanced security erase completion
91	O	V	Current advanced power management value
92	O	V	Master Password Revision Code
93			Hardware reset result. The contents of bits (12:0) of this word will change only during the execution of a hardware reset.
		F	15 Will be cleared to zero
		F	14 Will be set to one
		V	13 1 = device detected CBLID – above V_{iH} 0 = device detected CBLID – below V_{iL}
			12–8 Device 1 hardware reset result. Device 0 will clear these bits to zero. Device 1 will set these bits as follows:

(*Continued on following page*)

Table 3.6 – IDENTIFY DEVICE Data (*Continued*)

Word	O/M	F/V	Description
		F	12 Reserved
		V	11 0 = Device 1 did not assert PDIAG- 1 = Device 1 asserted PDIAG-
		V	10–9 These bits indicate how Device 1 determined the device number: 00 = Reserved 01 = A jumper was used 10 = The CSEL signal was used 11 = Some other method was used or the method is unknown
			8 Will be set to one
			7–0 Device 0 hardware reset result. Device 1 will clear these bits to zero. Device 0 will set these bits as follows:
		F	7 Reserved
		F	6 0 = Device 0 does not respond when Device 1 is selected 1 = Device 0 responds when Device 1 is selected
		V	5 0 = Device 0 did not detect the assertion of DASP- 1 = Device 0 detected the assertion of DASP-
		V	4 0 = Device 0 did not detect the assertion of PDIAG- 1 = Device 0 detected the assertion of PDIAG-
		V	3 0 = Device 0 failed diagnostics 1 = Device 0 passed diagnostics
		V	2–1 These bits indicate how Device 0 determined the device number: 00 = Reserved 01 = A jumper was used 10 = The CSEL signal was used 11 = Some other method was used or the method is unknown
		F	0 Will be set to one
94	O	V V	15–8 Vendor's recommended acoustic management value 7–0 Current automatic acoustic management value
95		F	Stream Minimum Request Size
96		V	Streaming Transfer Time – DMA
97		V	Streaming Access Latency – DMA and PIO
98–99		F	Streaming Performance Granularity
100–103	O	V	Maximum user LBA for 48-bit Address feature set
104	O	V	Streaming Transfer Time – PIO
105		F	Reserved
106	O	 F F F F F	Physical Sector Size / Logical Sector Size 15 Will be cleared to zero 14 Will be set to one 13 1 = Device has multiple logical sectors per physical sector. 12 1= Device Logical Sector Longer than 256 Words 11–4 Reserved 3–0 2^X logical sectors per physical sector
107	O	F	Inter-seek delay for ISO-7779 acoustic testing in microseconds
108	O	F	15–12 NAA (3:0) 11–0 IEEE OUI (23:12)

(*Continued on following page*)

Table 3.6 – IDENTIFY DEVICE Data (*Continued*)

Word	O/M	F/V	Description
109	O	F	15–4 IEEE OUI (11:0) 3–0 Unique ID (35:32)
110	O	F	15–0 Unique ID (31:16)
111	O	F	15–0 Unique ID (15:0)
112–115	O	F	Reserved for worldwide name extension to 128 bits
116	O	V	Reserved for technical report-
117–118	O	F	Words per Logical Sector
119–126		F	Reserved
127	O	 F F	Removable Media Status Notification feature set support 15–2 Reserved 1–0 00 = Removable Media Status Notification feature set not supported 01 = Removable Media Status Notification feature supported 10 = Reserved 11 = Reserved
128	O	 F V F F V V V V F	Security status 15–9 Reserved 8 Security level 0 = High, 1 = Maximum 7–6 Reserved 5 1 = Enhanced security erase supported 4 1 = Security count expired 3 1 = Security frozen 2 1 = Security locked 1 1 = Security enabled 0 1 = Security supported
129–159		X	Vendor specific
160	O	 F F F V F	CFA power mode 1 15 Word 160 supported 14 Reserved 13 CFA power mode 1 is required for one or more commands implemented by the device 12 CFA power mode 1 disabled 11–0 Maximum current in ma
161–175		X	Reserved for assignment by the CompactFlash™ Association
176–205	O	V	Current media serial number
206–254		F	Reserved
255	M	X	Integrity word 15–8 Checksum 7–0 Signature

Key:
O/M = Mandatory/optional requirement.
M = Support of the word is mandatory.
O = Support of the word is optional.
F/V = Fixed/variable content.
F = The content of the word is fixed and does not change. For removable media devices, these values may change when media is removed or changed.
V = The contents of the word is variable and may change depending on the state of the device or the commands executed by the device.
X = The content of the word may be fixed or variable.

IDENTIFY PACKET DEVICE

Command code

A1h

Feature set

PACKET Command feature set

- Use prohibited for devices not implementing the PACKET Command feature set.
- Mandatory for devices implementing the PACKET Command feature set.

Protocol

PIO data-in

Inputs

Register	7	6	5	4	3	2	1	0
Features	na							
Sector Count	na							
LBA Low	na							
LBA Mid	na							
LBA High	na							
Device	obs	na	obs	DEV	na	na	na	na
Command	A1h							

Device register –
DEV will specify the selected device.

Normal outputs

Register	7	6	5	4	3	2	1	0
Error	na							
Sector Count	na							
LBA Low	na							
LBA Mid	na							
LBA High	na							
Device	obs	na	obs	DEV	na	na	na	na
Status	BSY	DRDY	DF	na	DRQ	na	na	ERR

Device register –
DEV will indicate the selected device.

Status register –
BSY will be cleared to zero, indicating command completion.
DRDY will be set to one.
DF (Device Fault) will be cleared to zero.
DRQ will be cleared to zero.
ERR will be cleared to zero.

Error outputs

The device will return command aborted if the device does not implement this command; otherwise, the device will not report an error.

Prerequisites

This command will be accepted, regardless of the state of DRDY.

Description

The IDENTIFY PACKET DEVICE command enables the host to receive parameter information from a device that implements the PACKET Command feature set.

Some devices may have to read the media in order to complete this command.

When the command is issued, the device sets the BSY bit to one, prepares to transfer the 256 words of device identification data to the host, sets the DRQ bit to one, clears the BSY bit to zero, and asserts INTRQ if nIEN is cleared to zero. The host may then transfer the data by reading the Data register. Table 3.7 defines the arrangement and meanings of the parameter words in the buffer. All reserved bits or words will be zero.

References to parallel implementation bus signals (e.g., DMACK, DMARQ, etc) apply only to parallel implementations. Some register bits (e.g., nIEN, SRST, etc.) have different requirements in the serial implementation.

Some parameters are defined as a group of bits. A word that is defined as a set of bits is transmitted with indicated bits on the respective data bus bit (e.g., bit 15 appears on DD15).

Some parameters are defined as a 16-bit value. A word that is defined as a 16-bit value places the most significant bit of the value on bit DD15 and the least significant bit on bit DD0.

Some parameters are defined as 32-bit values (e.g., words (61:60)). Such fields are transferred using two word transfers. The device will first transfer the least significant bits, bits (15:0) of the value, on bits DD(15:0), respectively. After the least significant bits have been transferred, the most significant bits, bits (31:16) of the value, will be transferred on DD(15:0), respectively.

Table 3.7 – IDENTIFY PACKET DEVICE Data

Word	O/M	F/V	Description		
0	M		General configuration bit – significant information:		
		F		15–14	10 = ATAPI device
		F			11 = Reserved
		F		13	Reserved
		F		12–8	Field indicates command packet set used by device
		F		7	1 = removable media device
		F		6–5	00 = Device will set DRQ to one within 3 ms of receiving PACKET command
					01 = Obsolete
					10 = Device will set DRQ to one within 50 μs of receiving PACKET command
					11 = Reserved
		F		4–3	Reserved
		V		2	Incomplete response
		F		1–0	00 = 12-byte command packet
					01 = 16-byte command packet
					1x = Reserved
1		F	Reserved		

(Continued on following page)

Table 3.7 – IDENTIFY PACKET DEVICE Data (*Continued*)

Word	O/M	F/V	Description
2		V	Unique configuration
3–9		F	Reserved
10–19	M	F	Serial number (20 ASCII characters)
20–22		F	Reserved
23–26	M	F	Firmware revision (8 ASCII characters)
27–46	M	F	Model number (40 ASCII characters)
47–48		F	Reserved
49	M		Capabilities
		F	15 1 = Interleaved DMA supported Devices that require the DMADIR bit in the Packet command will clear this bit to 0
		F	14 1 = Command queuing supported
		F	13 1 = Overlap operation supported
		F	12 1 = ATA software reset required (Obsolete)
		F	11 1 = IORDY supported
		F	10 1 = IORDY may be disabled
		F	9 Will be set to one
		F	8 1 = DMA supported Devices that require the DMADIR bit in the Packet command will clear this bit to 0
		X	7–0 Vendor specific
50	O		Capabilities
		F	15 Will be cleared to zero
		F	14 Will be set to one
		F	13–2 Reserved
		X	1 Obsolete
		F	0 Will be set to one to indicate a device specific Standby timer value minimum
51–52		X	Obsolete
53	M	F	15–3 Reserved
		F	2 1 = The fields reported in word 88 are valid 0 = The fields reported in word 88 are not valid
		F	1 1 = The fields reported in words (70:64) are valid 0 = The fields reported in words (70:64) are not valid
		X	0 Obsolete
54–61		F	Reserved
62	M	F	15 1 = DMADIR bit in the Packet command is required for DMA transfers 0 = DMADIR bit in Packet command is not required for DMA transfers
		F	14–11 Reserved
		F	10 1 = DMA is supported
		F	9 1 = Multiword DMA mode 2 is supported
		F	8 1 = Multiword DMA mode 1 is supported
		F	7 1 = Multiword DMA mode 0 is supported
		F	6 1 = Ultra DMA mode 6 and below are supported
		F	5 1 = Ultra DMA mode 5 and below are supported
		F	4 1 = Ultra DMA mode 4 and below are supported
		F	3 1 = Ultra DMA mode 3 and below are supported
		F	2 1 = Ultra DMA mode 2 and below are supported
		F	1 1 = Ultra DMA mode 1 and below are supported
		F	0 1 = Ultra DMA mode 0 is supported

(*Continued on following page*)

Table 3.7 – IDENTIFY PACKET DEVICE Data (*Continued*)

Word	O/M	F/V	Description
63	M	F	15–11 Reserved
		V	10 1 = Multiword DMA mode 2 is selected 0 = Multiword DMA mode 2 is not selected
		V	9 1 = Multiword DMA mode 1 is selected 0 = Multiword DMA mode 1 is not selected
		V	8 1 = Multiword DMA mode 0 is selected 0 = Multiword DMA mode 0 is not selected
		F	7–3 Reserved
		F	2 1 = Multiword DMA mode 2 and below are supported Devices that require the DMADIR bit in the Packet command will clear this bit to 0
		F	1 1 = Multiword DMA mode 1 and below are supported Devices that require the DMADIR bit in the Packet command will clear this bit to 0
		F	0 1 = Multiword DMA mode 0 is supported Multiword DMA mode selected Devices that require the DMADIR bit in the Packet command will clear this bit to 0
64	M	F	15–8 Reserved
		F	7–0 PIO transfer modes supported
65	M		Minimum Multiword DMA transfer cycle time per word
		F	15–0 Cycle time in nanoseconds
66	M		Manufacturer's recommended Multiword DMA transfer cycle time
		F	15–0 Cycle time in nanoseconds
67	M		Minimum PIO transfer cycle time without flow control
		F	15–0 Cycle time in nanoseconds
68	M		Minimum PIO transfer cycle time with IORDY flow control
		F	15–0 Cycle time in nanoseconds
69–70		F	Reserved (for future command overlap and queuing)
71	O	F	Typical time in ns from receipt of PACKET command to bus release
72	O	F	Typical time in ns from receipt of SERVICE command to BSY cleared to zero
73–74		F	Reserved
75	O		Queue depth
		F	15–5 Reserved
		F	4–0 Maximum queue depth supported – 1
76–79		R	Reserved for Serial ATA
80	M		Major version number 0000h or FFFFh = device does not report version
		F	15 Reserved
		F	14 Reserved for ATA/ATAPI-14
		F	13 Reserved for ATA/ATAPI-13
		F	12 Reserved for ATA/ATAPI-12
		F	11 Reserved for ATA/ATAPI-11
		F	10 Reserved for ATA/ATAPI-10
		F	9 Reserved for ATA/ATAPI-9
		F	8 Reserved for ATA/ATAPI-8
		F	7 1 = Supports ATA/ATAPI-7
		F	6 1 = supports ATA/ATAPI-6
		F	5 1 = supports ATA/ATAPI-5
		F	4 1 = supports ATA/ATAPI-4

(*Continued on following page*)

Table 3.7 – IDENTIFY PACKET DEVICE Data (*Continued*)

Word	O/M	F/V	Description
		F	3 Obsolete
		X	2 Obsolete
		X	1 Obsolete
		F	0 Reserved
81	M		Minor version number 0000h or FFFFh = device does not report version 0001h-FFFEh = Version number
82	M		Command set supported. If words (83:82) = 0000h or FFFFh command set notification not supported.
		X	15 Obsolete
		F	14 1 = NOP command supported
		F	13 1 = READ BUFFER command supported
		F	12 1 = WRITE BUFFER command supported
		X	11 Obsolete
		F	10 1 = Host Protected Area feature set supported
		F	9 1 = DEVICE RESET command supported
		F	8 1 = SERVICE interrupt supported
		F	7 1 = release interrupt supported
		F	6 1 = look-ahead supported
		F	5 1 = write cache supported
		F	4 Will be set to one, indicating the PACKET Command feature set is supported
		F	3 1 = Power Management feature set supported
		F	2 1 = Removable Media feature set supported
		F	1 1 = Security Mode feature set supported
		F	0 1 = SMART feature set supported
83	M		Command sets supported If words (83:82) = 0000h or FFFFh command set notification not supported
		F	15 Will be cleared to zero
		F	14 Will be set to one
		F	13 Reserved
		F	12 1 = FLUSH CACHE command supported
		F	11 1 = Device Configuration Overlay feature set supported
		F	10 Reserved
		F	9 1 = AUTOMATIC Acoustic Management feature set supported
		F	8 1 = SET MAX security extension supported
		F	7 See Address Offset Reserved Area Boot, INCITS TR27:2001
		F	6 1 = SET FEATURES subcommand required to spinup after power-up
		F	5 1 = Power-Up In Standby feature set supported
		F	4 1 = Removable Media Status Notification feature set supported
		F	3-1 Reserved
		F	0 1 = DOWNLOAD MICROCODE command supported
84	M		Command set/feature supported extension. If words 82, 83, and 84 = 0000h or FFFFh command set notification extension is not supported.
		F	15 Will be cleared to zero
		F	14 Will be set to one
		F	13–0 Reserved
85	M		Command set/feature enabled. If words 85, 86, and 87 = 0000h or FFFFh command set enabled notification is not supported.
		X	15 Obsolete
		F	14 1 = NOP command enabled
		F	13 1 = READ BUFFER command enabled

(*Continued on following page*)

Table 3.7 – IDENTIFY PACKET DEVICE Data (*Continued*)

Word	O/M	F/V	Description	
		F	12	1 = WRITE BUFFER command enabled
		X	11	Obsolete
		V	10	1 = Host Protected Area feature set enabled
		F	9	1 = DEVICE RESET command enabled
		V	8	1 = SERVICE interrupt enabled
		V	7	1 = release interrupt enabled
		V	6	1 = look-ahead enabled
		V	5	1 = write cache enabled
		F	4	Will be set to one, indicating the PACKET Command feature set is supported
		F	3	1 = Power Management feature set enabled
		V	2	1 = Removable Media feature set enabled
		V	1	1 = Security Mode feature set enabled
		V	0	1 = SMART feature set enabled
86	M		Command set/feature enabled. If words 85, 86, and 87 = 0000h or FFFFh command set enabled notification is not supported.	
		F	15–13	Reserved
		V	12	1 = FLUSH CACHE command supported
		F	11	1 = Device Configuration Overlay feature set supported
		F	10	Reserved
		V	9	1 = Automatic Acoustic Management feature set enabled
		V	8	1 = SET MAX security extension enabled by a SET MAX SET PASSWORD
		V	7	See Address Offset Reserved Area Boot, INCITS TR27:2001
		F	6	1 = SET FEATURES subcommand required to spinup after power-up
		V	5	1 = Power-Up In Standby feature set enabled
		V	4	1 = Removable Media Status Notification feature set enabled via the SET FEATURES command
		F	3–1	Reserved
		F	0	1 = DOWNLOAD MICROCODE command enabled
87	M		Command set/feature default. If words 85, 86, and 87 = 0000h or FFFFh command set default notification is not supported.	
		F	15	Will be cleared to zero
		F	14	Will be set to one
		F	13–0	Reserved
88	M	F	15	Reserved
			14	1 = Ultra DMA mode 6 is selected 0 = Ultra DMA mode 6 is not selected
		V	13	1 = Ultra DMA mode 5 is selected 0 = Ultra DMA mode 5 is not selected
		V	12	1 = Ultra DMA mode 4 is selected 0 = Ultra DMA mode 4 is not selected
		V	11	1 = Ultra DMA mode 3 is selected 0 = Ultra DMA mode 3 is not selected
		V	10	1 = Ultra DMA mode 2 is selected 0 = Ultra DMA mode 2 is not selected
		V	9	1 = Ultra DMA mode 1 is selected 0 = Ultra DMA mode 1 is not selected
		V	8	1 = Ultra DMA mode 0 is selected 0 = Ultra DMA mode 0 is not selected
		F	7	Reserved
		F	6	1 = Ultra DMA mode 6 and below are supported Devices that require the DMADIR bit in the Packet command will clear this bit to 0

(*Continued on following page*)

Table 3.7 – IDENTIFY PACKET DEVICE Data (*Continued*)

Word	O/M	F/V	Description
		F	5 1 = Ultra DMA mode 5 and below are supported Devices that require the DMADIR bit in the Packet command will clear this bit to 0
		F	4 1 = Ultra DMA mode 4 and below are supported Devices that require the DMADIR bit in the Packet command will clear this bit to 0
		F	3 1 = Ultra DMA mode 3 and below are supported Devices that require the DMADIR bit in the Packet command will clear this bit to 0
		F	2 1 = Ultra DMA mode 2 and below are supported Devices that require the DMADIR bit in the Packet command will clear this bit to 0
		F	1 1 = Ultra DMA mode 1 and below are supported Devices that require the DMADIR bit in the Packet command will clear this bit to 0
		F	0 1 = Ultra DMA mode 0 is supported Devices that require the DMADIR bit in the Packet command will clear this bit to 0
89–92		F	Reserved
93			Hardware reset result. The contents of bits (12:0) of this word will change only during the execution of a hardware reset.
		F	15 Will be cleared to zero
		F	14 Will be set to one
		V	13 1 = device detected CBLID – above V_{iH} 0 = device detected CBLID – below V_{iL}
			12–8 Device 1 hardware reset result. Device 0 will clear these bits to zero. Device 1 will set these bits as follows:
		F	12 Reserved
		V	11 0 = Device 1 did not assert PDIAG- 1 = Device 1 asserted PDIAG-
		V	10–9 These bits indicate how Device 1 determined the device number: 00 = Reserved 01 = a jumper was used 10 = the CSEL signal was used 11 = some other method was used or the method is unknown
		F	8 Will be set to one
			7–0 Device 0 hardware reset result. Device 1 will clear these bits to zero. Device 0 will set these bits as follows:
		F	7 Reserved
		F	6 0 = Device 0 does not respond when Device 1 is selected 1 = Device 0 responds when Device 1 is selected
		V	5 0 = Device 0 did not detect the assertion of DASP- 1 = Device 0 detected the assertion of DASP-
		V	4 0 = Device 0 did not detect the assertion of PDIAG- 1 = Device 0 detected the assertion of PDIAG-
		V	3 0 = Device 0 failed diagnostics 1 = Device 0 passed diagnostics
			2–1 These bits indicate how Device 0 determined the device number:
		F	00 = Reserved
		V	01 = A jumper was used
		V	10 = The CSEL signal was used

(*Continued on following page*)

Table 3.7 – IDENTIFY PACKET DEVICE Data (*Continued*)

Word	O/M	F/V	Description
		V F	11 = Some other method was used or the method is unknown 0 Will be set to one
94	O	V V	15–8 Vendor's recommended acoustic management value 7–0 Current automatic acoustic management value
95–124		F	Reserved
125	M	F	ATAPI byte count = 0 behavior
126		X	Obsolete
127	O	 F F	Removable Media Status Notification feature set support 15–2 Reserved 1–0 00 = Removable Media Status Notification feature set not supported 01 = Removable Media Status Notification feature set supported 10 = Reserved 11 = Reserved
128	O	 F V F F V V V V F	Security status 15–9 Reserved 8 Security level 0 = High, 1 = Maximum 7–6 Reserved 5 1 = Enhanced security erase supported 4 1 = Security count expired 3 1 = Security frozen 2 1 = Security locked 1 1 = Security enabled 0 1 = Security supported
129–159		X	Vendor specific
160–175		F	Reserved for assignment by the CompactFlash™ Association
176–254	O	F	Reserved
255	O	X	Integrity word
			15–8 Checksum
			7–0 Signature

Key:
O/M = Mandatory/optional requirement.
M = Support of the word is mandatory.
O = Support of the word is optional.
F/V = Fixed/variable content.
F = The content of the word is fixed and does not change. For removable media devices, these values may change when media is removed or changed.
V = The contents of the word is variable and may change depending on the state of the device or the commands executed by the device.
X = The content of the word may be fixed or variable.

PACKET

Command code

A0h

Feature set

PACKET Command feature set

- Use prohibited for devices not implementing the PACKET Command feature set.
- Mandatory for devices implementing the PACKET Command feature set.

Protocol

Packet

Inputs

Register	7	6	5	4	3	2	1	0
Features	na	na	na	na	na	DMADIR	OVL	DMA
Sector Count	Tag					na		
LBA Low	na							
Byte Count Low	Byte Count limit (7:0)							
Byte Count High	Byte Count limit (15:8)							
Device	obs	na	obs	DEV	na	na	na	na
Command	A0h							

Features register –

DMADIR – This bit indicates Packet DMA direction and is used only for devices that implement the Packet Command feature set with a Serial ATA bridge that requires direction indication from the host. Support for this bit is determined by reading bit 15 of word 62 in the IDENTIFY PACKET DEVICE data. If bit 15 of word 62 is set to 1, the device requires the use of the DMADIR bit for Packet DMA commands.

If the device requires the DMADIR bit to be set for Packet DMA operations and the current operations is DMA (i.e., bit 0, the DMA bit, is set), this bit indicates the direction of data transfer (0 = transfer to the device; 1 = transfer to the host). If the device requires the DMADIR bit to be set for Packet DMA operations but the current operations is PIO (i.e., bit 0, the DMA bit, is cleared), this bit is ignored.

Since the data transfer direction will be set by the host as the command is constructed, the DMADIR bit should not conflict with the data transfer direction of the command. If a conflict between the command transfer direction and the DMADIR bit occurs, the device should return with an ABORTED command and the sense key set to ILLEGAL REQUEST.

If the device does not require the DMADIR bit for Packet DMA operations, this bit should be cleared to 0.

A device that does not support the DMADIR feature may abort a command if the DMADIR bit is set to 1.

OVL – This bit is set to one to inform the device that the PACKET command is to be overlapped.

DMA – This bit is set to one to inform the device that the data transfer (not the command packet transfer) associated with this command is via Multiword DMA or Ultra DMA mode.

Sector Count register –

Tag – If the device supports command queuing, this field contains the command Tag for the command being delivered. A Tag may have any value between 0 and 31, regardless of the queue depth supported. If queuing is not supported, this field is not applicable.

Byte Count low and Byte Count high registers –

These registers are written by the host with the maximum byte count that is to be transferred in any single DRQ assertion for PIO transfers. The byte count does not apply to the command packet transfer. If the PACKET command does not transfer data, the byte count is ignored.

If the PACKET command results in a data transfer:

1. the host should not set the byte count limit to zero. If the host sets the byte count limit to zero, the contents of IDENTIFY PACKET DEVICE data word 125 determines the expected behavior;
2. the value set into the byte count limit will be even if the total requested data transfer length is greater than the byte count limit;
3. the value set into the byte count limit may be odd if the total requested data transfer length is equal to or less than the byte count limit;
4. the value FFFFh is interpreted by the device as though the value were FFFEh.

Device register –
DEV will specify the selected device.

Normal outputs

Awaiting command

When the device is ready to accept the command packet from the host, the register content will be as shown below.

Register	7	6	5	4	3	2	1	0
Error	na							
Interrupt reason	Tag					REL	I/O	C/D
LBA Low	na							
Byte Count Low	Byte Count (7:0)							
Byte Count High	Byte Count (15:8)							
Device	obs	na	obs	DEV	na	na	na	na
Status	BSY	na	DMRD	SERV	DRQ	na	na	CHK

Byte Count High/Low – will reflect the value set by the host when the command was issued.

Interrupt reason register –
Tag – If the device supports command queuing and overlap is enabled, this field contains the command Tag for the command. A Tag value may be any value between 0 and 31, regardless of the queue depth supported. If the device does not support command queuing or overlap is disabled, this field is not applicable.
REL – will be cleared to zero.
I/O – will be cleared to zero, indicating transfer to the device.
C/D – will be set to one, indicating the transfer of a command packet.

Device register –
DEV will indicate the selected device.

Status register –
BSY – will be cleared to zero.
DMRD (DMA ready) – will be cleared to zero.
SERV (Service) – will be set to one if another command is ready to be serviced. If overlap is not supported, this bit is command specific.
DRQ – will be set to one.
CHK – will be cleared to zero.

Data transmission

If overlap is not supported or not specified by the command, data transfer will occur after the receipt of the command packet. If overlap is supported and the command specifies that the command may be overlapped, data transfer may occur after receipt of the command packet or may occur after the receipt of a SERVICE command. When the device is ready to transfer data requested by a data transfer command, the device sets the following register content to initiate the data transfer.

Register	7	6	5	4	3	2	1	0
Error	na							
Interrupt reason	Tag					REL	I/O	C/D
LBA Low	na							
Byte Count Low	Byte Count (7:0)							
Byte Count High	Byte Count (15:8)							
Device	obs	na	obs	DEV	na	na	na	na
Status	BSY	na	DMRD	SERV	DRQ	na	na	CHK

Byte Count High/Low – If the transfer is to be in PIO mode, the byte count of the data to be transferred for this DRQ assertion will be presented.

Valid byte count values are as follows:

1. The byte count will be less than or equal to the byte count limit value from the host.
2. The byte count will not be zero.
3. The byte count will be less than or equal to FFFEh.
4. The byte count will be even except for the last transfer of a command.
5. If the byte count is odd, the last valid byte transferred is on DD(7:0) and the data on DD(15:8) is a pad byte of undefined value.
6. If the last transfer of a command has a pad byte, the byte count will be odd.

Interrupt reason register –

Tag – If the device supports command queuing and overlap is enabled, this field contains the command Tag for the command. A Tag value may be any value between 0 and 31, regardless of the queue depth supported. If the device does not support command queuing or overlap is disabled, this field is not applicable.

REL – will be cleared to zero.

I/O – will be cleared to zero if the transfer is to the device. I/O will be set to one if the transfer is to the host.

C/D – will be cleared to zero, indicating the transfer of data.

Device register –

DEV will indicate the selected device.

Status register –

BSY – will be cleared to zero.

DMRD (DMA ready) – will be set to one if the transfer is to be a DMA or Ultra DMA transfer and the device supports overlap DMA.

SERV (Service) – will be set to one if another command is ready to be serviced. If overlap is not supported, this bit is command specific.

DRQ – will be set to one.
CHK – will be cleared to zero.

Bus release (overlap feature set only)

After receiving the command packet, the device sets BSY to one and clears DRQ to zero. If the command packet requires a data transfer, the OVL bit is set to one, the Release interrupt is disabled, and the device is not prepared to immediately transfer data, the device may perform a bus release by placing the following register content in the Command Block registers. If the command packet requires a data transfer, the OVL bit is set to one, and the Release interrupt is enabled, the device will perform a bus release by setting the register content as follows.

Register	7	6	5	4	3	2	1	0
Error	na							
Interrupt reason	Tag					REL	I/O	C/D
LBA Low	na							
Byte Count Low	na							
Byte Count High	na							
Device	obs	na	obs	DEV	na	na	na	na
Status	BSY	DRDY	DMRD	SERV	DRQ	na	na	CHK

Byte Count High/Low – na.

Interrupt reason register –
Tag – If the device supports command queuing and overlap is enabled, this field contains the command Tag for the command. A Tag value may be any value between 0 and 31, regardless of the queue depth supported. If the device does not support command queuing or overlap is disabled, this field is not applicable.
REL – will be set to one.
I/O – will be cleared to zero.
C/D – will be cleared to zero.

Device register –
DEV will indicate the selected device.

Status register –
BSY – will be cleared to zero, indicating bus release.
DRDY – na.
DMRD (DMA ready) – will be cleared to zero.
SERV (Service) – will be set to one if another command is ready to be serviced. If overlap is not supported, this bit is command specific.
DRQ – will be cleared to zero.
CHK – will be cleared to zero.

Service request (overlap feature set only)

When the device is ready to transfer data or complete a command after the command has performed a bus release, the device will set the SERV bit and not change the state of any other register bit. When the SERVICE command is received, the device will set outputs as described in data transfer, successful command completion, or error outputs, depending on the service the device requires.

Successful command completion

When the device has command completion without error, the device sets the following register content.

Register	7	6	5	4	3	2	1	0
Error	na							
Interrupt reason	Tag					REL	I/O	C/D
LBA Low	na							
Byte Count Low	na							
Byte Count High	na							
Device	obs	na	obs	DEV	na	na	na	na
Status	BSY	DRDY	DMRD	SERV	DRQ	na	na	CHK

Byte Count High/Low – na.

Interrupt reason register –
Tag – If the device supports command queuing and overlap is enabled, this field contains the command Tag for the command. A Tag value may be any value between 0 and 31, regardless of the queue depth supported. If the device does not support command queuing or overlap is disabled, this field is not applicable.
REL – will be cleared to zero.
I/O – will be set to one.
C/D – will be set to one.

Device register –
DEV will indicate the selected device.

Status register –
BSY – will be cleared to zero, indicating command completion.
DRDY – will be set to one.
DMRD (DMA ready) – na.
SERV (Service) – will be set to one if another command is ready to be serviced. If overlap is not supported, this bit is command specific.
DRQ – will be cleared to zero.
CHK – will be cleared to zero.

Error outputs

The device will not terminate the PACKET command with an error before the last byte of the command packet has been written.

Register	7	6	5	4	3	2	1	0
Error	Sense key (see page 140)				na	ABRT	EOM	ILI
Interrupt reason	Tag					REL	I/O	C/D
LBA Low	na							
Byte Count Low	na							
Byte Count High	na							
Device	obs	na	obs	DEV	na	na	na	na
Status	BSY	DRDY	DF	SERV	DRQ	na	na	CHK

Error register –

Sense Key is a command packet set specific error indication.

ABRT will be set to one if the requested command has been command aborted because the command code or a command parameter is invalid. ABRT may be set to one if the device is not able to complete the action requested by the command.

EOM – The meaning of this bit is command set specific. See the appropriate command set standard for the definition of this bit.

ILI – The meaning of this bit is command set specific. See the appropriate command set standard for the definition of this bit.

Interrupt reason register –

Tag – If the device supports command queuing and overlap is enabled, this field contains the command Tag for the command. A Tag value may be any value between 0 and 31, regardless of the queue depth supported. If the device does not support command queuing or overlap is disabled, this field is not applicable.

REL – will be cleared to zero.

I/O – will be set to one.

C/D – will be set to one.

Device register –

DEV will indicate the selected device.

Status register –

BSY will be cleared to zero, indicating command completion.

DRDY will be set to one.

SERV (Service) – will be set to one if another command is ready to be serviced. If overlap is not supported, this bit is command specific.

DF (Device Fault) will be set to one if a device fault has occurred.

DRQ will be cleared to zero.

CHK will be set to one if an Error register sense key or code bit is set.

Prerequisites

This command will be accepted, regardless of the state of DRDY.

Description

The PACKET command is used to transfer a device command via a command packet. If the native form of the encapsulated command is shorter than the packet size reported in bits (1:0) of word 0 of the IDENTIFY PACKET DEVICE response, the encapsulated command will begin at byte 0 of the packet. Packet bytes beyond the end of the encapsulated command are reserved.

If the device supports overlap, the OVL bit is set to one in the Features register and the Release interrupt has been disabled via the SET FEATURES command, the device may or may not perform a bus release. If the device is ready for the data transfer, the device may begin the transfer immediately, as described in the non-overlapped protocol. If the data is not ready, the device may perform a bus release and complete the transfer after the execution of a SERVICE command.

SCSI Sense Keys

Sense Key	Description
0h	NO SENSE. Indicates that there is no specific sense key information to be reported. This may occur for a successful command or a command that received CHECK CONDITION status because one of the filemark, EOM, or ILI bits is set to one.
1h	RECOVERED ERROR. Indicates that the last command completed successfully with some recovery action performed by the device server. Details may be determinable by examining the additional sense bytes and the information field. When multiple recovered errors occur during one command, the choice of which error to report (first, last, most severe, etc.) is vendor specific.
2h	NOT READY. Indicates that the logical unit addressed cannot be accessed. Operator intervention may be required to correct this condition.
3h	MEDIUM ERROR. Indicates that the command terminated with a non-recovered error condition that was probably caused by a flaw in the medium or an error in the recorded data. This sense key may also be returned if the device server is unable to distinguish between a flaw in the medium and a specific hardware failure (sense key 4h).
4h	HARDWARE ERROR. Indicates that the device server detected a non-recoverable hardware failure (e.g., controller failure, device failure, parity error, etc.) while performing the command or during a self test.
5h	ILLEGAL REQUEST. Indicates that there was an illegal parameter in the command descriptor block (CDB) or in the additional parameters supplied as data for some commands (FORMAT UNIT, SEARCH DATA, etc.). If the device server detects an invalid parameter in the CDB, then it will terminate the command without altering the medium. If the device server detects an invalid parameter in the additional parameters supplied as data, then the device server may have already altered the medium.
6h	UNIT ATTENTION. Indicates that the removable medium may have been changed or the target has been reset. [See SAM-2 for more details (www.t10.org).]
7h	DATA PROTECT. Indicates that a command that reads or writes the medium was attempted on a block that is protected from this operation. The read or write operation is not performed.
8h	BLANK CHECK. Indicates that a write-once device or a sequential-access device encountered blank medium or format-defined end-of-data indication while reading, or a write-once device encountered a non-blank medium while writing.
9h	Vendor Specific. This sense key is available for reporting vendor-specific conditions.
Ah	COPY ABORTED. Indicates a COPY, COMPARE, or COPY AND VERIFY command was aborted due to an error condition on the source device, the destination device, or both.
Bh	ABORTED COMMAND. Indicates that the device server aborted the command. The application client may be able to recover by trying the command again.

READ BUFFER

Command code

E4h

Feature set

General feature set

- Optional for devices not implementing the PACKET Command feature set.
- Use prohibited for devices implementing the PACKET Command feature set.

Protocol

PIO data-in

Inputs

Register	7	6	5	4	3	2	1	0
Features	na							
Sector Count	na							
LBA Low	na							
LBA Mid	na							
LBA High	na							
Device	obs	na	obs	DEV	na	na	na	na
Command	E4h							

Device register –
DEV will specify the selected device.

Normal outputs

Register	7	6	5	4	3	2	1	0
Error	na							
Sector Count	na							
LBA Low	na							
LBA Mid	na							
LBA High	na							
Device	obs	na	obs	DEV	na	na	na	na
Status	BSY	DRDY	DF	na	DRQ	na	na	ERR

Device register –
DEV will indicate the selected device.
Status register –
BSY will be cleared to zero, indicating command completion.
DRDY will be set to one.
DF (Device Fault) will be cleared to zero.
DRQ will be cleared to zero.
ERR will be cleared to zero.

Error outputs

The device will return command aborted if the command is not supported.

Register	7	6	5	4	3	2	1	0
Error	na	na	na	na	na	ABRT	na	na
Sector Count	na							
LBA Low	na							
LBA Mid	na							
LBA High	na							
Device	obs	na	obs	DEV	na	na	na	na
Status	BSY	DRDY	DF	na	DRQ	na	na	ERR

Error register –
ABRT will be set to one if this command is not supported. ABRT may be set to one if the device is not able to complete the action requested by the command.
Device register –
DEV will indicate the selected device.
Status register –
BSY will be cleared to zero, indicating command completion.
DRDY will be set to one.
DF (Device Fault) will be set to one if a device fault has occurred.
DRQ will be cleared to zero.
ERR will be set to one if an Error register bit is set to one.

Prerequisites

DRDY set to one. The command prior to a READ BUFFER command will be a WRITE BUFFER command.

Description

The READ BUFFER command enables the host to read the current contents of the device's sector buffer.

The READ BUFFER and WRITE BUFFER commands will be synchronized such that sequential WRITE BUFFER and READ BUFFER commands access the same 512 bytes within the buffer.

READ DMA

Command code

C8h

Feature set

General feature set
- Mandatory for devices not implementing the PACKET Command feature set.
- Use prohibited for devices implementing the PACKET Command feature set.

Protocol

DMA

Inputs

Register	7	6	5	4	3	2	1	0
Features	na							
Sector Count	Sector count							
LBA Low	LBA (7:0)							
LBA Mid	LBA (15:8)							
LBA High	LBA (23:16)							
Device	obs	LBA	obs	DEV	LBA (27:24)			
Command	C8h							

Sector Count –
Number of sectors to be transferred. A value of 00h specifies that 256 sectors are to be transferred.
LBA Low – starting LBA bits (7:0).
LBA Mid – starting LBA bits (15:8).
LBA High – starting LBA bits (23:16).
Device –
The LBA bit will be set to one to specify the address is an LBA.
DEV will specify the selected device.
Bits (3:0) will be starting LBA bits (27:24).

Normal outputs

Register	7	6	5	4	3	2	1	0
Error	na							
Sector Count	na							
LBA Low	na							
LBA Mid	na							
LBA High	na							
Device	obs	na	obs	DEV	na	na	na	na
Status	BSY	DRDY	DF	na	DRQ	na	na	ERR

Device register –
DEV will indicate the selected device.
Status register –
BSY will be cleared to zero, indicating command completion.
DRDY will be set to one.
DF (Device Fault) will be cleared to zero.
DRQ will be cleared to zero.
ERR will be cleared to zero.

Error outputs

An unrecoverable error encountered during the execution of this command results in the termination of the command. The Command Block registers contain the address of the sector where the first unrecoverable error occurred. The amount of data transferred is indeterminate.

Register	7	6	5	4	3	2	1	0
Error	ICRC	UNC	MC	IDNF	MCR	ABRT	NM	obs
Sector Count	na							
LBA Low	LBA (7:0)							
LBA Mid	LBA (15:8)							
LBA High	LBA (23:16)							
Device	obs	na	obs	DEV	LBA (27:24)			
Status	BSY	DRDY	DF	na	DRQ	na	na	ERR

Error register –

ICRC will be set to one if an interface CRC error has occurred during an Ultra DMA data transfer. The content of this bit is not applicable for Multiword DMA transfers.

UNC will be set to one if data is uncorrectable

MC will be set to one if the media in a removable media device changed since the issuance of the last command. The device will clear the device internal media change detected state.

IDNF will be set to one if a user-accessible address could not be found. IDNF will be set to one if an address outside of the range of user-accessible addresses is requested if command aborted is not returned.

MCR will be set to one if a media change request has been detected by a removable media device. This bit is only cleared by a GET MEDIA STATUS or a media access command.

ABRT will be set to one if this command is not supported or if an error, including an ICRC error, has occurred during an Ultra DMA data transfer. ABRT may be set to one if the device is not able to complete the action requested by the command. ABRT will be set to one if an address outside of the range of user-accessible addresses is requested if IDNF is not set to one.

NM will be set to one if no media is present in a removable media device.

LBA Low, LBA Mid, LBA High, Device –

Will be written with the address of the first unrecoverable error.

Device register –

DEV will indicate the selected device.

Status register –

BSY will be cleared to zero, indicating command completion.

DRDY will be set to one.

DF (Device Fault) will be set to one if a device fault has occurred.

DRQ will be cleared to zero.

ERR will be set to one if an Error register bit is set to one; however, if SE is set to one, ERR will be cleared to zero.

Prerequisites

DRDY set to one. The host will initialize the DMA channel.

Description

The READ DMA command allows the host to read data using the DMA data transfer protocol.

READ DMA EXT

Command code

25h

Feature set

48-bit Address feature set

- Mandatory for devices implementing the 48-bit Address feature set.
- Use prohibited for devices implementing the PACKET Command feature set.

Protocol

DMA

Inputs

Register		7	6	5	4	3	2	1	0
Features	Current	Reserved							
	Previous	Reserved							
Sector Count	Current	Sector count (7:0)							
	Previous	Sector count (15:8)							
LBA Low	Current	LBA (7:0)							
	Previous	LBA (31:24)							
LBA Mid	Current	LBA (15:8)							
	Previous	LBA (39:32)							
LBA High	Current	LBA (23:16)							
	Previous	LBA (47:40)							
Device		obs	LBA	obs	DEV	Reserved			
Command		25h							
NOTE – The value indicated as Current is the value most recently written to the register. The value indicated as Previous is the value that was in the register before the most recent write to the register.									

Sector Count Current –
Number of sectors to be transferred low order, bits (7:0).
Sector Count Previous –
Number of sectors to be transferred high order, bits (15:8). 0000h in the Sector Count register specifies that 65,536 sectors are to be transferred.
LBA Low Current – LBA (7:0).
LBA Low Previous – LBA (31:24).
LBA Mid Current – LBA (15:8).
LBA Mid Previous – LBA (39:32).
LBA High Current – LBA (23:16).
LBA High Previous – LBA (47:40).
Device –
DEV will specify the selected device.
LBA bit will be set to 1.

Normal outputs

Register		7	6	5	4	3	2	1	0
Error		na							
Sector Count	HOB = 0	Reserved							
	HOB = 1	Reserved							
LBA Low	HOB = 0	Reserved							
	HOB = 1	Reserved							
LBA Mid	HOB = 0	Reserved							
	HOB = 1	Reserved							
LBA High	HOB = 0	Reserved							
	HOB = 1	Reserved							
Device		obs	na	obs	DEV	Reserved			
Status		BSY	DRDY	DF	na	DRQ	na	na	ERR
NOTE – HOB = 0 indicates the value read by the host when the HOB bit of the Device Control register is cleared to zero. HOB = 1 indicates the value read by the host when the HOB bit of the Device Control register is set to one.									

Device register –
DEV will indicate the selected device.
Status register –
BSY will be cleared to zero, indicating command completion.
DRDY will be set to one.
DF (Device Fault) will be cleared to zero.
DRQ will be cleared to zero.
ERR will be cleared to zero.

Error outputs

An unrecoverable error encountered during the execution of this command results in the termination of the command. The Command Block registers contain the address of the sector where the first unrecoverable error occurred. The amount of data transferred is indeterminate.

Register		7	6	5	4	3	2	1	0
Error		ICRC	UNC	MC	IDNF	MCR	ABRT	NM	obs
Sector Count	HOB = 0	Reserved							
	HOB = 1	Reserved							
LBA Low	HOB = 0	LBA (7:0)							
	HOB = 1	LBA (31:24)							
LBA Mid	HOB = 0	LBA (15:8)							
	HOB = 1	LBA (39:32)							
LBA High	HOB = 0	LBA (23:16)							
	HOB = 1	LBA (47:40)							
Device		obs	na	obs	DEV	Reserved			
Status		BSY	DRDY	DF	na	DRQ	na	na	ERR
NOTE – HOB = 0 indicates the value read by the host when the HOB bit of the Device Control register is cleared to zero. HOB = 1 indicates the value read by the host when the HOB bit of the Device Control register is set to one.									

Error register –

- ICRC will be set to one if an interface CRC error has occurred during an Ultra DMA data transfer. The content of this bit is not applicable for Multiword DMA transfers.
- UNC will be set to one if data is uncorrectable.
- MC will be set to one if the media in a removable media device changed since the issuance of the last command. The device will clear the device internal media change detected state.
- IDNF will be set to one if a user-accessible address could not be found. IDNF will be set to one if an address outside of the range of user-accessible addresses is requested if command aborted is not returned.
- MCR will be set to one if a media change request has been detected by a removable media device. This bit is only cleared by a GET MEDIA STATUS or a media access command.
- ABRT will be set to one if this command is not supported or if an error, including an ICRC error, has occurred during an Ultra DMA data transfer. ABRT may be set to one if the device is not able to complete the action requested by the command. ABRT will be set to one if an address outside of the range of user-accessible addresses is requested if IDNF is not set to one.
- NM will be set to one if no media is present in a removable media device.

LBA Low – LBA (7:0) of the address of the first unrecoverable error when read with Device Control register HOB bit cleared to zero.
LBA (31:24) of the address of the first unrecoverable error when read with Device Control register HOB bit set to one.
LBA Mid – LBA (15:8) of the address of the first unrecoverable error when read with Device Control register HOB bit cleared to zero.
LBA (39:32) of the address of the first unrecoverable error when read with Device Control register HOB bit set to one.
LBA High – LBA (23:16) of the address of the first unrecoverable error when read with Device Control register HOB bit cleared to zero.
LBA (47:40) of the address of the first unrecoverable error when read with Device Control register HOB bit set to one.
Device register –
DEV will indicate the selected device.
Status register –
BSY will be cleared to zero, indicating command completion.
DRDY will be set to one.
DF (Device Fault) will be set to one if a device fault has occurred.
DRQ will be cleared to zero.
ERR will be set to one if an Error register bit is set to one; however, if SE is set to one, ERR will be cleared to zero.

Prerequisites

DRDY set to one. The host will initialize the DMA channel.

Description

The READ DMA EXT command allows the host to read data using the DMA data transfer protocol.

READ DMA QUEUED

Command code

C7h

Feature set

Overlapped feature set
- Mandatory for devices implementing the Overlapped feature set and not implementing the PACKET Command feature set.
- Use prohibited for devices implementing the PACKET command feature set.

Protocol

DMA QUEUED

Inputs

Register	7	6	5	4	3	2	1	0
Features	Sector Count							
Sector Count	Tag					na	na	na
LBA Low	LBA (7:0)							
LBA Mid	LBA (15:8)							
LBA High	LBA (23:16)							
Device	obs	LBA	obs	DEV	LBA (27:24)			
Command	C7h							

Features –
Number of sectors to be transferred. A value of 00h specifies that 256 sectors are to be transferred.
Sector count –
If the device supports command queuing, bits (7:3) contain the Tag for the command being delivered. A Tag value may be any value between 0 and 31, regardless of the queue depth supported. If queuing is not supported, this register will be set to the value 00h.
LBA Low – Starting LBA bits (7:0).
LBA Mid – Starting LBA bits (15:8).
LBA High – Starting LBA bits (23:16).
Device –
The LBA bit will be set to one to specify the address is an LBA.
DEV will specify the selected device.
Bits (3:0) starting LBA bits (27:24).

Normal outputs

Data transmission

Data transfer may occur after receipt of the command or may occur after the receipt of a SERVICE command. When the device is ready to transfer data requested by a data transfer command, the device sets the following register content to initiate the data transfer.

Register	7	6	5	4	3	2	1	0
Error	na							
Sector Count	Tag					REL	I/O	C/D
LBA Low	na							
LBA Mid	na							
LBA High	na							
Device	obs	na	obs	DEV	na	na	na	na
Status	BSY	DRDY	DF	SERV	DRQ	na	na	CHK

Sector Count register –
Tag – This field contains the command Tag for the command. A Tag value may be any value between 0 and 31, regardless of the queue depth supported. If the device does not support command queuing or overlap is disabled, this register will be set to the value 00h.
REL – will be cleared to zero.
I/O – will be set to one, indicating the transfer is to the host.
C/D – will be cleared to zero, indicating the transfer of data.

Device register –
DEV will indicate the selected device.
Status register –
BSY – will be cleared to zero.
DRDY – will be set to one.
DF (Device Fault) – will be cleared to zero.
SERV (Service) – will be set to one if another command is ready to be serviced.
DRQ – will be set to one.
CHK – will be cleared to zero.

Bus release

If the device performs a bus release before transferring data for this command, the register content upon performing a bus release will be as shown below.

Register	7	6	5	4	3	2	1	0
Error	na							
Sector Count	Tag					REL	I/O	C/D
LBA Low	na							
LBA Mid	na							
LBA High	na							
Device	obs	na	obs	DEV	na			
Status	BSY	DRDY	DF	SERV	DRQ	na	na	ERR

Sector Count register –
Tag – If the device supports command queuing, this field will contain the Tag of the command being bus released. If the device does not support command queuing, this field will be set to the value 00h.
REL will be set to one.
I/O will be zero.
C/D will be zero.
Device register –
DEV will indicate the selected device.
Status register –
BSY will be cleared to zero, indicating bus release.
DRDY will be set to one.
SERV (Service) will be cleared to zero when no other queued command is ready for service. SERV will be set to one when another queued command is ready for service. SERV will be set to one when the device has prepared this command for service.
DF (Device Fault) will be cleared to zero.
DRQ bit will be cleared to zero.
ERR bit will be cleared to zero.

Service request

When the device is ready to transfer data or complete a command after the command has performed a bus release, the device will set the SERV bit and not change the state of any other register bit. When the SERVICE command is received, the device will set outputs as described in data transfer, command completion, or error outputs depending on the service the device requires.

Command completion

When the transfer of all requested data has occurred without error, the register content will be as shown below.

Register	7	6	5	4	3	2	1	0
Error	00h							
Sector Count	Tag					REL	I/O	C/D
LBA Low	na							
LBA Mid	na							
LBA High	na							
Device	obs	na	obs	DEV	na			
Status	BSY	DRDY	DF	SERV	DRQ	na	na	ERR

Sector Count register –
Tag – If the device supports command queuing, this field will contain the Tag of the completed command. If the device does not support command queuing, this field will be set to the value 00h.
REL will be cleared to zero.
I/O will be set to one.
C/D will be set to one.
Device register –
DEV will indicate the selected device.
Status register –
BSY will be cleared to zero, indicating command completion.
DRDY will be set to one.
SERV (Service) will be cleared to zero when no other queued command is ready for service. SERV will be set to one when another queued command is ready for service.
DF (Device Fault) will be cleared to zero.
DRQ bit will be cleared to zero.
ERR bit will be cleared to zero.

Error outputs

The Sector Count register contains the Tag for this command if the device supports command queuing. The device will return command aborted if the command is not supported or if the device has not had overlapped interrupt enabled. The device will return command aborted if the device supports command queuing and the Tag is invalid. An unrecoverable error encountered during the execution of this command results in the termination of the command and the Command Block registers contain the sector where the first unrecoverable error occurred. If a queue existed, the unrecoverable error will cause the queue to abort.

Register	7	6	5	4	3	2	1	0
Error	ICRC	UNC	MC	IDNF	MCR	ABRT	NM	obs
Sector Count	Tag					REL	I/O	C/D
LBA Low	LBA (7:0)							
LBA Mid	LBA (15:8)							
LBA High	LBA (23:16)							
Device	obs	na	obs	DEV	LBA (27:24)			
Status	BSY	DRDY	DF	SERV	DRQ	na	na	ERR

Error register –
ICRC will be set to one if an interface CRC error has occurred during an Ultra DMA data transfer. The content of this bit is not applicable for Multiword DMA transfers.
UNC will be set to one if data is uncorrectable.
MC will be set to one if the media in a removable media device changed since the issuance of the last command. The device will clear the device internal media change detected state.
IDNF will be set to one if a user-accessible address could not be found. IDNF will be set to one if an address outside of the range of user-accessible addresses is requested if ABRT is not set to one.
MCR will be set to one if a media change request has been detected by a removable media device. This bit is only cleared by a GET MEDIA STATUS or a media access command.
ABRT will be set to one if this command is not supported or if an error, including an ICRC error, has occurred during an Ultra DMA data transfer. ABRT may be set to one if the device is not able to complete the action requested by the command. ABRT will be set to one if an address outside of the range of user-accessible addresses is requested if IDNF is not set to one.
NM will be set to one if no media is present in a removable media device.
Sector Count register –
Tag – If the device supports command queuing, this field will contain the Tag of the completed command. If the device does not support command queuing, this field will be set to the value 00h.
REL will be cleared to zero.
I/O will be set to one.
C/D will be set to one.
LBA Low, LBA Mid, LBA High, Device –
Will be written with the address of first unrecoverable error.
DEV will indicate the selected device.
Status register –
BSY will be cleared to zero, indicating command completion.
DRDY will be set to one.
DF (Device Fault) will be set to one if a device fault has occurred.
SERV (Service) will be cleared to zero when no other queued command is ready for service. SERV will be set to one when another queued command is ready for service.
DRQ will be cleared to zero.
ERR will be set to one if an Error register bit is set to one.

Prerequisites

DRDY set to one. The host will initialize the DMA channel.

Description

This command executes in a similar manner to a READ DMA command. The device may perform a bus release or may execute the data transfer without performing a bus release if the data is ready to transfer.

READ DMA QUEUED EXT

Command code

26h

Feature set

48-bit Address feature set

- Mandatory for devices implementing the Overlapped feature set and the 48-bit Address feature set and not implementing the PACKET Command feature set.
- Use prohibited for devices implementing the PACKET command feature set.

Protocol

DMA QUEUED

Inputs

Register		7	6	5	4	3	2	1	0
Features	Current	Sector count (7:0)							
	Previous	Sector count (15:8)							
Sector Count	Current	Tag					Reserved		
	Previous	Reserved							
LBA Low	Current	LBA (7:0)							
	Previous	LBA (31:24)							
LBA Mid	Current	LBA (15:8)							
	Previous	LBA (39:32)							
LBA High	Current	LBA (23:16)							
	Previous	LBA (47:40)							
Device		obs	LBA	obs	DEV	Reserved			
Command		26h							
NOTE – The value indicated as Current is the value most recently written to the register. The value indicated as Previous is the value that was in the register before the most recent write to the register.									

Features Current –
Number of sectors to be transferred low order, bits (7:0).
Features Previous –
Number of sectors to be transferred high order, bits (15:8). 0000h in the Features register specifies that 65,536 sectors are to be transferred.
Sector Count Current –
If the device supports command queuing, bits (7:3) contain the Tag for the command being delivered. A Tag value may be any value between 0 and 31, regardless of the queue depth supported. If queuing is not supported, this register will be set to the value 00h.
Sector Count Previous –
Reserved
LBA Low Current – LBA (7:0).
LBA Low Previous – LBA (31:24).
LBA Mid Current – LBA (15:8).
LBA Mid Previous – LBA (39:32).
LBA High Current – LBA (23:16).
LBA High Previous – LBA (47:40).

Device –
DEV will specify the selected device.
LBA will be set to one

Normal outputs

Data transmission

Data transfer may occur after receipt of the command or may occur after the receipt of a SERVICE command. When the device is ready to transfer data requested by a data transfer command, the device sets the following register content to initiate the data transfer.

Register		7	6	5	4	3	2	1	0
Error		na							
Sector Count	HOB = 0	Tag					REL	I/O	C/D
	HOB = 1	Reserved							
LBA Low	HOB = 0	Reserved							
	HOB = 1	Reserved							
LBA Mid	HOB = 0	Reserved							
	HOB = 1	Reserved							
LBA High	HOB = 0	Reserved							
	HOB = 1	Reserved							
Device		obs	na	obs	DEV	Reserved			
Status		BSY	DRDY	DF	na	DRQ	na	na	ERR
NOTE – HOB = 0 indicates the value read by the host when the HOB bit of the Device Control register is cleared to zero. HOB = 1 indicates the value read by the host when the HOB bit of the Device Control register is set to one.									

Sector Count (when the HOB bit of the Device Control register is cleared to zero) –
Tag – This field contains the command Tag for the command. A Tag value may be any value between 0 and 31, regardless of the queue depth supported. If the device does not support command queuing or overlap is disabled, this register will be set to the value 00h.
REL – will be cleared to zero.
I/O – will be set to one, indicating the transfer is to the host.
C/D – will be cleared to zero, indicating the transfer of data.
Device register –
DEV will indicate the selected device.
Status register –
BSY will be cleared to zero, indicating command completion.
DRDY will be set to one.
DF (Device Fault) will be cleared to zero.
DRQ will be cleared to zero.
ERR will be cleared to zero.

Bus release

If the device performs a bus release before transferring data for this command, the register content upon performing a bus release will be as shown below.

Register		7	6	5	4	3	2	1	0
Error		na							
Sector Count	HOB = 0	Tag					REL	I/O	C/D
	HOB = 1	Reserved							
LBA Low	HOB = 0	Reserved							
	HOB = 1	Reserved							
LBA Mid	HOB = 0	Reserved							
	HOB = 1	Reserved							
LBA High	HOB = 0	Reserved							
	HOB = 1	Reserved							
Device		obs	na	obs	DEV	Reserved			
Status		BSY	DRDY	DF	SERV	DRQ	na	na	ERR
NOTE – HOB = 0 indicates the value read by the host when the HOB bit of the Device Control register is cleared to zero. HOB = 1 indicates the value read by the host when the HOB bit of the Device Control register is set to one.									

Sector Count (when the HOB bit of the Device Control register is cleared to zero) –

Tag – This field contains the command Tag for the command. A Tag value may be any value between 0 and 31, regardless of the queue depth supported. If the device does not support command queuing or overlap is disabled, this register will be set to the value 00h.

REL – will be set to one.

I/O – will be set to one, indicating the transfer is to the host.

C/D – will be cleared to zero, indicating the transfer of data.

Device register –

DEV will indicate the selected device.

Status register –

BSY will be cleared to zero, indicating command completion.

DRDY will be set to one.

DF (Device Fault) will be cleared to zero.

SERV (Service) will be cleared to zero when no other queued command is ready for service. SERV will be set to one when another queued command is ready for service. SERV will be set to one when the device has prepared this command for service.

DRQ will be cleared to zero.

ERR will be cleared to zero.

Service request

When the device is ready to transfer data or complete a command after the command has performed a bus release, the device will set the SERV bit and not change the state of any other register bit. When the SERVICE command is received, the device will set outputs as described in data transfer, command completion, or error outputs depending on the service the device requires.

Command completion

When the transfer of all requested data has occurred without error, the register content will be as shown below.

Register		7	6	5	4	3	2	1	0
Error		na							
Sector Count	HOB = 0	Tag					REL	I/O	C/D
	HOB = 1	Reserved							
LBA Low	HOB = 0	Reserved							
	HOB = 1	Reserved							
LBA Mid	HOB = 0	Reserved							
	HOB = 1	Reserved							
LBA High	HOB = 0	Reserved							
	HOB = 1	Reserved							
Device		obs	na	obs	DEV	Reserved			
Status		BSY	DRDY	DF	SERV	DRQ	na	na	ERR
NOTE – HOB = 0 indicates the value read by the host when the HOB bit of the Device Control register is cleared to zero. HOB = 1 indicates the value read by the host when the HOB bit of the Device Control register is set to one.									

Sector Count (when the HOB bit of the Device Control register is cleared to zero) –

Tag – This field contains the command Tag for the command. A Tag value may be any value between 0 and 31, regardless of the queue depth supported. If the device does not support command queuing or overlap is disabled, this register will be set to the value 00h.

REL – will be cleared to zero.

I/O – will be set to one.

C/D – will be set to one.

Device register –

DEV will indicate the selected device.

Status register –

BSY will be cleared to zero, indicating command completion.

DRDY will be set to one.

DF (Device Fault) will be cleared to zero.

SERV (Service) will be cleared to zero when no other queued command is ready for service. SERV will be set to one when another queued command is ready for service.

DRQ will be cleared to zero.

ERR will be cleared to zero.

Error outputs

The Sector Count register contains the Tag for this command if the device supports command queuing. The device will return command aborted if the command is not supported or if the device has not had overlapped interrupt enabled. The device will return command aborted if the device supports command queuing and the Tag is invalid. An unrecoverable error encountered during the execution of this command results in the termination of the command and the Command Block registers contain the sector where the first unrecoverable error occurred. If a queue existed, the unrecoverable error will cause the queue to abort.

Register		7	6	5	4	3	2	1	0
Error		ICRC	UNC	MC	IDNF	MCR	ABRT	NM	obs
Sector Count	HOB = 0	Tag					REL	I/O	C/D
	HOB = 1	Reserved							
LBA Low	HOB = 0	LBA (7:0)							
	HOB = 1	LBA (31:24)							
LBA Mid	HOB = 0	LBA (15:8)							
	HOB = 1	LBA (39:32)							
LBA High	HOB = 0	LBA (23:16)							
	HOB = 1	LBA (47:40)							
Device		obs	na	obs	DEV	Reserved			
Status		BSY	DRDY	DF	SERV	DRQ	na	na	ERR
NOTE – HOB = 0 indicates the value read by the host when the HOB bit of the Device Control register is cleared to zero. HOB = 1 indicates the value read by the host when the HOB bit of the Device Control register is set to one.									

Error register –

ICRC will be set to one if an interface CRC error has occurred during an Ultra DMA data transfer. The content of this bit is not applicable for Multiword DMA transfers.

UNC will be set to one if data is uncorrectable.

MC will be set to one if the media in a removable media device changed since the issuance of the last command. The device will clear the device internal media change detected state.

IDNF will be set to one if a user-accessible address could not be found. IDNF will be set to one if an address outside of the range of user-accessible addresses is requested if command aborted is not returned.

MCR will be set to one if a media change request has been detected by a removable media device. This bit is only cleared by a GET MEDIA STATUS or a media access command.

ABRT will be set to one if this command is not supported or if an error, including an ICRC error, has occurred during an Ultra DMA data transfer. ABRT may be set to one if the device is not able to complete the action requested by the command. ABRT will be set to one if an address outside of the range of user-accessible addresses is requested if IDNF is not set to one.

NM will be set to one if no media is present in a removable media device.

Sector Count (when the HOB bit of the Device Control register is cleared to zero) –

Tag – This field contains the command Tag for the command. A Tag value may be any value between 0 and 31, regardless of the queue depth supported. If the device does not support command queuing or overlap is disabled, this register will be set to the value 00h.

REL – will be cleared to zero.

I/O – will be set to one.

C/D – will be set to one.

LBA Low –

LBA (7:0) of the address of the first unrecoverable error when read with Device Control register HOB bit cleared to zero.

LBA (31:24) of the address of the first unrecoverable error when read with Device Control register HOB bit set to one.

LBA Mid –

LBA (15:8) of the address of the first unrecoverable error when read with Device Control register HOB bit cleared to zero.

LBA (39:32) of the address of the first unrecoverable error when read with Device Control register HOB bit set to one.

LBA High –

LBA (23:16) of the address of the first unrecoverable error when read with Device Control register HOB bit cleared to zero.

LBA (47:40) of the address of the first unrecoverable error when read with Device Control register HOB bit set to one.

Device register –

DEV will indicate the selected device.

Status register –

BSY will be cleared to zero, indicating command completion.

DRDY will be set to one.

DF (Device Fault) will be set to one if a device fault has occurred.

DRQ will be cleared to zero.

ERR will be set to one if an Error register bit is set to one.

Prerequisites

DRDY set to one. The host will initialize the DMA channel.

Description

This command executes in a similar manner to a READ DMA command. The device may perform a bus release or may execute the data transfer without performing a bus release if the data is ready to transfer.

READ MULTIPLE

Command code

C4h

Feature set

General feature set

- Mandatory for devices not implementing the PACKET Command feature set.
- Use prohibited for devices implementing the PACKET Command feature set.

Protocol

PIO data-in

Inputs

Register	7	6	5	4	3	2	1	0
Features	na							
Sector Count	Sector count							
LBA Low	LBA (7:0)							
LBA Mid	LBA (15:0)							
LBA High	LBA (23:16)							
Device	obs	LBA	obs	DEV	LBA (27:24)			
Command	C4h							

Sector Count –
Number of sectors to be transferred. A value of 00h specifies that 256 sectors are to be transferred.
Device –
DEV will specify the selected device.
Bits (3:0) starting LBA bits (27:24).

Normal outputs

Register	7	6	5	4	3	2	1	0
Error	na							
Sector Count	na							
LBA Low	na							
LBA Mid	na							
LBA High	na							
Device	obs	na	obs	DEV	na	na	na	na
Status	BSY	DRDY	DF	na	DRQ	na	na	ERR

Device register –
DEV will indicate the selected device.
Status register –
BSY will be cleared to zero, indicating command completion.
DRDY will be set to one.
DF (Device Fault) will be cleared to zero.
DRQ will be cleared to zero.
ERR will be cleared to zero.

Error outputs

An unrecoverable error encountered during the execution of this command results in the termination of the command. The Command Block registers contain the address of the sector where the first unrecoverable error occurred. The amount of data transferred is indeterminate.

Register	7	6	5	4	3	2	1	0
Error	na	UNC	MC	IDNF	MCR	ABRT	NM	obs
Sector Count	na							
LBA Low	LBA (7:0)							
LBA Mid	LBA (15:8)							
LBA High	LBA (23:16)							
Device	obs	na	obs	DEV	LBA (27:24)			
Status	BSY	DRDY	DF	na	DRQ	na	na	ERR

Error register –
UNC will be set to one if data is uncorrectable.
MC will be set to one if the media in a removable media device changed since the issuance of the last command. The device will clear the device internal media change detected state.
IDNF will be set to one if a user-accessible address could not be. IDNF will be set to one if an address outside of the range of user-accessible addresses is requested if command aborted is not returned.

MCR will be set to one if a media change request has been detected by a removable media device. This bit is only cleared by a GET MEDIA STATUS or a media access command.
ABRT will be set to one if this command is not supported or if an error, including an ICRC error, has occurred during an Ultra DMA data transfer. ABRT may be set to one if the device is not able to complete the action requested by the command. ABRT will be set to one if an address outside of the range of user-accessible addresses is requested if IDNF is not set to one.
NM will be set to one if no media is present in a removable media device.

LBA Low, LBA Mid, LBA High, Device –
will be written with the address of first unrecoverable error.
DEV will indicate the selected device.

Status register –
BSY will be cleared to zero, indicating command completion.
DRDY will be set to one.
DF (Device Fault) will be set to one if a device fault has occurred.
DRQ will be cleared to zero.
ERR will be set to one if an Error register bit is set to one.

Prerequisites

DRDY set to one. If bit 8 of IDENTIFY DEVICE data word 59 is cleared to zero, a successful SET MULTIPLE MODE command will precede a READ MULTIPLE command.

Description

This command reads the number of sectors specified in the Sector Count register.
The number of sectors per block is defined by the content of word 59 in the IDENTIFY DEVICE data. The device will interrupt for each DRQ block transferred.

When the READ MULTIPLE command is issued, the Sector Count register contains the number of sectors (not the number of blocks) requested.

If the number of requested sectors is not evenly divisible by the block count, as many full blocks as possible are transferred, followed by a final, partial block transfer. The partial block transfer will be for n sectors, where n = remainder (sector count/ block count).

If the READ MULTIPLE command is received when READ MULTIPLE commands are disabled, the READ MULTIPLE operation will be rejected with command aborted.

Device errors encountered during READ MULTIPLE commands are posted at the beginning of the block or partial block transfer, but the DRQ bit is still set to one and the data transfer will take place, including transfer of corrupted data, if any. The contents of the Command Block Registers following the transfer of a data block that had a sector in error are undefined. The host should retry the transfer as individual requests to obtain valid error information.

Subsequent blocks or partial blocks are transferred only if the error was a correctable data error. All other errors cause the command to stop after transfer of the block that contained the error.

READ SECTOR(S)

Command code

20h

Feature set

General feature set
- Mandatory for all devices.
- PACKET Command feature set devices.

Protocol

PIO data-in

Inputs

Register	7	6	5	4	3	2	1	0
Features	na							
Sector Count	Sector count							
LBA Low	LBA (7:0)							
LBA Mid	LBA (15:8)							
LBA High	LBA (23:16)							
Device	obs	LBA	obs	DEV	LBA (27:24)			
Command	20h							

Sector Count –
Number of sectors to be transferred. A value of 00h specifies that 256 sectors are to be transferred.
Device –
DEV will specify the selected device.
Bits (3:0) starting LBA bits (27:24).

Outputs

Normal outputs

Register	7	6	5	4	3	2	1	0
Error	na							
Sector Count	na							
LBA Low	na							
LBA Mid	na							
LBA High	na							
Device	obs	na	obs	DEV	na	na	na	na
Status	BSY	DRDY	DF	na	DRQ	na	na	ERR

Device register –
DEV will indicate the selected device.
Status register –
BSY will be cleared to zero, indicating command completion.
DRDY will be set to one.
DF (Device Fault) will be cleared to zero.
DRQ will be cleared to zero.
ERR will be cleared to zero.

Outputs for PACKET Command feature set devices

In response to this command, devices that implement the PACKET Command feature set will post command aborted and place the PACKET Command feature set signature in the LBA High and the LBA Mid register.

Error outputs

An unrecoverable error encountered during the execution of this command results in the termination of the command. The Command Block registers contain the address of the sector where the first unrecoverable error occurred. The amount of data transferred is indeterminate.

Register	7	6	5	4	3	2	1	0
Error	na	UNC	MC	IDNF	MCR	ABRT	NM	obs
Sector Count	na							
LBA Low	LBA (7:0)							
LBA Mid	LBA (15:8)							
LBA High	LBA (23:16)							
Device	obs	na	obs	DEV	LBA (27:24)			
Status	BSY	DRDY	DF	na	DRQ	na	na	ERR

Error register –
UNC will be set to one if data is uncorrectable.
MC will be set to one if the media in a removable media device changed since the issuance of the last command. The device will clear the device internal media change detected state.
IDNF will be set to one if a user-accessible address could not be found. IDNF will be set to one if an address outside of the range of user-accessible addresses is requested if command aborted is not returned.
MCR will be set to one if a media change request has been detected by a removable media device. This bit is only cleared by a GET MEDIA STATUS or a media access command.
ABRT will be set to one if this command is not supported or if an error, including an ICRC error, has occurred during an Ultra DMA data transfer. ABRT may be set to one if the device is not able to complete the action requested by the command. ABRT will be set to one if an address outside of the range of user-accessible addresses is requested if IDNF is not set to one.
NM will be set to one if no media is present in a removable media device.

LBA Low, LBA Mid, LBA High, Device –
Will be written with the address of first unrecoverable error.
DEV will indicate the selected device.

Status register –
BSY will be cleared to zero, indicating command completion.
DRDY will be set to one.
DF (Device Fault) will be set to one if a device fault has occurred.
DRQ will be cleared to zero.
ERR will be set to one if an Error register bit is set to one.

Prerequisites

DRDY set to one.

Description

This command reads from 1 to 256 sectors, as specified in the Sector Count register. A sector count of 0 requests 256 sectors. The transfer will begin at the sector specified in the LBA Low, LBA Mid, LBA High, and Device registers. The device will interrupt for each DRQ block transferred.

The DRQ bit is always set to one prior to data transfer, regardless of the presence or absence of an error condition.

SERVICE

Command code

A2h

Feature set

Overlap and Queued feature sets

- Mandatory when the Overlapped feature set is implemented.

Protocol

PACKET or READ/WRITE DMA QUEUED

Inputs

Register	7	6	5	4	3	2	1	0
Features	na							
Sector Count	na							
LBA Low	na							
LBA Mid	na							
LBA High	na							
Device	obs	na	obs	DEV	na	na	na	na
Command	A2h							

Device register –

DEV will specify the selected device.

Outputs

Outputs as a result of a SERVICE command are described in the command description for the command for which SERVICE is being requested.

Prerequisites

The device will have performed a bus release for a previous overlap PACKET, READ DMA QUEUED, READ DMA QUEUED EXT, WRITE DMA QUEUED, or WRITE DMA QUEUED EXT command and will have set the SERV bit to one to request the SERVICE command be issued to continue data transfer and/or provide command status.

Description

The SERVICE command is used to provide data transfer and/or status of a command that was previously bus released.

SET MULTIPLE MODE

Command code

C6h

Feature set

General feature set

- Mandatory for devices not implementing the PACKET Command feature set.
- Use prohibited for devices implementing the PACKET Command feature set.

Protocol

Non-data

Inputs

If the content of the Sector Count register is not zero, then the Sector Count register contains the number of sectors per block for the device to be used on all following READ/WRITE MULTIPLE commands. The content of the Sector Count register will be less than or equal to the value in bits (7:0) in word 47 in the IDENTIFY DEVICE data. The host should set the content of the Sector Count register to 1, 2, 4, 8, 16, 32, 64, or 128.

If the content of the Sector Count register is zero and the SET MULTIPLE command completes without error, then the device will respond to any subsequent READ MULTIPLE or WRITE MULTIPLE command with command aborted until a subsequent successful SET MULTIPLE command completion where the Sector Count register is not set to zero.

Register	7	6	5	4	3	2	1	0
Features	na							
Sector Count	Sectors per block							
LBA Low	na							
LBA Mid	na							
LBA High	na							
Device	obs	na	obs	DEV	na			
Command	C6h							

Device register –

DEV will specify the selected device.

Normal outputs

Register	7	6	5	4	3	2	1	0
Error	na							
Sector Count	na							
LBA Low	na							
LBA Mid	na							
LBA High	na							
Device	obs	na	obs	DEV	na			
Status	BSY	DRDY	DF	na	DRQ	na	na	ERR

Device register –

DEV will indicate the selected device.

Status register –

BSY will be cleared to zero, indicating command completion.

DRDY will be set to one.

DF (Device Fault) will be cleared to zero.
DRQ will be cleared to zero.
ERR will be cleared to zero.

Error outputs

If a block count is not supported, the device will return command aborted.

Register	7	6	5	4	3	2	1	0
Error	na	na	na	na	na	ABRT	na	na
Sector Count	na							
LBA Low	na							
LBA Mid	na							
LBA High	na							
Device	obs	na	obs	DEV	na			
Status	BSY	DRDY	DF	na	DRQ	na	na	ERR

Error register –
ABRT will be set to one if the block count is not supported. ABRT may be set to one if the device is not able to complete the action requested by the command.
Device register –
DEV will indicate the selected device.
Status register –
BSY will be cleared to zero, indicating command completion.
DRDY will be set to one.
DF (Device Fault) will be set to one if a device fault has occurred.
DRQ will be cleared to zero.
ERR will be set to one if an Error register bit is set to one.

Prerequisites

DRDY set to one.

Description

This command establishes the block count for READ MULTIPLE, READ MULTIPLE EXT, WRITE MULTIPLE, and WRITE MULTIPLE EXT commands.

Devices will support the block size specified in the IDENTIFY DEVICE parameter word 47, bits (7:0), and may also support smaller values.

Upon receipt of the command, the device checks the Sector Count register. If the content of the Sector Count register is not zero, the Sector Count register contains a valid value, and the block count is supported, then the value in the Sector Count register is used for all subsequent READ MULTIPLE, READ MULTIPLE EXT, WRITE MULTIPLE, and WRITE MULTIPLE EXT commands and their execution is enabled. If the content of the Sector Count register is zero, the device may:

1. disable multiple mode and respond with command aborted to all subsequent READ MULTIPLE, READ MULTIPLE EXT, WRITE MULTIPLE, and WRITE MULTIPLE EXT commands;
2. respond with command aborted to the SET MULTIPLE MODE command;
3. retain the previous multiple mode settings.

After a successful SET MULTIPLE command, the device will report the valid value set by that command in bits (7:0) in word 59 in the IDENTIFY DEVICE data.

After a power-on or hardware reset, if bit 8 is set to one and bits (7:0) are cleared to zero in word 59 of the IDENTIFY DEVICE data, a SET MULTIPLE command is required before issuing a READ MULTIPLE, READ MULTIPLE EXT, WRITE MULTIPLE, or WRITE MULTIPLE EXT command. If bit 8 is set to one and bits (7:0) are not cleared to zero, a SET MULTIPLE command may be issued to change the multiple value required before issuing a READ MULTIPLE, READ MULTIPLE EXT, WRITE MULTIPLE, or WRITE MULTIPLE EXT command.

WRITE BUFFER

Command code

E8h

Feature set

General feature set
- Optional for devices not implementing the PACKET Command feature set.
- Use prohibited for devices implementing the PACKET Command feature set.

Protocol

PIO data-out

Inputs

Register	7	6	5	4	3	2	1	0
Features	na							
Sector Count	na							
LBA Low	na							
LBA Mid	na							
LBA High	na							
Device	obs	na	obs	DEV	na	na	na	na
Command	E8h							

Device register –
DEV will specify the selected device.

Normal outputs

Register	7	6	5	4	3	2	1	0
Error	na							
Sector Count	na							
LBA Low	na							
LBA Mid	na							
LBA High	na							
Device	obs	na	obs	DEV	na	na	na	na
Status	BSY	DRDY	DF	na	DRQ	na	na	ERR

Device register –
DEV will indicate the selected device.
Status register –
BSY will be cleared to zero, indicating command completion.
DRDY will be set to one.
DF (Device Fault) will be cleared to zero.
DRQ will be cleared to zero.
ERR will be cleared to zero.

Error outputs

The device will return command aborted if the command is not supported.

Register	7	6	5	4	3	2	1	0
Error	na	na	na	na	na	ABRT	na	na
Sector Count	na							
LBA Low	na							
LBA Mid	na							
LBA High	na							
Device	obs	na	obs	DEV	na			
Status	BSY	DRDY	DF	na	DRQ	na	na	ERR

Error register –
ABRT will be set to one if this command is not supported. ABRT may be set to one if the device is not able to complete the action requested by the command.
Device register –
DEV will indicate the selected device.
Status register –
BSY will be cleared to zero, indicating command completion.
DRDY will be set to one.
DF (Device Fault) will be set to one if a device fault has occurred.
DRQ will be cleared to zero.
ERR will be set to one if an Error register bit is set to one.

Prerequisites

DRDY set to one.

Description

This command enables the host to write the contents of one sector in the device's buffer.

The READ BUFFER and WRITE BUFFER commands will be synchronized within the device such that sequential WRITE BUFFER and READ BUFFER commands access the same 512 bytes within the buffer.

WRITE DMA

Command code

CAh

Feature set

General feature set

- Mandatory for devices not implementing the PACKET Command feature set.
- Use prohibited for devices implementing the PACKET Command feature set.

Protocol

DMA

Inputs

The LBA Mid, LBA High, Device, and LBA Low specify the starting sector address to be written. The Sector Count register specifies the number of sectors to be transferred.

Register	7	6	5	4	3	2	1	0
Features	na							
Sector Count	Sector count							
LBA Low	LBA (7:0)							
LBA Mid	LBA (15:8)							
LBA High	LBA (23:16)							
Device	obs	LBA	obs	DEV	LBA (27:24)			
Command	CAh							

Sector Count –
Number of sectors to be transferred. A value of 00h specifies that 256 sectors are to be transferred.
Device –
The LBA bit will be set to one to specify the address is an LBA.
DEV will specify the selected device.
Bits (3:0) starting LBA bits (27:24).

Normal outputs

Register	7	6	5	4	3	2	1	0
Error	na							
Sector Count	na							
LBA Low	na							
LBA Mid	na							
LBA High	na							
Device	obs	na	obs	DEV	na	na	na	na
Status	BSY	DRDY	DF	na	DRQ	na	na	ERR

Device register –
DEV will indicate the selected device.
Status register –
BSY will be cleared to zero, indicating command completion.
DRDY will be set to one.
DF (Device Fault) will be cleared to zero.
DRQ will be cleared to zero.
ERR will be cleared to zero.

Error outputs

An unrecoverable error encountered during the execution of this command results in the termination of the command. The Command Block registers contain the address of the sector where the first unrecoverable error occurred. The amount of data transferred is indeterminate.

Register	7	6	5	4	3	2	1	0
Error	ICRC	WP	MC	IDNF	MCR	ABRT	NM	obs
Sector Count	na							
LBA Low	LBA (7:0)							
LBA Mid	LBA (15:8)							
LBA High	LBA (23:16)							
Device	obs	na	obs	DEV	LBA (27:24)			
Status	BSY	DRDY	DF	na	DRQ	na	na	ERR

Error register –

ICRC will be set to one if an interface CRC error has occurred during an Ultra DMA data transfer. The content of this bit is not applicable for Multiword DMA transfers.

WP will be set to one if the media in a removable media device is write protected.

MC will be set to one if the media in a removable media device changed since the issuance of the last command. The device will clear the device internal media change detected state.

IDNF will be set to one if a user-accessible address could not be found. IDNF will be set to one if an address outside of the range of user-accessible addresses is requested if command aborted is not returned.

MCR will be set to one if a media change request has been detected by a removable media device. This bit is only cleared by a GET MEDIA STATUS or a media access command.

ABRT will be set to one if this command is not supported or if an error, including an ICRC error, has occurred during an Ultra DMA data transfer. ABRT may be set to one if the device is not able to complete the action requested by the command. ABRT will be set to one if an address outside of the range of user-accessible addresses is requested if IDNF is not set to one.

NM will be set to one if no media is present in a removable media device.

LBA Low, LBA Mid, LBA High, Device –

Will be written with address of first unrecoverable error.

DEV will indicate the selected device.

Status register –

BSY will be cleared to zero, indicating command completion.

DRDY will be set to one.

DF (Device Fault) will be set to one if a device fault has occurred.

DRQ will be cleared to zero.

ERR will be set to one if an Error register bit is set to one; however, if SE is set to one, ERR will be cleared to zero.

Prerequisites

DRDY set to one. The host will initialize the DMA channel.

Description

The WRITE DMA command allows the host to write data using the DMA data transfer protocol.

WRITE DMA EXT

Command code

35h

Feature set

48-bit Address feature set

- Mandatory for devices implementing the 48-bit Address feature set.
- Use prohibited for devices implementing the PACKET Command feature set.

Protocol

DMA

Inputs

Register		7	6	5	4	3	2	1	0
Features	Current	Reserved							
	Previous	Reserved							
Sector Count	Current	Sector count (7:0)							
	Previous	Sector count (15:8)							
LBA Low	Current	LBA (7:0)							
	Previous	LBA (31:24)							
LBA Mid	Current	LBA (15:8)							
	Previous	LBA (39:32)							
LBA High	Current	LBA (23:16)							
	Previous	LBA (47:40)							
Device		obs	LBA	obs	DEV	Reserved			
Command		35h							
NOTE – The value indicated as Current is the value most recently written to the register. The value indicated as Previous is the value that was in the register before the most recent write to the register.									

Sector Count Current –
Number of sectors to be transferred low order, bits (7:0).

Sector Count Previous –
Number of sectors to be transferred high order, bits (15:8). 0000h in the Sector Count register specifies that 65,536 sectors are to be transferred.

LBA Low Current – LBA (7:0).

LBA Low Previous – LBA (31:24).

LBA Mid Current – LBA (15:8).

LBA Mid Previous – LBA (39:32).

LBA High Current – LBA (23:16).

LBA High Previous – LBA (47:40).

Device –
The LBA bit will be set to one to specify the address is an LBA.
DEV will specify the selected device.

Normal outputs

Register		7	6	5	4	3	2	1	0
Error		na							
Sector Count	HOB = 0	Reserved							
	HOB = 1	Reserved							
LBA Low	HOB = 0	Reserved							
	HOB = 1	Reserved							
LBA Mid	HOB = 0	Reserved							
	HOB = 1	Reserved							
LBA High	HOB = 0	Reserved							
	HOB = 1	Reserved							
Device		obs	Na	obs	DEV	Reserved			
Status		BSY	DRDY	DF	na	DRQ	na	na	ERR
NOTE – HOB = 0 indicates the value read by the host when the HOB bit of the Device Control register is cleared to zero. HOB = 1 indicates the value read by the host when the HOB bit of the Device Control register is set to one.									

Device register –
DEV will indicate the selected device.
Status register –
BSY will be cleared to zero, indicating command completion.
DRDY will be set to one.
DF (Device Fault) will be cleared to zero.
DRQ will be cleared to zero.
ERR will be cleared to zero.

Error outputs

An unrecoverable error encountered during the execution of this command results in the termination of the command. The Command Block registers contain the address of the sector where the first unrecoverable error occurred. The amount of data transferred is indeterminate.

Register		7	6	5	4	3	2	1	0
Error		ICRC	WP	MC	IDNF	MCR	ABRT	NM	obs
Sector Count	HOB = 0	Reserved							
	HOB = 1	Reserved							
LBA Low	HOB = 0	LBA (7:0)							
	HOB = 1	LBA (31:24)							
LBA Mid	HOB = 0	LBA (15:8)							
	HOB = 1	LBA (39:32)							
LBA High	HOB = 0	LBA (23:16)							
	HOB = 1	LBA (47:40)							
Device		obs	na	obs	DEV	Reserved			
Status		BSY	DRDY	DF	na	DRQ	na	na	ERR
NOTE – HOB = 0 indicates the value read by the host when the HOB bit of the Device Control register is cleared to zero. HOB = 1 indicates the value read by the host when the HOB bit of the Device Control register is set to one.									

Error register –
ICRC will be set to one if an interface CRC error has occurred during an Ultra DMA data transfer. The content of this bit is not applicable for Multiword DMA transfers.
WP will be set to one if the media in a removable media device is write protected.
MC will be set to one if the media in a removable media device changed since the issuance of the last command. The device will clear the device internal media change detected state.
IDNF will be set to one if a user-accessible address could not be found IDNF will be set to one if an address outside of the range of user-accessible addresses is requested if command aborted is not returned.
MCR will be set to one if a media change request has been detected by a removable media device. This bit is only cleared by a GET MEDIA STATUS or a media access command.
ABRT will be set to one if this command is not supported or if an error, including an ICRC error, has occurred during an Ultra DMA data transfer. ABRT may be set to one if the device is not able to complete the action requested by the command. ABRT will be set to one if an address outside of the range of user-accessible addresses is requested if IDNF is not set to one.
NM will be set to one if no media is present in a removable media device.
LBA Low –
LBA (7:0) of the address of the first unrecoverable error when read with Device Control register HOB bit cleared to zero.
LBA (31:24) of the address of the first unrecoverable error when read with Device Control register HOB bit set to one.
LBA Mid –
LBA (15:8) of the address of the first unrecoverable error when read with Device Control register HOB bit cleared to zero.
LBA (39:32) of the address of the first unrecoverable error when read with Device Control register HOB bit set to one.
LBA High –
LBA (23:16) of the address of the first unrecoverable error when read with Device Control register HOB bit cleared to zero.
LBA (47:40) of the address of the first unrecoverable error when read with Device Control register HOB bit set to one.
Device register –
DEV will indicate the selected device.
Status register –
BSY will be cleared to zero, indicating command completion.
DRDY will be set to one.
DF (Device Fault) will be set to one if a device fault has occurred.
DRQ will be cleared to zero.
ERR will be set to one if an Error register bit is set to one. However, if SE is set to one, ERR will be cleared to zero.

Prerequisites

DRDY set to one. The host will initialize the DMA channel.

Description

The WRITE DMA EXT command allows the host to write data using the DMA data transfer protocol.

WRITE DMA QUEUED

Command code

CCh

Feature set

Overlapped feature set

- Mandatory for devices implementing the Overlapped feature set and not implementing the PACKET Command feature set.
- Use prohibited for devices implementing the PACKET Command feature set.

Protocol

DMA QUEUED

Inputs

Register	7	6	5	4	3	2	1	0
Features	Sector Count							
Sector Count	Tag					na		
LBA Low	LBA (7:0)							
LBA Mid	LBA (15:8)							
LBA High	LBA (23:16)							
Device	obs	LBA	obs	DEV	LBA (27:24)			
Command	CCh							

Features –
Number of sectors to be transferred. A value of 00h specifies that 256 sectors are to be transferred.
Sector count –
If the device supports command queuing, bits (7:3) contain the Tag for the command being delivered. A Tag value may be any value between 0 and 31, regardless of the queue depth supported. If queuing is not supported, this field is not applicable.
Device –
The LBA bit will be set to one to specify the address is an LBA.
DEV will specify the selected device.
Bits (3:0) starting LBA bits (27:24).

Normal outputs

Data transmission

Data transfer may occur after receipt of the command or may occur after the receipt of a SERVICE command. When the device is ready to transfer data requested by a data transfer command, the device sets the following register content to initiate the data transfer.

Register	7	6	5	4	3	2	1	0
Error	na							
Sector Count	Tag					REL	I/O	C/D
LBA Low	na							
LBA Mid	na							
LBA High	na							
Device	obs	na	obs	DEV	na	na	na	na
Status	BSY	DRDY	DF	SERV	DRQ	na	na	CHK

Interrupt reason register –

Tag – This field contains the command Tag for the command. A Tag value may be any value between 0 and 31, regardless of the queue depth supported. If the device does not support command queuing or overlap is disabled, this field is not applicable.

REL – will be cleared to zero.

I/O – will be cleared to zero, indicating the transfer is from the host.

C/D – will be cleared to zero, indicating the transfer of data.

Device register –

DEV – will indicate the selected device.

Status register –

BSY – will be cleared to zero.

DRDY – will be set to one.

DF (Device Fault) – will be cleared to zero.

SERV (Service) – will be set to one if another command is ready to be serviced.

DRQ – will be set to one.

CHK – will be cleared to zero.

Bus release

If the device performs a bus release before transferring data for this command, the register content upon performing a bus release will be as shown below.

Register	7	6	5	4	3	2	1	0
Error	na							
Sector Count	Tag					REL	I/O	C/D
LBA Low	na							
LBA Mid	na							
LBA High	na							
Device	obs	na	obs	DEV	na			
Status	BSY	DRDY	DF	SERV	DRQ	na	na	ERR

Sector Count register –

Tag – If the device supports command queuing, this field will contain the Tag of the command being bus released. If the device does not support command queuing, this field will be zeros.

REL bit will be set, indicating that the device has bus released an overlap command.

I/O will be cleared to zero.

C/D will be cleared to zero.

Device register –
DEV will indicate the selected device.
Status register –
BSY will be cleared to zero, indicating bus release.
DRDY will be set to one.
SERV (Service) will be cleared to zero if no other queued command is ready for service. SERV will be set to one when another queued command is ready for service. This bit will be set to one when the device has prepared this command for service.
DF (Device Fault) will be cleared to zero.
DRQ bit will be cleared to zero.
ERR bit will be cleared to zero.

Service request

When the device is ready to transfer data or complete a command after the command has performed a bus release, the device will set the SERV bit and not change the state of any other register bit. When the SERVICE command is received, the device will set outputs as described in data transfer, command completion, or error outputs depending on the service the device requires.

Command completion

When the transfer of all requested data has occurred without error, the register content will be as shown below.

Register	7	6	5	4	3	2	1	0
Error	00h							
Sector Count	Tag					REL	I/O	C/D
LBA Low	Na							
LBA Mid	Na							
LBA High	Na							
Device	obs	na	obs	DEV	na			
Status	BSY	DRDY	DF	SERV	DRQ	na	na	ERR

Sector Count register –
Tag – If the device supports command queuing, this field will contain the Tag of the completed command. If the device does not support command queuing, this field will be zeros.
REL will be cleared to zero.
I/O will be set to one.
C/D will be set to one.
Device register –
DEV will indicate the selected device.
Status register –
BSY will be cleared to zero, indicating command completion.
DRDY will be set to one.
SERV (Service) will be cleared to zero when no other queued command is ready for service. SERV will be set to one when another queued command is ready for service.
DF (Device Fault) will be cleared to zero.
DRQ bit will be cleared to zero.
ERR bit will be cleared to zero.

Error outputs

The Sector Count register contains the Tag for this command if the device supports command queuing. The device will return command aborted if the command is not supported. The device will return command aborted if the device supports command queuing and the Tag is invalid. An unrecoverable error encountered during the execution of this command results in the termination of the command and the Command Block registers contain the sector where the first unrecoverable error occurred. If a queue existed, the unrecoverable error will cause the queue to abort. The device may remain BSY for some time when responding to these errors.

Register	7	6	5	4	3	2	1	0
Error	ICRC	WP	MC	IDNF	MCR	ABRT	NM	na
Sector Count	Tag					REL	I/O	C/D
LBA Low	LBA (7:0)							
LBA Mid	LBA (15:8)							
LBA High	LBA (23:16)							
Device	obs	na	obs	DEV	LBA (27:24)			
Status	BSY	DRDY	DF	SERV	DRQ	na	na	ERR

Error register –

ICRC will be set to one if an interface CRC error has occurred during an Ultra DMA data transfer. The content of this bit is not applicable for Multiword DMA transfers.

WP will be set to one if the media in a removable media device is write protected.

MC will be set to one if the media in a removable media device changed since the issuance of the last command. The device will clear the device internal media change detected state.

IDNF will be set to one if a user-accessible address could not be. IDNF will be set to one if an address outside of the range of user-accessible addresses is requested if command aborted is not returned.

MCR will be set to one if a media change request has been detected by a removable media device. This bit is only cleared by a GET MEDIA STATUS or a media access command.

ABRT will be set to one if this command is not supported or if an error, including an ICRC error, has occurred during an Ultra DMA data transfer. ABRT may be set to one if the device is not able to complete the action requested by the command. ABRT will be set to one if an address outside of the range of user-accessible addresses is requested if IDNF is not set to one.

NM will be set to one if no media is present in a removable media device.

Sector Count register –

Tag – If the device supports command queuing, this field will contain the Tag of the completed command. If the device does not support command queuing, this field will be zeros.

REL will be cleared to zero.

I/O will be set to one.

C/D will be set to one.

LBA Low, LBA Mid, LBA High, Device –

Will be written with the address of the first unrecoverable error.

DEV will indicate the selected device.

Status register –

BSY will be cleared to zero, indicating command completion.

DRDY will be set to one.

DF (Device Fault) will be set to one if a device fault has occurred.

SERV (Service) will be cleared to zero when no other queued command is ready for service. SERV will be set to one when another queued command is ready for service.

DRQ will be cleared to zero.
ERR will be set to one if an Error register bit is set to one.

Prerequisites

DRDY set to one. The host will initialize the DMA channel.

Description

This command executes in a similar manner to a WRITE DMA command. The device may perform a bus release or may execute the data transfer without performing a bus release if the data is ready to transfer.

If the device performs a bus release, the host will reselect the device using the SERVICE command.

Once the data transfer is begun, the device will not perform a bus release until the entire data transfer has been completed.

WRITE DMA QUEUED EXT

Command code

36h

Feature set

Overlapped feature set and 48-bit Address feature set

- Mandatory for devices implementing the Overlapped feature set and the 48-bit Address feature set and not implementing the PACKET Command feature set.
- Use prohibited for devices implementing the PACKET Command feature set.

Protocol

DMA QUEUED

Inputs

<table>
<tr><th colspan="2">Register</th><th>7</th><th>6</th><th>5</th><th>4</th><th>3</th><th>2</th><th>1</th><th>0</th></tr>
<tr><td>Features</td><td>Current</td><td colspan="8">Sector count (7:0)</td></tr>
<tr><td></td><td>Previous</td><td colspan="8">Sector count (15:8)</td></tr>
<tr><td>Sector Count</td><td>Current</td><td colspan="5">Tag</td><td colspan="3">Reserved</td></tr>
<tr><td></td><td>Previous</td><td colspan="8">Reserved</td></tr>
<tr><td>LBA Low</td><td>Current</td><td colspan="8">LBA (7:0)</td></tr>
<tr><td></td><td>Previous</td><td colspan="8">LBA (31:24)</td></tr>
<tr><td>LBA Mid</td><td>Current</td><td colspan="8">LBA (15:8)</td></tr>
<tr><td></td><td>Previous</td><td colspan="8">LBA (39:32)</td></tr>
<tr><td>LBA High</td><td>Current</td><td colspan="8">LBA (23:16)</td></tr>
<tr><td></td><td>Previous</td><td colspan="8">LBA (47:40)</td></tr>
<tr><td colspan="2">Device</td><td>obs</td><td>LBA</td><td>obs</td><td>DEV</td><td colspan="4">Reserved</td></tr>
<tr><td colspan="2">Command</td><td colspan="8">36h</td></tr>
<tr><td colspan="10">NOTE – The value indicated as Current is the value most recently written to the register. The value indicated as Previous is the value that was in the register before the most recent write to the register.</td></tr>
</table>

Features Current –
Number of sectors to be transferred low order, bits (7:0).
Features Previous –
Number of sectors to be transferred high order, bits (15:8). 0000h in the Features register specifies that 65,536 sectors are to be transferred.
Sector Count Current –
If the device supports command queuing, bits (7:3) contain the Tag for the command being delivered. A Tag value may be any value between 0 and 31, regardless of the queue depth supported. If queuing is not supported, this register will be set to the value 00h.
Sector Count Previous – Reserved
LBA Low Current – LBA (7:0).
LBA Low Previous – LBA (31:24).
LBA Mid Current – LBA (15:8).
LBA Mid Previous – LBA (39:32).
LBA High Current – LBA (23:16).
LBA High Previous – LBA (47:40).
Device –
The LBA bit will be set to one to specify the address is an LBA.
DEV will specify the selected device.

Normal outputs

Data transmission

Data transfer may occur after receipt of the command or may occur after the receipt of a SERVICE command. When the device is ready to transfer data requested by a data transfer command, the device sets the following register content to initiate the data transfer.

Register		7	6	5	4	3	2	1	0
Error		na							
Sector Count	HOB = 0	Tag					REL	I/O	C/D
	HOB = 1	Reserved							
LBA Low	HOB = 0	Reserved							
	HOB = 1	Reserved							
LBA Mid	HOB = 0	Reserved							
	HOB = 1	Reserved							
LBA High	HOB = 0	Reserved							
	HOB = 1	Reserved							
Device		obs	na	obs	DEV	Reserved			
Status		BSY	DRDY	DF	na	DRQ	na	na	ERR
NOTE – HOB = 0 indicates the value read by the host when the HOB bit of the Device Control register is cleared to zero. HOB = 1 indicates the value read by the host when the HOB bit of the Device Control register is set to one.									

Sector Count (when HOB of the Device Control register is cleared to zero) –
Tag – This field contains the command Tag for the command. A Tag value may be any value between 0 and 31, regardless of the queue depth supported. If the device does not support command queuing or overlap is disabled, this register will be set to the value 00h.
REL – will be cleared to zero.
I/O – will be cleared to zero, indicating the transfer is from the host.

C/D – will be cleared to zero, indicating the transfer of data.

Device register –

DEV will indicate the selected device.

Status register –

BSY will be cleared to zero.

DRDY will be set to one.

DF (Device Fault) will be cleared to zero.

DRQ will be cleared to zero.

ERR will be cleared to zero.

Bus release

If the device performs a bus release before transferring data for this command, the register content upon performing a bus release will be as shown below.

Register		7	6	5	4	3	2	1	0
Error		na							
Sector Count	HOB = 0	Tag					REL	I/O	C/D
	HOB = 1	Reserved							
LBA Low	HOB = 0	Reserved							
	HOB = 1	Reserved							
LBA Mid	HOB = 0	Reserved							
	HOB = 1	Reserved							
LBA High	HOB = 0	Reserved							
	HOB = 1	Reserved							
Device		obs	na	obs	DEV	Reserved			
Status		BSY	DRDY	DF	SERV	DRQ	na	na	ERR
NOTE – HOB = 0 indicates the value read by the host when the HOB bit of the Device Control register is cleared to zero. HOB = 1 indicates the value read by the host when the HOB bit of the Device Control register is set to one.									

Sector Count (when HOB of the Device Control register is cleared to zero) –

Tag – This field contains the command Tag for the command. A Tag value may be any value between 0 and 31, regardless of the queue depth supported. If the device does not support command queuing or overlap is disabled, this register will be set to the value 00h.

REL – will be set to one.

I/O – will be cleared to zero.

C/D – will be cleared to zero, indicating the transfer of data.

Device register –

DEV will indicate the selected device.

Status register –

BSY will be cleared to zero.

DRDY will be set to one.

DF (Device Fault) will be cleared to zero.

SERV (Service) will be cleared to zero when no other queued command is ready for service. SERV will be set to one when another queued command is ready for service. SERV will be set to one when the device has prepared this command for service.

DRQ will be cleared to zero.

ERR will be cleared to zero.

Service request

When the device is ready to transfer data or complete a command after the command has performed a bus release, the device will set the SERV bit to one and not change the state of any other register bit. When the SERVICE command is received, the device will set outputs as described in data transfer, command completion, or error outputs, depending on the service the device requires.

Command completion

When the transfer of all requested data has occurred without error, the register content will be as shown below.

Register		7	6	5	4	3	2	1	0
Error		na							
Sector Count	HOB = 0	Tag					REL	I/O	C/D
	HOB = 1	Reserved							
LBA Low	HOB = 0	Reserved							
	HOB = 1	Reserved							
LBA Mid	HOB = 0	Reserved							
	HOB = 1	Reserved							
LBA High	HOB = 0	Reserved							
	HOB = 1	Reserved							
Device		obs	na	Obs	DEV	Reserved			
Status		BSY	DRDY	DF	SERV	DRQ	na	na	ERR
NOTE – HOB = 0 indicates the value read by the host when the HOB bit of the Device Control register is cleared to zero. HOB = 1 indicates the value read by the host when the HOB bit of the Device Control register is set to one.									

Sector Count (when HOB of the Device Control register is cleared to zero) –
- Tag – This field contains the command Tag for the command. A Tag value may be any value between 0 and 31, regardless of the queue depth supported. If the device does not support command queuing or overlap is disabled, this register will be set to the value 00h.
- REL – will be cleared to zero.
- I/O – will be set to one.
- C/D – will be set to one.

Device register –
- DEV will indicate the selected device.

Status register –
- BSY will be cleared to zero, indicating command completion.
- DRDY will be set to one.
- DF (Device Fault) will be cleared to zero.
- SERV (Service) will be cleared to zero when no other queued command is ready for service. SERV will be set to one when another queued command is ready for service.
- DRQ will be cleared to zero.
- ERR will be cleared to zero.

Error outputs

The Sector Count register contains the Tag for this command if the device supports command queuing. The device will return command aborted if the command is not supported or if the device has not had overlapped interrupt enabled. The device will return command aborted if the device supports command queuing and the Tag is invalid. An unrecoverable error encountered during the execution of this command results in the termination of the command and the Command Block registers contain the sector where the first unrecoverable error occurred. If a queue existed, the unrecoverable error will cause the queue to abort.

Register		7	6	5	4	3	2	1	0
Error		ICRC	WP	MC	IDNF	MCR	ABRT	NM	obs
Sector Count	HOB = 0	Tag					REL	I/O	C/D
	HOB = 1	Reserved							
LBA Low	HOB = 0	LBA (7:0)							
	HOB = 1	LBA (31:24)							
LBA Mid	HOB = 0	LBA (15:8)							
	HOB = 1	LBA (39:32)							
LBA High	HOB = 0	LBA (23:16)							
	HOB = 1	LBA (47:40)							
Device		obs	na	obs	DEV	Reserved			
Status		BSY	DRDY	DF	SERV	DRQ	na	na	ERR
NOTE – HOB = 0 indicates the value read by the host when the HOB bit of the Device Control register is cleared to zero. HOB = 1 indicates the value read by the host when the HOB bit of the Device Control register is set to one.									

Error register –

ICRC will be set to one if an interface CRC error has occurred during an Ultra DMA data transfer. The content of this bit is not applicable for Multiword DMA transfers.

WP will be set to one if the media in a removable media device is write protected.

MC will be set to one if the media in a removable media device changed since the issuance of the last command. The device will clear the device internal media change detected state.

IDNF will be set to one if a user-accessible address could not be found. IDNF will be set to one if an address outside of the range of user-accessible addresses is requested if command aborted is not returned.

MCR will be set to one if a media change request has been detected by a removable media device. This bit is only cleared by a GET MEDIA STATUS or a media access command.

ABRT will be set to one if this command is not supported or if an error, including an ICRC error, has occurred during an Ultra DMA data transfer. ABRT may be set to one if the device is not able to complete the action requested by the command. ABRT will be set to one if an address outside of the range of user-accessible addresses is requested if IDNF is not set to one.

NM will be set to one if no media is present in a removable media device.

Sector Count (when HOB of the Device Control register is cleared to zero) –

Tag – This field contains the command Tag for the command. A Tag value may be any value between 0 and 31, regardless of the queue depth supported. If the device does not support command queuing or overlap is disabled, this register will be set to the value 00h.

REL – will be cleared to zero.

I/O – will be set to one.

C/D – will be set to one.

LBA Low –

LBA (7:0) of the address of the first unrecoverable error when read with Device Control register HOB bit cleared to zero.

LBA (31:24) of the address of the first unrecoverable error when read with Device Control register HOB bit set to one.

LBA Mid –

LBA (15:8) of the address of the first unrecoverable error when read with Device Control register HOB bit cleared to zero.

LBA (39:32) of the address of the first unrecoverable error when read with Device Control register HOB bit set to one.

LBA High –

LBA (23:16) of the address of the first unrecoverable error when read with Device Control register HOB bit cleared to zero.

LBA (47:40) of the address of the first unrecoverable error when read with Device Control register HOB bit set to one.

Device register –

DEV will indicate the selected device.

Status register –

BSY will be cleared to zero, indicating command completion.

DRDY will be set to one.

DF (Device Fault) will be set to one if a device fault has occurred.

DRQ will be cleared to zero.

ERR will be set to one if an Error register bit is set to one.

Prerequisites

DRDY set to one. The host will initialize the DMA channel.

Description

This command executes in a similar manner to a WRITE DMA EXT command. The device may perform a bus release or may execute the data transfer without performing a bus release if the data is ready to transfer.

If the device performs a bus release, the host will reselect the device using the SERVICE command.

Once the data transfer is begun, the device will not perform a bus release until the entire data transfer has been completed.

WRITE MULTIPLE

Command code

C5h

Feature set

General feature set

- Mandatory for devices not implementing the PACKET Command feature set.
- Use prohibited for devices implementing the PACKET Command feature set.

Protocol

PIO data-out

Inputs

The LBA Mid, LBA High, Device, and LBA Low specify the starting sector address to be written. The Sector Count register specifies the number of sectors to be transferred.

Register	7	6	5	4	3	2	1	0
Features	na							
Sector Count	Sector count							
LBA Low	LBA (7:0)							
LBA Mid	LBA (15:8)							
LBA High	LBA (23:16)							
Device	obs	LBA	obs	DEV	LBA (27:24)			
Command	C5h							

Sector Count –
Number of sectors to be transferred. A value of 00h specifies that 256 sectors will be transferred.
Device –
The LBA bit will be set to one to specify the address is an LBA.
DEV will specify the selected device.
Bits (3:0) starting LBA bits (27:24).

Normal outputs

Register	7	6	5	4	3	2	1	0
Error	na							
Sector Count	na							
LBA Low	na							
LBA Mid	na							
LBA High	na							
Device	obs	na	obs	DEV	na	na	na	na
Status	BSY	DRDY	DF	na	DRQ	na	na	ERR

Device register –
DEV will indicate the selected device.
Status register –
BSY will be cleared to zero, indicating command completion.
DRDY will be set to one.
DF (Device Fault) will be cleared to zero.
DRQ will be cleared to zero.
ERR will be cleared to zero.

Error outputs

An unrecoverable error encountered during the execution of this command results in the termination of the command. The Command Block registers contain the address of the sector where the first unrecoverable error occurred. The amount of data transferred is indeterminate.

Register	7	6	5	4	3	2	1	0
Error	na	WP	MC	IDNF	MCR	ABRT	NM	na
Sector Count	na							
LBA Low	LBA (7:0)							
LBA Mid	LBA (15:8)							
LBA High	LBA (23:16)							
Device	obs	na	obs	DEV	LBA (27:24)			
Status	BSY	DRDY	DF	na	DRQ	na	na	ERR

Error register –
WP will be set to one if the media in a removable media device is write protected.
MC will be set to one if the media in a removable media device changed since the issuance of the last command. The device will clear the device internal media change detected state.
IDNF will be set to one if a user-accessible address could not be found. IDNF will be set to one if an address outside of the range of user-accessible addresses is requested if command aborted is not returned.
MCR will be set to one if a media change request has been detected by a removable media device. This bit is only cleared by a GET MEDIA STATUS or a media access command.
ABRT will be set to one if this command is not supported or if an error, including an ICRC error, has occurred during an Ultra DMA data transfer. ABRT may be set to one if the device is not able to complete action requested by the command. ABRT will be set to one if an address outside of the range of user-accessible addresses is requested if IDNF is not set to one.
NM will be set to one if no media is present in a removable media device.
LBA Low, LBA Mid, LBA High, Device –
Will be written with the address of the first unrecoverable error.
DEV will indicate the selected device.
Status register –
BSY will be cleared to zero, indicating command completion.
DRDY will be set to one.
DF (Device Fault) will be set to one if a device fault has occurred.
DRQ will be cleared to zero.
ERR will be set to one if an Error register bit is set to one.

Prerequisites

DRDY set to one. If bit 8 of IDENTIFY DEVICE data word 59 is cleared to zero, a successful SET MULTIPLE MODE command will proceed a WRITE MULTIPLE command.

Description

This command writes the number of sectors specified in the Sector Count register.

The number of sectors per block is defined by the content of word 59 of the IDENTIFY DEVICE response. The device will interrupt for each DRQ block transferred.

When the WRITE MULTIPLE command is issued, the Sector Count register contains the number of sectors (not the number of blocks) requested.

If the number of requested sectors is not evenly divisible by the block count, as many full blocks as possible are transferred, followed by a final, partial block transfer. The partial block transfer is for n sectors, where:

n = Remainder (sector count/block count)

If the WRITE MULTIPLE command is received when WRITE MULTIPLE commands are disabled, the Write Multiple operation will be rejected with command aborted.

Device errors encountered during WRITE MULTIPLE commands are posted after the attempted device write of the block or partial block transferred. The command ends with the sector in error, even if the error was in the middle of a block. Subsequent blocks are not transferred in the event of an error.

The contents of the Command Block Registers following the transfer of a data block that had a sector in error are undefined. The host should retry the transfer as individual requests to obtain valid error information. Interrupt pending is set when the DRQ bit is set to one at the beginning of each block or partial block.

WRITE MULTIPLE EXT

Command code

39h

Feature set

48-bit Address feature set

- Mandatory for devices implementing the 48-bit Address feature set.
- Use prohibited for devices implementing the PACKET Command feature set.

Protocol

PIO data-out

Inputs

Register		7	6	5	4	3	2	1	0
Features	Current	Reserved							
	Previous	Reserved							
Sector Count	Current	Sector count (7:0)							
	Previous	Sector count (15:8)							
LBA Low	Current	LBA (7:0)							
	Previous	LBA (31:24)							
LBA Mid	Current	LBA (15:8)							
	Previous	LBA (39:32)							
LBA High	Current	LBA (23:16)							
	Previous	LBA (47:40)							
Device		obs	LBA	obs	DEV	Reserved			
Command		39h							
NOTE – The value indicated as Current is the value most recently written to the register. The value indicated as Previous is the value that was in the register before the most recent write to the register.									

Sector Count Current –

Number of sectors to be transferred low order, bits (7:0).

Sector Count Previous –

Number of sectors to be transferred high order, bits (15:8). 0000h in the Sector Count register specifies that 65,536 sectors are to be transferred.

Device –

The LBA bit will be set to one to specify the address is an LBA.

DEV will specify the selected device.

Normal outputs

Register		7	6	5	4	3	2	1	0
Error		na							
Sector Count	HOB = 0	Reserved							
	HOB = 1	Reserved							
LBA Low	HOB = 0	Reserved							
	HOB = 1	Reserved							
LBA Mid	HOB = 0	Reserved							
	HOB = 1	Reserved							
LBA High	HOB = 0	Reserved							
	HOB = 1	Reserved							
Device		obs	na	obs	DEV	Reserved			
Status		BSY	DRDY	DF	na	DRQ	na	na	ERR
NOTE – HOB = 0 indicates the value read by the host when the HOB bit of the Device Control register is cleared to zero. HOB = 1 indicates the value read by the host when the HOB bit of the Device Control register is set to one.									

Device register –

DEV will indicate the selected device.

Status register –

BSY will be cleared to zero, indicating command completion.

DRDY will be set to one.

DF (Device Fault) will be cleared to zero.

DRQ will be cleared to zero.

ERR will be cleared to zero.

Error outputs

An unrecoverable error encountered during the execution of this command results in the termination of the command. The Command Block registers contain the address of the sector where the first unrecoverable error occurred. The amount of data transferred is indeterminate.

Register		7	6	5	4	3	2	1	0
Error		na	WP	MC	IDNF	MCR	ABRT	NM	obs
Sector Count	HOB = 0	Reserved							
	HOB = 1	Reserved							
LBA Low	HOB = 0	LBA (7:0)							
	HOB = 1	LBA (31:24)							
LBA Mid	HOB = 0	LBA (15:8)							
	HOB = 1	LBA (39:32)							
LBA High	HOB = 0	LBA (23:16)							
	HOB = 1	LBA (47:40)							
Device		obs	na	obs	DEV	Reserved			
Status		BSY	DRDY	DF	na	DRQ	na	na	ERR
NOTE – HOB = 0 indicates the value read by the host when the HOB bit of the Device Control register is cleared to zero. HOB = 1 indicates the value read by the host when the HOB bit of the Device Control register is set to one.									

Error register –

WP will be set to one if the media in a removable media device is write protected.

MC will be set to one if the media in a removable media device changed since the issuance of the last command. The device will clear the device internal media change detected state.

IDNF will be set to one if a user-accessible address could not be found. IDNF will be set to one if an address outside of the range of user-accessible addresses is requested if command aborted is not returned.

MCR will be set to one if a media change request has been detected by a removable media device. This bit is only cleared by a GET MEDIA STATUS or a media access command.

ABRT will be set to one if this command is not supported. ABRT may be set to one if the device is not able to complete the action requested by the command. ABRT will be set to one if an address outside of the range of user-accessible addresses is requested if IDNF is not set to one.

NM will be set to one if no media is present in a removable media device.

LBA Low –

LBA (7:0) of the address of the first unrecoverable error when read with Device Control register HOB bit cleared to zero.

LBA (31:24) of the address of the first unrecoverable error when read with Device Control register HOB bit set to one.

LBA Mid –

LBA (15:8) of the address of the first unrecoverable error when read with Device Control register HOB bit cleared to zero.

LBA (39:32) of the address of the first unrecoverable error when read with Device Control register HOB bit set to one.

LBA High –

LBA (23:16) of the address of the first unrecoverable error when read with Device Control register HOB bit cleared to zero.

LBA (47:40) of the address of the first unrecoverable error when read with Device Control register HOB bit set to one.

Device register –

DEV will indicate the selected device.

Status register –

BSY will be cleared to zero, indicating command completion.

DRDY will be set to one.

DF (Device Fault) will be set to one if a device fault has occurred.

DRQ will be cleared to zero.

ERR will be set to one if an Error register bit is set to one.

Prerequisites

DRDY set to one. If bit 8 of IDENTIFY DEVICE data word 59 is cleared to zero, a successful SET MULTIPLE MODE command will proceed a WRITE MULTIPLE EXT command.

Description

This command writes the number of sectors specified in the Sector Count register.

The number of sectors per block is defined by the content of word 59 in the IDENTIFY DEVICE response. The device will interrupt for each DRQ block transferred.

When the WRITE MULTIPLE EXT command is issued, the Sector Count register contains the number of sectors (not the number of blocks) requested.

If the number of requested sectors is not evenly divisible by the block count, as many full blocks as possible are transferred, followed by a final, partial block transfer. The partial block transfer is for n sectors, where:

n = Remainder (sector count/block count)

If the WRITE MULTIPLE EXT command is received when WRITE MULTIPLE EXT commands are disabled, the Write Multiple operation will be rejected with command aborted.

Device errors encountered during WRITE MULTIPLE EXT commands are posted after the attempted device write of the block or partial block transferred. The command ends with the sector in error, even if the error was in the middle of a block. Subsequent blocks are not transferred in the event of an error.

The contents of the Command Block Registers following the transfer of a data block that had a sector in error are undefined. The host should retry the transfer as individual requests to obtain valid error information. Interrupt pending is set when the DRQ bit is set to one at the beginning of each block or partial block.

WRITE SECTOR(S)

Command code

30h

Feature set

General feature set
- Mandatory for devices not implementing the PACKET Command feature set.
- Use prohibited for devices implementing the PACKET Command feature set.

Protocol

PIO data-out

Inputs

The LBA Mid, LBA High, Device, and LBA Low specify the starting sector address to be written. The Sector Count register specifies the number of sectors to be transferred.

Register	7	6	5	4	3	2	1	0
Features	na							
Sector Count	Sector count							
LBA Low	LBA (7:0)							
LBA Mid	LBA (15:8)							
LBA High	LBA (23:16)							
Device	obs	LBA	obs	DEV	LBA (27:24)			
Command	30h							

Sector Count –

Number of sectors to be transferred. A value of 00h specifies that 256 sectors are to be transferred.

Device –

The LBA bit will be set to one to specify the address is an LBA.

DEV will specify the selected device.

Bits (3:0) starting LBA bits (27:24).

Normal outputs

Register	7	6	5	4	3	2	1	0
Error	na							
Sector Count	na							
LBA Low	na							
LBA Mid	na							
LBA High	na							
Device	obs	na	obs	DEV	na	na	na	na
Status	BSY	DRDY	DF	na	DRQ	na	na	ERR

Device register –
 DEV will indicate the selected device.
Status register –
 BSY will be cleared to zero, indicating command completion.
 DRDY will be set to one.
 DF (Device Fault) will be cleared to zero.
 DRQ will be cleared to zero.
 ERR will be cleared to zero.

Error outputs

An unrecoverable error encountered during the execution of this command results in the termination of the command. The Command Block registers contain the address of the sector where the first unrecoverable error occurred. The amount of data transferred is indeterminate.

Register	7	6	5	4	3	2	1	0
Error	na	WP	MC	IDNF	MCR	ABRT	NM	na
Sector Count	na							
LBA Low	LBA (7:0)							
LBA Mid	LBA (15:8)							
LBA High	LBA (23:16)							
Device	obs	na	obs	DEV	LBA (27:24)			
Status	BSY	DRDY	DF	na	DRQ	na	na	ERR

Error register –
 WP will be set to one if the media in a removable media device is write protected.
 MC will be set to one if the media in a removable media device changed since the issuance of the last command. The device will clear the device internal media change detected state.
 IDNF will be set to one if a user-accessible address could not be found. IDNF will be set to one if an address outside of the range of user-accessible addresses is requested if command aborted is not returned.
 MCR will be set to one if a media change request has been detected by a removable media device. This bit is only cleared by a GET MEDIA STATUS or a media access command.
 ABRT will be set to one if this command is not supported or if an error, including an ICRC error, has occurred during an Ultra DMA data transfer. ABRT may be set to one if the device is not able to complete the action requested by the command. ABRT will be set to one if an address outside of the range of user-accessible addresses is requested if IDNF is not set to one.
 NM will be set to one if no media is present in a removable media device.

LBA Low, LBA Mid, LBA High, Device –
Will be written with the address of the first unrecoverable error.
DEV will indicate the selected device.
Status register –
BSY will be cleared to zero, indicating command completion.
DRDY will be set to one.
DF (Device Fault) will be set to one if a device fault has occurred.
DRQ will be cleared to zero.
ERR will be set to one if an Error register bit is set to one.

Prerequisites

DRDY set to one.

Description

This command writes from 1 to 256 sectors, as specified in the Sector Count register. A sector count of 0 requests 256 sectors. The device will interrupt for each DRQ block transferred.

WRITE SECTOR(S) EXT

Command code

34h

Feature set

48-bit Address feature set
- Mandatory for devices implementing the 48-bit Address feature set.
- Use prohibited for devices implementing the PACKET Command feature set.

Protocol

PIO data-out

Inputs

Register		7	6	5	4	3	2	1	0
Features	Current	Reserved							
	Previous	Reserved							
Sector Count	Current	Sector count (7:0)							
	Previous	Sector count (15:8)							
LBA Low	Current	LBA (7:0)							
	Previous	LBA (31:24)							
LBA Mid	Current	LBA (15:8)							
	Previous	LBA (39:32)							
LBA High	Current	LBA (23:16)							
	Previous	LBA (47:40)							
Device		obs	LBA	obs	DEV	Reserved			
Command		34h							
NOTE – The value indicated as Current is the value most recently written to the register. The value indicated as Previous is the value that was in the register before the most recent write to the register.									

Sector Count Current –
Number of sectors to be transferred low order, bits (7:0).
Sector Count Previous –
Number of sectors to be transferred high order, bits (15:8).
Device –
The LBA bit will be set to one to specify the address is an LBA.
DEV will specify the selected device.

Normal outputs

Register		7	6	5	4	3	2	1	0
Error		na							
Sector Count	HOB = 0	Reserved							
	HOB = 1	Reserved							
LBA Low	HOB = 0	Reserved							
	HOB = 1	Reserved							
LBA Mid	HOB = 0	Reserved							
	HOB = 1	Reserved							
LBA High	HOB = 0	Reserved							
	HOB = 1	Reserved							
Device		obs	na	obs	DEV	Reserved			
Status		BSY	DRDY	DF	na	DRQ	na	na	ERR
NOTE – HOB = 0 indicates the value read by the host when the HOB bit of the Device Control register is cleared to zero. HOB = 1 indicates the value read by the host when the HOB bit of the Device Control register is set to one.									

Device register –
DEV will indicate the selected device.
Status register –
BSY will be cleared to zero, indicating command completion.
DRDY will be set to one.
DF (Device Fault) will be cleared to zero.
DRQ will be cleared to zero.
ERR will be cleared to zero.

Error outputs

An unrecoverable error encountered during the execution of this command results in the termination of the command. The Command Block registers contain the address of the sector where the first unrecoverable error occurred. The amount of data transferred is indeterminate.

Register		7	6	5	4	3	2	1	0
Error		na	WP	MC	IDNF	MCR	ABRT	NM	obs
Sector Count	HOB = 0	Reserved							
	HOB = 1	Reserved							
LBA Low	HOB = 0	LBA (7:0)							
	HOB = 1	LBA (31:24)							
LBA Mid	HOB = 0	LBA (15:8)							
	HOB = 1	LBA (39:32)							
LBA High	HOB = 0	LBA (23:16)							
	HOB = 1	LBA (47:40)							
Device		obs	na	obs	DEV	Reserved			
Status		BSY	DRDY	DF	na	DRQ	na	na	ERR
NOTE – HOB = 0 indicates the value read by the host when the HOB bit of the Device Control register is cleared to zero. HOB = 1 indicates the value read by the host when the HOB bit of the Device Control register is set to one.									

Error register –

WP will be set to one if the media in a removable media device is write protected.

MC will be set to one if the media in a removable media device changed since the issuance of the last command. The device will clear the device internal media change detected state.

IDNF will be set to one if a user-accessible address could not be found. IDNF will be set to one if an address outside of the range of user-accessible addresses is requested if command aborted is not returned.

MCR will be set to one if a media change request has been detected by a removable media device. This bit is only cleared by a GET MEDIA STATUS or a media access command.

ABRT will be set to one if this command is not supported. ABRT may be set to one if the device is not able to complete the action requested by the command. ABRT will be set to one if an address outside of the range of user-accessible addresses is requested if IDNF is not set to one.

NM will be set to one if no media is present in a removable media device.

LBA Low – LBA (7:0) of the address of the first unrecoverable error when read with Device Control register HOB bit cleared to zero.

LBA (31:24) of the address of the first unrecoverable error when read with Device Control register HOB bit set to one.

LBA Mid – LBA (15:8) of the address of the first unrecoverable error when read with Device Control register HOB bit cleared to zero.

LBA (39:32) of the address of the first unrecoverable error when read with Device Control register HOB bit set to one.

LBA High – LBA (23:16) of the address of the first unrecoverable error when read with Device Control register HOB bit cleared to zero.

LBA (47:40) of the address of the first unrecoverable error when read with Device Control register HOB bit set to one.

Device register –

DEV will indicate the selected device.

Status register –

BSY will be cleared to zero, indicating command completion.

DRDY will be set to one.

DF (Device Fault) will be set to one if a device fault has occurred.

DRQ will be cleared to zero.

ERR will be set to one if an Error register bit is set to one.

Prerequisites

DRDY set to one.

Description

This command writes from 1 to 65,536 sectors, as specified in the Sector Count register. A sector count value of 0000h requests 65,536 sectors. The device will interrupt for each DRQ block transferred.

3.4 Review Questions

1. Which organization is responsible for the maintenance and development of Serial ATA Specifications?

   (a) INCITS T-10
   (b) SATA Trade Association
   (c) SATA-IO
   (d) INCITS T-13

2. Which organization is responsible for the maintenance and development of industry Standards?

   (a) INCITS T-10
   (b) FCIA
   (c) SATA-IO
   (d) INCITS T-13

3. Which Standards provide details about the ATA command sets?

   (a) AST-8
   (b) ACS-8
   (c) AAM-8
   (d) APT-8

4. What ATA/ATAPI-8 Standard defines the serial transport layers?

   (a) AST-8
   (b) ACS-8
   (c) AAM-8
   (d) APT-8

5. What is the operation code of the READ DMA command?

   (a) 2Ch
   (b) 3Ah
   (c) C8h
   (d) 25h

6. What is the difference between a command that ends with EXT and commands that don't?

   (a) EXT is used for external commands
   (b) EXT Commands are only used for queuing
   (c) EXT Commands use 48-bit addressing
   (d) EXT Commands use 24-bit addressing

7. What command subsection determines if data is transferred and how it will be transferred?

(a) Features sets
(b) Command code
(c) Inputs
(d) Protocol

8. What command can be used to determine which link speeds a SATA device supports?

(a) IDENTIFY DEVICE
(b) READ LOGS
(c) READ FEATURES
(d) INQUIRY

Chapter 4

SATA Transport Layer

Objectives

- Protocol characteristics of the SATA Transport layer
- All Frame Information Structures (FIS) format and use cases
- Detailed FIS field definitions
- FIS Status and Error codes
- Enhancement to transport protocol to reduce data processing overhead, including First-party DMA
- SATA II registers for discovery, configuration, queuing, and asynchronous events
- Command processing protocol examples for all protocols (PIO, DMA, ATAPI, etc.)
- Native Command Queuing (NCQ)
  - Protocol changes from original SATA queuing model
  - Queuing commands and Set Device Bits FIS examples
  - Queuing management commands

4.1 SATA Transport Layer

This section will provide all the details of the SATA Transport layer. The Transport layer includes the framing information structures (FISes) and the Transport layer protocol.

The Transport layer need not be cognizant of how frames are transmitted and received. The Transport layer simply constructs FISes for transmission and decomposes received FISes. Host and device Transport layer states differ in that the source of the FIS content is different. The Transport layer maintains no context in terms of ATA commands or previous FIS content.

4.1.1 Transport Layer Services

The SATA Transport Layer is responsible for the management of FISes.

At the request of the Application layer, the Transport layer will

- gather FIS content appropriate for the FIS type;
- format the FIS;
- make frame transmission requests of the Link layer;
- pass FIS contents to the Link layer;
- request Link Layer flow control, if required; and
- receive transmission status from the Link layer and report to the Application layer.

4.1.2 Frame Information Structure (FIS)

A frame is a group of Dwords that convey information between the host and the device, as described previously. Primitives are used to define the boundaries of the frame and may be inserted to control the rate of the information flow. This section will focus on and describe the actual information content of the frame—referred as a payload—and assumes the reader is aware of the Primitives that are needed to support the information content.

- A FIS is a mechanism to transfer information between host and device Application layers
  - Shadow Register Block contents
  - ATA commands
  - Command completion status
  - Data movement setup information
  - Read and write data
  - Self-test activation
- Each FIS type is identified by a Type Code
- Maximum FIS length of 8192 bytes
  - 2048 Dwords

4.1.3 FIS Types

Table 4.1 shows the FIS types defined by the SATA 1.0 Specification. FISes fall into three basic categories:

- Host to Device
- Device to Host
- Bi-directional (i.e., may be sent by either host or device)

4.1.4 Register – Host to Device FIS

The Register – Host to Device FIS is used to transfer the contents of the Shadow Register Block from the host to the device. This is the mechanism for issuing legacy ATA commands to the device.

- Transmission of a Register – Host to Device FIS is initiated by a write operation to either the Command register or to the Control register with a value different from what is currently in the Control register in the host adapter's Shadow Register Block.

Table 4.1 FIS Type Codes and Descriptions

Type field value	Description	Direction
27h	Register FIS – Host to Device	H>D
34h	Register FIS – Device to Host	D>H
39h	DMA Activate FIS – Device to Host	D>H
41h	DMA Setup FIS – Bi-directional	H< >D
46h	Data FIS – Bi-directional	H< >D
58h	BIST Activate FIS – Bi-directional	H< >D
5Fh	PIO Setup FIS – Device to Host	D>H
A1h	Set Device Bits FIS – Device to Host	D>H
A6h	Reserved for future Serial ATA definition	
B8h	Reserved for future Serial ATA definition	
BFh	Reserved for future Serial ATA definition	
C7h	Vendor specific	
D4h	Vendor specific	
D9h	Reserved for future Serial ATA definition	

- Upon initiating transmission, the current contents of the Shadow Register Block are transmitted and the C bit in the FIS is set according to whether the transmission was a result of the Command register or the Control register being written.
- The host adapter shall set the BSY bit in the shadow Status register to one within 400 ns of the write operation to the Command register that initiated the transmission.
- The host adapter shall set the BSY bit in the shadow Status register to one within 400 ns of the write operation to the Control register if the write to the Control register changes the state of the SRST bit from 0 to 1, but it shall not set the BSY bit in the shadow Status register for writes to the Control register that do not change the state of the SRST bit from 0 to 1.

It is important to note that Serial ATA host adapters enforce the same access control to the Shadow Register Block as legacy (parallel) ATA devices enforce to the Command Block Registers.

- Specifically, the host is prohibited from writing the Features, Sector Count, Sector Number, Cylinder Low, Cylinder High, or Device/Head registers when either BSY or DRQ is set in the Status Register.
- Any write to the Command register when BSY or DRQ is set is ignored unless the write is to issue a Device Reset command.

Upon reception of a valid Register – Host to Device FIS, the device updates its local copy of the Command and Control Block Register contents and either initiates execution of the command indicated in the Command register or initiates execution of the control request indicated in the Control register, depending on the state of the C bit in the FIS.

Host to Device FIS Example

	byte n+3	byte n+2	byte n+1	byte n
SOF	D23.1	D23.1	D21.5	K28.3
Data dword 0	Features 7:0	Command	c | r | r | r | PM Port	FIS Type (**27h**)
Data dword 1	Device [Dev/Head]	LBA 23:16 [Cyl High]	LBA 15:8 [Cyl Low]	LBA 7:0 [Sector Number]
Data dword 2	Features 15:8	LBA 47:40 [Cyl High (exp)]	LBA 39:32 [Cyl Low (exp)]	LBA 31:24 [Sector # (exp)]
Data dword 3	Control	***ICC***	Count 15:8 [Sector Cnt (exp)]	Count 7:0 [Sector Count]
Data dword 4	Reserved	Reserved	Reserved	Reserved
CRC	Dxx.y	Dxx.y	Dxx.y	Dxx.y
EOF	D21.6	D21.6	D21.5	K28.3

Note: c = Command bit; r = reserved; ICC = Isochronous Command Completion.

ICC Isochronous Command Completion (ICC) contains a value that is set by the host to inform device of a time limit. If a command does not define the use of this field, it shall be reserved. This capability is new in ***SATA-3***. (***All SATA-3 changes will be set in bold/italics.***)

[text] Represents legacy SATA-1 or SATA-2 byte definitions.

PM Port: A four bit field that determines which port multiplier port is being addressed

c This bit is set to one when the register transfer is due to an update of the Command register. The bit is cleared to zero when the register transfer is due to an update of the Device Control register. Setting C bit to one and SRST bit to one in the Device Control Field is invalid and results in indeterminate behaviour.

Command: Contains the contents of the Command register (code) of the Shadow Register Block.

Device: Contains the contents of the Device register of the Shadow Register Block.

4.1.5 Register – Device to Host FIS

The Register – Device to Host FIS is used by the device to update the contents of the host adapter's Shadow Register Block. This is the mechanism by which devices indicate command completion status or otherwise change the contents of the host adapter's Shadow Register Block.

Transmission of a Register – Device to Host FIS is initiated by the device in order to update the contents of the host adapter's Shadow Register Block. Transmission of the Register – Device to Host FIS is typically as a result of command completion by the device.

- Upon reception of a valid Register – Device to Host FIS, the received register contents are transferred to the host adapter's Shadow Register Block. Transmission of a Register – Device to Host FIS is initiated by the device in order to update the contents of the host adapter's Shadow Register Block.

- Transmission of the Register – Device to Host FIS is typically as a result of command completion by the device.
- If the BSY bit and DRQ bit in the shadow Status Register are both cleared when a Register – Device to Host FIS is received by the host adapter, then the host adapter shall discard the contents of the received FIS and not update the contents of any shadow register.

Device to Host FIS Example

	byte n+3	byte n+2	byte n+1	byte n
SOF	D23.1	D23.1	D21.5	K28.3
Data dword 0	Error	Status	r | **i** | r | r | PM Port	FIS Type (**34h**)
Data dword 1	Device [Dev/Head]	LBA 23:16 [Cyl High]	LBA 15:8 [Cyl Low]	LBA 7:0 [Sector Number]
Data dword 2	Reserved	LBA 47:40 [Cyl High (exp)]	LBA 39:32 [Cyl Low (exp)]	LBA 31:24 [Sector # (exp)]
Data dword 3	Reserved	Reserved	Count 15:8 [Sector Cnt (exp)]	Count 7:0 [Sector Count]
Data dword 4	Reserved	Reserved	Reserved	Reserved
CRC	Dxx.y	Dxx.y	Dxx.y	Dxx.y
EOF	D21.6	D21.6	D21.5	K28.3

i Interrupt: This bit reflects the interrupt bit line of the device. Devices shall not modify the behavior of this bit based on the state of the nIEN bit received in Register Host to Device FISes.

Status Field

Bit	Description
7	Device Busy (BSY) The Busy bit is transport dependent. In legacy devices, this bit informed the host that it was busy.
6	Device Ready (DRDY) The Data Ready bit is transport dependent. (a) Refer to the appropriate transport standard for the usage of the Data Ready bit. (b) In ATA/ATAPI-7, this bit was documented as the DRDY bit. (c) It was used by a device when it was ready to send read data.
5	Device Fault/Stream Error (SE or DF) The Stream Error bit shall be set to one if an error occurred during the processing of a command in the Streaming feature set, and either the Read Continuous (RC) bit is set to one in a READ STREAM command or the Write Continuous (WC) bit is set to one in a WRITE STREAM command. If the device enters a condition in which continued operation may affect user data integrity (e.g., failure to spin-up without error, or no spares remaining for reallocation), then the device shall set the Device Fault bit to one and no longer accept commands. This condition is only cleared by power cycling the device.
4	Deferred Write Error (DWE) (a) The Deferred Write Error bit shall be set to one if an error was detected in a deferred write to the media for a previous WRITE STREAM DMA EXT command or WRITE STREAM EXT command. (b) If the Deferred Write Error bit is set to one, the location of the deferred error is only reported in the Write Stream Error Log. (c) In ATA/ATAPI-7, this bit was documented as the DWE bit.
3	Data Request (DRQ) The Data Request bit is transport dependent. (d) Refer to the appropriate transport standard for the usage of the Data Request bit. (e) In ATA/ATAPI-7, this bit was documented as the DRQ bit. (f) It was used by a device when it was ready to receive write data.
2	Alignment Error or obsolete (AE) The Alignment Error bit shall be set to one if: (a) IDENTIFY DEVICE data word 106 bit 13 is set to one; (c) IDENTIFY DEVICE data word 69 bit 13 is set to one; (c) IDENTIFY DEVICE data word 49 bits (1:0) are 01b or 10b; and (d) the device successfully processes a write command where: (1) the first byte of data transfer does not begin at the first byte of a physical sector (see IDENTIFY DEVICE data word 209 bits (13:0)); or (2) the last byte of data transfer does not end at the last byte of a physical sector (see IDENTIFY DEVICE data word 209 bits (13:0)).
1	Sense Data Available or obsolete The Sense Data Available bit shall be set to one if: (a) IDENTIFY DEVICE data word 119 bit 6 is set to one; (b) the device has successfully processed enable sense data reporting with sense data available reporting set to one; and (c) the device has sense data to report after processing any command.
0	Check Condition or Error (legacy) The Check Condition bit shall be set to one if an Error sense key is greater than zero or any Error bit is set to one.

Error Field

Bit	Description
7:4	Sense Key from SCSI Standards
7	Interface CRC (ICRC) shall be set to one if an interface CRC error has occurred during an Ultra DMA data transfer.
6	Uncorrectable Error (UE) shall be set to one if data is uncorrectable.
5	Obsolete
4	ID Not Found (IDNF) shall be set to one if an address outside of the range of user-accessible addresses is requested if command aborted is not returned
3	Obsolete
2	Abort (ABRT) shall be set to one if the device aborted the command. The Abort bit shall be cleared to zero if the device did not abort the command.
1	End of Media
0	Illegal Length Indicator or Command Completion Time Out or Media Error or Attempted Partial Range Removal or Insufficient NV Cache space or Insufficient LBA Range Entries Remaining
If the Error field is set to 7Fh and the Sense Data Reporting feature set is enabled with Error field reporting, then see Sense Data Reporting feature set for a description of Error field reporting.	

Register D⇨H and H⇨D FIS Block Addressing

The format of the Register – Host to Device and Register – Device to Host FISes accommodates the registers required for the 48 bit LBA addressing scheme. The command code indicates when the expanded registers are valid (48-bit command have special command codes).

- Register FISes are formatted to accommodate 48-bit addressing.
- The "base" and "expanded" FIS fields represent the 2-byte deep FIFO used by PATA 48-bit addressing mode.
- The use of 48-bit addressing is not mandatory for SATA.
  - 28-bit addressing is often used.
  - 48-bit addressing is required to access beyond 128GB.
- If 48-bit addressing is not used, the "expanded" register fields are set to zero.
  - But the expanded fields are always carried by the FIS

48-Bit Addressing for All FIS Examples

	byte n+3	byte n+2	byte n+1	byte n
Data dword 1	Device [Dev/Head]	LBA 23:16 [Cyl High]	LBA 15:8 [Cyl Low]	LBA 7:0 [Sector Number]
Data dword 2	FIS type dependent	LBA 47:40 [Cyl High (exp)]	LBA 39:32 [Cyl Low (exp)]	LBA 31:24 [Sector # (exp)]
Data dword 3	FIS type dependent	FIS type dependent	Count 15:8 [Sector Cnt (exp)]	Count 7:0 [Sector Count]

4.1.6 Set Device Bits FIS

The Set Device Bits – Device to Host FIS is used by the device to load Shadow Register Block bits for which the device has exclusive write access.

- These bits are the eight bits of the Error register and six of the eight bits of the Status register.
- This FIS does not alter bit 7, BSY, or bit 3, DRQ, of the Status register.

The FIS includes a bit to signal the host adapter to generate an interrupt if the BSY bit and the DRQ bit in the shadow Status Register are both cleared to zero when this FIS is received.

- Some Serial ATA to parallel ATA Bridge solutions may elect to not support this FIS based on the requirements of their target markets.
- Upon reception, such devices will process this FIS as if it were an invalid FIS type and return the R_ERR end of frame handshake.

The device transmits a Set Device Bits – Device to Host to alter one or more bits in the Error register or in the Status register in the Shadow Register Block.

- This FIS should be used by the device to set the SERV bit in the Status register to request service for a bus released command.
- When used for this purpose, the device shall set the Interrupt bit to one.

Upon receiving a Set Device Bits – Device to Host:

- The host adapter shall load the data from the Error field into the shadow Error register, the data from the Status-Hi field into bits 6, 5, and 4 of the shadow Status register, and the data from the Status-Lo field into bits 2, 1, and 0 of the shadow Status register.
- Bit 7, BSY, and bit 3, DRQ, of the shadow Status register shall not be changed.
- If the "i" bit in the FIS is set to one, and if both the BSY bit and the DRQ bit in the Shadow status register are cleared to zero when this FIS is received, then the host adapter shall enter an interrupt-pending state.

Set Device Bits FIS Example

<table>
<tr><th></th><th>byte n+3</th><th colspan="4">byte n+2</th><th colspan="5">byte n+1</th><th>byte n</th></tr>
<tr><td>SOF</td><td>D23.1</td><td colspan="4">D23.1</td><td colspan="5">D21.5</td><td>K28.3</td></tr>
<tr><td>Data dword 0</td><td>Error</td><td>r</td><td>Status high</td><td>r</td><td>Status low</td><td>n</td><td>i</td><td>r</td><td>r</td><td>PM Port</td><td>FIS Type (A1h)</td></tr>
<tr><td>Data dword 1</td><td colspan="11">SActive (NCQ) or Notify (SNodification) or Reserved (SATA-1)</td></tr>
<tr><td>CRC</td><td>Dxx.y</td><td colspan="4">Dxx.y</td><td colspan="5">Dxx.y</td><td>Dxx.y</td></tr>
<tr><td>EOF</td><td>D21.6</td><td colspan="4">D21.6</td><td colspan="5">D21.5</td><td>K28.3</td></tr>
</table>

Note: r = reserved; n = Notification; i = Interrupt.

4.1.7 First-Party DMA Setup FIS

The First-party DMA Setup – Device to Host or Host to Device FIS is the mechanism by which First-party DMA access to host memory is initiated.

- This FIS is used to request the host or device to program its DMA controller before transferring data. The FIS allows the actual host memory regions to be abstracted (depending on implementation) by having memory regions referenced via a base memory descriptor representing a memory region that the host has granted the device access to.
- The specific implementation for the memory descriptor abstraction is not defined.
- Serial ATA 1.0 devices do not implement the First-party DMA architecture.
- This FIS is modified for "Native SATA Command Queuing" and Auto Activate features (described shortly).

First-Party DMA Setup FIS Example

<table>
<tr><th></th><th>byte n+3</th><th>byte n+2</th><th colspan="5">byte n+1</th><th>byte n</th></tr>
<tr><td>SOF</td><td>D23.1</td><td>D23.1</td><td colspan="5">D21.5</td><td>K28.3</td></tr>
<tr><td>Data dword 0</td><td>Reserved</td><td>Reserved</td><td>a</td><td>i</td><td>d</td><td>r</td><td>PM Port</td><td>FIS Type (41h)</td></tr>
<tr><td>Data dword 1</td><td colspan="8">DMA Buffer ID Low</td></tr>
<tr><td>Data dword 2</td><td colspan="8">DMA Buffer ID High</td></tr>
<tr><td>Data dword 3</td><td colspan="8">Reserved</td></tr>
<tr><td>Data dword 4</td><td colspan="8">DMA Buffer Offset</td></tr>
<tr><td>Data dword 5</td><td colspan="8">DMA Transfer Count</td></tr>
<tr><td>Data dword 6</td><td colspan="8">Reserved</td></tr>
<tr><td>CRC</td><td>Dxx.y</td><td>Dxx.y</td><td colspan="5">Dxx.y</td><td>Dxx.y</td></tr>
<tr><td>EOF</td><td>D21.6</td><td>D21.6</td><td colspan="5">D21.5</td><td>K28.3</td></tr>
</table>

4.1.8 DMA Activate FIS

The DMA Activate – Device to Host FIS is used by the device to signal the host to proceed with a DMA transfer of data from the host to the device (Write Operation).

- This is the mechanism by which a legacy device signals its readiness to receive DMA data from the host.

A situation may arise where the host needs to send multiple Data FISes in order to complete the overall data transfer request. The host shall wait for a successful reception of a DMA Activate FIS before sending each of the Data FISes that are needed.

The device transmits a DMA Activate – Device to Host to the host in order to initiate the flow of DMA data from the host to the device as part of the data transfer portion of a corresponding DMA write command.

- When transmitting this FIS, the device shall be prepared to subsequently receive a Data – Host to Device FIS from the host with the DMA data for the corresponding command.

Upon receiving a DMA Activate – Device to Host

- If the host adapter's DMA controller has been programmed and armed, the host adapter shall initiate the transmission of a Data FIS and shall transmit in this FIS the data corresponding to the host memory regions indicated by the DMA controller's context.
- If the host adapter's DMA controller has not yet been programmed and armed, the host adapter shall set an internal state indicating that the DMA controller has been activated by the device, and as soon as the DMA controller has been programmed and armed, a Data FIS shall be transmitted to the device with the data corresponding to the host memory regions indicated by the DMA controller context.

DMA Activate FIS Example

	byte n+3	byte n+2	byte n+1	byte n
SOF	D23.1	D23.1	D21.5	K28.3
Data dword 0	Reserved	Reserved	r | r | r | r | PM Port	FIS Type (**39h**)
CRC	Dxx.y	Dxx.y	Dxx.y	Dxx.y
EOF	D21.6	D21.6	D21.5	K28.3

4.1.9 SATA II Changes to DMA Transport

This section briefly describes the changes outlined made to the second generation of Serial ATA Specifications.

Auto-Activate in DMA Setup FIS

In the Serial ATA 1.0a Specification, First-party DMA transfers from the host to the device require transmission of both the DMA Setup FIS and a subsequent DMA Activate FIS in order to trigger the host transfer of data to the device (i.e., Write operations). Because the device can elect to submit the DMA Setup FIS only when it is already prepared to receive the subsequent Data FIS from the host, the extra transaction for the DMA Activate FIS can be eliminated by merely having the DMA Setup FIS automatically activate the DMA controller.

In the Serial ATA 1.0a Specification, the high-order bit of byte 1 (the second byte) of the DMA Setup FIS is reserved and cleared to zero. In order to eliminate the need for the DMA Activate FIS immediately following a DMA Setup FIS for a host to device transfer, this bit is defined as the Auto-Activate bit. The modified definition of the DMA Setup FIS is illustrated in the SATA-2 FIS Changes tables provided in this section.

DMA Auto-Activate Bit Summary

SATA II provides extension to "DMA setup" FIS.

- This eliminates the requirement to use the "DMA activate" FIS.
  - SATA originally required the use of the DMA Setup FIS, followed by the DMA Activate FIS.
- The "Auto-Activate" field is introduced into the DMA Setup FIS.
  - This field was "reserved" in prior versions of the SATA Specifications.
- This results in overhead reduction for DMA transfers due to one less FIS transaction required during data transfers.

SATA-2 DMA Setup FIS Changes for Auto-Activate

	byte n+3	byte n+2	byte n+1	byte n
SOF	D23.1	D23.1	D21.5	K28.3
Data dword 0	Reserved	Reserved	a | **i** | d | r | PM Port	FIS Type (**41h**)
Data dword 1	DMA Buffer ID Low			
Data dword 2	DMA Buffer ID High			
Data dword 3	Reserved			
Data dword 4	DMA Buffer Offset			
Data dword 5	DMA Transfer Count			
Data dword 6	Reserved			
CRC	Dxx.y	Dxx.y	Dxx.y	Dxx.y
EOF	D21.6	D21.6	D21.5	K28.3

a – For a DMA Setup with transfer, direction from host to device (write) indicates whether the host should immediately proceed with the data transfer without awaiting a subsequent DMA Activate to start the transfer. For a DMA Setup with transfer direction from device to host (read), this bit shall be zero.

Changes to DMA Setup FIS for Queuing

- When Read and Write commands are queued, memory regions or buffers must be allocated to receive the data when it is subsequently transferred
  - Buffers in Host for Read operation
  - Buffers in Device for Write operation
- Each buffer requires an identifier so that the DMA engine can direct data to the correct buffer
  - Command Tag is used as a Buffer ID
- Command Tag is transmitted in the DMA setup FIS
  - DMA setup must be transmitted for each queued command prior to data transfer
    - Multiple DMA Setup FISes may be required for long data transfers

DMA Setup FIS Changes for Queuing

	byte n+3	byte n+2	byte n+1	byte n
SOF	D23.1	D23.1	D21.5	K28.3
Data dword 0	Reserved	Reserved	a / i / d / r / PM Port	FIS Type (**41h**)
Data dword 1	DMA Buffer ID Low			TAG
Data dword 2	DMA Buffer ID High			
Data dword 3	Reserved			
Data dword 4	DMA Buffer Offset			
Data dword 5	DMA Transfer Count			
Data dword 6	Reserved			
CRC	Dxx.y	Dxx.y	Dxx.y	Dxx.y
EOF	D21.6	D21.6	D21.5	K28.3

FIS Type — Set to a value of 41h. Defines the rest of the FIS fields. Defines the total length of the FIS as seven Dwords.

d — Indicates whether subsequent data transferred after this FIS is from transmitter to receiver or from receiver to transmitter. Since the DMA Setup FIS is only issued by the device for the queuing model defined here, the value in the field is defined as:

1 = device to host transfer (write to host memory)
0 = host to device transfer (read from host memory)

a — For a DMA Setup with transfer, direction from host to device indicates whether the host should immediately proceed with the data transfer without awaiting a subsequent DMA Activate to start the transfer. For DMA Setup with transfer direction from device to host, this bit shall be zero.

TAG — This field is used to identify the DMA buffer region in host memory to select for the data transfer. The low order 5 bits of the DMA Buffer Identifier Low field shall be set to the TAG value corresponding to the command TAG for which data is being transferred. The remaining bits of the DMA Buffer Identifier Low/High shall be cleared to zero. The 64-bit Buffer Identifier field is used to convey a TAG value that occupies the five least-significant bits of the field.

DMA Buffer Offset — This is the byte offset into the buffer for the transfer. Bits <1:0> shall be zero. The device may specify a nonzero value in this field only if the host indicates support for it through the Set Features mechanism. Data is transferred to/from sequentially increasing logical addresses starting at the specified offset in the specified buffer.

DMA Transfer Count	This is the number of bytes that will be transferred. Bit zero must be zero and the value must accurately reflect the length of the data transfer to follow. Devices shall not set this field to "0"; a value of "0" for this field is illegal and will result in indeterminate behavior.
i Interrupt	The queuing model defined here does not make use of an interrupt following the data transfer phase (after the transfer count is exhausted). The I bit shall be cleared to zero.

4.1.10 Data FIS

The Data – Host to Device and the Data – Device to Host FISes are used for transporting payload data, such as the data read from or written to a number of sectors on a hard drive.

- The FIS may either be generated by the device to transmit data to the host (Read Operations) or may be generated by the host to transmit data to the device (Write Operations).
- This FIS is generally only one element of a sequence of transactions leading up to a data transmission, and the transactions leading up to and following the Data FIS establish the proper context for both the host and device.

The byte count of the payload is not an explicit parameter; rather it is inferred by counting the number of Dwords between the SOF and EOF primitives, and discounting the FIS type and CRC Dwords.

- The payload size shall be no more than 2048 Dwords (8192 bytes) for native implementations. Parallel ATA to Serial ATA bridge implementations should not transmit more than 2048 Dwords (8192 bytes) in any one Data FIS.
  - However, separate bridge hardware is not required to break up PIO or DMA transfers from parallel ATA into more than one Serial ATA Data FIS and may therefore produce payloads longer than 2048 Dwords.
  - Receivers shall tolerate reception of data payloads longer than 2048 Dwords without error.
- Non-packet devices, with or without bridges, should report a SET MULTIPLE limit of 16 sectors or less in word 47 of their IDENTIFY DEVICE information.
- In the case that the transfer length represents an odd number of words, the last word shall be placed in the low order word (word 0) of the final Dword, and the high order word (word 1) of the final Dword shall be padded with zeros before transmission.

The device transmits a Data – Device to Host FIS to the host during the data transfer phase of legacy mode PIO reads, DMA reads, and First-party DMA writes to host memory.

- The device shall precede a Data FIS with any necessary context-setting transactions as appropriate for the particular command sequence.
- For example, a First-party DMA host memory write must be preceded by a First-party DMA Setup – Device to Host FIS to establish a proper context for the Data FIS that follows.

Data FIS Example

	byte n+3	byte n+2	byte n+1	byte n
SOF	D23.1	D23.1	D21.5	K28.3
Data dword 0	Reserved			FIS Type (**46h**)
Data dword 1	Data dword			
Data dword 2	(minimum 1 dword – maximum 2048 dwords)			
Data dword 3	Data dword			
Data dword 4	Data dword			
Data dword -	Data dword			
Data dword n	Data dword			
CRC	Dxx.y	Dxx.y	Dxx.y	Dxx.y
EOF	D21.6	D21.6	D21.5	K28.3

- The host transmits a Data – Host to Device FIS to the device during the data transfer phase of PIO writes, DMA writes, and First-party DMA reads of host memory. The FIS shall be preceded by any necessary context-setting transactions as appropriate for the particular command sequence. For example, a legacy mode DMA write to the device is preceded by a DMA Activate – Device to Host FIS with the DMA context having been pre-established by the host.
- When used for transferring data for DMA operations, multiple Data – Host to Device or Device to Host FISes can follow in either direction. Segmentation can occur when the transfer count exceeds the maximum Data – Host to Device or Device to Host transfer length or if a data transfer is interrupted.
- When used for transferring data in response to a PIO Setup, all of the data must be transmitted in a single Data FIS.
- In the event that a transfer is broken into multiple FISes, all intermediate FISes must contain an integral number of full Dwords. If the total data transfer is for an odd number of words, then the high order word (word 1) of the last Dword of the last FIS shall be padded with zeros before transmission and discarded on reception.
- The Serial ATA protocol does not permit for the transfer of an odd number of bytes.
- Neither the host nor the device is expected to buffer an entire Data FIS in order to check the CRC of the FIS before processing the data. Incorrect data reception for a Data FIS should be reflected in the overall command completion status.

4.1.11 BIST Activate FIS

The BIST Activate FIS is used to place the receiver in one of "n" loopback modes.

The BIST Activate FIS is a bi-directional request in that it can be sent by either the host or the device. The sender and receiver have distinct responsibilities in order to ensure proper cooperation between the two parties. The state machines for transmission and reception of the FIS are symmetrical.

The state machines for the transmission of the FIS do not attempt to specify the actions the sender takes once successful transmission of the request has been performed. After the Application layer is notified of the successful transmission of the FIS, the sender's Application layer will prepare its own Application, Transport, and Physical layers into the appropriate states that support the transmission of a stream of data. The FIS shall not be considered successfully transmitted until the receiver has acknowledged reception of the FIS as per normal FIS transfers documented in various sections of this specification. The transmitter of the BIST Activate FIS should transmit continuous SYNC primitives after reception of the R_OK primitive until such a time that it is ready to interact with the receiver in the BIST exchange.

Similarly, the state machines for the reception of the FIS do not specify the actions of the receiver's Application layer. Once the FIS has been received, the receiver's Application layer must place its own Application, Transport, and Physical layers into states that will perform the appropriate retransmission of the sender's data. The receiver shall not enter the BIST state until after it has properly received a good BIST Activate FIS (good CRC), indicated a successful transfer of the FIS to the transmitting side via the R_OK primitive, and has received at least one good SYNC primitive. Once in the self-test mode, a receiver shall continue to allow processing of the COMINIT or COMRESET signals in order to exit from the self-test mode.

BIST Activate FIS Example

<table>
<tr><th></th><th>byte n+3</th><th colspan="8">byte n+2</th><th colspan="2">byte n+1</th><th>byte n</th></tr>
<tr><td>SOF</td><td>D23.1</td><td colspan="8">D23.1</td><td colspan="2">D21.5</td><td>K28.3</td></tr>
<tr><td>Data dword 0</td><td>Reserved</td><td>T</td><td>A</td><td>S</td><td>L</td><td>F</td><td>P</td><td>R</td><td>V</td><td>Reserved</td><td>PM Port</td><td>FIS Type (58h)</td></tr>
<tr><td>Data dword 1</td><td>Data 31:24</td><td colspan="8">Data 23:16</td><td colspan="2">Data 15:8</td><td>Data 7:0</td></tr>
<tr><td>Data dword 2</td><td>Data 31:24</td><td colspan="8">Data 23:16</td><td colspan="2">Data 15:8</td><td>Data 7:0</td></tr>
<tr><td>CRC</td><td>Dxx.y</td><td colspan="8">Dxx.y</td><td colspan="2">Dxx.y</td><td>Dxx.y</td></tr>
<tr><td>EOF</td><td>D21.6</td><td colspan="8">D21.6</td><td colspan="2">D21.5</td><td>K28.3</td></tr>
</table>

T: The Far-End Transmit Mode
A: ALIGNP sequence bypass mode
S: The Bypass Scrambling Mode
L: The Far End Retimed Loopback Mode
F: The Far End Analog (Analog Front End—AFE)
P: The transmit primitives bit
R: Reserved
V: The vendor-unique mode

BIST Activate FIS Modes and Bit Settings

Note: The BIST mode is intended for Inspection/Observation Testing, as well as support for conventional laboratory equipment, rather than for in-system automated testing.

The setting of the F, L, and T bits is mutually exclusive. It is the responsibility of the sender of the BIST Activate FIS to ensure that only one of these bits is set. Refer to the table below for valid bit settings within a BIST Activate FIS.

BIST Test Mode	F	L	T	P	A	S	V
Far End Analog Loopback	1	0	0	0	0	0	0
Far End Retimed Loopback	0	1	0	0	0	0	0
Far End Transmit with ALIGNs, scrambled data	0	0	1	0	0	0	0
Far End Transmit with ALIGNs, unscrambled data	0	0	1	0	0	1	0
Far End Transmit without ALIGNs, scrambled data	0	0	1	0	1	0	0
Far End Transmit without ALIGNs, unscrambled data	0	0	1	0	1	1	0
Far End Transmit primitives with ALIGNs	0	0	1	1	0	na	0
Far End Transmit primitives without ALIGNs	0	0	1	1	1	na	0
Vendor Specific	na	na	na	na	na	na	1
Key: 0 – bit shall be cleared to zero 1 – bit shall be set to one							

4.1.12 PIO Setup FIS

The PIO Setup – Device to Host FIS is used by the device to provide the host adapter with sufficient information regarding a PIO data phase to allow the host adapter to efficiently handle PIO data transfers. For PIO data transfers, the device shall send to the host a PIO Setup – Device to Host FIS just before each and every data transfer FIS that is required to complete the data transfer. Data transfers from Host to Device as well as data transfers from Device to Host shall follow this algorithm. Because of the stringent timing constraints in the ATA standard, the PIO Setup FIS includes both the starting and ending status values. These are used by the host adapter to first signal to host software regarding readiness for PIO write data (BSY deasserted and DRQ asserted), and, following the PIO write burst, to properly signal the host software by deasserting DRQ and possibly raising BSY.

PIO Setup FIS Example

	byte n+3	byte n+2	byte n+1	byte n
SOF	D23.1	D23.1	D21.5	K28.3
Data dword 0	Error	Status	r | i | d | r | PM Port	FIS Type (**5Fh**)
Data dword 1	Device [Dev/Head]	LBA 23:16 [Cyl High]	LBA 15:8 [Cyl Low]	LBA 7:0 [Sector Number]
Data dword 2	Reserved	LBA 47:40 [Cyl High (exp)]	LBA 39:32 [Cyl Low (exp)]	LBA 31:24 [Sector # (exp)]
Data dword 3	E_Status	Reserved	Count 15:8 [Sector Cnt (exp)]	Count 7:0 [Sector Count]
Data dword 4	Reserved	Reserved	Transfer Count (bytes)	
CRC	Dxx.y	Dxx.y	Dxx.y	Dxx.y
EOF	D21.6	D21.6	D21.5	K28.3

d – Direction 0 = Out (Write) 1 = In (Read)
E_Status – Contains the new value of the Status register of the Command Block at the conclusion of the subsequent Data FIS.
Transfer Count – Holds the number of bytes to be transferred in the subsequent Data FIS. The Transfer Count value shall be nonzero and the low order bit shall be zero (even number of bytes transferred).

4.1.13 Changes to FISes for Port Multipliers

All FISes will be modified, as seen below, to accommodate the addressing of Port Multipliers. The PM Port field is in the first Dword of all FIS types. The PM Port field corresponds to reserved bits common to all FIS types in prior SATA Specification. The first Dword of all FIS types is shown below.

Standard FIS without PM Field

	31	30	29	28	27	26	25	24	23	22	21	20	19	18	17	16	15	14	13	12	11	10	9	8	7	6	5	4	3	2	1	0
Dword0	See specification								See specification								x	x	x	Reserved (0)					FIS Type (xxh)							

FIS with PM Field

	31	30	29	28	27	26	25	24	23	22	21	20	19	18	17	16	15	14	13	12	11	10	9	8	7	6	5	4	3	2	1	0
Dword0	See specification								See specification								x	x	x	x	PM Port				FIS Type (xxh)							

FIS Type
- Defines FIS Type–specific fields and FIS length

PM Port
- Specifies the port address that the FIS should be delivered to or is received from
- Ports 0 to 14 are device ports
  0000 = Port 0
  0001 = Port 1
  0010 = Port 2
  0011 = Port 3
  0100 = Port 4
  0101 = Port 5
  0110 = Port 6
  0111 = Port 7
  1000 = Port 8
  1001 = Port 9
  1010 = Port A (10)
  1011 = Port B (11)
  1100 = Port C (12)
  1101 = Port D (13)
  1110 = Port E (14)
  1111 = Port F (15) Control Port

4.2 Serial Interface Host Adapter Registers Overview

Serial implementations of ATA host adapters include an additional block of registers.

- These registers are mapped separately and independently from the ATA Command Block Registers for reporting additional status and error information and to allow control of capabilities unique to the serial implementation of ATA.
- These additional registers are referred to as the SStatus and SControl Registers (SCRs).
  - They consist of and are organized as 16 contiguous 32-bit registers.
  - NCQ adds an additional SCR to those defined in the standard.
  - There is an additional SCR that will be added to address Notification Events.

4.2.1 SCR Mapping and Organization

The base address and mapping scheme for these registers is defined by the specific host adapter implementation—for example, PCI controller implementations may map the SCRs using the PCI mapping capabilities.

- The table below illustrates the overall organization of the SStatus and SControl registers.
- Parallel implementation ATA software does not make use of the serial interface SStatus and SControl registers.
- The SStatus and SControl register are associated with the serial interface and are independent of any Device 0/Device 1 emulation the host adapter may implement.

SStatus and SControl Registers

SATA register	0	SATA Status/Control
SATA register	1	SATA Status/Control
...	...	SATA Status/Control
SATA register	14	SATA Status/Control
SATA register	15	SATA Status/Control

4.2.2 SStatus, SError, and SControl Registers

The serial implementation of ATA provides an additional block of registers to control the interface and to retrieve interface state information.

- There are 16 contiguous registers allocated, of which the first three are defined and the remaining 13 are reserved for future definition.
- The table below defines the serial implementation of SATA Status and Control registers.

SCR[0]	SStatus Register
SCR[1]	SError Register
SCR[2]	SControl Register
SCR[3]	SActive Register (used during Native Command Queuing)
SCR[4]	SNotification Register (SATA II)
....	....
SCR[15]	Reserved

SStatus Register

The SStatus register is a 32-bit read-only register that conveys the current state of the interface and host adapter.

- The register conveys the interface state at the time it is read and is updated continuously and asynchronously by the host adapter.
- Writes to the register have no effect.

Bits	31	30	29	28	27	26	25	24	23	22	21	20	19	18	17	16	15	14	13	12	11	10	9	8	7	6	5	4	3	2	1	0
SCR0	Reserved (0)																				IPM				SPD				DET			

DET The DET value indicates the interface device detection and Phy state.

0000 No device detected and Phy communication not established
0001 Device presence detected but Phy communication not established
0011 Device presence detected and Phy communication established
0100 Phy in offline mode as a result of the interface being disabled or running in a BIST loopback mode
All other values reserved

SPD The SPD value indicates the negotiated interface communication speed established

0000 No negotiated speed (device not present or communication not established)
0001 Generation 1 communication rate negotiated
0010 Generation 2 communication rate negotiated
0011 Generation 3 communication rate negotiated
All other values reserved

IPM The IPM value indicates the current interface power management state

0000 Device not present or communication not established
0001 Interface in active state
0010 Interface in PARTIAL power management state
0110 Interface in SLUMBER power management state
All other values reserved

Reserved All reserved fields will be cleared to zero.

Note: The interface must be in the active state for the interface device detection value (DET field) to be accurate.

- When the interface is in the partial or slumber state, no communication between the host and target is established, resulting in a DET value corresponding to no device present or no communication established.
- As a result, the insertion or removal of a device may not be accurately detected under all conditions, such as when the interface is quiescent as a result of being in the sleep or slumber state.
- This field alone may, therefore, be insufficient to satisfy all the requirements for device attach or detach detection during all possible interface states.

SControl Register

The SControl register is a 32-bit read-write register that provides the interface by which software controls the serial ATA interface capabilities.

- Writes to the SControl register result in an action being taken by the host adapter or interface.
- Reads from the register return the last value written to it.

Bits	31	30	29	28	27	26	25	24	23	22	21	20	19	18	17	16	15	14	13	12	11	10	9	8	7	6	5	4	3	2	1	0
SCR2	Reserved (0)												PMP				SPM				IPM				SPD				DET			

DET The DET field controls the host adapter device detection and interface initialization.

0000b No device detection or initialization action requested.

0001b Perform interface communication initialization sequence to establish communication. This is functionally equivalent to a hard reset and results in the interface being reset and communications reinitialized. Upon a write to the SControl register that sets the LSB of the DET field to one, the host shall transition to the HP1:HR Reset state and shall remain in that state until the LSB of the DET field is cleared to zero by a subsequent write to the SControl register.

0100b Disable the interface and put Phy in offline mode.

All other values reserved.

SPD The SPD field represents the highest allowed communication speed the interface is allowed to negotiate when interface communication speed is established.

0000b No speed negotiation restrictions.

0001b Limit speed negotiation to a rate not > Generation 1 communication rate.

0010b Limit speed negotiation to a rate not > Generation 2 communication rate.

0011b Limit speed negotiation to a rate not > Generation 3 communication rate.

All other values reserved.

IPM The IPM field represents the enabled interface power management states that can be invoked via the serial interface power management capabilities.

0000b No interface power management state restrictions.

0001b Transitions to the PARTIAL power management state disabled.

0010b Transitions to the SLUMBER power management state disabled.

0011b Transitions to both the PARTIAL and SLUMBER power management states disabled.

All other values reserved.

SPM A nonzero value written to this field shall cause the power management state specified to be initiated. A value written to this field is treated as a one-shot. This field shall be read as 0000b.

0000b No power management state transition requested.

0001b Transition to the Partial power management state initiated.

0010b Transition to the Slumber power management state initiated.

0100b Transition to the active power management state initiated.

All other values reserved.

PMP The Port Multiplier Port (PMP) field represents the 4-bit value to be placed in the PM Port field of all transmitted FISes.

This field is "0000" upon power-up.

This field is optional, and an HBA implementation may choose to ignore this field if the FIS to be transmitted is constructed via an alternative method.

Reserved All reserved fields shall be cleared to zero.

SError Register

The SError register is a 32-bit register that conveys supplemental interface error information to complement the error information available in the Shadow Error register.

- The register represents all the detected errors accumulated since the last time the SError register was cleared (regardless of whether it was recovered by the interface).
- Set bits in the error register are explicitly cleared by a write operation to the SError register, or a reset operation.
- The value written to clear set error bits shall have 1's encoded in the bit positions corresponding to the bits that are to be cleared.
- Host software should clear the Interface SError register at appropriate checkpoints in order to best isolate error conditions and the commands they impact.

Bits	31	30	29	28	27	26	25	24	23	22	21	20	19	18	17	16	15	14	13	12	11	10	9	8	7	6	5	4	3	2	1	0
SCR1	DIAG																ERR															
	R				A	X	F	T	S	H	C	D	B	W	I	N	R	R	R	R	E	P	C	T	R	R	R	R	R	R	M	I

R = Reserved

	Error
E	Internal error – reset required
P	Protocol error – reset required
C	Persistent communication error – device may have failed
T	Nonrecovered transient error – host should retry operation
M	Recovered communication error – no action required
I	Recovered data integrity error – host may wish to log

	Diagnostic
F	Unrecognized FIS type
T	Transport state transition error
S	Link state transition error
H	Handshake error – one or more R_ERR received
C	CRC error
D	8b/10b disparity error
B	10b to 8b decode error
W	COMWAKE detected
I	PHY internal error
N	PHYRDY change
X	Exchange (SATA II: Extensions to SATA 1.0a)
A	Port Selector Presence Detected

ERR The ERR field contains error information for use by host software in determining the appropriate response to the error condition.

C Non-recovered persistent communication or data integrity error: A communication error that was not recovered occurred that is expected to be persistent. Since the error condition is expected to be persistent, the operation need not be retried by host software. Persistent communications errors may arise from faulty interconnect with the device, from a device that has been removed or has failed, or a number of other causes.

E Internal error: The host bus adapter experienced an internal error that caused the operation to fail and may have put the host bus adapter into an error state. Host software should reset the interface before retrying the operation. If the condition persists, the host bus adapter may suffer from a design issue, rendering it incompatible with the attached device.

I Recovered data integrity error: A data integrity error occurred that was recovered by the interface through a retry operation or other recovery action. This can arise from a noise burst in the transmission, a voltage supply variation, or from other causes. No action is required by host software since the operation ultimately succeeded; however, host software may elect to track such recovered errors in order to gauge overall communications integrity and step down the negotiated communication speed.

M Recovered communications error: Communications between the device and host was temporarily lost but was re-established. This can arise from a device temporarily being removed, from a temporary loss of Phy synchronization, or from other causes and may be derived from the PhyNRdy signal between the Phy and Link layers. No action is required by the host software since the operation ultimately succeeded; however, host software may elect to track such recovered errors in order to gauge overall communications integrity and step down the negotiated communication speed.

P Protocol error: A violation of the serial implementation of ATA protocol was detected. This can arise from invalid or poorly formed FISes being received, from invalid state transitions, or from other causes. Host software should reset the interface and retry the corresponding operation. If such an error persists, the attached device may have a design issue, rendering it incompatible with the host bus adapter.

R Reserved bit for future use: Shall be cleared to zero.

T Non-recovered transient data integrity error: A data integrity error occurred that was not recovered by the interface. Since the error condition is not expected to be persistent, the operation should be retried by the host software.

DIAG The DIAG field contains diagnostic error information for use by diagnostic software in validating correct operation or isolating failure modes.

B 10b to 8b Decode error: When set to a one, this bit indicates that one or more 10b to 8b decoding errors occurred since the bit was last cleared.

C CRC Error: When set to one, this bit indicates that one or more CRC errors occurred with the Link layer since the bit was last cleared.

D Disparity Error: When set to one, this bit indicates that incorrect disparity was detected one or more times since the last time the bit was cleared.

F Unrecognized FIS type: When set to one, this bit indicates that since the bit was last cleared, one or more FISes were received by the Transport layer with good CRC but had a type field that was not recognized.

I Phy Internal Error: When set to one, this bit indicates that the Phy detected some internal error since the last time this bit was cleared.

N PhyRdy change: When set to one, this bit indicates that the PhyRdy signal changed state since the last time this bit was cleared.

H Handshake error: When set to one, this bit indicates that one or more R_ERR handshake response was received in response to frame transmission. Such errors may be the result of a CRC error detected by the recipient, a disparity or 10b/8b decoding error, or other error condition leading to a negative handshake on a transmitted frame.

R Reserved bit for future use: Shall be cleared to zero.

S Link Sequence Error: When set to one, this bit indicates that one or more Link state machine error conditions was encountered since the last time this bit was cleared. The Link layer state machine defines the conditions under which the link layer detects an erroneous transition.

T Transport state transition error: When set to one, this bit indicates that an error has occurred in the transition from one state to another within the Transport layer since the last time this bit was cleared.

W When set to one, this bit indicates that a COMWAKE signal was detected by the Phy since the last time this bit was cleared.

X Exchange: When set to one, this bit indicates that device presence has changed since the last time this bit was cleared. The means by which the implementation determines that the device presence has changed is vendor specific. This bit may be set anytime a Phy reset initialization sequence occurs, as determined by reception of the COMINIT signal whether in response to a new device being inserted, in response to a COMRESET having been issued, or in response to power-up.

A Port Selector presence detected: This bit is set to one when COMWAKE is received while the host is in state HP2: HR_AwaitCOMINIT. On power-up reset, this bit is cleared to 0. The bit is cleared to 0 when the host writes a “1” to this bit location.

Exchange Bit Description: There is no direct indication in Serial ATA 1.0a of a device change condition, and in a hot-plug scenario, in which devices may be removed and inserted, host notification of a device change condition is essential in order to ensure that data intended for one device is not inadvertently written to another device when the operator changes the devices.

- In order to facilitate a means for notifying host software of the potential of a device having been changed, an additional bit was added to the DIAG field of the SError superset register.

SActive Register

The SActive register is used during Native Command Queuing (NCQ). This register can be written to when a Set Device Bits FIS is received from a device. Prior to NCQ, the second word in the Set Device Bits FIS was reserved. This word will contain the contents of the SActive register, as seen below.

Bits	31	30	29	28	27	26	25	24	23	22	21	20	19	18	17	16	15	14	13	12	11	10	9	8	7	6	5	4	3	2	1	0
SCR3	SActive																															
	31	30	29	28	27	26	25	24	23	22	21	20	19	18	17	16	15	14	13	12	11	10	9	8	7	6	5	4	3	2	1	0

SActive Register Characteristics

Communication between the host and drive about which commands are outstanding is handled through the 32-bit register in the host called SActive.

- The SActive register has one bit allocated to each possible tag value (0–31), that is, bit x shows the status of the command with tag x.

- If a bit in the SActive register is set, it means that a command with that tag is outstanding in the drive (or a command with that tag is about to be issued to the drive).
- If a bit in the SActive register is cleared, it means that a command with that tag is not outstanding in the drive.
- The host and drive work together to make sure that the SActive register is accurate at all times.

Host Sets Bits

The host can set bits in the SActive register, whereas the device can clear bits in the SActive register.

- This ensures that updates to the SActive register require no synchronization between the host and the drive.
- Before issuing a command, the host sets the bit corresponding to the tag of the command it is about to issue.
- When the drive successfully completes a command, it will clear the bit corresponding to the tag of the command it just finished.

Device Clears Bits

The drive clears bits in the SActive register using the Set Device Bits FIS, covered in the NCQ section.

- The SActive field of the Set Device Bits FIS is used to convey successful (or unsuccessful) status to the host.
- When a bit is set in the SActive field of the FIS, it means that the command with the corresponding tag has completed successfully.
- The host controller will clear bits in the SActive register corresponding to bits that are set to one in the SActive field of a received Set Device Bits FIS.

SNotification Register

- The Serial ATA interface notification register—SNotification—is a 32-bit register that conveys the devices that have sent the host a Set Device Bits FIS with the Notification bit set.
- When the host receives a Set Device Bits FIS with the Notification bit set to one, the host shall set the bit in the SNotification register corresponding to the value of the PM Port field in the received FIS; the PM Port field is defined in the Port Multiplier 1.0 Specification.
  - For example, if the PM Port field is set to 7, then the host shall set bit 7 in the SNotification register to 1.
  - After setting the bit in the SNotification register, the host shall generate an interrupt if the I bit is set to one in the FIS and interrupts are enabled.

Bits	31	30	29	28	27	26	25	24	23	22	21	20	19	18	17	16	15	14	13	12	11	10	9	8	7	6	5	4	3	2	1	0
SCR4	Reserved (0)																**Notify**															
	0	0	0	0	0	0	0	0	0	0	0	0	0	0	0	0	15	14	13	12	11	10	9	8	7	6	5	4	3	2	1	0

Notify

- The field represents whether a particular device with the corresponding PM Port number has sent a Set Device Bits FIS to the host with the Notification bit set.

Set bits in the SNotification register are explicitly cleared by a write operation to the SNotification register, or a power-on reset operation.

- The register is not cleared due to a COMRESET; software is responsible for clearing the register, as appropriate.
- The value written to clear set bits shall have 1's encoded in the bit positions corresponding to the bits that are to be cleared.

Asynchronous notification is a mechanism for a device to send a notification to the host that the device requires attention.

- A few examples of how this mechanism could be used include indicating media has been inserted in an ATAPI device or indicating that a hot-plug event has occurred on a Port Multiplier port.
- The mechanism that the host uses to determine the event and the action that is required is outside the scope of the specification; refer to the appropriate command set specification for the specific device for more information.
- The Serial ATA II Specification added:
    - ATAPI Notification – An ATAPI device shall indicate whether it supports asynchronous notification in Word 78 of IDENTIFY PACKET DEVICE.
    - The feature is enabled by using SET FEATURES.

Event Example (Informative)

An example of an event that may cause the device to generate an asynchronous notification to the host to request attention is the Media Change Event. The Media Change Event occurs when an ATAPI device has detected a change in device state—media has either been inserted or removed.

4.2.3 Command Processing Examples

The following descriptions show how commands are processed, what registers and bits are set or cleared and when they are set or cleared, which FISes are used to move registers between the host and device, and the overall transport protocol of the SATA architecture.

4.2.4 Legacy DMA Read by Host from Device

1. Prior to the command being issued to the device, the host driver software programs the host adapter's DMA controller with the memory address pointer(s) and the transfer direction, and arms the DMA controller (enables the "run" flag).
2. The host driver software issues the command to the device by writing the Shadow Command Block and Shadow Control Block (Command register last).
    - In response to the Shadow Command Register being written, the host adapter sets the BSY bit in the Shadow Status register and transmits a Register – Host to Device FIS to the device with the Shadow Command Block and Shadow Control Block contents.

3. When the device has processed the command and is ready, it transmits the read data to the host in the form of one or more Data FISes.
   - This transfer proceeds in response to flow control signals/readiness.
4. The host adapter recognizes that the incoming frame is a Data FIS and the DMA controller is programmed, and it directs the incoming data to the host adapter's DMA controller, which forwards the incoming data to the appropriate host memory locations.
5. Upon completion of the transfer, the device transmits a Register – Device to Host FIS to indicate ending status for the command, clearing the BSY bit in the Status register, and if the interrupt flag is set in the header, an interrupt is asserted to the host.

Legacy DMA Read by Host from Device Protocol Example

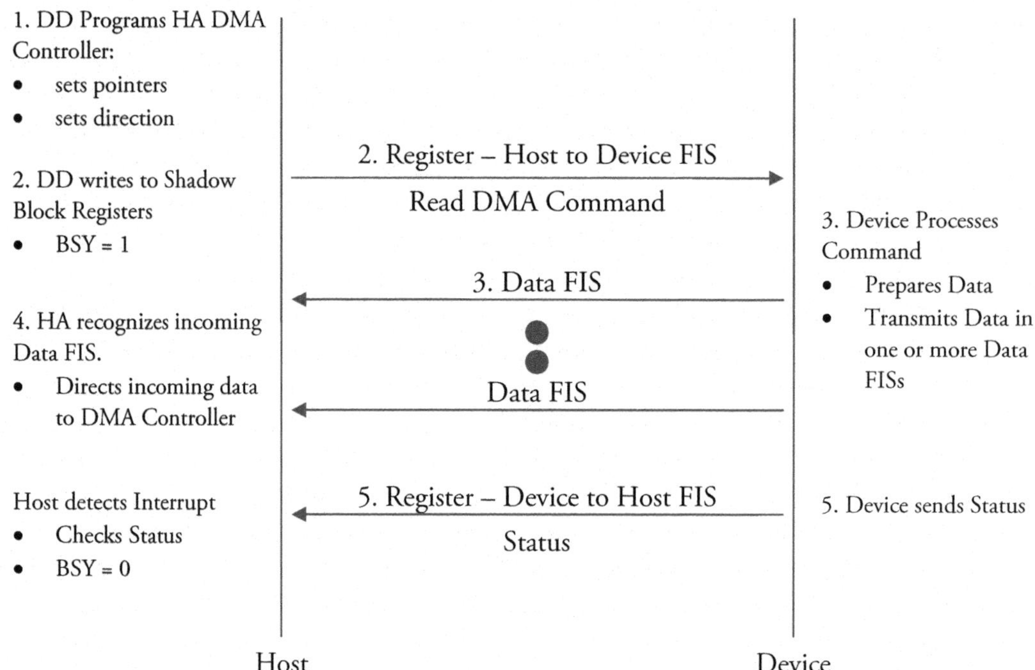

4.2.5 Legacy DMA Write by Host to Device

1. Prior to the command being issued to the device, the host driver software programs the host adapter's DMA controller with the memory address pointer(s) and the transfer direction, and arms the DMA controller (enables the "run" flag). As a result, the DMA controller becomes armed but remains paused pending a signal from the device to proceed with the data transfer.
2. The host driver software issues the command to the device by writing the Shadow Command Block and Shadow Control Block (Command register last).
   - In response to the Shadow command Register being written, the host adapter sets the BSY bit in the Shadow Status register and transmits a Register – Host to Device FIS to the device with the Shadow Command Block and Shadow Control Block contents.
3. When the device is ready to receive the data from the host, the device transmits a DMA Activate FIS to the host, which activates the armed DMA controller.
4. The DMA controller transmits the write data to the device in the form of one or more Data FIS.

a. If more than one Data FIS is required to complete the overall data transfer request, a DMA Activate FIS will be sent prior to each and every one of the subsequent Data FISes.
b. The amount of data transmitted to the device is determined by the transfer count programmed into the host adapter's DMA controller by the host driver software during the command setup phase.

5. Upon completion of the transfer, the device transmits a Register – Device to Host FIS to indicate ending status for the command, clearing the BSY bit in the Status register, and if the interrupt flag is set in the header, an interrupt is asserted to the host.

Legacy DMA Write by Host to Device Protocol Example

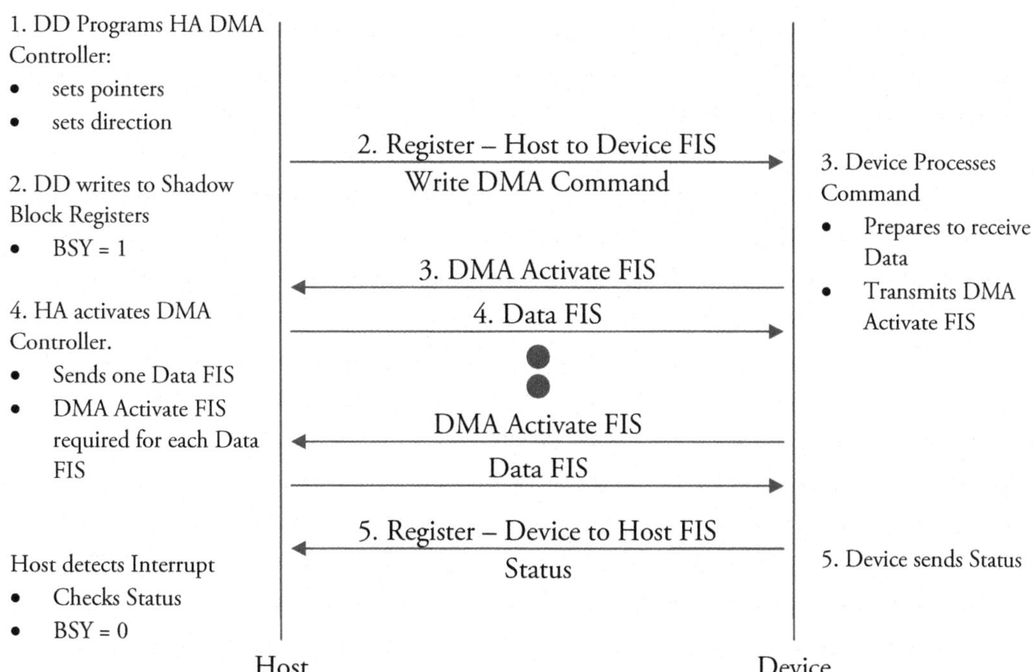

4.2.6 PIO Data Read from the Device

1. The host driver software issues a PIO read command to the device by writing the Shadow Command Block and Shadow Control Block (Command register last).
   - In response to the Command register being written, the host adapter sets the BSY bit in the Shadow Status register and transmits a Register – Host to Device FIS to the device with the Shadow Command Block and Shadow Control Block contents.
2. When the device has processed the command and is ready to begin transferring data to the host, it first transmits a PIO Setup FIS to the host.
3. Upon receiving the PIO Setup FIS, the host adapter holds the FIS contents in a temporary holding buffer.
   - When the Data FIS is received and while holding the PIO Setup FIS, the host adapter transfers the register contents from the PIO Setup FIS into the Shadow Command Block and Shadow Control Block including the initial status value, resulting in DRQ getting set and BSY getting cleared in the Status register.
   - In addition, if the interrupt flag is set, an interrupt is generated to the host.
4. The device follows the PIO Setup FIS with a Data – Device to Host FIS.

5. The host adapter receives the incoming data that is part of the Data FIS into a speed matching FIFO that is conceptually attached to the Shadow Data Register.
   a. As a result of the issued interrupt and DRQ being set in the Status register, host software does an IN Operation on the data register and pulls data from the head of the speed matching FIFO while the serial link is adding data to the tail of the FIFO.
   b. The flow control scheme handles data throttling to avoid underflow/overflow of the receive speed matching FIFO that feeds the Data Shadow Register.

When the number of words read by host software from the Data Shadow register reaches the value indicated in the PIO Setup FIS, the host transfers the ending status value from the earlier PIO Setup FIS E_Status field into the Shadow Status register, resulting in DRQ being cleared and the ending status reported.

- If there are more data blocks to be transferred, the ending status from the E_Status field will indicate BSY, and the process will repeat from the device sending the PIO Setup FIS to the host.
- No Status is returned unless an error occurs.

PIO Read Command Protocol Example

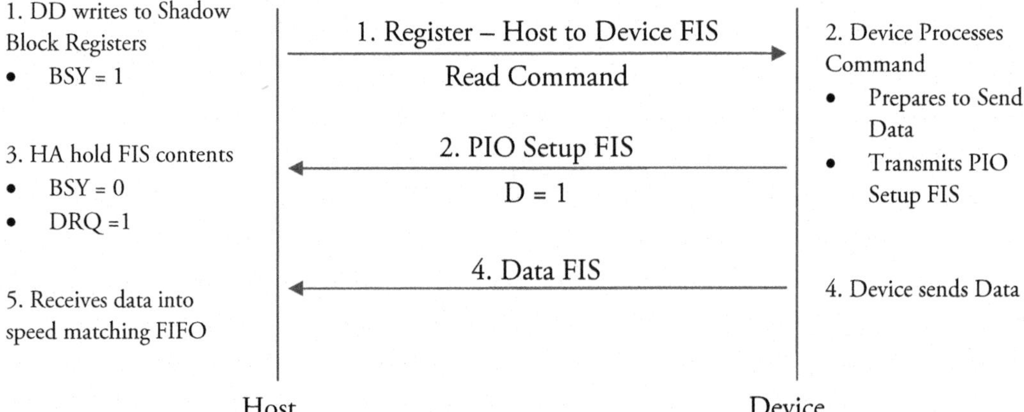

4.2.7 PIO Data Write to the Device

1. The host driver software issues a PIO write command to the device by writing the Shadow Command Block and Shadow Control Block (Command register last).
   - In response to the Command register being written, the host adapter sets the BSY bit in the Shadow Status register and transmits a Register – Host to Device FIS to the device with the Shadow Command Block and Shadow Control Block contents.
2. When the device is ready to receive the PIO write data, it transmits a PIO Setup FIS to the host to indicate that the target is ready to receive PIO data and the number of words of data that are to be transferred.
3. In response to a PIO Setup FIS with the D bit indicating a write to the device, the host transfers the beginning Status register contents from the PIO Setup FIS into the Shadow Status register, resulting in DRQ getting set, BSY cleared. Also, if the interrupt flag is set, an interrupt is generated to the host.
4. As a result of DRQ being set in the Shadow Status register, the host driver software starts an OUT operation to the data register.

- The data written to the data register is placed in an outbound speed matching FIFO and is transmitted to the device as a Data – Host to Device FIS. The OUT operation pushes data onto the tail of the FIFO and the serial link pulls data from the head.
- The flow control scheme handles data throttling to avoid underflow of the transmit FIFO.

5. When all words indicated in the PIO Setup FIS have been written to the transmit FIFO, the host adapter transfers the final status value indicated in the PIO Setup frame into the Shadow Status register, resulting in DRQ being cleared, and closes the frame with a CRC and EOF.
   - If additional sectors of data are to be transferred, the ending status from the E_Status field value transferred to the Shadow Status register would have the BSY bit set, and the state is the same as immediately after the command was first issued to the device.
   - If there are more data blocks to be transferred, the ending status from the E_Status field will indicate BSY, and the process will repeat from the device sending the PIO Setup FIS to the host.
6. When all sectors have been transferred, the device shall send a Register – Device to Host FIS with the command complete interrupt and not BSY status.
   - In the case of a write error, the device may, on any sector boundary, include error status and a command complete interrupt in the PIO setup FIS, and there is no need to send the Register – Device to Host FIS.

PIO Write Command Protocol Example

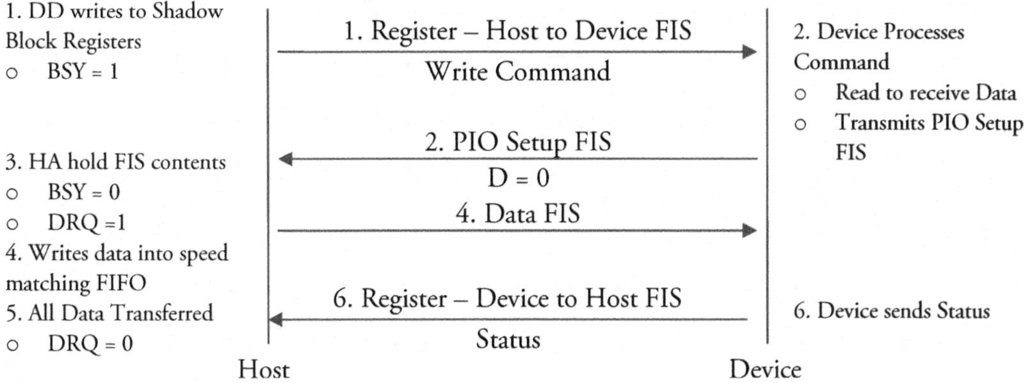

READ DMA QUEUED Example

Note: The serial implementation of ATA devices may choose to not implement the parallel implementation of ATA queuing in favor of a more efficient serial implementation queuing mechanism.

1. Prior to the command being issued to the device, the host driver software programs the host-side DMA controller with the memory address pointer(s) and the transfer direction, and arms the DMA controller (enables the "run" flag).
2. The host driver software issues the command to the device by writing the Shadow Command Block and Shadow Control Block (Command register last).
   - In response to the Shadow command Register being written, the host adapter sets the BSY bit in the Shadow Status register and transmits a Register – Host to Device FIS to the device with the Shadow Command Block and Shadow Control Block contents.
3. When the device has queued the command and wishes to release the bus, it transmits a Register FIS to the host, resulting in the BSY bit being cleared and the REL bit being set in the Status register.

4. When the device is ready to complete the transfer for the queued command, it transmits a Set Device Bits FIS to the host, resulting in the SERV bit being set in the Status register. If no other command is active (i.e., BSY set to one), then an interrupt is also generated.
5. In response to the service request, the host software deactivates the DMA controller (if activated) and issues a SERVICE command to the device by writing the Shadow Command Block and Shadow Control Block, resulting in the BSY bit getting set and a register FIS being transmitted to the device.
6. In response to the SERVICE request, the device transmits a Register FIS to the host conveying the TAG value to the host and clearing the BSY bit and setting the DRQ bit.
7. When the DRQ bit is set, the host software reads the TAG value from the Shadow Command Block and restores the DMA controller context appropriate for the command that is completing.
8. The device transmits the read data to the host in the form of one or more Data FISes.
   - This transfer proceeds in response to flow control signals/readiness.
   - Any DMA data arriving before the DMA controller has its context restored will back up into the inbound speed matching FIFO until the FIFO is filled and will thereafter be flow controlled to throttle the incoming data until the DMA controller has its context restored by the host software.
9. The host adapter recognizes that the incoming packet is a Data FIS and the DMA controller is programmed, and directs the incoming data to the host adapter's DMA controller, which forwards the incoming data to the appropriate host memory locations.
10. Upon completion of the transfer, the target transmits a Register – Device to Host FIS to indicate ending status for the command, clearing the BSY bit in the Status register, and if the interrupt flag is set in the header, an interrupt is asserted to the host.

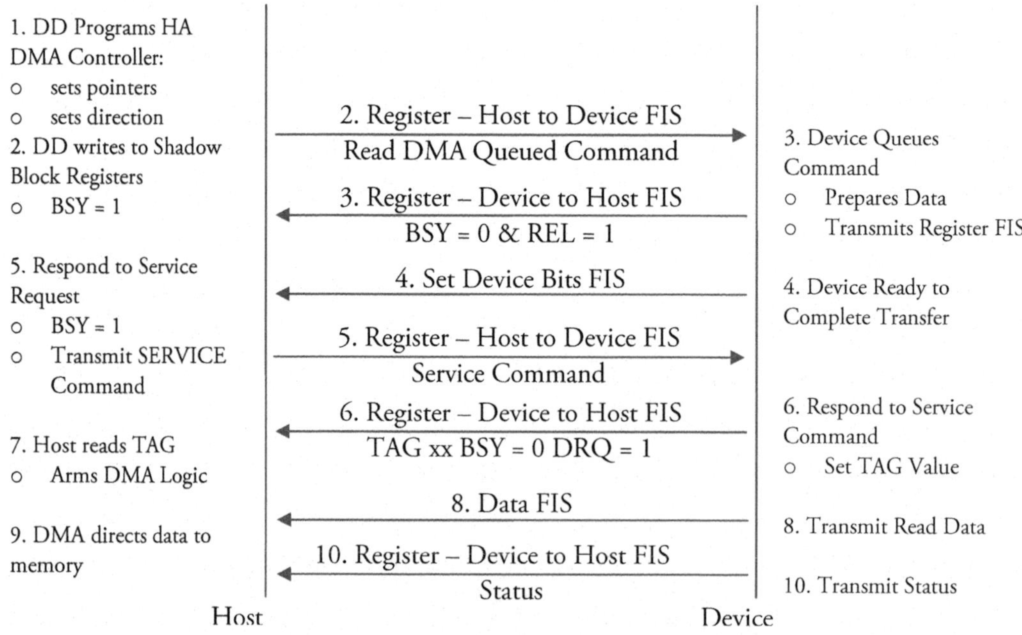

4.2.8 WRITE DMA QUEUED Example

Note: Serial implementations of ATA devices may choose to not implement the parallel implementation of ATA queuing in favor of a more efficient serial implementation queuing mechanism.

1. Prior to the command being issued to the device, the host driver software programs the host-side DMA controller with the memory address pointer(s) and the transfer direction, and arms the DMA controller (enables the "run" flag).
2. The host driver software issues the command to the device by writing the Shadow Command Block and Shadow Control Block (Command register last).
   - In response to the Shadow Command Register being written, the host adapter sets the BSY bit in the Shadow Status register and transmits a Register – Host to Device FIS to the device with the Shadow Command Block and Shadow Control Block contents.
3. When the device has queued the command and wishes to release the bus, it transmits a register FIS to the host, resulting in the BSY bit being cleared and the REL bit being set in the Status register.
4. When the device is ready to complete the transfer for the queued command, it transmits a Set Device Bits FIS to the host, resulting in the SERV bit being set in the Status register. If no other command is active (i.e., BSY set to one), then an interrupt is also generated.
5. In response to the service request, the host software deactivates the DMA controller (if activated) and issues a SERVICE command to the device by writing the Shadow Command Block and Shadow Control Block, resulting in the BSY bit getting set and a register FIS being transmitted to the device.
6. In response to the SERVICE request, the device transmits a register FIS to the host conveying the TAG value to the host, clearing the BSY bit, and setting the DRQ bit.
7. When the DRQ bit is set, the host software reads the TAG value from the Shadow Command Block and restores the DMA controller context appropriate for the command that is completing.
8. When the device is ready to receive the data from the host, the device transmits a DMA Activate FIS to the host, which activates the armed DMA controller.
9. The DMA controller transmits the write data to the device in the form of one or more Data – Host to Device FISes.
   - If more than one Data FIS is required to complete the overall data transfer request, a DMA Activate FIS shall be sent prior to each and every one of the subsequent Data FISes.
   - The amount of data transmitted to the device is determined by the transfer count programmed into the host's DMA controller by the host driver software during the command setup phase.
   - If the DMA Activate FIS arrives at the host prior to the host software restoring the DMA context, the DMA Activate FIS results in the DMA controller starting the transfer as soon as the host software completes programming it (i.e., the controller is already activated, and the transfer starts as soon as the context is restored).
10. Upon completion of the transfer, the target transmits a Register – Host to Device FIS to indicate ending status for the command, clearing the BSY bit in the Status register, and if the interrupt flag is set in the header, an interrupt is asserted to the host.

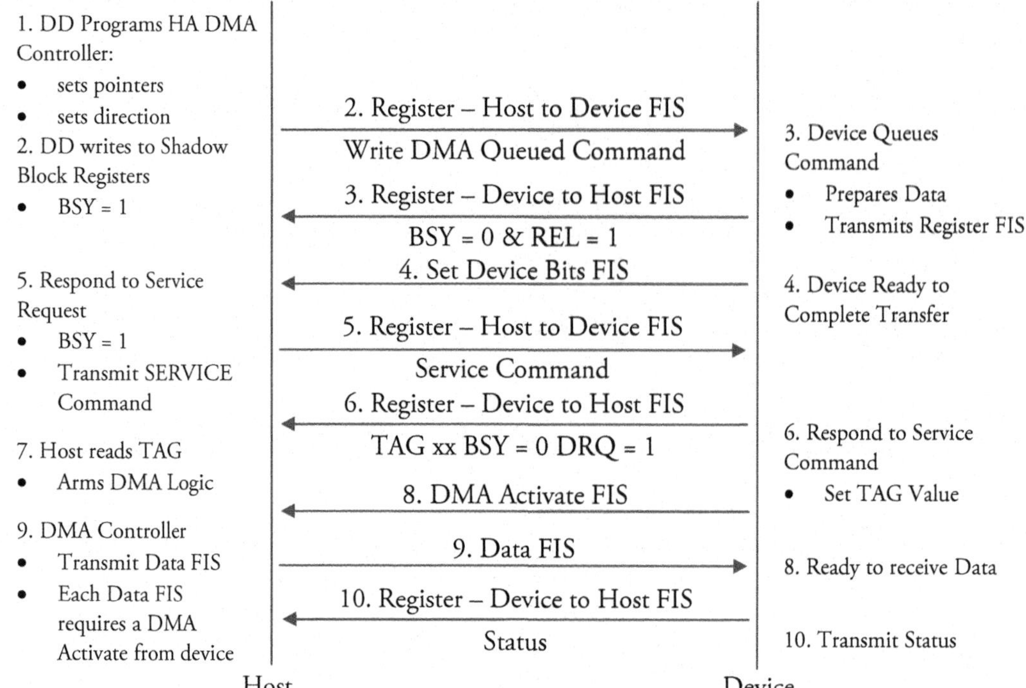

4.2.9 ATAPI PACKET Commands with PIO Data-In

1. The host driver software issues a PACKET command to the device by writing the Shadow Command Block and Shadow Control Block (Command register last).
   - In response to the Command register being written, the host adapter sets the BSY bit in the Shadow Status register and transmits a Register – Host to Device FIS to the device with the Shadow Command Block and Shadow Control Block contents.
2. When the device is ready to receive the ATAPI command packet, it transmits a PIO Setup – Device to Host FIS to the host to indicate that the target is ready to receive PIO data and the number of words of data that are to be transferred.
3. The host transfers the beginning Status register contents from the PIO Setup FIS into the Shadow Status register, resulting in BSY getting negated and DRQ getting asserted.
   - As a result of BSY getting cleared and DRQ being set in the Shadow Status register, the host driver software writes the command packet to the Shadow Data register.
4. The data written to the data register is placed in an outbound speed matching FIFO and is transmitted to the device as a Data – Host to Device FIS. The writes to the data register push data onto the tail of the FIFO and the serial link pulls data from the head. The flow control scheme handles data throttling to avoid underflow of the transmit FIFO.
   - When the number of words indicated in the PIO Setup FIS have been written to the transmit FIFO, the host adapter transfers the final status value indicated in the PIO setup frame into the Shadow Status register, resulting in DRQ being cleared and BSY being set, and closes the frame with a CRC and EOF. This completes the transmission of the command packet to the device.

5. When the device has processed the command and is ready to begin transferring data to the host, it first transmits a PIO Setup – Device to Host FIS to the host.
   - Upon receiving the PIO Setup – Device to Host FIS, the host adapter holds the FIS contents in a temporary holding buffer.
6. The device follows the PIO Setup – Device to Host FIS with a Data – Device to Host FIS. Upon receiving the Data FIS while holding the PIO Setup FIS context, the host adapter transfers the register contents from the PIO Setup FIS into the Shadow Command Block and Shadow Control Block, including the initial status value, resulting in DRQ getting set and BSY getting cleared in the Status register. Also, if the interrupt flag is set, an interrupt is generated to the host.
7. The host adapter receives the incoming data that is part of the Data FIS into a speed matching FIFO that is conceptually attached to the Data Shadow Register.
   - As a result of the issued interrupt and DRQ being set in the Status register, the host software reads the byte count and does a IN operation on the data register to pull data from the head of the speed matching FIFO while the serial link is adding data to the tail of the FIFO. The flow control scheme handles data throttling to avoid underflow/overflow of the receive speed matching FIFO that feeds the Data Shadow Register.
8. When the number of words received in the Data FIS reaches the value indicated in the PIO Setup FIS and the host FIFO is empty, the host transfers the ending status from the E_Status field value from the earlier PIO Setup into the Shadow Status register, resulting in DRQ being cleared and BSY being set.
9. The device transmits final ending status by sending a Register Device to Host FIS with BSY cleared to zero and the command completion status for the command and the interrupt flag set.
10. The host detects an incoming register frame that contains BSY cleared to zero and command completion status for the command and the interrupt flag set, and it places the frame content into the Shadow Command Block and Shadow Control Block to complete the command.

ATAPI PACKET Commands with PIO Data-IN Protocol Example

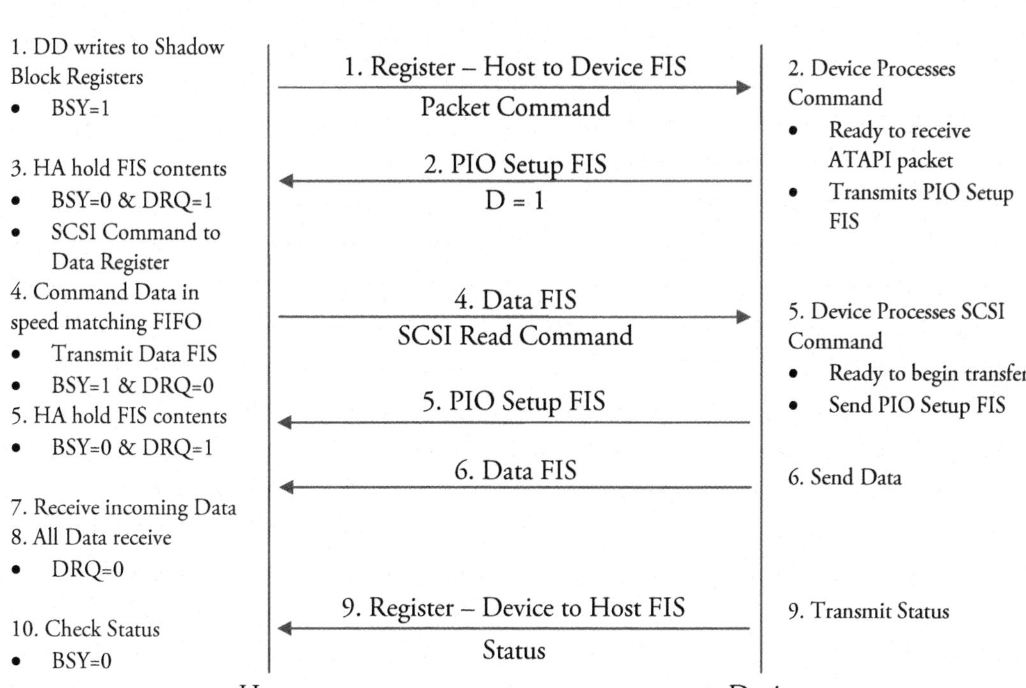

4.2.10 ATAPI PACKET Commands with PIO Data-Out

1. The host driver software issues a PACKET command to the device by writing the Shadow Command Block and Shadow Control Block (Command register last).
   - In response to the Command register being written, the host adapter sets the BSY bit in the Shadow Status register and transmits a Register – Host to Device FIS to the device with the Shadow Command Block and Shadow Control Block contents.
2. When the device is ready to receive the ATAPI command packet, it transmits a PIO Setup – Device to Host FIS to the host to indicate that the target is ready to receive PIO data and the number of words of data that are to be transferred.
3. The host transfers the beginning Status register contents from the PIO Setup – Device to Host FIS into the Shadow Status register, resulting in BSY getting negated and DRQ getting asserted.
   - As a result of BSY getting cleared and DRQ being set in the Shadow Status register, the host driver software writes the command packet to the Shadow Data register.
4. The data written to the data register is placed in an outbound speed matching FIFO and is transmitted to the device as a Data – Host to Device FIS. The writes to the data register push data onto the tail of the FIFO and the serial link pulls data from the head. The flow control scheme handles data throttling to avoid underflow of the transmit FIFO.
   - When the number of words indicated in the PIO Setup FIS have been written to the transmit FIFO, the host adapter transfers the final status value indicated in the PIO Setup frame into the Shadow Status register, resulting in DRQ being cleared and BSY being set, and closes the frame with a CRC and EOF. This completes the transmission of the command packet to the device.
5. When the device has processed the command and is ready to receive the PIO write data, it transmits a PIO Setup – Device to Host FIS to the host to indicate that the target is ready to receive PIO data and the number of words of data that are to be transferred.
6. In response to a PIO Setup FIS with the D bit indicating a write to the device, the host transfers the beginning Status register contents from the PIO Setup FIS into the Shadow Status register, resulting in DRQ getting set. Also, if the interrupt flag is set, an interrupt is generated to the host.
   - As a result of DRQ being set in the Shadow Status register, the host driver software reads the byte count and starts an OUT operation to the data register.
7. The data written to the data register is placed in an outbound speed matching FIFO and is transmitted to the device as a Data – Host to Device FIS. The OUT operation pushes data onto the tail of the FIFO and the serial link pulls data from the head. The flow control scheme handles data throttling to avoid underflow of the transmit FIFO.
   - When the number of words indicated in the PIO Setup FIS have been written to the transmit FIFO, the host adapter transfers the final status value indicated in the PIO Setup FIS into the Shadow Status register, resulting in DRQ being cleared, BSY being set, and closes the frame with a CRC and EOF.
8. The device transmits command completion status by sending a Register – Device to Host FIS with BSY cleared to zero and the ending status for the command and the interrupt flag set.
9. The host detects an incoming Register – Device to Host FIS that contains BSY cleared to zero, command completion status for the command, and the interrupt flag set, and it places the frame content into the Shadow Command Block and Shadow Control Block to complete the command.

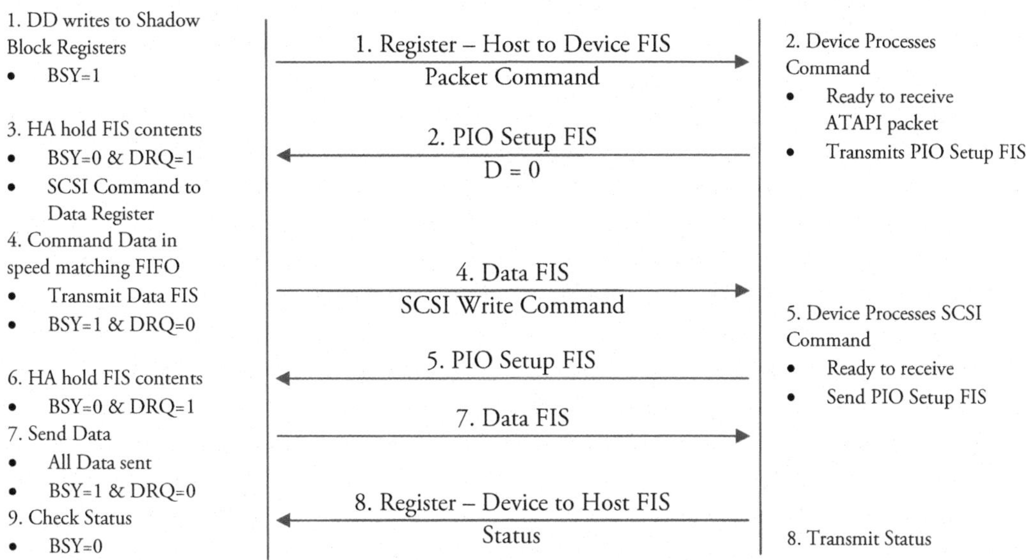

4.2.11 ATAPI PACKET Commands with DMA Data-In

1. Prior to the command being issued to the device, the host driver software programs the host-side DMA controller with the memory address pointer(s) and the transfer direction, and arms the DMA controller (enables the "run" flag).
2. The host driver software issues a PACKET command to the device by writing the Shadow Command Block and Shadow Control Block (Command register last).
   - In response to the Command register being written, the host adapter sets the BSY bit in the Shadow Status register and transmits a Register – Host to Device FIS to the device with the Shadow Command Block and Shadow Control Block contents.
3. When the device is ready to receive the ATAPI command packet, it transmits a PIO Setup – Device to Host FIS to the host to indicate that the target is ready to receive PIO data and the number of words of data that are to be transferred.
4. The host transfers the beginning Status register contents from the PIO Setup FIS into the Shadow Status register, resulting in BSY getting negated and DRQ getting asserted.
   - As a result of BSY getting cleared and DRQ being set in the Shadow Status register, the host driver software writes the command packet to the Shadow Data register.
5. The data written to the data register is placed in an outbound speed matching FIFO and is transmitted to the device as a Data – Host to Device FIS. The writes to the data register push data onto the tail of the FIFO and the serial link pulls data from the head. The flow control scheme handles data throttling to avoid underflow of the transmit FIFO.
6. When the number of words indicated in the PIO Setup FIS have been written to the transmit FIFO, the host adapter transfers the final status value indicated in the PIO setup frame into the Shadow Status register, resulting in DRQ being cleared and BSY being set, and closes the frame with a CRC and EOF.
   - This completes the transmission of the command packet to the device.

7. When the device has processed the command and is ready, it transmits the read data to the host in the form of a single Data FIS.
   - This transfer proceeds in response to flow control signals/readiness.
8. The host adapter recognizes that the incoming packet is a Data FIS and the DMA controller is programmed, and directs the incoming data to the host adapter's DMA controller, which forwards the incoming data to the appropriate host memory locations.
9. Upon completion of the transfer, the target transmits a Register – Device to Host FIS to indicate command completion status, clearing the BSY bit in the Status register, and if the interrupt flag is set in the header, an interrupt is asserted to the host.

ATAPI PACKET Commands with DMA Data-IN Protocol Example

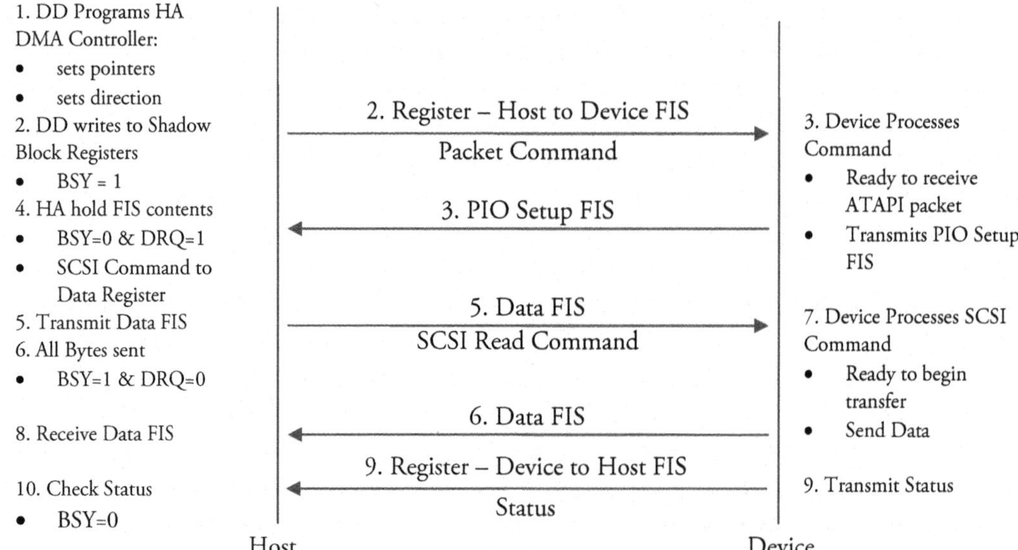

4.2.12 ATAPI PACKET Commands with DMA Data-Out

1. Prior to the command being issued to the device, the host driver software programs the host-side DMA controller with the memory address pointer(s) and the transfer direction, and arms the DMA controller (enables the "run" flag). As a result, the DMA controller becomes armed but remains paused pending a signal from the device to proceed with the data transfer.
2. The host driver software issues a PACKET command to the device by writing the Shadow Command Block and Shadow Control Block (Command register last).
   - In response to the Command register being written, the host adapter sets the BSY bit in the Shadow Status register and transmits a Register – Host to Device FIS to the device with the Shadow Command Block and Shadow Control Block contents.
3. When the device is ready to receive the ATAPI command packet, it transmits a PIO Setup – Device to Host FIS to the host to indicate that the target is ready to receive PIO data and the number of words of data that are to be transferred.
4. The host transfers the beginning Status register contents from the PIO Setup FIS into the Shadow Status register, resulting in BSY getting negated and DRQ getting asserted.
   - As a result of BSY getting cleared and DRQ being set in the Shadow Status register, the host driver software writes the command packet to the Shadow Data register.

5. The data written to the data register is placed in an outbound speed matching FIFO and is transmitted to the device as a Data – Host to Device FIS. The writes to the data register push data onto the tail of the FIFO and the serial link pulls data from the head. The flow control scheme handles data throttling to avoid underflow of the transmit FIFO.
6. When the number of words indicated in the PIO Setup FIS have been written to the transmit FIFO, the host adapter transfers the final status value indicated in the PIO Setup frame into the Shadow Status register, resulting in DRQ being cleared and BSY being set, and closes the frame with a CRC and EOF. This completes the transmission of the command packet to the device.
7. When the device is ready to receive the data from the host, the device transmits a DMA Activate FIS to the host, which activates the armed DMA controller.
8. The DMA controller transmits the write data to the device in the form of one or more Data FIS. The transfer proceeds in response to flow control signals/readiness. The amount of data transmitted to the device is determined by the transfer count programmed into the host's DMA engine by the host driver software during the command setup phase.
9. Upon completion of the transfer, the device transmits a Register – Device to Host FIS to indicate ending status for the command, clearing the BSY bit in the Status register, and if the interrupt flag is set in the header, an interrupt is asserted to the host.

ATAPI PACKET Commands with DMA Data-OUT Protocol Example

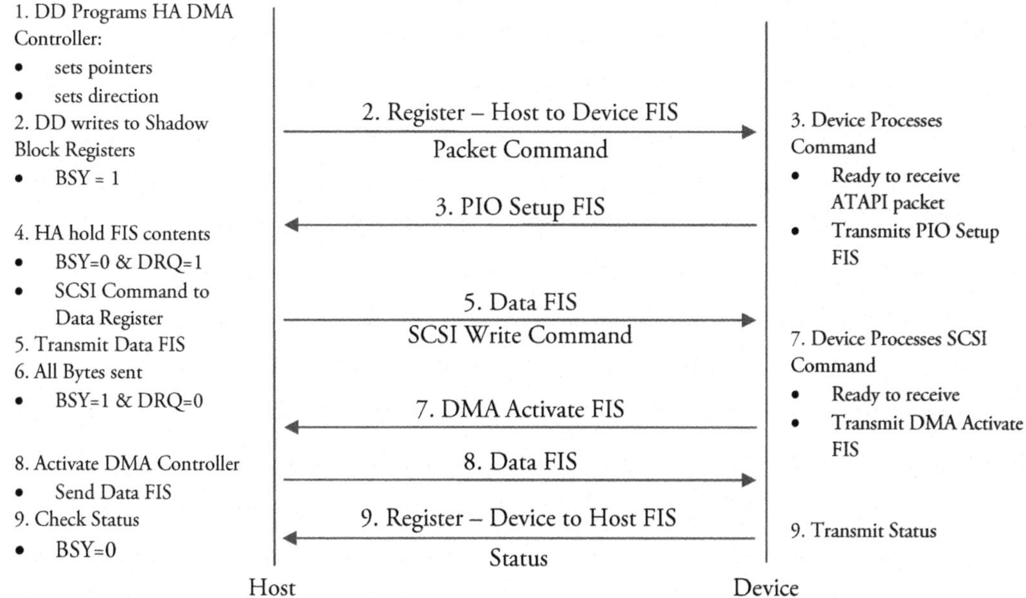

4.2.13 First-Party DMA Read of Host Memory by Device

This section only outlines the basic First-party DMA read transaction and does not outline command sequences that utilize this capability.

1. As a result of some condition and the target having adequate knowledge of host memory (presumably some prior state has been conveyed to the target), the target signals a DMA read request for a given address and given transfer count to the host by transmitting a First-party DMA Setup FIS.

2. Upon receiving the First-party DMA Setup FIS, the host adapter transfers the appropriate memory address, count value, and transfer direction into the host-side DMA controller and arms and activates the DMA controller (enables the "run" flag).
3. In response to being set up, armed, and activated, the DMA controller retrieves data from host memory and transmits it to the device in the form of one or more Data – Host to Device FISes. The transfer proceeds in response to flow control signals/readiness.

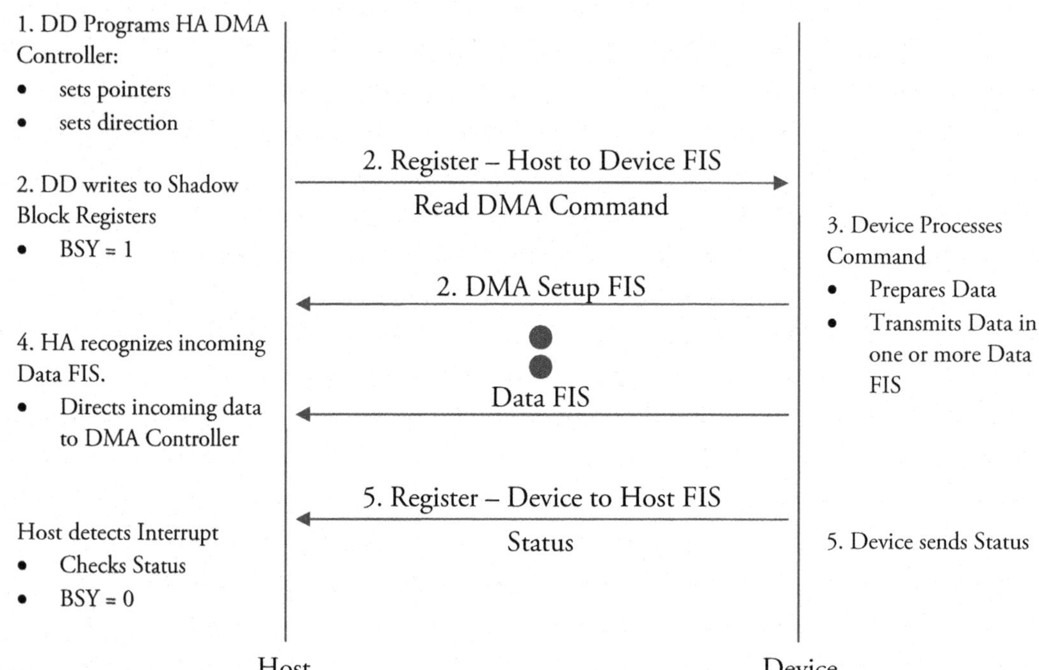

4.2.14 First-Party DMA Write of Host Memory by Device

This section only outlines the basic First-party DMA write transaction and does not outline command sequences that utilize this capability.

1. As a result of some condition and the target having adequate knowledge of host memory (presumably some prior state has been conveyed to the target), the target signals a DMA write request for a given address and given transfer count to the host by transmitting a First-party DMA Setup FIS.
2. Upon receiving the First-party DMA Setup FIS, the host adapter transfers the appropriate memory address, count value, and transfer direction into the host-side DMA controller and arms the DMA controller (enables the "run" flag).
3. The device then transmits the data to the host in the form of one or more Data – Device to Host FISes. This DMA transfer proceeds in response to flow control signals/readiness.
4. The host adapter recognizes that the incoming packet is a Data FIS and directs the incoming data to the host adapter's DMA controller, which forwards the incoming data to the appropriate host memory locations.

4.2.15 PIO Data Read from the Device – Odd Word Count

- In response to decoding and processing a PIO read command with a transfer count for an odd number of 16-bit words, the device transmits the corresponding data to the host in the form of a single Data FIS. The device pads the upper 16-bits of the final 32-bit DWORD of the last transmitted FIS in order to close the FIS. The CRC value transmitted at the end of the FIS is computed over the entire FIS, including any pad bytes in the final transmitted symbol.
- Host driver software responsible for retrieving the PIO data is aware of the number of words of data it expects to retrieve from the Data register and performs an IN operation for an odd number of repetitions.
- Upon exhaustion of the IN operation by the host driver software, the receive FIFO that interfaces with the data register has one 16-bit word of received data remaining in it that corresponds to the pad that the device included at the end of the transmitted frame. This remaining word of data left in the data register FIFO is flushed upon the next write of the Command register or upon the receipt of the next Data FIS from the device.

4.2.16 PIO Data Write to the Device – Odd Word Count

- In response to decoding and processing a PIO write command with a transfer count for an odd number of 16-bit words, the device transmits a PIO Setup FIS to the host indicating it is ready to receive the PIO data and indicating the transfer count. The conveyed transfer count is for an odd number of 16-bit word quantities.
- Host driver software responsible for transmitting the PIO data is aware of the number of words of data it will write to the Data register and performs an OUT operation for an odd number of repetitions.
- After the final write by the software driver to the Data register, the transfer count indicated in the PIO Setup packet is exhausted, which signals the host adapter to close the FIS. Since the transfer count was odd, the upper 16 bits of the final 32-bit DWORD of data to transmit remains zeroed (pad), and the host adapter closes the FIS after transmitting this final padded DWORD. The CRC value transmitted at the end of the FIS is computed over the entire FIS, including any pad bytes in the final transmitted symbol.
- Having awareness of the command set and having decoded the current command, the device that receives the transmitted data has knowledge of the expected data transfer length. Upon receiving the data from the host, the device removes the 16-bit pad data in the upper 16 bits of the final 32-bit DWORD of received data.

4.2.17 First-Party DMA Read of Host Memory by Device – Odd Word Count

- When the device wishes to initiate a First-party DMA read of host memory for an odd word count, it constructs a First-party DMA Setup FIS that includes the base address and the transfer count (for an odd number of words).
- Upon receiving the First-party DMA Setup FIS, the host adapter transfers the starting address and (odd) transfer count to the DMA controller and activates the DMA controller to initiate a transfer to the device of the requested data.
- The host transmits the data to the device in the form of one or more Data FIS. Because the transfer count is odd, the DMA controller completes its data transfer from host memory to the transmit FIFO after filling only the low order 16-bits of the last DWORD in the FIFO, leaving

the upper 16 bits zeroed. This padded final DWORD is transmitted as the final symbol in the data frame. The CRC value transmitted at the end of the FIS is computed over the entire FIS, including any pad bytes in the final transmitted symbol.

- Having awareness of the amount of data requested, the device that receives the transmitted data has knowledge of the expected data transfer length. Upon receiving the data from the host, the device removes the 16-bit pad data in the upper 16 bits of the final 32-bit DWORD of received data.

4.2.18 First-Party DMA Write of Host Memory by Device – Odd Word Count

- When the device wishes to initiate a First-party DMA write to host memory for an odd word count, it constructs a First-party DMA Setup FIS that includes the base address and transfer count (for an odd number of words).
- Upon receiving the First-party DMA Setup FIS, the host adapter transfers the starting address and (odd) transfer count to the DMA controller and activates the DMA controller.
- The device transmits the data to the host in the form of one or more Data FIS. Because the transfer count is odd, the last 32-bit DWORD transmitted to the host has the high order 16 bits padded with zeroes. The CRC value transmitted at the end of the FIS is computed over the entire FIS, including any pad bytes in the final transmitted symbol.
- The host adapter receives the incoming data, and the DMA controller directs the received data from the receive FIFO to the appropriate host memory locations. The DMA controller has a transfer granularity of a 16-bit word.
- Upon receiving the final 32-bit DWORD of receive data, the DMA controller transfers the first half (low order 16 bits) to the corresponding final memory location at which point the DMA controller's transfer count is exhausted. The DMA controller drops the high order 16 bits of the final received DWORD since it represents data received beyond the end of the requested DMA transfer. The dropped 16 high order bits corresponds with the 16 bits of transmission pad inserted by the sender.

4.3 Native Command Queuing

This section details the operational characteristics of Native Command Queuing (NCQ) and how it is implemented in a SATA environment.

- NCQ was among the advanced and most anticipated features introduced in the Serial ATA II: Extensions to Serial ATA 1.0 Specification.
- NCQ is a powerful interface/disk technology designed to increase performance and endurance by allowing the drive to internally optimize the execution order of workloads.

Intelligent reordering of commands within the drive's internal command queue

- helps improve performance of queued workloads by minimizing mechanical positioning latencies on the drive;
- could prolong the life of a disc drive by minimizing or decreasing actuator movements.

This section will provide a basis for understanding how the features of NCQ apply to complete storage solutions and how software developers can enhance their applications to take advantage of Serial ATA NCQ, thereby creating higher performance applications.

4.3.1 Introduction to NCQ

Accessing media on mass storage devices, such as hard disc drives (HDD), can have a negative impact on overall system performance.

- Unlike other purely electrical components in a modern system, HDDs are still largely mechanical devices.
- Drives are hampered by the inertia of their mechanical components, which effectively limits the speed of media access and retrieval of data.
- Mechanical performance can be physically improved only up to a certain point, and these performance improvements usually come at an increased cost of the mechanical components.

However, intelligent internal management of the sequence of mechanical processes can greatly improve the efficiency of the entire workflow. The operative words are intelligent and internal, meaning that the drive itself has to assess the location of the target logical block addresses (LBAs) and then make appropriate decisions on the order in which commands should be executed to achieve the highest performance.

Native Command Queuing:

- NCQ is a command protocol in Serial ATA that allows multiple commands to be outstanding within a drive at the same time.
- Drives that support NCQ have an internal queue wherein outstanding commands can be dynamically rescheduled or reordered, along with the necessary tracking mechanisms, for outstanding and completed portions of the workload.
- NCQ also has a mechanism that allows the host to issue additional commands to the drive while the drive is seeking data for another command.

Operating systems such as Microsoft Windows™ and Linux™ are increasingly taking advantage of multithreaded software or processor-based Hyper-Threading Technology. These features have a high potential to create workloads where multiple commands are outstanding to the drive at the same time. By utilizing NCQ, the potential disc performance is increased significantly for these workloads.

HDD Basics

Hard drives are electromechanical devices and, therefore, hybrids of electronics and mechanical components. The mechanical portions of drives are subject to wear and tear and are also the critical limiting factor for performance. To understand the mechanical limitations, a short discussion of how data is laid out on a drive may be helpful.

Data is written to the drive in concentric circles, called tracks, starting from the outer diameter of the bottom platter, disc 0, and the first read/write head, head 0. When one complete circle on one side of the disc, track 0 on head 0, is complete, the drive starts writing to the next head on the other side of the disc, track 0 and head 1. When the track is complete on head 1, the drive starts writing to the next head, track 0 and head 2, on the second disc. This process continues until the last head on the last side of the final disc has completed the first track. The drive then will start writing the second track, track 1, with head 0 and continues with the same process as it did when writing track 0. This process results in concentric circles wherein, as writing continues, the data moves closer and closer to the inner diameter of the discs. A particular track on all heads, or sides of the discs, is collectively called a cylinder. Thus, data is laid out across the discs sequentially in cylinders starting from the outer diameter of the drive.

One of the major mechanical challenges is that applications rarely request data in the order that it is written to the disc. Rather, applications tend to request data scattered throughout all portions of the drive. The mechanical movement required to position the appropriate read/write head to the right track in the right rotational position is nontrivial.

The mechanical overheads that affect drive performance most are seek latencies and rotational latencies. Both seek and rotational latencies need to be addressed in a cohesive optimization algorithm.

The best-known algorithm to minimize both seek and rotational latencies is called Rotational Position Ordering. Rotational Position Ordering (or Sorting) allows the drive to select the order of command execution at the media in a manner that minimizes access time to maximize performance. Access time consists of both seek time to position the actuator and latency time to wait for the data to rotate under the head. Both seek time and rotational latency time can be several milliseconds in duration.

Earlier algorithms simply minimized seek distance to minimize seek time. However, a short seek may result in a longer overall access time if the target location requires a significant rotational latency period to wait for the location to rotate under the head. Rotational Position Ordering considers the rotational position of the disc as well as the seek distance when considering the order to execute commands. Commands are executed in an order that results in the shortest overall access time—the combined seek and rotational latency time—to increase performance.

Native Command Queuing allows a drive to take advantage of Rotational Position Ordering to optimally reorder commands to maximize performance.

Seek Latency Optimization

Seek latencies are caused by the time it takes the read/write head to position and settle over the correct track containing the target Logical Block Addressing (LBA). To satisfy several commands, the drive will need to access all target LBAs. Without queuing, the drive will have to access the target LBAs in the order that the commands are issued. However, if all of the commands are outstanding to the drive at the same time, the drive can satisfy the commands in the optimal order. The optimal order to reduce seek latencies would be the order that minimizes the amount of mechanical movement.

One rather simplistic analogy would be an elevator. If all stops were approached in the order in which the buttons were pressed, the elevator would operate in a very inefficient manner and waste an enormous amount of time going back and forth between the different target locations.

As trivial as it may sound, most of today's hard drives in the desktop environment still operate exactly in this fashion. Elevators have evolved to understand that reordering the targets will result in a more economic and, by extension, faster mode of operation. With Serial ATA, not only is reordering from a specific starting point possible but the reordering scheme is dynamic, meaning that at any given time, additional commands can be added to the queue. These new commands are either incorporated into an ongoing thread or postponed for the next series of command execution, depending on how well they fit into the outstanding workload.

To translate this into HDD technology, reducing mechanical overhead in a drive can be accomplished by accepting the queued commands (floor buttons pushed) and reordering them to efficiently deliver the data the host is asking for. While the drive is executing one command, a new command may enter the queue and be integrated in the outstanding workload. If the new command happens to be the most mechanically efficient to process, it will then be next in line to complete.

4.3.2 Benefits of Native Command Queuing (NCQ)

It is clear that there is a need for reordering outstanding commands in order to reduce mechanical overhead and consequently improve input/output (I/O) latencies.

- It is also clear, however, that simply collecting commands in a queue is not worth the silicon they are stored on.
- Efficient reordering algorithms take both the linear and the angular position of the target data into account and will optimize for both in order to yield the minimal total service time.
- This process is referred to as "command reordering based on seek and rotational optimization" or tagged command queuing.
- A side effect of command queuing and the reduced mechanical workload will be less mechanical wear, providing the additional benefit of improved endurance.
- Serial ATA II provides an efficient protocol implementation of tagged command queuing called Native Command Queuing (NCQ).

Native Command Queuing achieves high performance and efficiency through efficient command reordering. In addition, there are three new capabilities that are built into the Serial ATA protocol to enhance NCQ performance, including race-free status return, interrupt aggregation, and First-party DMA.

Race-Free Status Return Mechanism

This feature allows status to be communicated about any command at any time.

- There is no "handshake" required with the host for this status return to take place.
- The drive may issue command completions for multiple commands back-to-back or even at the same time.

Interrupt Aggregation

Generally, the drive interrupts the host each time it completes a command. The more interrupts, the bigger the host processing burden.

- However, with NCQ, the average number of interrupts per command can be less than one.
- If the drive completes multiple commands in a short time span—a frequent occurrence with a highly queued workload—the individual interrupts may be aggregated.
- In that case, the host controller only has to process one interrupt for multiple commands.

First-Party DMA (FPDMA)

Native Command Queuing has a mechanism that lets the drive set up the Direct Memory Access (DMA) operation for a data transfer without host software intervention. This mechanism is called First-party DMA (FPDMA).

- The drive selects the DMA context by sending a DMA Setup FIS to the host controller.
- This FIS specifies the tag of the command for which the DMA is being set up. Based on the tag value, the host controller will load the PRD table pointer for that command into the DMA engine, and the transfer can proceed without any software intervention.
- This is the means by which the drive can effectively reorder commands since it can select the buffer to transfer on its own initiative.

4.3.3 *Detailed Description of NCQ*

There are three main components to Native Command Queuing:

1. Building a queue of commands in the drive
2. Transferring data for each command
3. Returning status for the commands that were completed

Building a Queue

The drive must know when it receives a particular command whether it should queue the command or whether it should execute that command immediately. In addition, the drive must understand the protocol to use for a received command; the command protocol could be NCQ, DMA, PIO, and so forth.

1. The drive determines this information by the particular command opcode that is issued.
2. Therefore, in order to take advantage of NCQ, commands that are specifically for NCQ were defined.
3. There are two NCQ commands that were added as part of the NCQ definition in Serial ATA II:
   - Read FPDMA Queued
   - Write FPDMA Queued

The Write FPDMA Queued command inputs are shown in the table below.

1. The commands are extended LBA and sector count commands to accommodate the large capacities in today's drives.
2. The commands also contain a force unit access (FUA) bit for high availability applications.
   - When the FUA bit is set for a Write FPDMA Queued command, the drive will commit the data to media before returning success for the command.
   - By using the FUA bit as necessary on writes, the host can manage the amount of data that has not been committed to media within the drive's internal cache.

Write FPDMA Queued Inputs

<table>
<tr><th>Register</th><th>7</th><th>6</th><th>5</th><th>4</th><th>3</th><th>2</th><th>1</th><th>0</th></tr>
<tr><td>Features 7:0</td><td colspan="8">Sector Count 7:0</td></tr>
<tr><td>Features 15:8</td><td colspan="8">Sector Count 15:8</td></tr>
<tr><td>Count 7:0</td><td colspan="5">TAG</td><td colspan="3">Reserved</td></tr>
<tr><td>Count 15:8</td><td colspan="2">PRIO(1:0)</td><td colspan="6">Reserved</td></tr>
<tr><td>LBA Low</td><td colspan="8">LBA (7:0)</td></tr>
<tr><td>LBA Low</td><td colspan="8">LBA (31:24)</td></tr>
<tr><td>LBA Mid</td><td colspan="8">LBA (15:8)</td></tr>
<tr><td>LBA Mid</td><td colspan="8">LBA (39:32)</td></tr>
<tr><td>LBA High</td><td colspan="8">LBA (23:16)</td></tr>
<tr><td>LBA High</td><td colspan="8">LBA (47:40)</td></tr>
<tr><td>ICC</td><td colspan="8">ICC (7:0)</td></tr>
<tr><td>Device</td><td>FUA</td><td>1</td><td>0</td><td>0</td><td colspan="4">Reserved</td></tr>
<tr><td>Command</td><td colspan="8">61h</td></tr>
</table>

One interesting field is the TAG field in the Sector Count register.

- Each queued command issued has a tag associated with it.
- The tag is a shorthand mechanism used between the host and the device to identify a particular outstanding command.
- Tag values can be between 0 and 31, although the drive can report support for a queue depth less than 32. In this case, tag values are limited to the maximum tag value the drive supports.
- Having tag values limited to be between 0 and 31 has some nice advantages, including that status for all commands can be reported in one 32-bit value.
- Each outstanding command must have a unique tag value.

PRIO The Priority (PRIO) value is assigned by the host based on the priority of the command issued. The device should complete high priority requests in a more timely fashion than normal and isochronous requests. The device should complete isochronous requests prior to its associated deadline.
00b Normal Priority
01b Isochronous—deadline-dependent priority
10b High priority
11b Reserved

ICC The Isochronous Command Completion (ICC) field is valid when PRIO is set to a value of 01b. It is assigned by the host based on the intended deadline associated with the command issued. When a deadline has expired, the device shall continue to complete the command as soon as possible. This behavior may be modified by the host if the device supports the NCQ QUEUE MANAGEMENT command and supports the Deadline Handling subcommand. This subcommand allows the host to set whether the device shall abort (or continue processing) commands that have exceeded the time set in ICC. There are several parameters encoded in the ICC field: Fine or Coarse timing, Interval and the Max Time. The Interval indicates the time units of the Time Limit parameter.

If ICC Bit 7 is cleared to zero, then:

- The time interval is fine-grained
- Interval = 10 msec
- Time Limit = (ICC[6:0] + 1) * 10 msec
- Max Fine Time = 128 * 10 msec = 1.28 sec

If ICC Bit 7 is set to one (coarse encoding), then:

- The time interval is coarse-grained
- Interval = 0.5 sec;
- Time Limit = (ICC[6:0] + 1) * 0.5 sec
- Max Coarse Time = 128 * 0.5 sec = 64 sec

Read FPDMA Queued Inputs

Register	7	6	5	4	3	2	1	0
Features 7:0	Sector Count 7:0							
Features 15:8	Sector Count 15:8							
Count 7:0	TAG					Reserved		
Count 15:8	PRIO(1:0)		Reserved					
LBA Low	LBA (7:0)							
LBA Low	LBA (31:24)							
LBA Mid	LBA (15:8)							
LBA Mid	LBA (39:32)							
LBA High	LBA (23:16)							
LBA High	LBA (47:40)							
ICC	ICC (7:0)							
Device	FUA	1	0	0	Reserved			
Command	60h							

The Read and Write FPDMA Queued commands are issued just like any other command would be—that is, the taskfile is written with the particular register values and then the Command register is written with the command opcode. The difference between queued and non-queued commands is what happens after the command is issued.

- If a non-queued command was issued, the drive would transfer the data for that command and then clear the BSY bit in the Status register to tell the host that the command was completed.
- When a queued command is issued, the drive will clear BSY immediately, before any data is transferred to the host.
  - In queuing, the BSY bit is not used to convey command completion.
  - Instead, the BSY bit is used to convey whether the drive is ready to accept a new command.
  - As soon as the BSY bit is cleared, the host can issue another queued command to the drive.
  - In this way, a queue of commands can be built within the drive.

Transferring Data

NCQ takes advantage of a feature called First-party DMA (FPDMA) to transfer data between the drive and the host.

- First-party DMA allows the drive to have control over programming the DMA engine for a data transfer.
- This is an important enhancement since only the drive knows the current angular and rotational position of the drive head.
- The drive can then select the next data transfer to minimize both seek and rotational latencies.
- The First-party DMA mechanism is effectively what allows the drive to reorder commands in the most optimal way.

As an additional optimization, the drive can also return data out-of-order to further minimize the rotational latency. First-party DMA allows the drive to

- return partial data for a command;
- send partial data for another command; and then
- finish sending the data for the first command, if this is the most efficient means for completing the data transfers.

To program the DMA engine for a data transfer, the drive issues a DMA Setup FIS to the host, as shown below. There are a few key fields in the DMA Setup FIS that are important for programming the DMA engine.

DMA Setup FIS

	byte n+3	byte n+2	byte n+1	byte n
SOF	D23.1	D23.1	D21.5	K28.3
Data dword 0	Reserved	Reserved	a / i / d / r / PM Port	FIS Type (**41h**)
Data dword 1	0h			TAG
Data dword 2	0h			
Data dword 3	Reserved			
Data dword 4	DMA Buffer Offset			
Data dword 5	DMA Transfer Count			
Data dword 6	Reserved			
CRC	Dxx.y	Dxx.y	Dxx.y	Dxx.y
EOF	D21.6	D21.6	D21.5	K28.3

The **TAG** field identifies the tag of the command that the DMA transfer is for. For host memory protection from a rogue device, it is important to not allow the drive to indiscriminately specify physical addresses to transfer data to and from in host memory.

- The tag acts as a handle to the physical memory buffer in the host such that the drive does not need to have any knowledge of the actual physical memory addresses.
- Instead, the host uses the tag to identify which PRD table to use for the data transfer and programs the DMA engine accordingly.

The **DMA Buffer Offset** field is used to support out-of-order data delivery, also referred to as nonzero buffer offset within the specification.

- Nonzero buffer offset allows the drive to transfer data out-of-order or in-order, but in multiple pieces.

The **DMA Transfer Count** Field identifies the number of bytes to be transferred.

- The D bit specifies the direction of the transfer (whether it is a read or a write).
- The A bit is an optimization for writes called Auto-Activate, which can eliminate one FIS transfer during a write command.

Important Note

HBA designers note that new commands cannot be issued between the DMA Setup FIS and the completion of the transfer of the data for that DMA Setup FIS.

- It is important that the drive is not interrupted while actively transferring data since taking a new command may cause a hiccup in the transfer of data.
    - Thus, this restriction was added explicitly in the NCQ definition.
- Analogously, drives cannot send a Set Device Bits FIS before the completion of the data transfer for that DMA Setup FIS.
    - There is one exemption to this restriction: If an error is encountered before all of the data is transferred, a drive may send a Set Device Bits to terminate the transfer with error status.

After the DMA Setup FIS is issued by the drive, data is transferred using the same FISes that are used in a non-queued DMA data transfer operation.

Status Return

Command status return is **race-free** and allows interrupts for multiple commands to be aggregated.

- The host and the drive work in concert to achieve race-free status return without handshakes between the host and drive.
- Communication between the host and drive about which commands are outstanding is handled through a 32-bit register in the host called SActive.
- The SActive register has one bit allocated to each possible tag—that is, bit x shows the status of the command with tag x.
- If a bit in the SActive register is set, it means that a command with that tag is outstanding in the drive (or a command with that tag is about to be issued to the drive).
- If a bit in the SActive register is cleared, it means that a command with that tag is not outstanding in the drive.
- The host and drive work together to make sure that the SActive register is accurate at all times.

The host can set bits in the SActive register, whereas the device can clear bits in the SActive register.

- This ensures that updates to the SActive register require no synchronization between the host and the drive.
- Before issuing a command, the host sets the bit corresponding to the tag of the command it is about to issue.
- When the drive successfully completes a command, it will clear the bit corresponding to the tag of the command it just finished.

The drive clears bits in the SActive register using the Set Device Bits FIS, as shown below.

- The SActive field of the Set Device Bits FIS is used to convey successful status to the host.
- When a bit is set in the SActive field of the FIS, it means that the command with the corresponding tag has completed successfully.
- The host controller will clear bits in the SActive register corresponding to bits that are set to one in the SActive field of a received Set Device Bits FIS.

Set Device Bits FIS

<table>
<tr><td></td><td colspan="8">3</td><td colspan="8">2</td><td colspan="8">1</td><td colspan="8">0</td></tr>
<tr><td>0</td><td colspan="8">Error</td><td>R</td><td colspan="3">Status High</td><td>R</td><td colspan="3">Status Low</td><td>R</td><td>I</td><td>R</td><td colspan="5">Reserved</td><td colspan="8">FIS Type (A1h)</td></tr>
<tr><td rowspan="2">1</td><td colspan="32">SActive</td></tr>
<tr><td>31</td><td>30</td><td>29</td><td>28</td><td>27</td><td>26</td><td>25</td><td>24</td><td>23</td><td>22</td><td>21</td><td>20</td><td>19</td><td>18</td><td>17</td><td>16</td><td>15</td><td>14</td><td>13</td><td>12</td><td>11</td><td>10</td><td>9</td><td>8</td><td>7</td><td>6</td><td>5</td><td>4</td><td>3</td><td>2</td><td>1</td><td>0</td></tr>
</table>

Another key feature is that the Set Device Bits FIS can convey that multiple commands have completed at the same time.

- This ensures that the host will only receive one interrupt for multiple command completions.
- For example, if the drive completes the command with tag 3 and the command with tag 7 very close together in time, the drive may elect to send one Set Device Bits FIS that has both bit 3 and bit 7 set to one.
- This will complete both commands successfully and is guaranteed to generate only one interrupt.

Since the drive can return a Set Device Bits FIS without a host handshake, it is possible to receive two Set Device Bits FISes very close together in time.

- If the second Set Device Bits FIS arrives before host software has serviced the interrupt for the first, then the interrupts are automatically aggregated.
- This means that the host effectively only services one interrupt rather than two, thus reducing overhead.

4.3.4 How Applications Take Advantage of Queuing

The advantages of queuing are only realized if a queue of requests is built into the drive.

- One major issue in current desktop workloads is that many applications ask for one piece of data at a time, and often only ask for the next piece of data once the previous piece of data has been received.
- In this type of scenario, the drive is only receiving one outstanding command at a time.
- When only one command is outstanding at a time, the drive can perform no reordering and all the benefits of queuing are lost.

Note that with the advent of Hyper-Threading Technology, it is possible to build a queue even if applications issue one request at a time.

- Hyper-Threading Technology allows significantly higher amounts of multithreading to occur such that multiple applications are more likely to have I/O requests pending at the same time.
- However, the best performance improvement can only be achieved if applications are slightly modified to take advantage of queuing.

The modifications to take advantage of queuing are actually fairly minor.

- Today most applications are written to use synchronous I/O, also called blocking I/O.
  - In synchronous I/O, the function call to read from or write to a file does not return until the actual read or write is complete.

- In the future, applications should be written to use asynchronous I/O.
  - Asynchronous I/O is non-blocking, meaning that the function call to read from or write to a file will actually return before the request is complete.
- The application determines whether the I/O has completed by checking for an event to be signaled or by receiving a callback.
- Since the call returns immediately, the application can continue to do useful work, including issuing more read or write file functions.

The preferred method for writing an application that needs to make several different file accesses is to issue all of the file accesses using non-blocking I/O calls.

- Then, the application can use events or callbacks to determine when individual calls have completed.
- If there are a large number of I/Os, on the order of four to eight, by issuing all of the I/Os at the same time, the total time to retrieve all of the data can be cut in half.

4.3.5 Conclusion

Native Command Queuing (NCQ) has the potential to offer significant performance advantages. The benefits of NCQ are realized when a queue of commands is built up in the drive such that the drive can optimally reorder the commands to reduce both seek and rotational latency. NCQ delivers an efficient solution through features in the Serial ATA protocol including race-free status return, interrupt aggregation, and First-party DMA.

The NCQ performance advantage can only be realized when a queue is built in the drive. Therefore, it is imperative that applications and operating systems use asynchronous I/O where possible and keep the drive queue active with multiple commands at a time. Independent Software Vendors (ISVs) and operating system providers are encouraged to start utilizing asynchronous I/O in order to take advantage of the benefits of NCQ.

4.3.6 NCQ Example

- In the following example (see Figures 4.1–4.3), a Host issues two ReadFPDMAQueued commands to a device
  - 1st Command Tag = 0
  - 2nd Command Tag = 5
- Device chooses to execute command Tag = 5 first
  - DMA setup and data transfer for command 5 occurs
  - Command 5 completes
- Device then executes command Tag = 0
  - DMA setup and data transfer for command 0 occurs
  - Command 0 completes

Commands Queued

Figure 4.1 shows two Read FPDMA Queued commands being sent to the device. As the host sends the commands, it sets bits in the SActive register that correspond to the TAGs that it wants to use. Upon

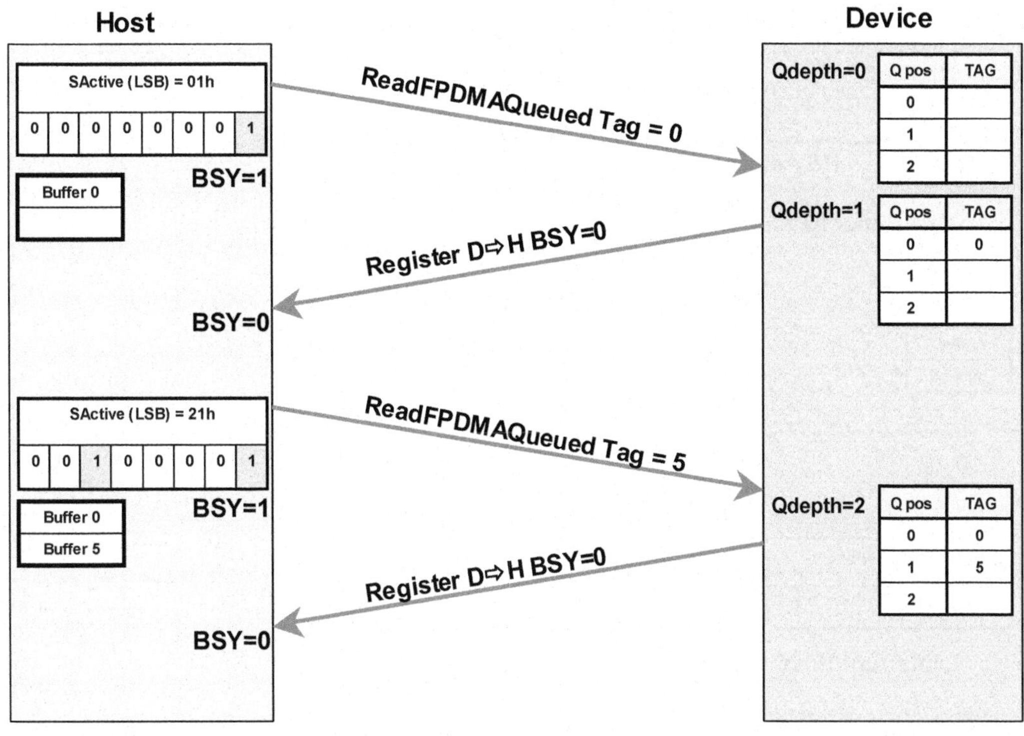

Figure 4.1 Host issues two Read FPDMAQueued commands.

queuing of the commands, the device returns a Register – Device to Host FIS, which clears the BSY bit in the host, thus allowing it to send another command.

Command 5 Executes

In Figure 4.2, the device reorders the two commands based on Rotational Positional Ordering and determines that the command with the TAG of 5 should be executed first. The device initiates the command by issuing a DMA Setup FIS to the Host. When the host receives the DMA Setup FIS, it determines the TAG value and activates the DMA control logic, which in turn programs the proper Data Pointers. Now the First-party DMA transfer can commence. The Device transmits the Data FISes required to complete the transfer and then sends the Set Device Bits FIS to clear the SActvie bit associated with the command that completed.

Command 0 Executes

The device initiates the second command, by issuing a DMA Setup FIS to the Host (see Figure 4.3). When the host receives the DMA Setup FIS, it determines the TAG value and activates the DMA control logic, which in turn programs the proper Data Pointers. Now, the First-party DMA transfer can commence. The Device transmits the Data FISes required to complete the transfer and then sends the Set Device Bits FIS to clear the SActvie bit associated with the command that completed (0).

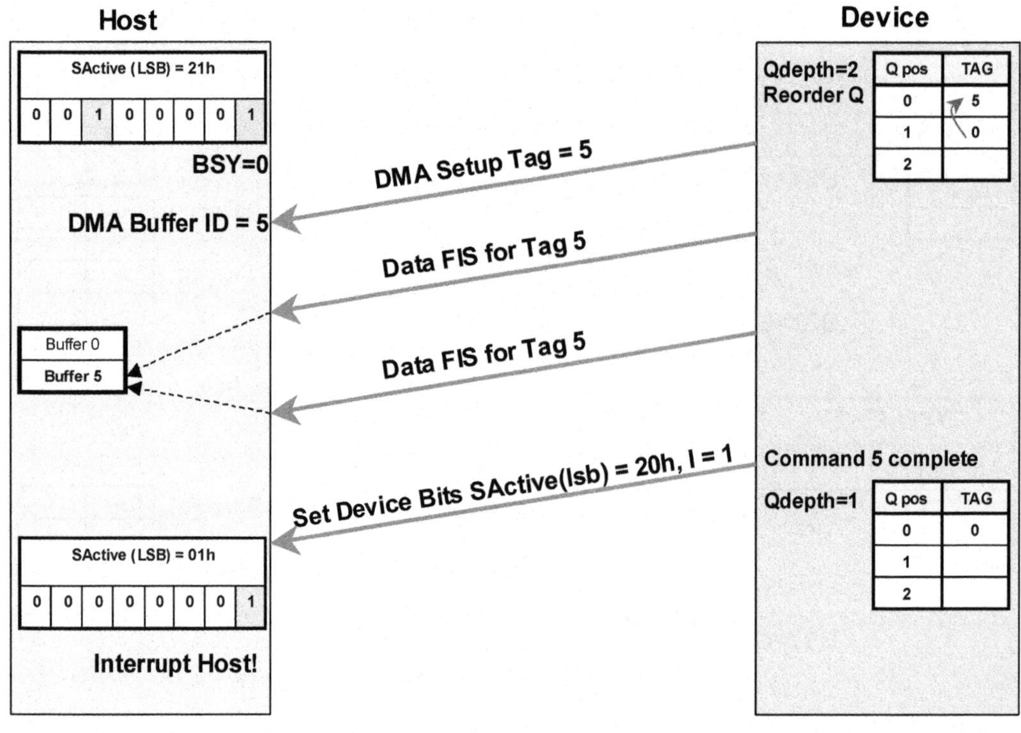

Figure 4.2 Device returns data for one Read FPDMAQueued command (Tag 5).

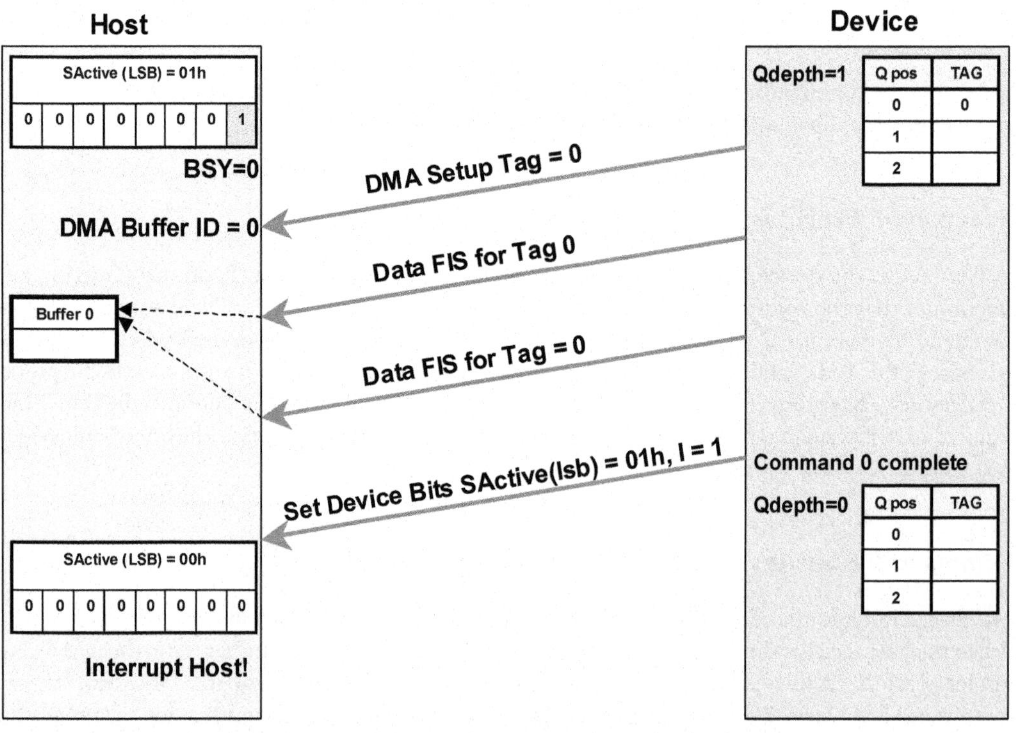

Figure 4.3 Device returns data for last Read FPDMAQueued command (Tag 0).

NCQ Example 2

- In the following example (see Figures 4.4–4.6), a host issues two WriteFPDMAQueued commands to a device
  - 1st Command Tag = 2
  - 2nd Command Tag = 3
  - DMA Auto-Activate enabled by the device for each command
    - Eliminates the need for DMA Activate FIS from the device before Data FIS
- Then the Host issues one ReadFPDMAQueued command to a device
  - 3rd Command Tag = 4
- Device chooses to execute command Tag = 4 first
  - DMA setup and data transfer for command 4 occurs
  - Command 4 completes
- Device then executes command Tag = 2
  - DMA setup and data transfer for command 2 occurs
- Device then executes command Tag = 3
  - DMA setup and data transfer for command 3 occurs
  - Commands 2 and 3 complete and are reported in a single FIS

In Figure 4.4, the commands are sent by the host and queued by the device. Each time the host sends a command, it sets a bit in the SActive register that corresponds to the TAG value it wants to set

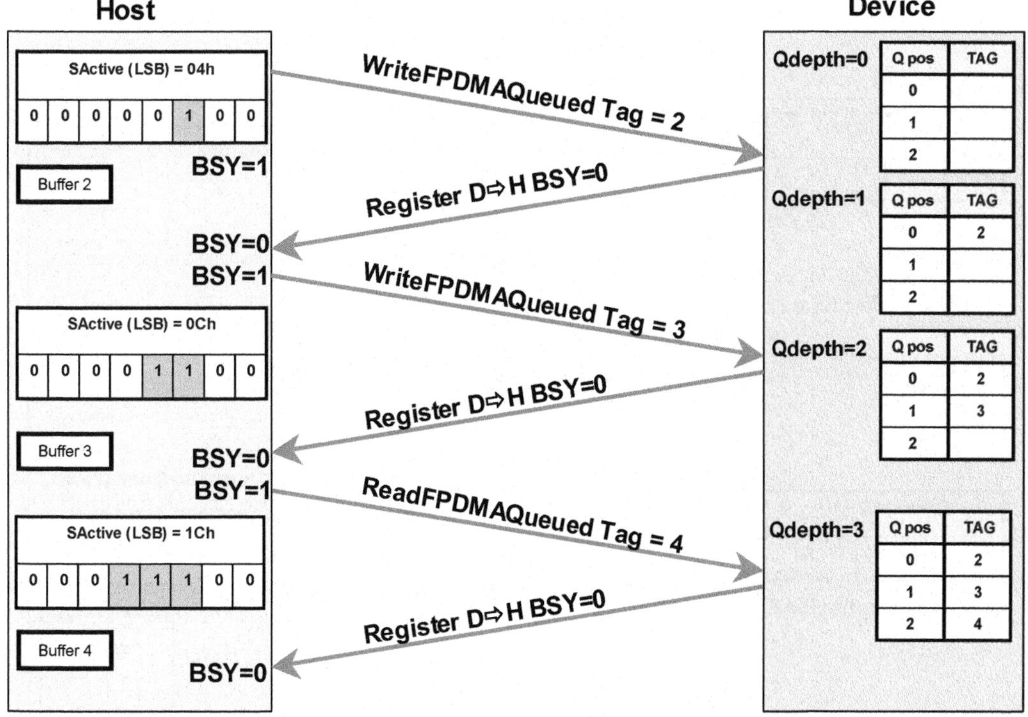

Figure 4.4 Host queues two Writes and one Read command.

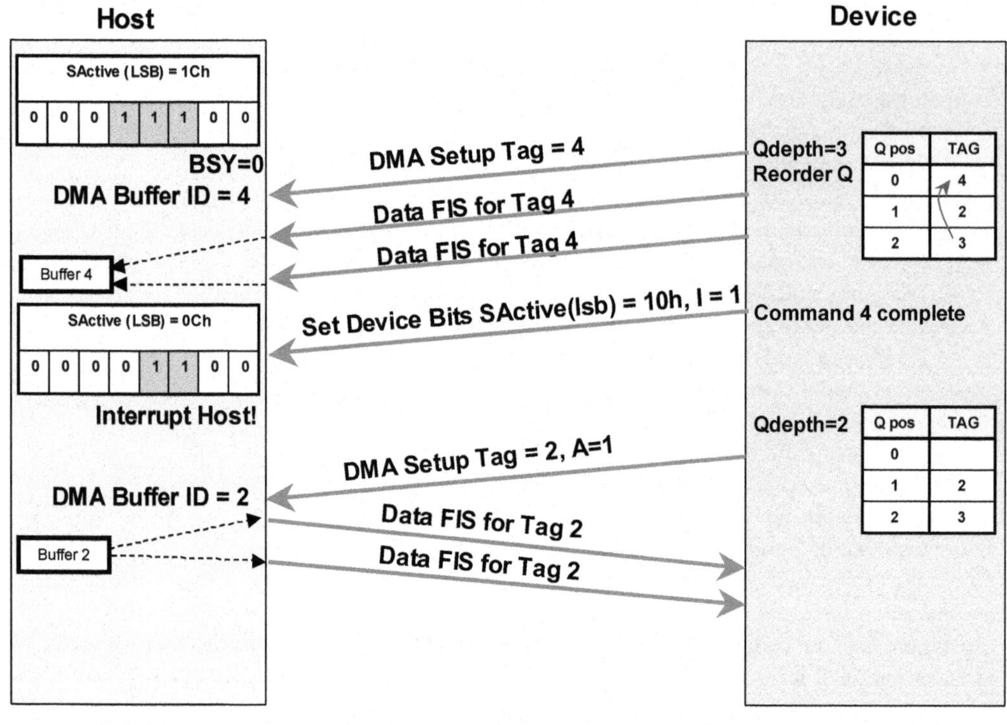

Figure 4.5 Device reorders commands—returns data for Tag 4 and requests data for Tag 2.

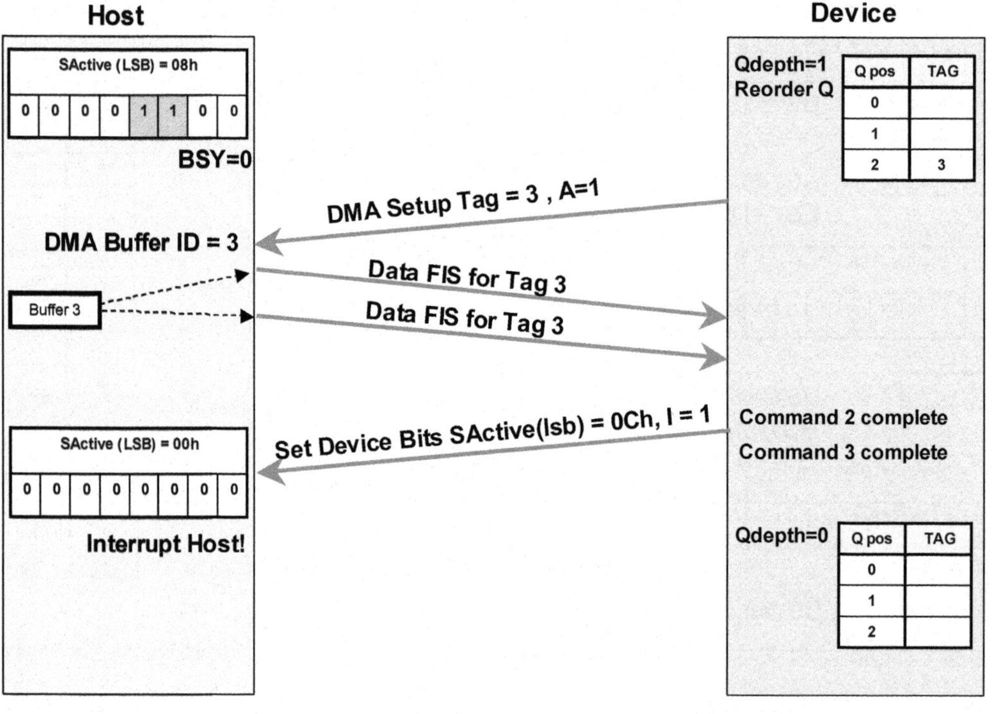

Figure 4.6 Device requests data for Tag 3.

in the Register – Host to Device FIS that carries the command. In this example, the host starts with a TAG value of 2 and increments the TAG value as it sends each command. For each command that is sent by the host, the device sends a Register – Device to Host FIS to clear the BSY bit so that another command can be queued.

Again, as shown in Figure 4.5, the device reordered the commands based on the Rotational Positional Ordering algorithm and determines that it will execute the Read command (TAG = 4) first. The device issues a DMA Setup FIS to the host to initiate the transfer. When the host receives the DMA Setup FIS, it determines the TAG value, activates the DMA control logic, and assigns the appropriate buffer pointers. When the host receives the Data FISes from the device, the DMA control logic sends the data to the appropriate buffer. The device completes the transfer by sending a Set Device Bits FIS to the host, which in turn generates an interrupt and clears the appropriate SActive value associated with the completed command.

Figure 4.5 also shows the next command in the queue being executed. In this instance, the Write operation associated with TAG 2 executes. Note that in this particular example, the Auto-Activate bits were set in the DMA Setup FIS (A = 1). The Auto-Activate bit will eliminate the need to send a DMA Activate FIS from the device to the host for write operations. If the Auto-Activate bit was not set then each Data FIS would require a DMA Activate FIS prior to the Data FIS from the host (an example of this can be seen later in the analyzer trace). In addition, note that no Set Device Bits FIS was sent at this time. This may be due to the fact that the device has yet to commit the data to the medium (physical platters).

In Figure 4.6, the device returns the data associated with command three. Again, the Auto-Activate bit is set in the DMA Setup FIS, thus eliminating the requirement to send DMA Activates prior to the Data FISes. When the DMA Setup FIS is received at the host, it determines the TAG value, activates the DMA control logic, and assigns the appropriate buffer pointers. When the host receives the Data FISes from the device, the DMA control logic sends the data to the appropriate buffer. The device completes the transfer by sending a Set Device Bits FIS to the host, which in turn generates an interrupt and clears the appropriate SActive value associated with the completed command. This example shows how the status for two outstanding commands can be sent in a single Set Device Bits FIS. This increases the performance from two perspectives—less overhead associated with issuing two Set Device Bits FISes and a decreased number of interrupts that need to be handled by the host.

4.4 NCQ Queue Management

The NCQ Queue Management feature allows the host to manage the outstanding NCQ commands and/or affect the processing of NCQ commands.

- The NCQ QUEUE MANAGEMENT command is a non-data NCQ command. Only specified NCQ QUEUE MANAGEMENT subcommands are executed as Immediate NCQ commands.
- If NCQ is disabled and an NCQ QUEUE MANAGEMENT command is issued to the device, then the device shall abort the command with the ERR bit set to one in the Status register and the ABRT bit set to one in the Error register.
- This command is prohibited for devices that implement the PACKET feature set.
- The queuing behavior of the device depends on which subcommand is specified.

NCQ Queue Management Inputs

Register	7	6	5	4	3	2	1	0
Features 7:0	Subcommand Specific				Subcommand			
Features 15:8	Reserved							
Count 7:0	TAG					Reserved		
Count 15:8	PRIO(1:0)		Reserved					
LBA Low (7:0)	Subcommand Specific (TTAG)					Reserved		
LBA Low (31:24)	Reserved							
LBA Mid (15:8)	Reserved							
LBA Mid (39:32)	Reserved							
LBA High (23:16)	Reserved							
LBA High (47:40)	Reserved							
Device	R	1	R	0	Reserved			
Command	63h							

Subcommand

0h	Abort NCQ queue
1h	Deadline handling
2h-Fh	Reserved

4.4.1 ABORT NCQ QUEUE Subcommand (0h)

A Subcommand set to 0h specifies the ABORT NCQ QUEUE subcommand.

- The ABORT NCQ QUEUE subcommand is an immediate NCQ command.
- Support for this subcommand is indicated in the NCQ Queue Management Log.
- The ABORT NCQ QUEUE subcommand shall affect only those NCQ commands for which the device has indicated command acceptance before accepting this NCQ QUEUE MANAGEMENT command.

Register	7	6	5	4	3	2	1	0
Features 7:0	Abort Type				0h			
Features 15:8	Reserved							
Count 7:0	TAG					Reserved		
Count 15:8	Reserved							
LBA Low (7:0)	TTAG					Reserved		
LBA Low (31:24)	Reserved							
LBA Mid (15:8)	Reserved							
LBA Mid (39:32)	Reserved							
LBA High (23:16)	Reserved							
LBA High (47:40)	Reserved							
Device	R	1	R	0	Reserved			
Command	63h							

NCQ QUEUE MANAGEMENT, Abort NCQ Queue – Command Definition

Type	Abort Type	Description
0h	Abort All	The device shall attempt to abort all outstanding NCQ commands.
1h	Abort Streaming	The device shall attempt to abort all outstanding NCQ streaming commands. All non-streaming NCQ commands shall be unaffected.
2h	Abort Non-Streaming	The device shall attempt to abort all outstanding NCQ non- streaming commands. All NCQ streaming commands shall be unaffected.
3h	Abort Selected	The device shall attempt to abort the outstanding NCQ command associated with the tag represented in TTAG field.
4h-Fh	Reserved	

4.4.2 NCQ Deadline Handling Subcommand (1h)

A subcommand set to 1h specifies the Deadline Handling Subcommand.

- This subcommand controls how NCQ Streaming commands are processed by the device.
- Support for this subcommand is indicated in the NCQ Queue Management Log.

NCQ QUEUE MANAGEMENT, Deadline Handling – Command Definition

Register	7	6	5	4	3	2	1	0
Features 7:0	Reserved		RDNC	WDNC	1h			
Features 15:8	Reserved							
Count 7:0	TAG					Reserved		
Count 15:8	Reserved							
LBA Low (7:0)	Reserved							
LBA Low (31:24)	Reserved							
LBA Mid (15:8)	Reserved							
LBA Mid (39:32)	Reserved							
LBA High (23:16)	Reserved							
LBA High (47:40)	Reserved							
Device	R	1	R	0	Reserved			
Command	63h							

WDNC (Write Data Not Continue)

- If the WDNC bit is cleared to zero, then the device may allow WRITE FPDMA QUEUED command completion times to exceed what the ICC parameter specified.
- If the WDNC bit is set to one, then all the WRITE FPDMA QUEUED commands shall be completed by the time specified by the ICC timer value, otherwise the device shall return command aborted for all outstanding commands.
- WDNC is only applicable to WRITE FPDMA QUEUED commands with PRIO set to 01b (Isochronous—deadline-dependent priority).

RDNC (Read Data Not Continue)

- If the RDNC bit is cleared to zero, then the device may allow READ FPDMA QUEUED command completion times to exceed what the ICC parameter specified.
- If the RDNC bit is set to one, then all READ FPDMA QUEUED commands shall be completed by the time specified by the ICC timer value; otherwise, the device shall return command aborted for all outstanding commands.
- RDNC is only applicable to READ FPDMA QUEUED commands with PRIO set to 01b (Isochronous—deadline-dependent priority).

The state of the WDNC and RDNC bits shall be preserved across software resets and COMRESETs (via Software Setting Preservations) and shall not be preserved across power cycles.

4.5 Chapter Review

1. Define FIS:

   a. Frame Information System
   b. Fragment Information Structure
   c. Frame Information Structure
   d. Frame Intelligent System

2. What is the FIS code for a device to host FIS?

   a. 27h
   b. 34h
   c. 39h
   d. 41h

3. The Register FIS – Host to Device FIS is used to transfer ___________ information to the device.

   a. Data
   b. Command
   c. Status
   d. DMA setup

4. The Register FIS – Device to Host FIS is used to transfer ___________ information to the host.

   a. Data
   b. Command
   c. Status
   d. DMA setup

The **DMA Activate** – Device to Host FIS is used by the device to signal the host to proceed with a DMA transfer of data from the host to the device (Write Operation).

The **First-party DMA Setup** – Device to Host or Host to Device FIS is the mechanism by which First-party DMA access to host memory is initiated.

The **Set Device Bits** – Device to Host FIS is used by the device to load Shadow Register Block bits for which the device has exclusive write access. These bits are the eight bits of the Error register and six of the eight bits of the Status register.

The **Data** – Host to Device and the Data – Device to Host FISes are used for transporting payload data, such as the data read from or written to a number of sectors on a hard drive.

The **BIST Activate** FIS is used to place the receiver in one of three loopback modes. The BIST Activate FIS is a bi-directional request in that it can be sent by either the host or the device.

The **SStatus register** conveys the current state of the interface and host adapter.

Bits	31	30	29	28	27	26	25	24	23	22	21	20	19	18	17	16	15	14	13	12	11	10	9	8	7	6	5	4	3	2	1	0
SCR0	Reserved (0)																				IPM				SPD				DET			

DET The DET value indicates the interface device detection and Phy state.

0000 No device detected and Phy communication not established

0001 Device presence detected but Phy communication not established

0011 Device presence detected and Phy communication established

0100 Phy in offline mode and is disabled or running in a BIST loopback mode

SPD The SPD value indicates the negotiated interface communication speed established

0000 No negotiated speed (device not present or communication not established)

0001 Generation 1 communication rate negotiated

0010 Generation 2 communication rate negotiated

0011 Generation 3 communication rate negotiated

IPM The IPM value indicates the current interface power management state

0000 Device not present or communication not established

0001 Interface in active state

0010 Interface in PARTIAL power management state

0110 Interface in SLUMBER power management state

The SControl register provides the interface by which software can control SATA interface capabilities.

- Writes to the SControl register result in an action being taken by the host adapter or interface.
- Reads from the register return the last value written to it.

Bits	31	30	29	28	27	26	25	24	23	22	21	20	19	18	17	16	15	14	13	12	11	10	9	8	7	6	5	4	3	2	1	0
SCR2	Reserved (0)												PMP				SPM				IPM				SPD				DET			

SPD Speed capabilities.

0000b No speed negotiation restrictions.

0001b Limit speed negotiation to a rate not > Generation 1 communication rate.

0010b Limit speed negotiation to a rate not > Generation 2 communication rate.

0011b Limit speed negotiation to a rate not > Generation 3 communication rate.

IPM Interface power management can be invoked via the interface power management capabilities.

0000b	No interface power management state restrictions.
0001b	Transitions to the PARTIAL power management state disabled.
0010b	Transitions to the SLUMBER power management state disabled.
0011b	Transitions to both the PARTIAL and SLUMBER power management states disabled.

SPM SATA Power Mode

0000b No power management state transition requested.
0001b Transition to the Partial power management state initiated.
0010b Transition to the Slumber power management state initiated.
0100b Transition to the active power management state initiated.

PMP The Port Multiplier Port (PMP) field represents the 4-bit value to be placed in the PM Port field of all transmitted FISes.

The **SError register** conveys supplemental interface error information to complement the error information available in the Shadow Error register.

Bits	31	30	29	28	27	26	25	24	23	22	21	20	19	18	17	16	15	14	13	12	11	10	9	8	7	6	5	4	3	2	1	0
SCR1	DIAG																ERR															
	R				A	X	F	T	S	H	C	D	B	W	I	N	R	R	R	R	E	P	C	T	R	R	R	R	R	R	M	I

R = Reserved

ERR fields

C	Non-recovered persistent communication or data integrity error
E	Internal error
I	Recovered data integrity error
M	Recovered communications error
P	Protocol error
T	Non-recovered transient data integrity error

DIAG The DIAG field contains diagnostic error information

A	Port Selector presence detected
B	10b to 8b Decode error
C	CRC Error
D	Disparity Error
F	Unrecognized FIS type
I	Phy Internal Error
N	PhyRdy change
H	Handshake error
S	Link Sequence Error
T	Transport state transition error
W	COMWAKE signal was detected
X	Exchange

The **SActive register** is used during Native Command Queuing (NCQ).

- This register can be written to when a Set Device Bits FIS is received from a device.
- Each bit position represents a Q Tag value (e.g. Tag 0, Tag 1, Tag 31)

Bits	31	30	29	28	27	26	25	24	23	22	21	20	19	18	17	16	15	14	13	12	11	10	9	8	7	6	5	4	3	2	1	0
SCR3	SActive																															
	31	30	29	28	27	26	25	24	23	22	21	20	19	18	17	16	15	14	13	12	11	10	9	8	7	6	5	4	3	2	1	0

Native Command Queuing

- NCQ was the major features introduced in the Serial ATA II.
- NCQ is a powerful interface/disk technology designed to increase performance and endurance by allowing the drive to internally optimize the execution order of workloads.

Intelligent reordering of commands within the drive's internal command queue

- Can improve performance of queued workloads by minimizing mechanical positioning latencies on the drive.
- Could prolong the life of a disc drive by minimizing or decreasing actuator movements.

Building Queues

The drive must know when it receives a particular command whether it should queue the command or whether it should execute that command immediately.

1. The drive must understand the protocol to use for a received command;
   - The command protocol could be NCQ, DMA, PIO, and so forth.
   - The drive determines this information by the particular command opcode that is issued.
2. Commands were specifically defined for NCQ.
   - Read FPDMA Queued
   - Write FPDMA Queued
3. Only FPDMA Queued commands can be executed when the device is queuing.

Chapter 5

SATA Link Layer

Objectives

This chapter will cover the Link layer functions that

- transmit and receive frames;
- transmit primitives based on control signals from the Transport layer;
- receive primitives from the Physical layer, which are converted to control signals to the Transport layer.

5.1 Link Layer Responsibilities

Figure 5.1 shows the key elements of the Link layer and outlines the major focal points of this section.

5.1.1 Frame Transmission Topics

When requested by the Transport layer to transmit a frame, the Link layer provides the following services:

1. Negotiates with its peer Link layer to transmit a frame; resolves arbitration conflicts if both host and device request transmission
2. Inserts a frame envelope around Transport layer data (i.e., SOF, CRC, EOF, etc.)
3. Receives data in the form of DWORDs from the Transport layer
4. Calculates CRC on Transport layer data
5. Transmits frames
6. Provides frame flow control in response to requests from the FIFO or the peer Link layer
7. Receives frame receipt acknowledgment from the peer Link layer
8. Reports good transmission or Link/Physical layer errors to the Transport layer
9. Performs 8b10b encoding
10. Scrambles (transforms) control and data DWORDs in such a way as to distribute the potential EMI emissions over a broader range

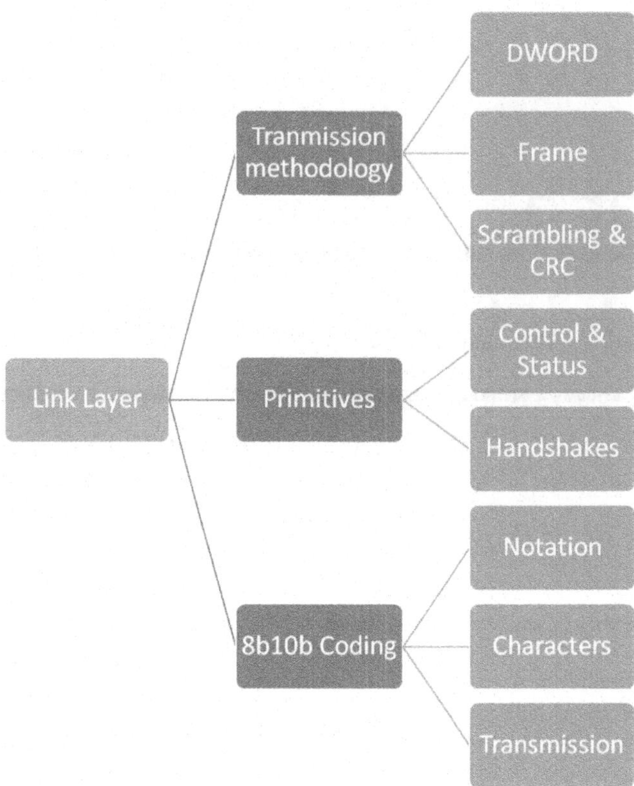

Figure 5.1 Link layer hierarchy.

5.1.2 Frame Reception Topics

When data is received from the Physical layer, the Link layer provides the following services:

1. Acknowledges to the peer Link layer readiness to receive a frame
2. Receives data in the form of encoded characters from the Physical layer
3. Decodes the encoded 8b10b character stream into aligned DWORDs of data
4. Removes the envelope around frames (i.e., SOF, CRC, EOF)
5. Calculates CRC on the received DWORDs
6. Provides frame flow control in response to requests from the FIFO or the peer Link layer
7. Compares the calculated CRC to the received CRC
8. Reports good reception or Link/Physical layer errors to the Transport layer and the peer Link layer
9. Descrambles (untransforms) the control and data DWORDs received from a peer Link layer

5.2 Transmission Methodology

The information on the serial line is a sequence of 8b10b encoded characters. The smallest unit of communication is known as a DWORD (see Chapter 2, Table 2.20—SATA Character Format). The contents of each DWORD provide link level control information to transfer information between hosts and devices.

5.2.1 SATA Structures

SATA uses two types of structures:

- Primitives (DWORD)
- Frames

5.2.2 DWORDs

In Chapter 2, we learned about the structure of DWORDs. The DWORD is made up of 4 characters, as seen in Table 5.1).

Table 5.1 SATA Character Format

<table>
<tr><td>31</td><td>30</td><td>29</td><td>28</td><td>27</td><td>26</td><td>25</td><td>24</td><td>23</td><td>22</td><td>21</td><td>20</td><td>19</td><td>18</td><td>17</td><td>16</td><td>15</td><td>14</td><td>13</td><td>12</td><td>11</td><td>10</td><td>09</td><td>08</td><td>07</td><td>06</td><td>05</td><td>04</td><td>03</td><td>02</td><td>01</td><td>00</td></tr>
<tr><td colspan="8">4th character</td><td colspan="8">3rd character</td><td colspan="8">2nd character</td><td colspan="8">1st character</td></tr>
<tr><td colspan="8">byte 3</td><td colspan="8">byte 2</td><td colspan="8">byte 1</td><td colspan="8">byte 0</td></tr>
<tr><td colspan="16">word 1</td><td colspan="16">word 0</td></tr>
<tr><td colspan="32">DWORD</td></tr>
</table>

Transmission characteristics include the following:

- Serial ATA is a word-oriented architecture.
  - Words are known as DWORDs.
  - DWORDs consists of 4 bytes (32 bits).
    - DWORD stands for double word.
- The minimum size data unit in SATA is the DWORD.
  - Not all fields are necessarily word size, but word boundaries are preserved on all data transmissions.
- There are two types of DWORDs:
  - **Data** DWORD
    - Used to express Transport layer information
  - **Primitive**
    - Used for low-level link control—that is, Link layer protocol
- Each 8-bit byte is encoded into a 10-bit character by the transmitter.
  - This is known as 8b10b encoding.
  - Each 10-bit character is decoded back into 8 bits by the receiver.
- All DWORDs transmitted on a SATA link are encoded.
  - A double word consists of 32 bits.
  - After encoding, a transmission word consists of 40 bits.
- This process is implemented in hardware to achieve the necessary speed.
  - 8b10b encoding was developed by IBM for use in its ESCON channel architecture.
  - 8b10b encoding is used on all of today's high-speed serial interfaces, including Fibre Channel, Gigabit Ethernet, Serial Attached SCSI, InfiniBand, and PCI Express.
- Link layers
  - The Link layer need not be cognizant of the content of frames.
  - Host and device Link layer state machines differ only in the fact that the ***host will back off*** in the event of a collision when attempting to transmit a frame—that is, X_RDY's being sent by both devices simultaneously.

5.2.3 *Primitive Structure*

A primitive consists of a single DWORD and is the smallest protocol unit of information that may be exchanged between a host and a device (see Figure 5.2).

  - Primitives are used primarily to convey real-time state information, to control the transfer of information and coordinate host/device communication.
  - All bytes in a primitive are constants, and the first byte is always a control character. Since all of the bytes are constants, a primitive cannot be used to convey variable information.
- SATA defines primitives for a variety of purposes:
  - To convey real-time state information
  - To control the transfer of information between device and host
  - To provide host/device coordination
- Primitives can be divided into categories:
  - Physical layer management
  - Frame delimiters
  - Frame transmission and flow control
  - Frame delivery status notification
  - Power management

4th character	3rd character	2nd character	1st character
Generic Primitive			
Dxx.y	Dxx.y	Dxx.y	K28.3

ALIGN Primitive			
D27.3	D10.2	D10.2	K28.5

All other SATA Primitives			
Dxx.y	Dxx.y	Dxx.y	K28.3
Defines primitive			Special Character

Data DWORD			
Dxx.y	Dxx.y	Dxx.y	Dxx.y

Figure 5.2 Primitive and Data DWORD format.

5.3 Frame Structure

A frame consists of multiple DWORDs:

- Always starts with an SOF primitive
- Followed by a user payload, called a Frame Information Structure (FIS), and a Cyclic Redundancy Check (CRC)
- Ends with an EOF primitive (see Figure 5.3)

					Frame Payload													
SYNC	SYNC	SYNC	SYNC	SOF	DW00	DW01	DW02	DW03	DW04	DW05	DWn-1	DWn	CRC	EOF	SYNC	SYNC	SYNC	SYNC
Primitives					Data Dwords (2048 max.)									Primitives				

Figure 5.3 Frame format.

Frame characteristics:

- The CRC is defined to be the last non-primitive DWORD immediately preceding the EOF primitive.
- Payload
    - Made up of Data DWORDS
    - Size is FIS dependent
    - Data frame payload accommodates 2048 DWORDS (8k)

Frames are preceded by primitives or scrambled DWORDs, as shown in Figure 5.4. Serial links require a continuous stream of transmission bits to keep the transmission lines charged and valid. This is unlike parallel interfaces, wherein the interface is only driven when information is being transmitted.

Some number of flow control primitives (HOLD or HOLDA, or a CONT stream to sustain a HOLD or HOLDA state) are allowed between the SOF and EOF primitives to throttle data flow for speed-matching purposes. HOLD primitives will be covered in detail in the section "HOLD/HOLDA Primitives" (in Section 5.4.2).

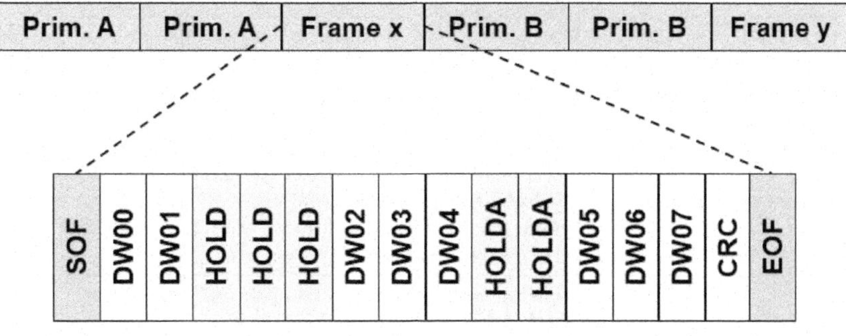

Figure 5.4 Frame transmission sequence.

5.4 Primitives

Table 5.2 defines and describes all SATA primitives.

Table 5.2 SATA Primitives and Their Descriptions

Primitive	Name	Description	Usage
ALIGN	Physical layer control	Upon receipt of an ALIGN, the physical layer readjusts internal operations as necessary to perform its functions correctly. This primitive is always sent in pairs; there is no condition where an odd number of ALIGN primitives will be sent (except as noted for retimed loopback).	Repeated
CONT	Continue repeating previous primitive	The CONT primitive allows long strings of repeated primitives to be eliminated. The CONT primitive is used to signal that the previously transmitted primitive pair is to be sustained until another primitive is received.	Single
DMAT	DMA terminate	This primitive is sent as a request to the transmitter to terminate a DMA data transmission early by computing a CRC on the data sent and ending with a EOF primitive. The transmitter context is assumed to remain stable after the EOF primitive has been sent.	Single
EOF	End of frame	EOF marks the end of a frame. The previous non-primitive DWORD is the CRC for the frame.	Single
HOLD	Hold data transmission	HOLD is transmitted in place of payload data within a frame when the transmitter does not have the next payload data ready for transmission. HOLD is also transmitted on the backchannel when a receiver is not ready to receive additional payload data.	Repeated*
HOLDA	Hold acknowledge	This primitive is sent by a transmitter as long as the HOLD primitive is received by its companion receiver.	Repeated*
PMACK	Power management acknowledge	Sent in response to a PMREQ_S or PMREQ_P when a receiving node is prepared to enter a power mode state.	Repeated
PMNAK	Power management denial	Sent in response to a PMREQ_S or PMREQ_P when a receiving node is not prepared to enter a power mode state or when power management is not supported.	Repeated
PMREQ_P	Power management request to partial	This primitive is sent continuously until PMACK or PMNAK is received. When PMACK is received, current node (host or device) will stop PMREQ_P and enters the partial power management state.	Repeated*
PMREQ_S	Power management request to Slumber	This primitive is sent continuously until PMACK or PMNAK is received. When PMACK is received, current node (host or device) will stop PMREQ_S and enters the Slumber power management state.	Repeated*
R_ERR	Reception error	Current node (host or device) detected error in received payload.	Repeated*
R_IP	Reception in progress	Current node (host or device) is receiving payload.	Repeated*
R_OK	Reception with no error	Current node (host or device) detected no error in received payload.	Repeated*
R_RDY	Receiver ready	Current node (host or device) is ready to receive payload.	Repeated*

(Continued on following page)

Table 5.2 SATA Primitives and Their Descriptions (*Continued*)

Primitive	Name	Description	Usage
SOF	Start of frame	Start of a frame. Payload and CRC follow to EOF.	Single
SYNC	Synchronization	Synchronizing primitive—always idle.	Repeated*
WTRM	Wait for frame termination	After transmission of any of the EOFs, the transmitter will transmit WTRM while waiting for reception status from receiver.	Repeated*
X_RDY	Transmission data ready	Current node (host or device) has payload ready for transmission	Repeated*
* This primitive is continued by use of the CONT primitive.			

5.4.1 Primitive Encoding

Table 5.3 shows the primitive encoding for all SATA primitives.

Table 5.3 Primitive Encoding for All SATA Primitives

Primitive name	Byte 3 contents	Byte 2 contents	Byte 1 contents	Byte 0 contents
ALIGN	D27.3	D10.2	D10.2	K28.5
CONT	D25.4	D25.4	D10.5	K28.3
DMAT	D22.1	D22.1	D21.5	K28.3
EOF	D21.6	D21.6	D21.5	K28.3
HOLD	D21.6	D21.6	D10.5	K28.3
HOLDA	D21.4	D21.4	D10.5	K28.3
PMACK	D21.4	D21.4	D21.4	K28.3
PMNAK	D21.7	D21.7	D21.4	K28.3
PMREQ_P	D23.0	D23.0	D21.5	K28.3
PMREQ_S	D21.3	D21.3	D21.4	K28.3
R_ERR	D22.2	D22.2	D21.5	K28.3
R_IP	D21.2	D21.2	D21.5	K28.3
R_OK	D21.1	D21.1	D21.5	K28.3
R_RDY	D10.2	D10.2	D21.4	K28.3
SOF	D23.1	D23.1	D21.5	K28.3
SYNC	D21.5	D21.5	D21.4	K28.3
WTRM	D24.2	D24.2	D21.5	K28.3
X_RDY	D23.2	D23.2	D21.5	K28.3

5.4.2 Detailed Primitives Descriptions

This section provides detailed descriptions of each SATA primitive.

ALIGN Primitive

One of the reasons for the ALIGN primitive is to compensate for the differences in crystal oscillator frequencies used to generate the clock that drives the bits on the interface and the clock internal buffers.

- Some bit rates will be slightly faster at the source and could cause a data overrun.
- Some bit rates could be slower and result in data underruns.
- SATA fixes this issue by sending two ALIGNS primitives every 256 DWORDs.
  - This protocol compensates for any differences, providing the manufacturer follows the standards recommendations.

This primitive is always sent in pairs: There is no condition wherein an odd number of ALIGN primitives are sent (except as noted for retimed loopback).

The Link layer will ignore reception of ALIGN primitives. The Physical layer may consume received ALIGN primitives. Implementations in which the PHY does not consume received ALIGN primitives will drop received ALIGN primitives at the input to the Link layer or will include Link layer processing that yields behavior equivalents to the behavior produced if all received ALIGN primitives are consumed by the PHY and not presented to the Link.

After communications have been established, the first and second words out of the Link layer will be the dual-ALIGN primitive sequence, followed by, at most, 254 non-ALIGN DWORDs. The cycle repeats, starting with another dual-consecutive ALIGN primitive sequence. The Link may issue more than one dual ALIGN primitive sequence but will not send an unpaired ALIGN primitive (i.e., ALIGN primitives are always sent in pairs), except as noted for retimed loopback.

Elasticity Buffer Management

Elasticity buffer circuitry may be required to absorb the slight differences in frequencies between the host and device.

- The greatest frequency difference results from a SSC-compliant device talking to a non-SSC device.
- The average frequency difference will be just over 0.25%, with excursions as much as 0.5%.

Since an elasticity buffer has a finite length, there needs to be a mechanism at the Physical layer protocol level that allows this receiver buffer to be reset without dropping or adding any bits to the data stream.

- This is especially important during reception of long continuous streams of data.
- This physical layer protocol must not only support oversampling architectures but must also accommodate unlimited frame sizes (the frame size is limited by the CRC polynomial).

The Link layer will keep track of a resettable counter that rolls over at most every 1024 transmitted characters (256 DWORDs).

- Prior to, or at the pre-roll-over point (all 1's), the Link layer will trigger the issuance of dual, consecutive ALIGN primitives, which will be included in the DWORD count.

CONT Primitive

The CONT primitive allows long strings of repeated primitives, as defined by Table 5.2, to be eliminated.

- When high-speed serial interfaces repeat bit patterns, electromagnetic interference (EMI) fields are created, which can lead to poor signal quality.
  - Excessive EMI leads to problems passing FCC Class-B tests.
- This simple protocol change eliminates this issue.
- This also enables the construction of lower cost connectors and cabling, due to eliminated emissions.

The CONT primitive causes the previously received repeated primitive to be sustained until another primitive is received. The reception of any primitive other than CONT or ALIGN will terminate the stream of repeated primitives and replace each primitive with a scrambled dword.

In order to accommodate EMI reductions, the scrambling of data is incorporated into the serial implementation of ATA. The scrambling of data is with a **linear feedback shift register** (LFSR), which is used in generating the scrambling pattern being reset at each SOF primitive, or rolling over every 2048 DWORDs. However, the scrambling of primitives is not as effective or simple because of the small number of control characters available. In order to accommodate EMI reductions, repeated primitives are eliminated through the use of the CONT primitive.

Any repeated primitive (see Table 5.2) may be sustained through the use of the CONT primitive. The recipient of the CONT primitive will ignore all data received after the CONT primitive until the reception of any primitive, excluding ALIGN. After transmitting the CONT character, the transmitter may send any sequence of data characters to the recipient, provided that no primitives are included. The transmitter will send a minimum of two identical repeated primitives (e.g., SYNC SYNC, HOLD HOLD, etc.) immediately preceding the CONT primitive, excluding ALIGNs. Valid CONT transmission sequences are shown in Table 5.4.

Table 5.4 Valid CONT Transmission Sequences

PRIM	PRIM	CONT	XXXX	XXXX	XXXX
PRIM	ALIGN	ALIGN	PRIM	CONT	XXXX
PRIM	PRIM	ALIGN	ALIGN	CONT	XXXX
PRIM	PRIM	CONT	ALIGN	ALIGN	XXXX
PRIM = Any primitive defined as repeated that may be continued in accordance with those noted in the primitive table. XXXX = Output of LFSR or any primitive other than CONT.					

To improve overall protocol robustness and avoid potential timeout situations caused by a reception error in a primitive, all repeated primitives will be transmitted a minimum of twice before a CONT primitive is transmitted. The first primitive correctly received is the initiator of any action within the receiver. This avoids scenarios, for example, where X_RDY is sent from the host, followed by a CONT, and the X_RDY is received improperly, resulting in the device not returning an R_RDY and causing the system to deadlock until a timeout/reset condition occurs. Reception of a CONT primitive when one of the two preceding DWORDs is not a valid repeated primitive results in undefined behavior.

The transmission of a CONT primitive is optional, but the ability to receive and properly process the CONT primitive is required.

Example of Primitive Scrambling

Step 1 – Host and device are both in IDLE State. After transmitting 2 SYNC primitives, they transmit a CONT primitive followed by scrambled Dwords.

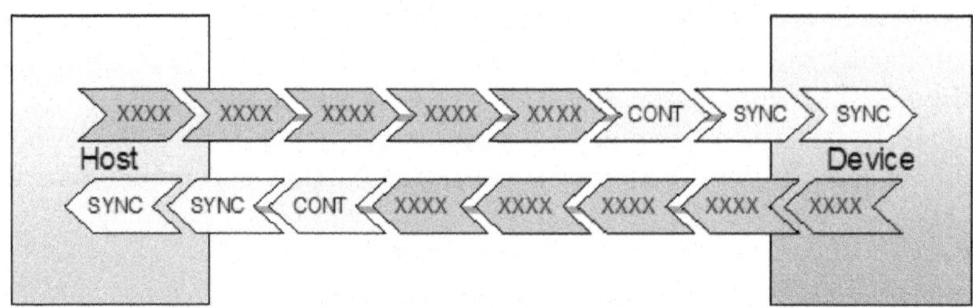

Step 2 – Host has data ready to transmit. Host sends 2 X_RDY primitives followed by a CONT and then by scrambled Dwords. Host waits for response from device.

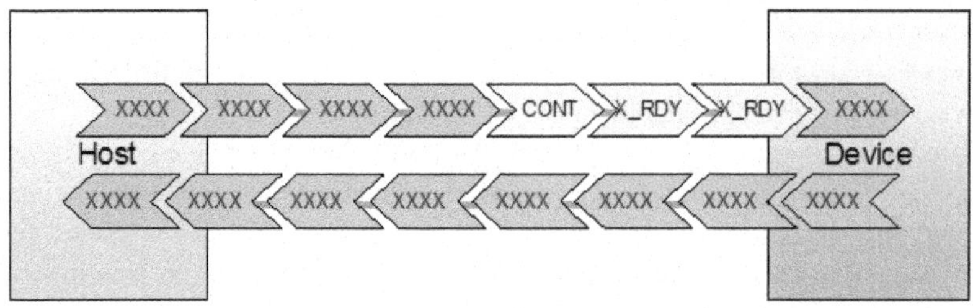

Step 3 – When the device receives X_RDY, it responds by sending 2 R_RDY primitives followed by a CONT and then scrambled Dwords indicating that it is ready to receive data.

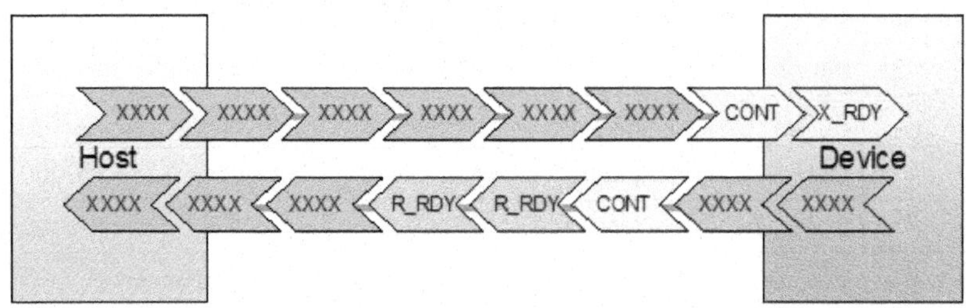

Step 4 – When the host receives R_RDY from the device, it starts transmitting a frame by sending an SOF primitive followed by data Dwords.

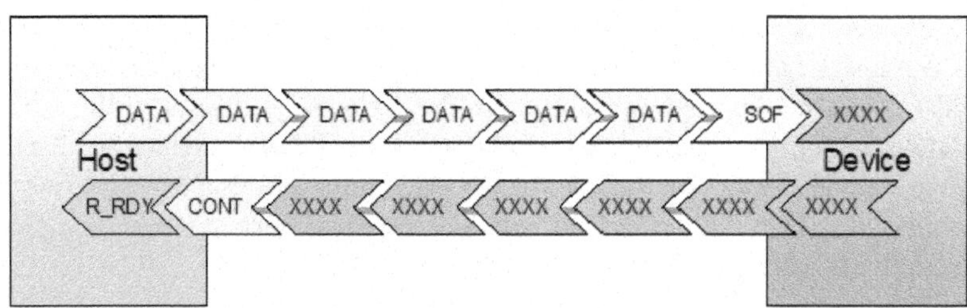

Step 5 – When the device receives the SOF it responds by sending 2 RIP primitives followed by a CONT and then scrambled Dwords indicating frame reception in progress.

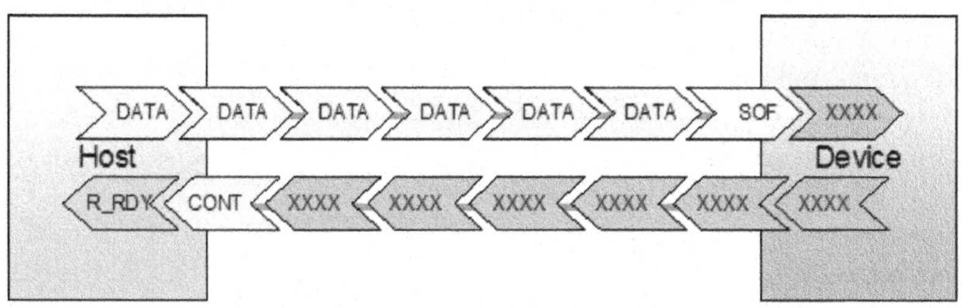

Figure 5.5 shows an entire frame transmission with suppressed primitive scrambling.

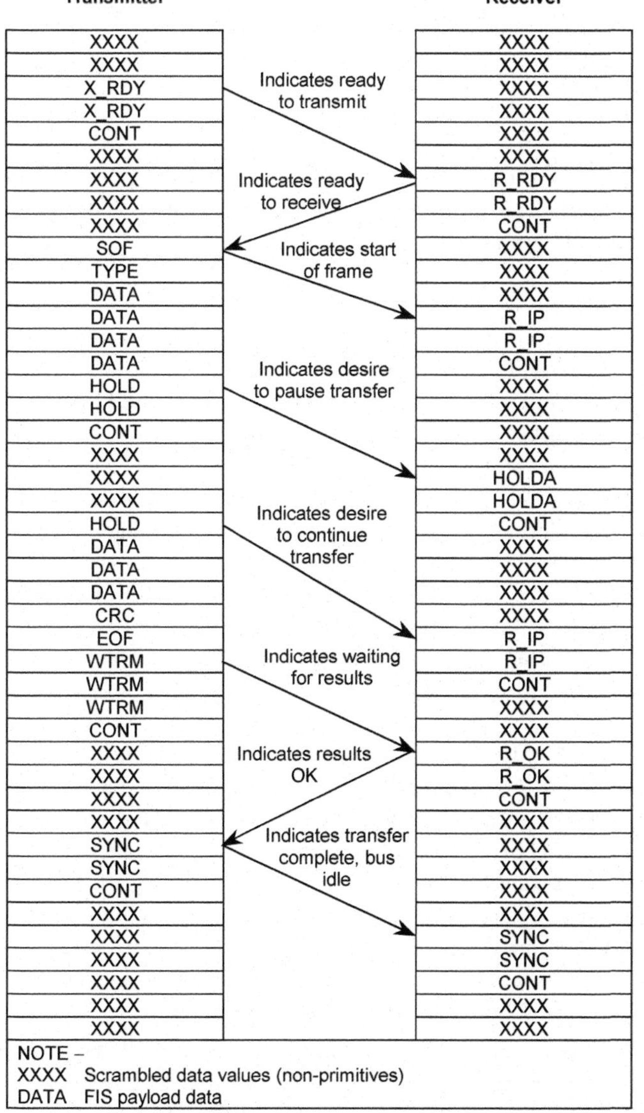

Figure 5.5 Protocol example of CONT Primitive sequence.

Scrambling of Data Following the CONT Primitive

The data following the CONT will be the output of an LFSR, which implements the same polynomial as is used to scramble FIS contents. That polynomial is defined in Section 5.5—Scrambling.

The resulting LFSR value will be encoded using the 8b10b rules for encoding data characters before transmission by the Link layer.

The LFSR used to supply data after the CONT primitive will be reset to the initial value upon detection of a COMINIT or COMRESET event.

Since the data following a CONT primitive is discarded by the Link layer, the value of the LFSR is undefined between CONT primitives. That is, the LFSR result used for CONT sequence N is not required to be continuous from the last LFSR result of CONT sequence N1.

The sequence of LFSR values used to scramble the payload contents of an FIS will not be affected by the scrambling of data used during repeated primitive suppression. That is, the data payload LFSR will not be advanced during repeated primitive suppression and will only be advanced for each payload data character that is scrambled using the data payload LFSR.

DMAT Primitive

This primitive is sent as a request to the transmitter to terminate a DMA data transmission early by computing a CRC on the data sent and ending with a EOF primitive. The DMA context is assumed to remain stable after the EOF primitive has been sent.

> *Note*: No consistent use of the DMA Terminate (DMAT) facility is defined, and its use may impact software compatibility. Implementations should tolerate reception of DMAT, as defined in the standard, but should avoid transmission of DMAT in order to minimize potential interaction problems. One valid response to reception of DMAT is to ignore it and complete the transfer.

In a parallel implementation of ATA, devices can abort a DMA transfer and post an error, by negating DMA Request, updating the Error and Status registers, and possibly setting the INTRQ line. This can be performed for both reads from and writes to the device.

A device may terminate a transfer with an EOF and send a Register Device to Host FIS to the host, with Error and Status registers updated appropriately. In the case of a DMA write to device, the device sends a DMA Activate FIS to the host, and then after receiving an SOF, accepts all data until receiving an EOF from the host. Since the device cannot terminate such a transfer once started, a DMAT primitive is used.

The DMAT primitive may be sent on the back channel during transmission of a Data FIS to signal the transmitter to terminate the transfer in progress. It may be used for both host to device transfers and for device to host transfers. If processed, reception of the DMAT signal will cause the recipient to close the current frame by inserting the CRC and EOF, and return to the idle state.

For host to device data transfers, upon receiving the DMAT signal, the host may terminate the transfer in progress by deactivating its DMA engine and closing the frame with valid CRC and EOF. The host DMA engine will preserve its state at the point it was deactivated so that the device may resume the transmission at a later time by transmitting another DMA Active FIS to reactivate the DMA engine. The device is responsible for either subsequently resuming the terminated transfer by transmitting another DMA Activate FIS or closing the affected command with appropriate status.

For device to host transfers, receipt of a DMAT signal by the device results in the permanent termination of the transfer and cannot be resumed. The device may terminate the transmission in progress and close the frame with a valid CRC and EOF, and will thereafter clean up the affected command by indicating appropriate status for that command. No facility for resuming a device to host transfer terminated with the DMAT signal is provided.

Some implementations may have an implementation-dependent latency associated with closing the affected Data FIS in response to the DMAT signal. For example, a host adapter may have a small transmit FIFO, and in order for the DMA engine to accurately reflect a state that can be resumed, the data already transferred by the DMA engine to the transmit FIFO may have to be transmitted prior to closing the affected Data FIS. Designs should minimize the DMAT response latency while being tolerant of other devices having a long latency.

HOLD/HOLDA Primitives

HOLD is transmitted in place of payload data within a frame when the transmitter does not have the next payload data ready for transmission. HOLD is also transmitted on the backchannel when a receiver is not ready to receive additional payload data.

The HOLDA primitive is sent by a transmitter as long the HOLD primitive is received by its companion receiver.

Flow Control Signaling Latency

There is a finite pipeline latency in a round-trip handshake across the serial interface. In order to accommodate efficient system design with sufficient buffering headroom to avoid buffer overflow in flow control situations, the maximum tolerable latency from when a receiver issues a HOLD signal until it receives the HOLDA signal from the transmitter is bounded. This allows the limit to be set in the receive FIFO so as to avoid buffer overflow while avoiding excessive buffering/FIFO space.

In the case where the receiver wants to flow control the incoming data, it transmits HOLD characters on the back channel. Some number of received DWORDs later, valid data ceases and HOLDA characters are received. The larger the latency between transmitting HOLD until receiving HOLDA, the larger the receive FIFO needs to be. The maximum allowed latency from the time the MSB of the HOLD primitive is on the wire until the MSB of the HOLDA is on the wire will be no more than 20 DWORD times. The LSB is transmitted first. A receiver will be able to accommodate the reception of 20 DWORDs of additional data after the time it transmits the HOLD flow control character to the transmitter, and the transmitter will respond with a HOLDA in response to receiving a HOLD character within 20 DWORD times.

The specified maximum latency figure is based on the layers and states described throughout this book. It is recognized that the Link layer may have two separate clock domains—the transmit clock domain and the receive clock domain. It is also recognized that a Link state machine could run at the DWORD clock rate, implying synchronizers between three potential clock domains. In practice, more efficient implementations would be pursued and the actual latencies may be less than indicated here. This accounting assumes a worst case synchronizer latency of 2.99 clocks in any clock domain and is rounded to three whole clocks. A one meter cable contains less than one-half DWORD and is therefore rounded to 0. Two DWORDs of pipeline delay are assumed for the Phy, and the FIFO is assumed to run at the Link state machine rate. No synchronization is needed between the two.

Table 5.5 outlines the origin of the 20 DWORD latency standard. The example illustrates the components of a round-trip delay when the receiver places a HOLD on the bus until reception of the HOLDA from the transmitter. This corresponds to the number of DWORDs that the receiver must be able to accept after transmitting a HOLD character.

Table 5.5 Latency Example

Receiver sends HOLD:		
	1 DWORD	Convert to 40-bit data
	1 DWORD	10b/8b conversion
	1 DWORD	Descrambling
	3 DWORDs	Synchronization between receive clock, and Link state machine clock
	1 DWORD	Link state machine is notified that primitive has been received
	1 DWORD	Link state machine takes action
	1 DWORD	FIFO is notified of primitive reception
	1 DWORD	FIFO stops sending data to Link layer
	1 DWORD	Link is notified to insert HOLDA
	1 DWORD	Link acts on notification and inserts HOLDA into data stream
	1 DWORD	Scrambling
	1 DWORD	8b10b conversion
	1 DWORD	Synchronize to transmit clock (three transmit clocks, which are four times the Link state machine rate)
	1 DWORD	Convert to 10-bit data
	2 DWORDs	PHY, transmit side
HOLDA on the cable.		

PMREQ_P, PMREQ_S, PMACK, and PMNAK Primitives

The PMREQ_P primitive is sent continuously to enter Partial power management state until PMACK or PMNAK is received. When PMACK is received, current node (host or device) will stop PMREQ_P and enters the Partial power management state.

The PMREQ_S primitive is sent continuously to enter Slumber power management state until PMACK or PMNAK is received. When PMACK is received, current node (host or device) will stop PMREQ_S and enters the Slumber power management state.

PMACK is sent in response to a PMREQ_S or PMREQ_P when a receiving node is prepared to enter a power mode state.

PMNAK is sent in response to a PMREQ_S or PMREQ_P when a receiving node is not prepared to enter a power mode state or when power management is not supported.

5.5 Scrambling

Scrambling is used to minimize the noise effects on signal quality due to long strings of repeated data characters or primitives. See Figure 5.6 for transmit and receive functions.

5.5.1 Frame Content Scrambling

The contents of a frame, that is, all data words between the SOF and EOF, including the CRC, will be scrambled before transmission by the physical layer.

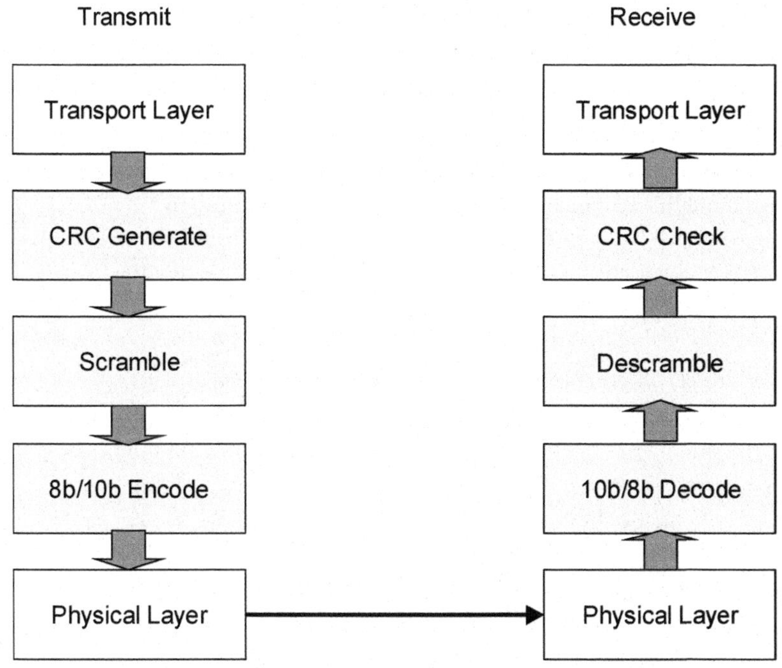

Figure 5.6 Link layer transmit/receive functions (CRC, Scrambling, Encoding).

Scrambling will be performed on DWORD quantities by XORing the data to be transmitted with the output of a LFSR. The shift register will implement the following polynomial:

$$G(X) = X^{16} + X^{15} + X^{13} + X^{4} + 1$$

The serial shift register will be initialized with a value of FFFFh before the first shifted output. The shift register will be initialized to the seed value before the SOF primitive.

Relationship between Scrambling and CRC

The order of the application of scrambling will be as follows. For a DWORD of data following the SOF primitive, the DWORD will be used in the calculation of the CRC. The same DWORD value will be XORed with the scrambler output, and the resulting DWORD submitted to the 8b10b encoder for transmission. Similarly, on reception, the DWORD will be decoded using a 10b/8b decoder, the scrambler output will be XORed with the resulting DWORD, and the resulting DWORD presented to the Link layer and subsequently used in calculating the CRC. The CRC DWORD will be scrambled according to the same rules.

Repeated Primitive Suppression

A second linear feedback shift register will be used to provide scrambled data for the suppression of repeated primitives with the CONT primitive. This LFSR will implement the same polynomial as the frame-scrambling LFSR but need not be reinitialized.

Relationship between Scrambling of FIS Data and Repeated Primitives

There are two separate scramblers used in the serial implementation of ATA. One scrambler is used for the data payload encoding and a separate scrambler is used for repeated primitive suppression. The scrambler used for data payload encoding will maintain consistent and contiguous context over the scrambled payload data characters of a frame (between SOF and EOF) and will not have its context affected by the scrambling of data used for repeated primitive suppression.

Scrambling is applied to all data (non-primitive) DWORDs. Primitives, including ALIGN, do not get scrambled and will not advance the data payload LFSR register. Similarly, the data payload LFSR

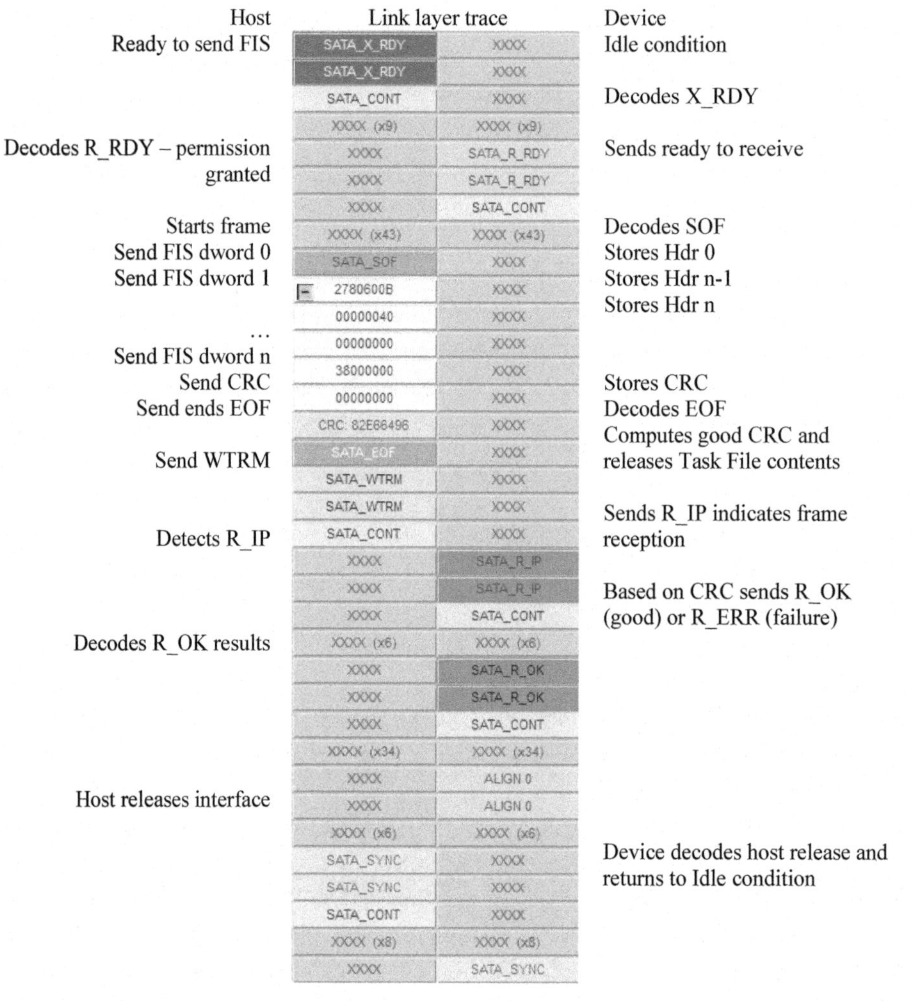

Figure 5.7 Link protocol trace of *Host* to *Device* FIS (frame shown in big-endian format).

Legend:

- xxxx = scrambled data
- (x6) = six instances of a DWORD or Primitive
- "SATA_" preceding all primitives is due to the fact that all SATA analyzers also decode SAS traces; the SAS standards require that all SATA primitives appear with a "SATA_" prefix.

will not be advanced during the transmission of DWORDs during repeated primitive suppression (i.e., after a CONT primitive). Since it is possible for a repeated primitive stream to occur in the middle of a data frame—multiple HOLD/HOLDA primitives are likely—care must be taken to ensure that the data payload LFSR is only advanced for each data payload data character that it scrambles and that it is not advanced for primitives or for data characters transmitted as part of repeated primitive suppression.

5.6 Link Layer Traces

The trace example below (see Figure 5.7) demonstrates the Link layer protocol to transmit a frame from the host to the device. In addition, notice the use of the CONT primitive for repeated primitives.

The trace example below (see Figure 5.8) demonstrates the Link layer protocol to transmit a frame from the device to the host.

Host	Link layer trace		Device
Idle condition	XXXX	SATA_X_RDY	Ready to send FIS
	XXXX	SATA_X_RDY	
Decodes X_RDY	XXXX	SATA_CONT	
	XXXX (x45)	XXXX (x45)	
Sends ready to receive	SATA_R_RDY	XXXX	Decodes R_RDY – permission
	SATA_R_RDY	XXXX	granted
	SATA_CONT	XXXX	
	XXXX (x10)	XXXX (x10)	Starts frame
Decodes SOF	XXXX	SATA_SOF	Send FIS dword 0
Stores Hdr 0	XXXX	34004000	Send FIS dword 1
Stores Hdr n-1	XXXX	00000040	
Stores Hdr n	XXXX	00000000	…
	XXXX	74000000	Send FIS dword n
Stores CRC	XXXX	00000000	Send CRC
Decodes EOF	XXXX	CRC: C4953B88	Send ends EOF
Computes good CRC and	XXXX	SATA_EOF	Send WTRM
releases Task File contents	XXXX	SATA_WTRM	
	XXXX	SATA_WTRM	
Sends R_IP indicates frame	XXXX	SATA_CONT	Detects R_IP
reception	XXXX (x34)	XXXX (x34)	
	SATA_R_IP	XXXX	
	SATA_R_IP	XXXX	
Based on CRC sends R_OK	SATA_CONT	XXXX	
(good) or R_ERR (failure)	XXXX (x6)	XXXX (x6)	Decodes R_OK results
	SATA_R_OK	XXXX	
	SATA_R_OK	XXXX	
	SATA_CONT	XXXX	
	XXXX (x8)	XXXX (x8)	Device releases interface
	XXXX	SATA_SYNC	
	XXXX	SATA_SYNC	
	XXXX	SATA_CONT	
	XXXX (x17)	XXXX (x17)	
Host decodes device release -	XXXX	ALIGN 0	
returns to Idle condition	XXXX	ALIGN 0	
	XXXX (x23)	XXXX (x23)	
	SATA_SYNC	XXXX	

Figure 5.8 Link protocol trace of *Device* to *Host* FIS (frame shown in big-endian format).

5.7 Encoding Method

Information to be transmitted over the serial interface will be

- encoded a byte (8 bits) at a time, along with a data or control character indicator, into a 10-bit encoded character; and then
- sent serially, bit by bit.

Information received over the serial interface will be

- collected 10 bits at a time;
- assembled into an encoded character; and
- decoded into the correct data characters and control characters.

The 8b10b code allows for the encoding of all 256 combinations of 8-bit data. A smaller subset of the control character set is utilized by SATA.

5.7.1 Notation and Conventions

The coding scheme uses a letter notation for describing data bits and control variables. A description of the translation process between these notations follows. This section also describes a convention used to differentiate data characters from control characters. Finally, translation examples for both a data character and a control character are presented.

5.7.2 Encoding Characteristics

- An unencoded byte of data is composed of 8 bits—A,B,C,D,E,F,G,H—and the control variable Z.
- The encoding process results in a 10-bit character—a,b,c,d,e,i,f,g,h,j.
- A bit is either a binary zero or binary one.
- The control variable, Z, has a value of D or K.
- When the control variable associated with a byte has the value D, the byte is referred to as a data character.
- When the control variable associated with a byte has the value K, the byte is referred to as a control character.

If a data byte is not accompanied by a specific control variable value, the control variable Z is assumed to be Z = D, and the data byte will be encoded as a data character.

Figure 5.9 illustrates the association between the numbered unencoded bits in a byte, the control variable, and the letter-labeled bits in the encoding scheme:

Data byte notation	7	6	5	4	3	2	1	0	Control variable
Unencoded bit notation	H	G	F	E	D	C	B	A	Z or D

Figure 5.9 Bit designations.

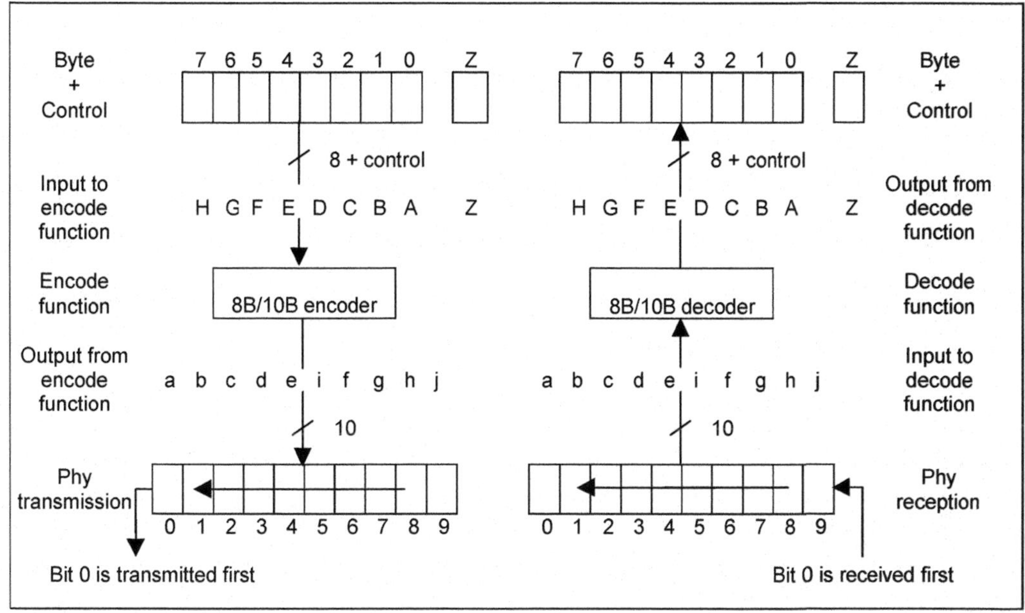

Figure 5.10 Nomenclature reference.

Each character is given a name Zxx.y

- where Z is the value of the control variable (D for a data character, K for a control character);
- xx is the decimal value of the binary number composed of the bits E, D, C, B, and A, in that order; and
- y is the decimal value of the binary number composed of the bits H, G, and F.

Figure 5.10 shows the relationship between the various representations.
Figure 5.11 shows conversions from byte notation to character notation for a control and data byte.

Byte notation	BCh, control character		4Ah, data character	
Bit notation	76543210	Control variable	76543210	Control variable
	10111100	K	01001010	D
Unencoded bit notation	HGF EDCBA	Z	HGF EDCBA	Z
	101 11100	K	010 01010	D
Bit notation reordered to conform with Zxx.y convention	Z	EDCBA HGF	Z	EDCBA HGF
	K	11100 101	D	01010 010
Character name	K 28	.5	D 10	.2

Figure 5.11 Byte notation computation.

Character Coding

The coding scheme translates unencoded data and control bytes to characters. The encoded characters are then transmitted by the physical layer over the serial line where they are received from the physical layer and decoded into the corresponding byte and control values.

The 8b10b coding process is defined in two stages. The first stage encodes the first five bits of the unencoded input byte into a 6-bit sub-block using a 5b6b encoder. The input to this stage includes the current Running Disparity value. The second stage uses a 3b4b encoder to encode the remaining three bits of the data byte, and the running disparity is modified by the 5b6b encoder into a 4-bit value.

In the derivations that follow (see Tables 5.6 and 5.7), the control variable (Z) is assumed to have a value of D, and thus is an implicit input.

Table 5.6 5b6b Encoding Scheme

Inputs		abcdei outputs		rd'	Inputs		abcdei outputs		rd'
Dx	EDCBA	rd+	rd-		Dx	EDCBA	rd+	rd-	
D0	00000	011000	100111		D16	10000	100100	011011	-rd
D1	00001	100010	011101	-rd	D17	10001	100011		
D2	00010	010010	101101		D18	10010	010011		
D3	00011	110001		rd	D19	10011	110010		rd
D4	00100	001010	110101	-rd	D20	10100	001011		
D5	00101	101001			D21	10101	101010		
D6	00110	011001		rd	D22	10110	011010		
D7	00111	000111	111000		D23	10111	000101	111010	-rd
D8	01000	000110	111001	-rd	D24	11000	001100	110011	
D9	01001	100101			D25	11001	100110		rd
D10	01010	010101			D26	11010	010110		
D11	01011	110100		rd	D27	11011	001001	110110	-rd
D12	01100	001101			D28	11100	001110		rd
D13	01101	101100			D29	11101	010001	101110	
D14	01110	011100			D30	11110	100001	011110	-rd
D15	01111	101000	010111	-rd	D31	11111	010100	101011	

Table 5.7 3b/4b Encoding Scheme

Inputs		fghj outputs		rd'
Dx.y	HGF	rd+	rd-	
Dx.0	000	0100	1011	-rd
Dx.1	001	1001		
Dx.2	010	0101		rd
Dx.3	011	0011	1100	
Dx.4	100	0010	1101	-rd
Dx.5	101	1010		rd
Dx.6	110	0110		
Dx.P7	111	0001	1110	-rd
Dx.A7	111	1000	0111	
Note: A7 replaces P7 if[(rd>0) and (e=i=0)] or [(rd<0) and (e=i=1)].				

Tables 5.6 and 5.7 describe the code and running disparity generation rules for each of the sub-blocks. The results can be used to generate the data in Table 5.8.

5.7.3 Running Disparity

Running Disparity is a binary parameter with either a negative (-) or positive (+) value.

After transmitting any encoded character, the transmitter will calculate a new value for its Running Disparity based on the value of the transmitted character.

After a COMRESET sequence, initial power-up, exiting any power management state, or exiting any diagnostic mode, the receiver will assume either the positive or negative value for its initial Running Disparity. Upon reception of an encoded character, the receiver will determine whether the encoded character is valid according to the following rules and tables and calculate a new value for its Running Disparity based on the contents of the received character.

The following rules will be used to calculate a new Running Disparity value for the transmitter after it sends an encoded character (transmitter's new Running Disparity) and for the receiver upon reception of an encoded character (receiver's new Running Disparity).

Running Disparity for an encoded character will be calculated on two sub-blocks, wherein the first 6 bits (abcdei) form one sub-block—the 6-bit sub-block. The last 4 bits (fghj) form the second sub-block—the 4-bit sub-block. Running Disparity at the beginning of the 6-bit sub-block is the Running Disparity at the end of the last encoded character or the initial conditions described above for the first encoded character transmitted or received. Running Disparity at the beginning of the 4-bit sub-block is the resulting Running Disparity from the 6-bit sub-block. Running Disparity at the end of the encoded character—and the initial Running Disparity for the next encoded character—is the Running Disparity at the end of the 4-bit sub-block.

Running Disparity for each of the sub-blocks will be calculated as follows:

> Running Disparity at the end of any sub-block is positive if the sub-block contains more ones than zeros. It is also positive at the end of the 6-bit sub-block if the value of the 6-bit sub-block is 000111, and is positive at the end of the 4-bit sub-block if the value of the 4-bit sub-block is 0011.
>
> Running Disparity at the end of any sub-block is negative if the sub-block contains more zeros than ones. It is also negative at the end of the 6-bit sub-block if the value of the 6-bit sub-block is 111000, and is negative at the end of the 4-bit sub-block if the value of the 4-bit sub-block is 1100.
>
> Otherwise, for any sub-block with an equal number of zeros and ones, the Running Disparity at the end of the sub-block is the same as at the beginning of the sub-block. Sub-blocks with an equal number of zeros and ones are said to have neutral disparity.

The 8b10b code restricts the generation of the 000111, 111000, 0011, and 1100 sub-blocks in order to limit the run length of zeros and ones between sub-blocks. Sub-blocks containing 000111 or 0011 are generated only when the running disparity at the beginning of the sub-block is positive, resulting in positive Running Disparity at the end of the sub-block. Similarly, sub-blocks containing 111000 or 1100 are generated only when the running disparity at the beginning of the sub-block is negative and the resulting Running Disparity is negative.

The rules for Running Disparity result in the generation of a character with disparity that is either the opposite of the previous character or neutral.

Sub-blocks with nonzero (non-neutral) disparity will be of alternating disparity.

Running Disparity Encoding Example

The coding examples in Figure 5.12 illustrate how the Running Disparity calculations are done.

The first conversion example completes the translation of the data byte value 4Ah (which is the character name of D10.2) into an encoded character value of "abcdei fghj" = "010101 0101." This value has special significance because (1) it is of neutral disparity, and (2) it also contains an alternating zero/one pattern that represents the highest data frequency that can be generated.

In the second example, the 8b10b character named D11.7 is encoded. Assuming a positive value for the incoming Running Disparity, this example shows the Dx.P7/Dx.A7 substitution. With an initial rd+ value, D10 translates to an abcdei value of 110100, with a resulting Running Disparity of positive for the 6-bit sub-block. Encoding the 4-bit sub-block triggers the substitution clause of Dx.A7 for Dx.P7 since [(rd>0) AND (e=i=0)].

Initial rd	Character name	abcdei output	6-bit sub-block rd	fghj output	4-bit sub-block rd	Encoded character	Ending rd
–	D10.2	010101	–	0101	–	010101 0101	–
+	D11.7	110100	+	1000	–	110100 1000	–

Figure 5.12 Coding example.

5.7.4 8b10b Valid Encoded Characters

Table 5.8 defines the valid data characters and valid control characters. This table will be used for generating encoded characters (encoding) for transmission. In the reception process, Table 5.8 is used to look up and verify the validity of received characters (decoding).

In this table, each data character and control character has two columns that represent two encoded characters. One column represents the output if the current Running Disparity is negative and the other is the output if the current Running Disparity is positive.

Table 5.8 Valid Data Characters

Name	byte	abcdei fghj output		Name	byte	abcdei fghj output	
		Current rd–	Current rd+			Current rd–	Current rd+
D0.0	00h	100111 0100	011000 1011	D0.1	20h	100111 1001	011000 1001
D1.0	01h	011101 0100	100010 1011	D1.1	21h	011101 1001	100010 1001
D2.0	02h	101101 0100	010010 1011	D2.1	22h	101101 1001	010010 1001
D3.0	03h	110001 1011	110001 0100	D3.1	23h	110001 1001	110001 1001
D4.0	04h	110101 0100	001010 1011	D4.1	24h	110101 1001	001010 1001
D5.0	05h	101001 1011	101001 0100	D5.1	25h	101001 1001	101001 1001
D6.0	06h	011001 1011	011001 0100	D6.1	26h	011001 1001	011001 1001
D7.0	07h	111000 1011	000111 0100	D7.1	27h	111000 1001	000111 1001

(Continued on following page)

Table 5.8 Valid Data Characters (*Continued*)

Name	byte	abcdei fghj output		Name	byte	abcdei fghj output	
		Current rd–	Current rd+			Current rd–	Current rd+
D8.0	08h	111001 0100	000110 1011	D8.1	28h	111001 1001	000110 1001
D9.0	09h	100101 1011	100101 0100	D9.1	29h	100101 1001	100101 1001
D10.0	0Ah	010101 1011	010101 0100	D10.1	2Ah	010101 1001	010101 1001
D11.0	0Bh	110100 1011	110100 0100	D11.1	2Bh	110100 1001	110100 1001
D12.0	0Ch	001101 1011	001101 0100	D12.1	2Ch	001101 1001	001101 1001
D13.0	0Dh	101100 1011	101100 0100	D13.1	2Dh	101100 1001	101100 1001
D14.0	0Eh	011100 1011	011100 0100	D14.1	2Eh	011100 1001	011100 1001
D15.0	0Fh	010111 0100	101000 1011	D15.1	2Fh	010111 1001	101000 1001
D16.0	10h	011011 0100	100100 1011	D16.1	30h	011011 1001	100100 1001
D17.0	11h	100011 1011	100011 0100	D17.1	31h	100011 1001	100011 1001
D18.0	12h	010011 1011	010011 0100	D18.1	32h	010011 1001	010011 1001
D19.0	13h	110010 1011	110010 0100	D19.1	33h	110010 1001	110010 1001
D20.0	14h	001011 1011	001011 0100	D20.1	34h	001011 1001	001011 1001
D21.0	15h	101010 1011	101010 0100	D21.1	35h	101010 1001	101010 1001
D22.0	16h	011010 1011	011010 0100	D22.1	36h	011010 1001	011010 1001
D23.0	17h	111010 0100	000101 1011	D23.1	37h	111010 1001	000101 1001
D24.0	18h	110011 0100	001100 1011	D24.1	38h	110011 1001	001100 1001
D25.0	19h	100110 1011	100110 0100	D25.1	39h	100110 1001	100110 1001
D26.0	1Ah	010110 1011	010110 0100	D26.1	3Ah	010110 1001	010110 1001
D27.0	1Bh	110110 0100	001001 1011	D27.1	3Bh	110110 1001	001001 1001
D28.0	1Ch	001110 1011	001110 0100	D28.1	3Ch	001110 1001	001110 1001
D29.0	1Dh	101110 0100	010001 1011	D29.1	3Dh	101110 1001	010001 1001
D30.0	1Eh	011110 0100	100001 1011	D30.1	3Eh	011110 1001	100001 1001
D31.0	1Fh	101011 0100	010100 1011	D31.1	3Fh	101011 1001	010100 1001
D0.2	40h	100111 0101	011000 0101	D0.3	60h	100111 0011	011000 1100
D1.2	41h	011101 0101	100010 0101	D1.3	61h	011101 0011	100010 1100
D2.2	42h	101101 0101	010010 0101	D2.3	62h	101101 0011	010010 1100
D3.2	43h	110001 0101	110001 0101	D3.3	63h	110001 1100	110001 0011
D4.2	44h	110101 0101	001010 0101	D4.3	64h	110101 0011	001010 1100
D5.2	45h	101001 0101	101001 0101	D5.3	65h	101001 1100	101001 0011
D6.2	46h	011001 0101	011001 0101	D6.3	66h	011001 1100	011001 0011
D7.2	47h	111000 0101	000111 0101	D7.3	67h	111000 1100	000111 0011
D8.2	48h	111001 0101	000110 0101	D8.3	68h	111001 0011	000110 1100
D9.2	49h	100101 0101	100101 0101	D9.3	69h	100101 1100	100101 0011
D10.2	4Ah	010101 0101	010101 0101	D10.3	6Ah	010101 1100	010101 0011
D11.2	4Bh	110100 0101	110100 0101	D11.3	6Bh	110100 1100	110100 0011
D12.2	4Ch	001101 0101	001101 0101	D12.3	6Ch	001101 1100	001101 0011
D13.2	4Dh	101100 0101	101100 0101	D13.3	6Dh	101100 1100	101100 0011

(*Continued on following page*)

Table 5.8 Valid Data Characters (*Continued*)

Name	byte	abcdei fghj output		Name	byte	abcdei fghj output	
		Current rd–	Current rd+			Current rd–	Current rd+
D14.2	4Eh	011100 0101	011100 0101	D14.3	6Eh	011100 1100	011100 0011
D15.2	4Fh	010111 0101	101000 0101	D15.3	6Fh	010111 0011	101000 1100
D16.2	50h	011011 0101	100100 0101	D16.3	70h	011011 0011	100100 1100
D17.2	51h	100011 0101	100011 0101	D17.3	71h	100011 1100	100011 0011
D18.2	52h	010011 0101	010011 0101	D18.3	72h	010011 1100	010011 0011
D19.2	53h	110010 0101	110010 0101	D19.3	73h	110010 1100	110010 0011
D20.2	54h	001011 0101	001011 0101	D20.3	74h	001011 1100	001011 0011
D21.2	55h	101010 0101	101010 0101	D21.3	75h	101010 1100	101010 0011
D22.2	56h	011010 0101	011010 0101	D22.3	76h	011010 1100	011010 0011
D23.2	57h	111010 0101	000101 0101	D23.3	77h	111010 0011	000101 1100
D24.2	58h	110011 0101	001100 0101	D24.3	78h	110011 0011	001100 1100
D25.2	59h	100110 0101	100110 0101	D25.3	79h	100110 1100	100110 0011
D26.2	5Ah	010110 0101	010110 0101	D26.3	7Ah	010110 1100	010110 0011
D27.2	5Bh	110110 0101	001001 0101	D27.3	7Bh	110110 0011	001001 1100
D28.2	5Ch	001110 0101	001110 0101	D28.3	7Ch	001110 1100	001110 0011
D29.2	5Dh	101110 0101	010001 0101	D29.3	7Dh	101110 0011	010001 1100
D30.2	5Eh	011110 0101	100001 0101	D30.3	7Eh	011110 0011	100001 1100
D31.2	5Fh	101011 0101	010100 0101	D31.3	7Fh	101011 0011	010100 1100
D0.4	80h	100111 0010	011000 1101	D0.5	A0h	100111 1010	011000 1010
D1.4	81h	011101 0010	100010 1101	D1.5	A1h	011101 1010	100010 1010
D2.4	82h	101101 0010	010010 1101	D2.5	A2h	101101 1010	010010 1010
D3.4	83h	110001 1101	110001 0010	D3.5	A3h	110001 1010	110001 1010
D4.4	84h	110101 0010	001010 1101	D4.5	A4h	110101 1010	001010 1010
D5.4	85h	101001 1101	101001 0010	D5.5	A5h	101001 1010	101001 1010
D6.4	86h	011001 1101	011001 0010	D6.5	A6h	011001 1010	011001 1010
D7.4	87h	111000 1101	000111 0010	D7.5	A7h	111000 1010	000111 1010
D8.4	88h	111001 0010	000110 1101	D8.5	A8h	111001 1010	000110 1010
D9.4	89h	100101 1101	100101 0010	D9.5	A9h	100101 1010	100101 1010
D10.4	8Ah	010101 1101	010101 0010	D10.5	AAh	010101 1010	010101 1010
D11.4	8Bh	110100 1101	110100 0010	D11.5	ABh	110100 1010	110100 1010
D12.4	8Ch	001101 1101	001101 0010	D12.5	ACh	001101 1010	001101 1010
D13.4	8Dh	101100 1101	101100 0010	D13.5	ADh	101100 1010	101100 1010
D14.4	8Eh	011100 1101	011100 0010	D14.5	AEh	011100 1010	011100 1010
D15.4	8Fh	010111 0010	101000 1101	D15.5	AFh	010111 1010	101000 1010
D16.4	90h	011011 0010	100100 1101	D16.5	B0h	011011 1010	100100 1010
D17.4	91h	100011 1101	100011 0010	D17.5	B1h	100011 1010	100011 1010
D18.4	92h	010011 1101	010011 0010	D18.5	B2h	010011 1010	010011 1010
D19.4	93h	110010 1101	110010 0010	D19.5	B3h	110010 1010	110010 1010

(*Continued on following page*)

Table 5.8 Valid Data Characters (*Continued*)

Name	byte	abcdei fghj output		Name	byte	abcdei fghj output	
		Current rd–	Current rd+			Current rd–	Current rd+
D20.4	94h	001011 1101	001011 0010	D20.5	B4h	001011 1010	001011 1010
D21.4	95h	101010 1101	101010 0010	D21.5	B5h	101010 1010	101010 1010
D22.4	96h	011010 1101	011010 0010	D22.5	B6h	011010 1010	011010 1010
D23.4	97h	111010 0010	000101 1101	D23.5	B7h	111010 1010	000101 1010
D24.4	98h	110011 0010	001100 1101	D24.5	B8h	110011 1010	001100 1010
D25.4	99h	100110 1101	100110 0010	D25.5	B9h	100110 1010	100110 1010
D26.4	9Ah	010110 1101	010110 0010	D26.5	BAh	010110 1010	010110 1010
D27.4	9Bh	110110 0010	001001 1101	D27.5	BBh	110110 1010	001001 1010
D28.4	9Ch	001110 1101	001110 0010	D28.5	BCh	001110 1010	001110 1010
D29.4	9Dh	101110 0010	010001 1101	D29.5	BDh	101110 1010	010001 1010
D30.4	9Eh	011110 0010	100001 1101	D30.5	BEh	011110 1010	100001 1010
D31.4	9Fh	101011 0010	010100 1101	D31.5	BFh	101011 1010	010100 1010
D0.6	C0h	100111 0110	011000 0110	D0.7	E0h	100111 0001	011000 1110
D1.6	C1h	011101 0110	100010 0110	D1.7	E1h	011101 0001	100010 1110
D2.6	C2h	101101 0110	010010 0110	D2.7	E2h	101101 0001	010010 1110
D3.6	C3h	110001 0110	110001 0110	D3.7	E3h	110001 1110	110001 0001
D4.6	C4h	110101 0110	001010 0110	D4.7	E4h	110101 0001	001010 1110
D5.6	C5h	101001 0110	101001 0110	D5.7	E5h	101001 1110	101001 0001
D6.6	C6h	011001 0110	011001 0110	D6.7	E6h	011001 1110	011001 0001
D7.6	C7h	111000 0110	000111 0110	D7.7	E7h	111000 1110	000111 0001
D8.6	C8h	111001 0110	000110 0110	D8.7	E8h	111001 0001	000110 1110
D9.6	C9h	100101 0110	100101 0110	D9.7	E9h	100101 1110	100101 0001
D10.6	CAh	010101 0110	010101 0110	D10.7	EAh	010101 1110	010101 0001
D11.6	CBh	110100 0110	110100 0110	D11.7	EBh	110100 1110	110100 1000
D12.6	CCh	001101 0110	001101 0110	D12.7	ECh	001101 1110	001101 0001
D13.6	CDh	101100 0110	101100 0110	D13.7	EDh	101100 1110	101100 1000
D14.6	CEh	011100 0110	011100 0110	D14.7	EEh	011100 1110	011100 1000
D15.6	CFh	010111 0110	101000 0110	D15.7	EFh	010111 0001	101000 1110
D16.6	D0h	011011 0110	100100 0110	D16.7	F0h	011011 0001	100100 1110
D17.6	D1h	100011 0110	100011 0110	D17.7	F1h	100011 0111	100011 0001
D18.6	D2h	010011 0110	010011 0110	D18.7	F2h	010011 0111	010011 0001
D19.6	D3h	110010 0110	110010 0110	D19.7	F3h	110010 1110	110010 0001
D20.6	D4h	001011 0110	001011 0110	D20.7	F4h	001011 0111	001011 0001
D21.6	D5h	101010 0110	101010 0110	D21.7	F5h	101010 1110	101010 0001
D22.6	D6h	011010 0110	011010 0110	D22.7	F6h	011010 1110	011010 0001
D23.6	D7h	111010 0110	000101 0110	D23.7	F7h	111010 0001	000101 1110
D24.6	D8h	110011 0110	001100 0110	D24.7	F8h	110011 0001	001100 1110
D25.6	D9h	100110 0110	100110 0110	D25.7	F9h	100110 1110	100110 0001

(*Continued on following page*)

Table 5.8 Valid Data Characters (*Continued*)

Name	byte	abcdei fghj output		Name	byte	abcdei fghj output	
		Current rd–	Current rd+			Current rd–	Current rd+
D26.6	DAh	010110 0110	010110 0110	D26.7	FAh	010110 1110	010110 0001
D27.6	DBh	110110 0110	001001 0110	D27.7	FBh	110110 0001	001001 1110
D28.6	DCh	001110 0110	001110 0110	D28.7	FCh	001110 1110	001110 0001
D29.6	DDh	101110 0110	010001 0110	D29.7	FDh	101110 0001	010001 1110
D30.6	DEh	011110 0110	100001 0110	D30.7	FEh	011110 0001	100001 1110
D31.6	DFh	101011 0110	010100 0110	D31.7	FFh	101011 0001	010100 1110

Control Characters

In the serial implementation of ATA, only the K28.3 and K28.5 control characters are valid and are always used as the first byte in a 4-byte primitive. The K28.3 control character is used to prefix all primitives other than the ALIGN primitive, whereas the K28.5 control character is used to prefix the ALIGN primitive. The encoding of characters within primitives follows the same rules as that applied to non-primitives, when calculating the running disparity between characters and between sub-blocks of each character within the primitive. The control characters K28.3 and K28.5 invert the current running disparity (see Table 5.9).

Table 5.9 Valid Control Characters

Name	abcdei fghj output		Description
	Current rd-	Current rd+	
K28.3	001111 0011	110000 1100	Occurs only at byte 0 of all primitives, except for the ALIGN primitive
K28.5	001111 1010	110000 0101	Occurs only at byte 0 of the ALIGN primitive

The running disparity at the end of the ALIGN primitive is the same as the running disparity at the beginning of the ALIGN primitive.

Bits within a Byte

The bits within an encoded character are labeled a,b,c,d,e,i,f,g,h,j. Bit "a" will be transmitted first, followed in order by "b," "c," "d," "e," "i," "f," "g," "h," and "j." Note that bit "i" is transmitted between bits "e" and "f," and that bit "j" is transmitted last, and not in the order that would be indicated by the letters of the alphabet.

Bytes within a DWORD

For all transmissions and receptions, the serial implementation of ATA organizes all values as DWORDs. Even when representing a 32-bit value, the DWORD will be considered a set of 4 bytes. The transmission order of the bytes within the DWORD will be from the least-significant byte (byte 0) to the most-significant byte (byte 3). This right-to-left transmission order differs from the Fibre Channel. Figure 5.13 illustrates how the bytes are arranged in a DWORD and the order in which bits are sent.

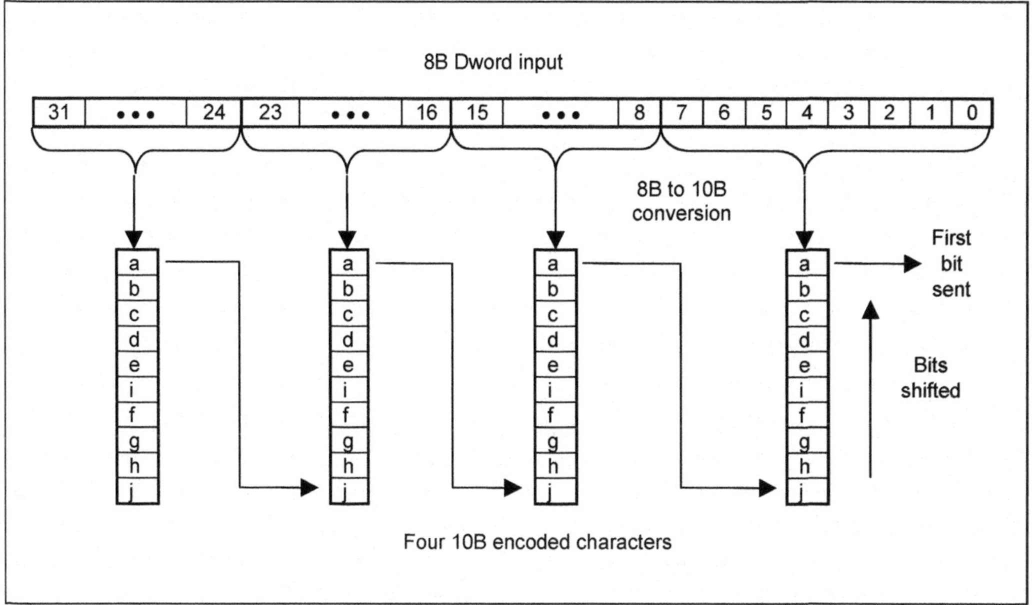

Figure 5.13 Dword bit ordering.

Disparity and the Detection of a Code Violation

Due to the propagation characteristics of the 8b10b code, it is possible that a single bit error might not be detected until several characters after the error is introduced. The following tables illustrate this effect. Table 5.10 shows a bit error being propagated two characters before being detected.

It is important to note that the serial implementation of ATA sends data in DWORD increments, but the transmitter and receiver operate in units of a byte (character). The example depicted in Table 5.9 don't show DWORD boundaries, so it is possible that an error in either of these cases could be deferred one full DWORD.

Table 5.10 Single Bit Error with Detection Delayed

	rd	Character	rd	Character	rd	Character	rd
Transmitted character stream	–	D21.1	–	D10.2	–	D23.5	+
Transmitted bit stream	–	101010 1001	–	010101 0101	–	111010 1010	+
Received bit stream	–	101010 1011 (See note 1)	+	010101 0101	+	111010 1010 (See note 2)	+
Decoded character stream	–	D21.0	+	D10.2	+	Code violation (See note 2)	+(See note 3)

NOTES –

1. Bit error introduced: 1001 → 1011
2. Sub-blocks with non-neutral disparity must alternate polarity (i.e., + → –). In this case, rd does not alternate (it stays positive for two sub-blocks in a row). The resulting encoded character does not exist in the rd+ column in the data or control code table, and so an invalid encoded character is recognized.
3. Running disparity is computed on the received character, regardless of the validity of the encoded character.

The frequency of disparity errors and code violations is an indicator of channel quality and corresponds directly to the bit error rate of the physical serial link between a host and a device. Implementations may elect to count such events and make them available to external firmware or software, although the method by which such counters are exposed is not defined in this standard.

Initial Running Disparity and the Running Disparity for each character is shown. In order to discover the errors, note that Running Disparity is actually computed at the end of each sub-block and is subsequently forwarded to the next sub-block. Footnotes indicate where the disparity error is detected. The error bit is underlined.

5.8 CRC Calculation

The Cyclic Redundancy Check (CRC) of a frame is a DWORD (32-bit) field that will follow the last DWORD of the contents of an FIS and precede the EOF primitive. The CRC calculation covers all of the FIS transport data between the SOF and EOF primitives, and excludes any intervening primitives and CONT stream contents. The CRC value will be computed on the contents of the FIS before encoding for transmission (scrambling) and after decoding upon reception.

The CRC will be calculated on DWORD quantities. If an FIS contains an odd number of words, the last word of the FIS will be padded with zeros to a full DWORD before the DWORD is used in the calculation of the CRC.

The CRC will be aligned on a DWORD boundary.

The CRC will be calculated using the following 32-bit generator polynomial:

$$G(X) = X^{32} + X^{26} + X^{23} + X^{22} + X^{16} + X^{12} + X^{11} + X^{10} + X^{8} + X^{7} + X^{5} + X^{4} + X^{2} + X + 1$$

The CRC value will be initialized with a value of 52325032h before the calculation begins.

The maximum number of DWORDs between the SOF primitive to the EOF primitive will not exceed 2064 DWORDs, including the FIS type and CRC.

5.9 Review Questions

1. How many bits in a DWORD?

   a. 8-bits
   b. 16-bits
   c. 32-bits
   d. 64-bits

2. What are the two types of DWORDs?

   a. Frame
   b. Data
   c. Special
   d. Primitive
   e. K28.3

3. What do all SATA primitives begin with?

   a. K28.5
   b. K28.3

c. K28.6
d. K28.7

4. What is a Data DWORD?

a. All Special characters
b. All Data characters
c. A mix of Special and Data characters
d. Special characters followed by three Data characters

5. What primitive is sent to pause frame transmission?

a. X_RDY
b. R_RDY
c. HOLD
d. R_IP
e. WTRM
f. R_OK

6. What primitive does the transmitting device send after the completion of frame transmission?

a. X_RDY
b. R_RDY
c. HOLD
d. R_IP
e. WTRM
f. R_OK

7. What primitive is sent to request permission to transmit a frame?

a. X_RDY
b. R_RDY
c. HOLD
d. R_IP
e. WTRM
f. R_OK

8. What primitive is sent if frame was properly received?

a. X_RDY
b. R_RDY
c. HOLD
d. R_OK
e. R_IP
f. WTRM

9. What primitive does the receiving device send when it is receiving a frame?

a. X_RDY
b. WTRM
c. R_RDY

d. HOLD
e. R_IP
f. R_OK

10. What primitive is sent to grant permission to transmit a frame?

a. X_RDY
b. HOLD
c. R_IP
d. R_RDY
e. WTRM
f. R_OK

11. What is the maximum payload size of a Data frame (in DWORDS)?

a. 1024
b. 2048
c. 4096
d. 8192

12. What two primitives are allowed between the SOF and EOF primitives during frame transmission?

a. X_RDY
b. HOLDA
c. HOLD
d. R_IP
e. WTRM

13. What is the function of the CONT primitive?

a. To maintain connectivity during idle link conditions
b. To minimize the EMI impact of long sequences of repeated primitives
c. To change the primitive
d. To start a frame

14. What primitive is sent every 254 DWORDs to compensate for clock tolerances?

a. ALIGN
b. HOLDA
c. HOLD
d. CONT

15. Inside frames, what is used to minimize the impact of repeated Data patterns and their impact on signal quality?

a. CRC
b. Encoding
c. Scrambling
d. Compression

16. What function do the PMxxx primitives serve?

    a. Preventive Maintenance
    b. Port Multipliers
    c. Power Management
    d. Preventive Management

17. What is the encoding scheme?

    a. 66b64b
    b. 8b10b
    c. 32b40b
    d. 128b130b

18. What is the symbol for the data value CFh?

    a. D13.6
    b. D14.6
    c. D15.6
    d. D16.6

Chapter 6

Physical Layer

Objectives

This chapter describes the Physical layer of the serial implementation of ATA. Unless otherwise described, the information is normative. The information that is provided and marked informative is provided to help the reader better understand the standards clauses and should be taken as examples only. Exact implementations may vary.

The following Physical layer functions and services are discussed:

- Out of band (OOB) signaling
  - COMWAKE
  - COMINIT
  - COMRESET
  - Initialization protocols
- SATA speed negotiation
- Resets and Signatures
- Power states (legacy)
- Built-in self-test
  - Loopback testing
- PHY logic diagram
- Electrical specifications (legacy)

6.1 PHY Layer Services

Following is a list of services provided by the Physical layer:

1. Transmit at up to 6 Gb/sec differential NRZ serial stream at specified voltage levels
2. Provide matched termination at the transmitter
3. Serialize a 10, 20, 40, or other width parallel input from the Link for transmission

4. Receive up to 6 Gb/sec differential NRZ serial stream
5. Provide a matched termination at the receiver
6. Extract data (and, optionally, clock) from the serial stream
7. Deserialize the serial stream
8. Detect the K28.5 comma character and provide a bit and word aligned 10, 20, 40, or other width parallel output
9. Provide specified OOB signal detection and transmission
10. Perform proper power-on sequencing and speed negotiation
11. Provide interface status to Link layer
12. Host/device absent/present
13. Host/device present but failed to negotiate communications
14. Optionally support power management modes
15. Optionally perform transmitter and receiver impedance calibration
16. Handle the input data rate frequency variation due to a spread spectrum transmitter clock
17. Accommodate request to go into Far-End retimed loopback test mode of operation when commanded

6.2 Out of Band (OOB) Signaling

Out of band (OOB) signaling is used to perform numerous Physical layer protocols. There are three OOB signals that are used (or detected) by the PHY, COMRESET, COMINIT, and COMWAKE.

- COMINIT, COMRESET, and COMWAKE OOB signaling is achieved by:
  - Transmission of a burst of ALIGN primitives
    - Each burst is 106.7ns in duration
    - Independent of link speed (i.e., all speeds perform same protocol sequence)

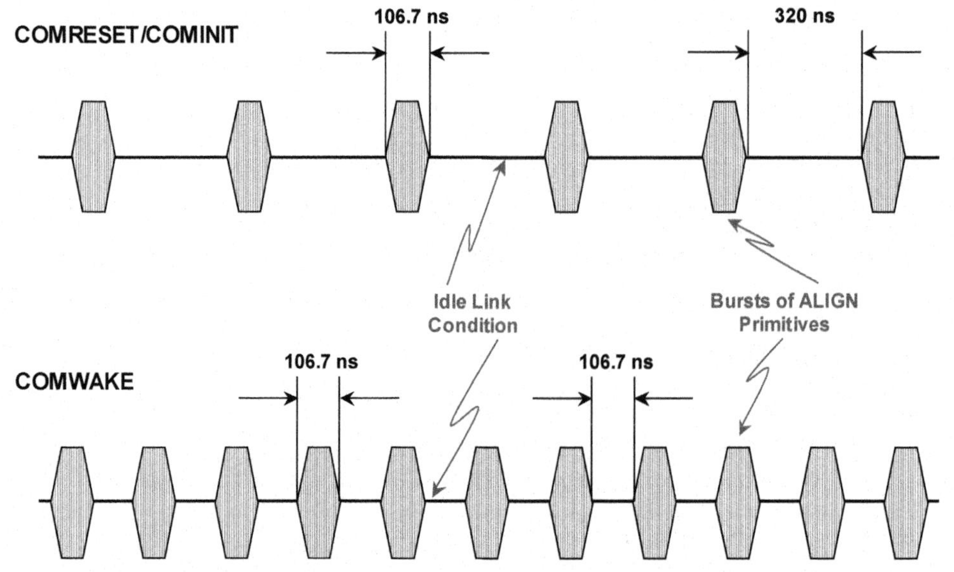

Figure 6.1 OOB signaling protocol.

Table 6.1 OOB Timing Values

Signal Characteristic	Nom	Min	Max	Unit	Comment
COMRESET/COMINIT detector on threshold	320	304	336	ns	Detector shall detect all bursts with spacings meeting this period
COMRESET/COMINIT transmit spacing	320.0	310.4	329.6	ns	As measured from 100mV differential cross points of last and first edges of bursts
COMWAKE detector off threshold		55	175	ns	Detector shall reject all bursts with spacings outside this spec.
COMWAKE detector on threshold	106.7	101.3	112	ns	Detector shall detect all burst spacings meeting this period
COMWAKE transmit spacing	106.7	103.5	109.9	ns	As measured from 100mV differential cross points from last to first edges of bursts
UI_{OOB}		646.67	686.67	ps	Operating data period during OOB burst transmission at 1.5 Gb/s
COMRESET/COMINIT detector off threshold		175	525	ns	Detector shall reject all bursts with spacings outside this spec.

  - Each burst is followed by an idle period
    - Idle periods are accomplished by using common-mode voltage levels
    - COMRESET/COMINT idle period is 320ns
    - COMWAKE idle period is 106.7ns
  - See Figure 6.1 below for a signaling example
- COMRESET always originates from the host controller
  - Forces a hard reset in the device
- The COMINIT signaling originates from the drive and requests a communication initialization
- COMWAKE signaling is used to bring the PHY out of a power-down state

Table 6.1 shows the specified minimum, maximum, and nominal values of the OOB signals.

During OOB signaling transmissions, the differential and common mode levels of the signal lines shall comply with the same electrical specifications as for in-band data transmission.

- OOB signals shall be observed by detecting the temporal spacing between adjacent bursts of activity on the differential pair.
- It is not required for a receiver to check the duration of an OOB burst.

6.2.1 OOB Signaling Protocol

Figure 6.2 shows the OOB protocol that takes place when a host and device power on or are instructed to perform a reset.

After OOB Initialization

Once OOB initialization has completed

1. the host and device perform Speed Negotiation; and
2. the device sends a ***Signature*** to the host via a Register Device to Host FIS.

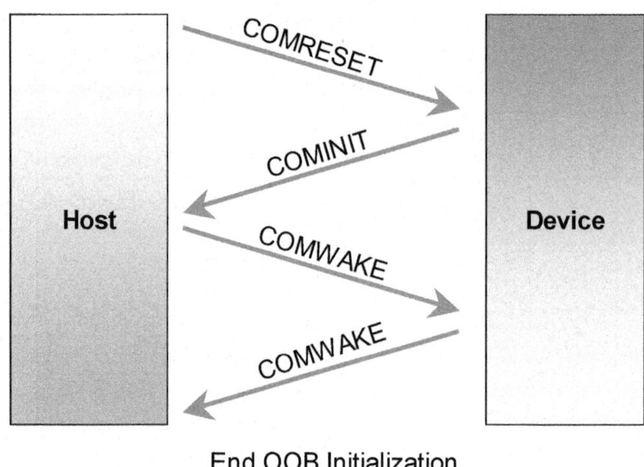

Figure 6.2 OOB initialization protocol.

6.2.2 Power-Up and COMRESET Timing

COMRESET Timings

COMRESET always originates from the host controller and forces a hard reset in the device. It is indicated by transmitting bursts separated by an idle bus condition. The COMINIT always originates at the device and forces a hard reset in the host. The OOB COMRESET signal shall consist of no less than six bursts (see Figure 6.3) and a multiple of six bursts, including inter-burst temporal spacing. The COMRESET signal will be:

COMINIT/COMRESET Sequence

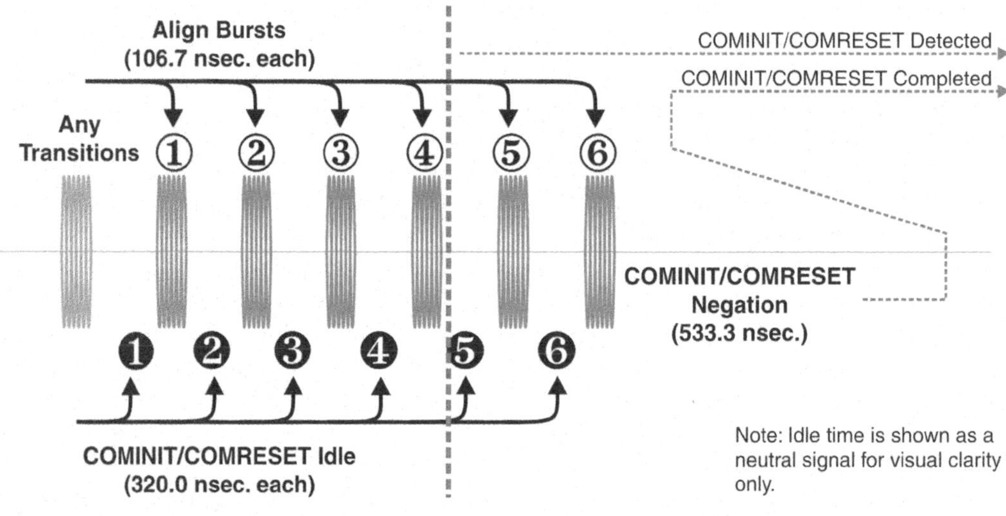

Figure 6.3 COMINIT/COMRESET timing sequence.

1. sustained/continued uninterrupted as long as the hard reset is asserted; or
2. started during the system hard reset and ended sometime after the deassertion of system hard reset; or
3. transmitted immediately following the deassertion of the system hard reset signal.

The host controller shall ignore any signal received from the device from the assertion of the hard reset signal until the COMRESET signal is transmitted.

Each burst shall be 160 Gen1 UI_{OOB}'s long (approximately 106.7 ns) and each inter-burst idle state shall be 480 Gen1 UI_{OOB}'s long (approximately 320 ns). A COMRESET detector will look for four consecutive bursts with 320 ns spacing (nominal).

Any spacing less than the COMRESET/COMINIT detector off threshold min time or greater than the COMRESET/COMINIT detector off threshold max time shall negate the COMRESET detector output.

- The COMRESET interface signal to the PHY layer will initiate the Reset sequence shown in Figure 6.3.
- The interface shall be held inactive for at least the COMRESET/COMINIT detector off threshold max time after the last burst to ensure that the far-end detector detects the deassertion properly.

COMWAKE

Any spacing less than or greater than the COMWAKE detector off threshold in Table 6.1 shall deassert the COMWAKE detector output.

- The COMWAKE OOB signaling is used to bring the PHY out of a power-down state (PARTIAL or SLUMBER).
- The interface shall be held inactive for at least the maximum COMWAKE detector off threshold after the last burst to ensure that the far-end detector detects the deassertion properly.
- The device shall hold the interface inactive no more than the maximum COMWAKE detector off threshold + 2 Gen1 DWORDs (approximately 228.3ns) at the end of a COMWAKE to prevent susceptibility to crosstalk.

COMWAKE can originate from either the host controller or the device. It is signaled by transmitting six bursts separated by an idle bus condition (see Figure 6.4).

- The OOB COMWAKE signaling shall consist of no less than six bursts, including inter-burst temporal spacing.
- Each burst shall be 160 Gen1 $UI_{OOB's}$ (approximately 106.7ns) long, and each inter-burst idle state shall be 160 Gen1 $UI_{OOB's}$ (approximately 106.7ns) long.
- A COMWAKE detector will look for four consecutive bursts with a COMWAKE transmit spacing time (approximately 106.7 ns nominal).

6.2.3 Link Speeds

The SATA serial links are designed to handle links speeds that will fuel the growth and performance of the technology. The SATA-IO has defined link speeds for the SATA architecture (see Table 6.2).

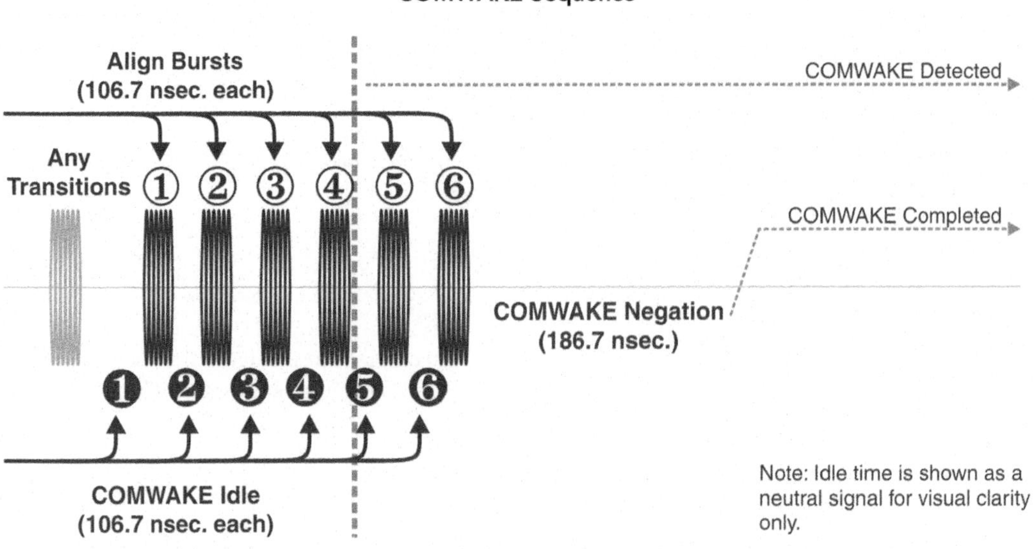

Figure 6.4 COMWAKE sequence.

Table 6.2 Link Speed Generations

Name	Nom	Date	Units	Description	Data Rate
G1	1.2		Gbits/s	1st Generation 8b data rate	
G1	1.5	2002	Gbits/s	1st Generation 10b bit rate	150 MB/s
G2	2.4		Gbits/s	2nd Generation 8b data rate	
G2	3.0	2007	Gbits/s	2nd Generation 10b bit rate	300 MB/s
G3	4.8		Gbits/s	3rd Generation 8b data rate	
G3	6.0	2012	Gbits/s	3rd Generation 10b bit rate	600 MB/s

- A SATA PHY reset sequence consists of
  - a SATA OOB sequence,
  - followed by a SATA speed negotiation sequence.
- SATA speed negotiation enables a SATA host and device to determine their highest mutually supported link speed.
  - SATA speed negotiation sequence is defined by the SATA standards but is hidden in the descriptions for COMINIT and COMWAKE.

SATA Speed Negotiation

Speed negotiation begins when a SATA device transmits COMWAKE (OOB signal).
Upon detecting COMWAKE:

- The SATA host begins transmitting the D10.2 character at its slowest supported link rate.
  - The host simultaneously steps through its supported receive speeds in an attempt to recognize incoming ALIGN (0)s.

- The SATA device begins stepping through its supported speeds.
  - It transmits a burst of 2,048 ALIGN (0)s at its highest speed.
  - It then steps down to its next highest speed and transmits another burst of 2,048 ALIGN (0)s, and so on.
- The SATA host detects ALIGN (0)s.
  - It then begins transmitting ALIGN (0) at the detected rate.
- The SATA device detects the ALIGN (0) and begins transmitting a non-ALIGN(0) signal.
  - Typically this will be the SYNC primitive or scrambled idle dwords.
- Finally, the SATA host transmits a NON-ALIGN (0) signal to complete the process

Figure 6.5 shows the speed negotiation sequence.

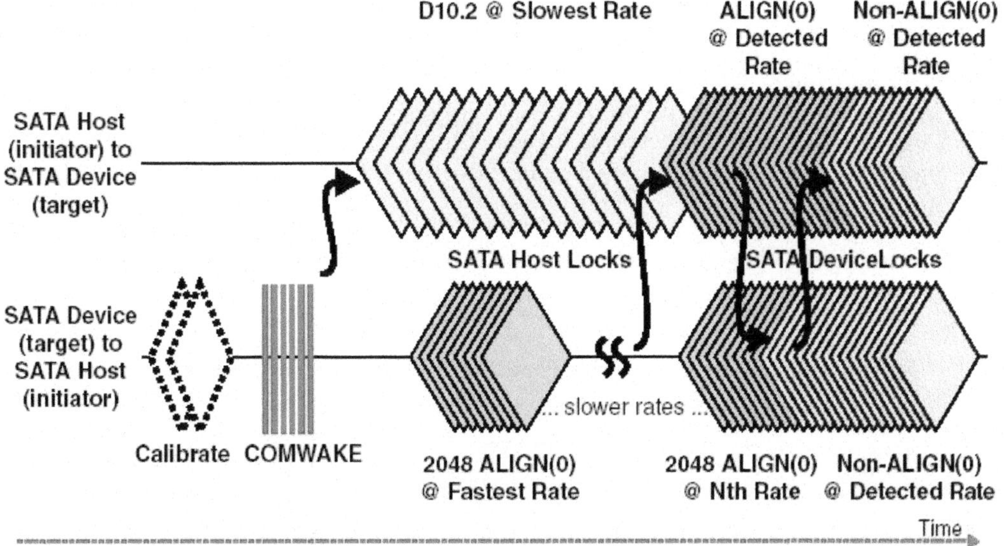

Figure 6.5 Speed negotiation diagram.

Speed Negotiation Timings

- There are two key timing values associated with SATA speed negotiation:
  - Await ALIGN Timeout
  - COMWAKE Response time
- Await ALIGN Timeout:
  - This value specifies how long a PHY waits for an ALIGN following recognition of COMWAKE completed.
  - This time is 1,310,720 out-of-band intervals (or 873.8 usec).
- COMWAKE Response time:
  - This value specifies the maximum amount of time that a SATA PHY may delay after recognizing COMWAKE before it must begin transmitting D10.2 characters.
  - This time is 533 ns.

6.2.4 Resets and Signatures

- There are a number of different types of reset a SATA device may perform
  - Power on reset.
  - OOB reset (COMRESET/COMINIT).
  - Software reset.
    - Host sends “Register Host to Device FIS,” with the Software Reset bit (SRST) set in the Device Control register.
  - Device reset.
    - Issued by Host ATA Application layer.
- All reset types have a similar effect:
  - All outstanding commands are terminated.
  - State machines at ***all*** architectural layers are reset.
  - Device sends “Register Device to Host FIS” containing device signature after completing reset.

Device Signatures

- Device type is indicated to host via a “signature.”
- Signature (see Table 6.3 for non-Packet and Table 6.4 for Packet device) is sent by device to host after a reset.

Table 6.3 Signature for a Device That Does Not Support the Packet Command

Register	Value
Sector Count	01h
Sector Number	01h
Cylinder Low	00h
Cylinder High	00h
Device/Head	00h
Error	01h
Status	00h–70h

Table 6.4 Signature for a Device That Supports the Packet Command

Register	Value
Sector Count	01h
Sector Number	01h
Cylinder Low	14h
Cylinder High	EBh
Device/Head	00h
Error	01h
Status	00h

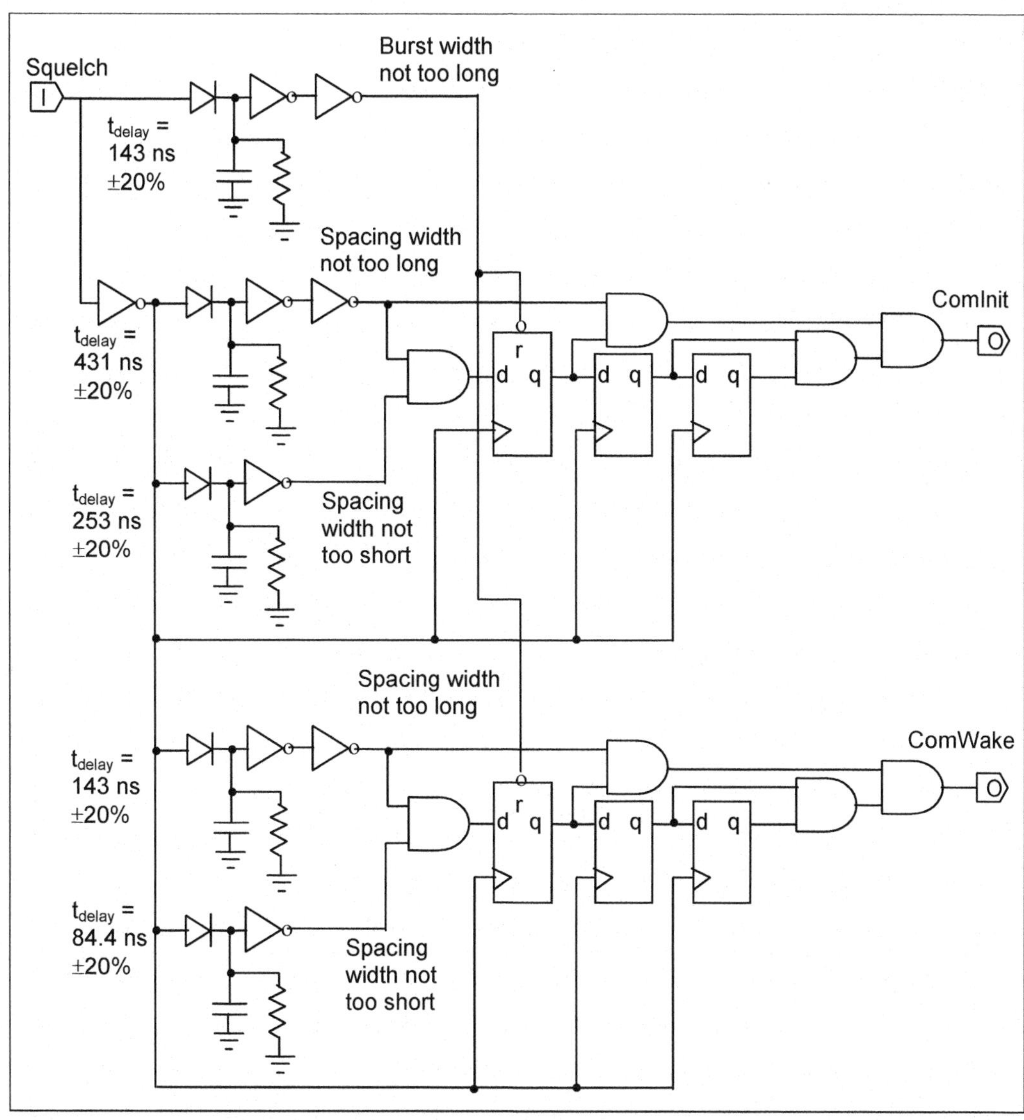

Figure 6.6 OOB signaling detector. (*Source:* www.t13.org)

6.2.5 OOB Signal Detection Logic

The example of OOB detection logic depicted in Figure 6.6 is directly from the SATA Standards (ATA Serial Transport (ATA8-AST) www.t13.org document number - d1697r0c-ATA8-AST).

- This logic cannot be modified from standard implementation due to timing frequencies and their relationship to industry standard defined OOB protocols.
- The squelch detector outputs a pulse when the signal goes away:
  - D.C. Idle
- Each detector has two timing circuits:

  - One detects if the spacing is too short.
  - One detects if the spacing is too long.
- If the spacing is correct, a counter is incremented:
  - If three consecutive cycles occur, the output is asserted

OBB Squelch Detector Description

The output of the squelch detector is fed into four frequency comparators (see Figure 6.7). When the period is within the window determined by the RC time constants for three consecutive cycles, the appropriate signal is asserted.

The Squelch detector makes use of a receiver with built-in hysteresis to filter out any signal not meeting the minimum amplitude. The squelch detector receiver should be a true differential to ensure that common-mode noise is rejected.

The full-swing output is fed into a pulse generator that charges up the capacitor through the diode. In the absence of a signal, a resistor discharges the capacitor to ground. The circuit outputs a true signal when the capacitor voltage is below the turn-on threshold of the Schmitt trigger buffer—indicating insufficient signal level. This circuit must be enabled in all power management states and should, therefore, be implemented with a small power budget.

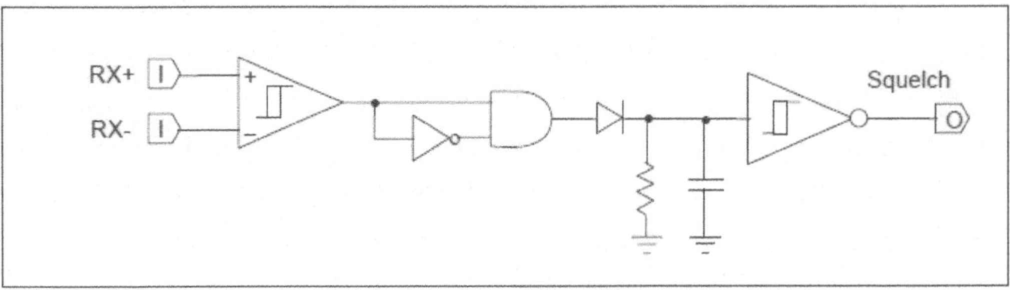

Figure 6.7 Squelch detector logic. (*Source:* www.t13.org)

Elasticity Buffer Management

Elasticity buffer circuitry may be required to absorb the slight differences in frequencies between the host and device.

- The greatest frequency differences result from a Spread Spectrum Clocking (SSC)–compliant device talking to a non-SSC device.
- The average frequency difference will be just over 0.25%, with excursions as much as 0.5%.

Since an elasticity buffer has a finite length, there needs to be a mechanism at the Physical layer protocol level that allows this receiver buffer to be reset without dropping or adding any bits to the data stream.

- This is especially important during reception of long continuous streams of data.
- This Physical layer protocol must not only support oversampling architectures but must also accommodate unlimited frame sizes (the frame size is limited by the CRC polynomial).

The Link layer shall keep track of a resettable counter that rolls over at every 1024 transmitted characters (256 dwords). Prior to, or at the pre-roll-over point (all 1's), the Link layer shall trigger the issuance of dual, consecutive ALIGN primitives, which shall be included in the dword count.

After communications have been established:

- The first and second words out of the Link layer shall be the dual-ALIGN primitive sequence, followed by, at most, 254 non-ALIGN dwords.
- The cycle repeats, starting with another dual-consecutive ALIGN primitive sequence.
- The Link may issue more than one dual ALIGN primitive sequence but shall not send an unpaired ALIGN primitive (i.e., ALIGN primitives are always sent in pairs), except as noted for retimed loopback.

6.2.6 Interface Power States

In the serial implementation of ATA, Interface Power States are controlled by the device and host adapter as defined in the Host PHY initialization and Device PHY initialization state diagrams and the Link Power Mode State Diagram. The Serial Interface Power States are defined in Table 6.5.

Table 6.5 Power States

READY	The PHY logic and main PLL are both on and active. The interface is synchronized and capable of receiving and sending data.
Partial	The PHY logic is powered, but is in a reduced power state. Both signal lines on the interface are at a neutral logic state (common mode voltage). The exit latency from this state shall be no longer than 10μs.
Slumber	The PHY logic is powered but is in a reduced power state. Both signal lines on the interface are at the neutral logic state (common mode voltage). The exit latency from this state shall be no longer than 10ms.

Power-On Sequence Timing Diagram

The following timing diagrams and descriptions (see Figures 6.8–6.10) are provided for clarity and are informative.

Description

1. Host/device power-off—Host and device power-off.
2. Power is applied—Host side signal conditioning pulls TX and RX pairs to neutral state (common mode voltage).
3. Host issues COMRESET sequence.
4. Once the power-on reset is released, the host puts the bus in a quiescent condition.
5. When the device detects the release of the COMRESET sequence, it responds with a COMINIT sequence. This is also the entry point if the device is late starting. The device may initiate communications at any time by issuing a COMINIT.
6. Host calibrates and issues a COMWAKE sequence.
7. Device responds—The device detects the COMWAKE sequence on its RX pair and calibrates its transmitter (optional). Following calibration, the device sends a six-burst COMWAKE sequence and then sends a continuous stream of the ALIGN sequence starting at the device's highest supported speed. After ALIGN DWORDs have been sent for 54.6μs (2048 nominal Gen1 DWORD times) without a response from the host, as determined by the detection of ALIGN

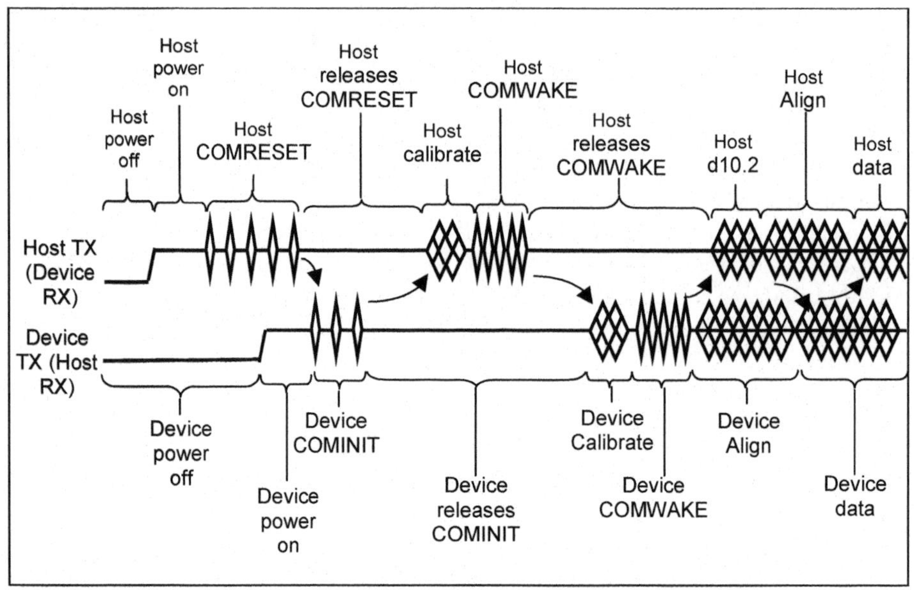

Figure 6.8 Power-on sequence. (*Source:* www.t13.org)

primitives received from the host, the device assumes that the host cannot communicate at that speed. If additional speeds are available, the device tries the next lower supported speed by sending ALIGN DWORDs at that rate for 54.6μs (2048 nominal Gen1 DWORD times). This step is repeated for as many legacy speeds as are supported. Once the lowest speed has been reached without response from the host, the device will enter an error state.

8. Host locks—After detecting the COMWAKE sequence, the host starts transmitting D10.2 characters at its lowest supported rate. Meanwhile, the host receiver locks to the ALIGN sequence and, when ready, returns the ALIGN sequence to the device at the same speed as received. A host must be designed such that it can acquire lock in 54.6μs (2048 nominal Gen1 DWORD times) at any given speed. The host should allow for at least 873.8μs (32768 Gen1 DWORD times) after detecting the release of COMWAKE to receive the first ALIGN. This will ensure interoperability with multi-generational and synchronous devices. If no is ALIGN is received within 873.8μs (32768 nominal Gen1 DWORD times), the host restarts the power-on sequence—repeating indefinitely until told to stop by the application layer.
9. Device locks—The device locks to the ALIGN sequence and, when ready, sends the SYNC primitive, indicating it is ready to start normal operation.
10. Upon receipt of three back-to-back non-ALIGN primitives, the communication link is established and normal operation may begin.

READY to Partial/Slumber

READY to Partial/Slumber is initiated by the transmission of the PMREQ_P or PMREQ_S and PMREQ_ACK primitives.

Partial/Slumber to READY

The host initiates a wakeup from the partial or slumber states by entering the initialization sequence at the HP5: HR_COMWAKE state in the Host PHY initialization state machine.

- Calibration and speed negotiation is bypassed since it has already been performed at power-on, and system performance depends on a quick resume latency.
- The device, therefore, transmits ALIGNs at the speed determined at power-on.

The device initiates a wakeup from the partial or slumber states by entering the initialization sequence at the DP6: DR_COMWAKE state in the Device PHY initialization state machine.

- Calibration and speed negotiation is bypassed since it has already been performed at power-on, and system performance depends on a quick resume latency.
- The device, therefore, transmits ALIGNs at the speed determined at power-on.

On to Partial/Slumber

Host Initiated

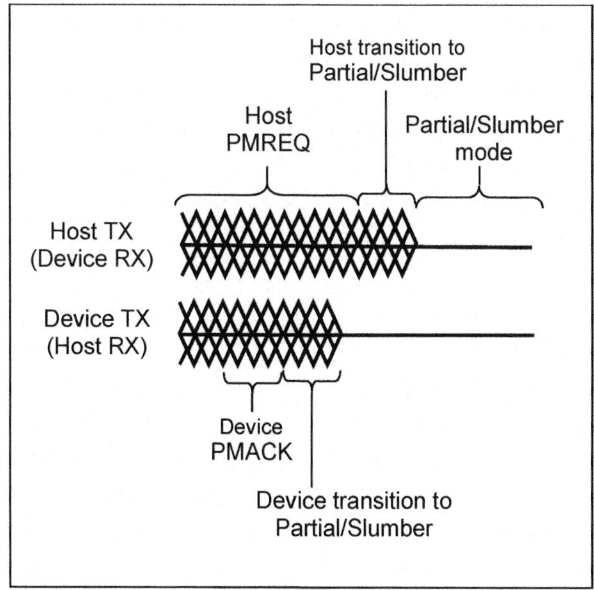

Figure 6.9 On to Partial/Slumber—host initiated. (*Source:* www.t13.org)

Detailed Sequence

a. Host Application layer sends request to host Transport layer.
b. Host Transport layer transmits request to host Link layer.
c. Host Link layer encodes request as PMREQ primitive and transmits it four times to host PHY layer.
d. Host PHY layer serializes PMREQ primitives and transmits them to device PHY layer.
e. Device PHY de-serializes PMREQ primitives and transmits them to device Link layer.
f. Device Link layer decodes PMREQ primitives and transmits request to device Transport layer.
g. Device Transport layer transmits request to device Application layer.
h. Device Application layer processes and accepts request. Issues accept to device Transport layer.
i. Device Transport layer transmits acceptance to device Link layer.

j. Device Link layer encodes acceptance as PMACK primitive and transmits it four times to device PHY layer.
k. Device PHY layer transmits four PMACK primitives to host PHY layer.
l. Device Link layer places device PHY layer in Partial/Slumber state.
m. Host PHY layer de-serializes PMACK primitives and transmits them to host Link layer.
n. Host Link layer decodes PMACK primitives and transmits acceptance to host Transport layer.
o. Host Link layer places host PHY layer in Partial/Slumber State.
p. Host Transport layer transmits acceptance to host Application layer.

Device Initiated

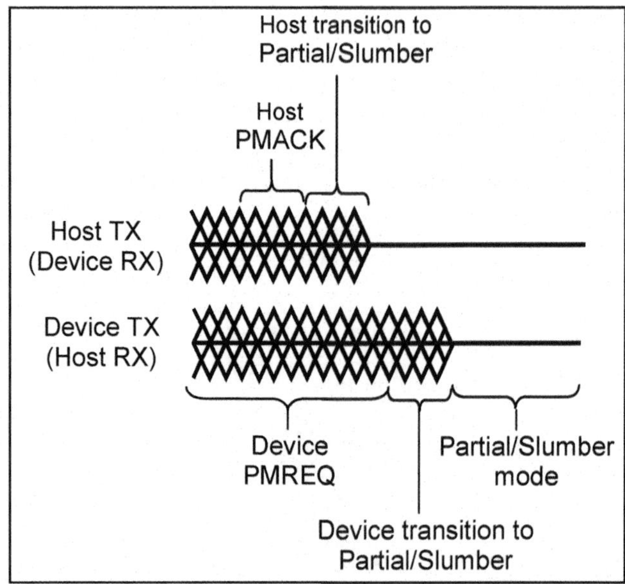

Figure 6.10 ON to Partial/Slumber—device initiated. (*Source:* www.t13.org)

Detailed Sequence

a. Device Application layer sends request to device Transport layer.
b. Device Transport layer transmits request to device Link layer.
c. Device Link layer encodes request as PMREQ primitive and transmits it to device PHY layer.
d. Device PHY layer serializes PMREQ primitives and transmits them to host PHY layer.
e. Host PHY de-serializes PMREQ primitives and transmits them to host Link layer.
f. Host Link layer decodes PMREQ primitives and transmits request to host Transport layer.
g. Host Transport layer transmits request to host Application layer.

Note: In this context, the host Application layer does not necessarily imply BIOS or other host CPU programming. Rather, the Application layer is the intelligent control section of the chipset logic.

h. Host Application layer processes and accepts request. Issues accept to host Transport layer.
i. Host Transport layer transmits acceptance to host Link layer.
j. Host Link layer encodes acceptance as PMACK primitive and transmits it four times to host PHY layer.
k. Host PHY layer transmits four PMACK primitives to device PHY layer.

l. Host Link layer asserts Partial/Slumber signal and places host PHY layer in Partial/Slumber state.
m. Host PHY layer negates Ready signal.
n. Device PHY layer de-serializes PMACK primitives and transmits them to device Link layer.
o. Device Link layer decodes PMACK primitives and transmits acceptance to device Transport layer.
p. Device Link layer asserts Partial/Slumber signal and places device PHY layer in Partial/Slumber State.
q. Device PHY layer negates Ready signal.
r. Device Transport layer transmits acceptance to device Application layer.

6.3 BIST (Built-In Self-Test)

BIST provides loopback testing of portions of the Physical layer. BIST is initiated by the BIST ACTIVATE FIS.

6.3.1 Loopback Testing

Three types of Loopback test schemes are defined.

a. Far-End Retimed – Mandatory
b. Far-End Analog – Optional
c. Near-End Analog (Effectively Retimed) – Optional

6.3.2 Loopback—Far End Retimed

Figure 6.11 illustrates the scope, at the architectural block diagram level, of the Far-End Retimed loopback. As this loopback scheme needs a specific action from the far-end connected interface, this mode shall be entered by way of the BIST Activate FIS.

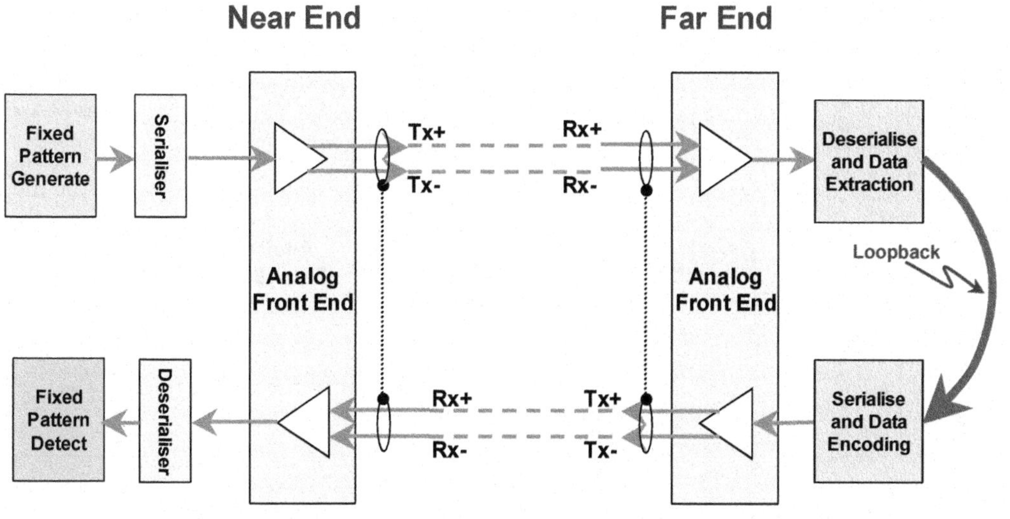

Figure 6.11 Far-End Retimed Loopback.

The Far-End Interface shall remain in this Far-End Retimed Loopback, until receipt of a COMRESET or COMINIT sequence.

As a minimum, Far-End Retimed Loopback shall involve far-end circuitry such that the datastream, at the Far-End interface, is extracted by the Serializer/Deserializer (SerDes) and data recovery circuit (DRC) before being sent back through the SerDes and transmitter with appropriately inserted retiming ALIGN primitives. The data may be decoded and descrambled in order to provide testing coverage for those portions of the host/device, provided the data is rescrambled using the same sequence of scrambler syndromes. The returned data shall be the same as the received data with the exception that the returned data may be encoded with different starting Running Disparity.

The initiator of the retimed loopback mode must account for the loopback host/device consuming up to two ALIGN primitives (one ALIGN sequence) every 256 dwords transmitted, and, if it requires any ALIGN primitives to be present in the returned data stream, it shall insert additional ALIGNs in the transmitted stream. The initiator shall transmit additional ALIGN sequences (Dual ALIGN primitives) in a single burst at the normal interval of every 256 dwords transmitted (as opposed to inserting ALIGN sequences at half the interval).

The loopback host/device may remove zero, one, or two ALIGN primitives from the received data. It may insert one or more ALIGN primitives if they are directly preceded or followed by the initiator-inserted ALIGN primitives (resulting in ALIGN sequences consisting of at least two ALIGN primitives) or it may insert two or more ALIGN primitives if not preceded or followed by the initiator's ALIGN primitives. One side effect of the loopback retiming is that the returned data stream may have instances of an odd number of ALIGN primitives; however, returned ALIGNs are always in bursts of two, and if the initiator transmitted dual ALIGN sequences (four consecutive ALIGNs), then the returned data stream shall include ALIGN bursts that are no shorter than two ALIGN primitives long (although the length of the ALIGN burst may be odd). The initiator of the retimed loopback mode shall not assume any relationship between the relative position of the ALIGNs returned by the loopback host/device and the relative position of the ALIGNs sent by the initiator.

In retimed loopback mode, the initiator shall transmit only valid 8b10b characters so the loopback host/device may 10b/8b decode it and re-encode it before retransmission. If the loopback host/device

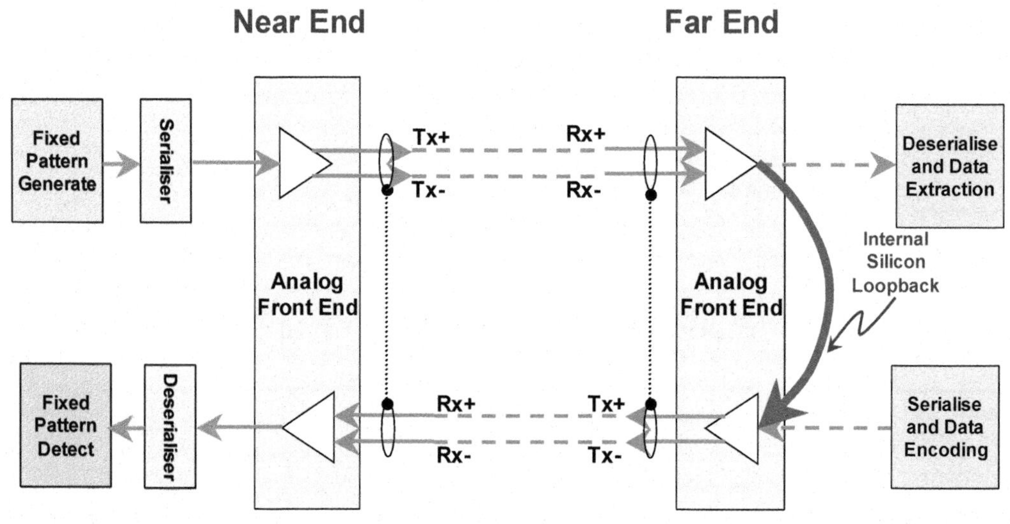

Figure 6.12 Far-End Analog.

descrambles incoming data, it is responsible for rescrambling it with the same sequence of scrambling syndromes in order to ensure the returned data is unchanged from the received data. The loopback host/device's running disparity for its transmitter and receiver are not guaranteed to be the same, and thus the loopback initiator shall 10b/8b decode the returned data rather than use the raw 10b returned stream for the purpose of data comparison. The loopback host/device shall return all received data unaltered and shall disregard protocol processing of primitives. Only the OOB signals and ALIGN processing is acted on by the loopback host/device, whereas all other data is retransmitted without interpretation.

6.3.3 Loopback—Far-End Analog (Optional)

Figure 6.12 illustrates the scope, at the architectural block diagram level, of the Far-End Analog loopback. As this loopback scheme needs a specific action from the far-end connected interface, this mode, if implemented, shall be entered by way of the BIST Activate FIS.

Once entered, the Far-End Interface shall remain in this Far-End Analog Loopback mode until receipt of the ComReset/ComInit sequence.

The implementation of Far-End AFE Loopback is optional due to the roundtrip characteristics of the test as well as the lack of retiming. This mode is intended to give a quick indication of connectivity, and test failure is not an indication of system failure.

6.3.4 Loopback—Near-End Analog (Optional)

Figure 6.13 illustrates the scope, at the architectural block diagram level, of the Near-End Analog loopback. This loopback scheme, if implemented, needs the far-end connected interface to be in a non-transmitting mode, such as Slumber mode or Partial mode.

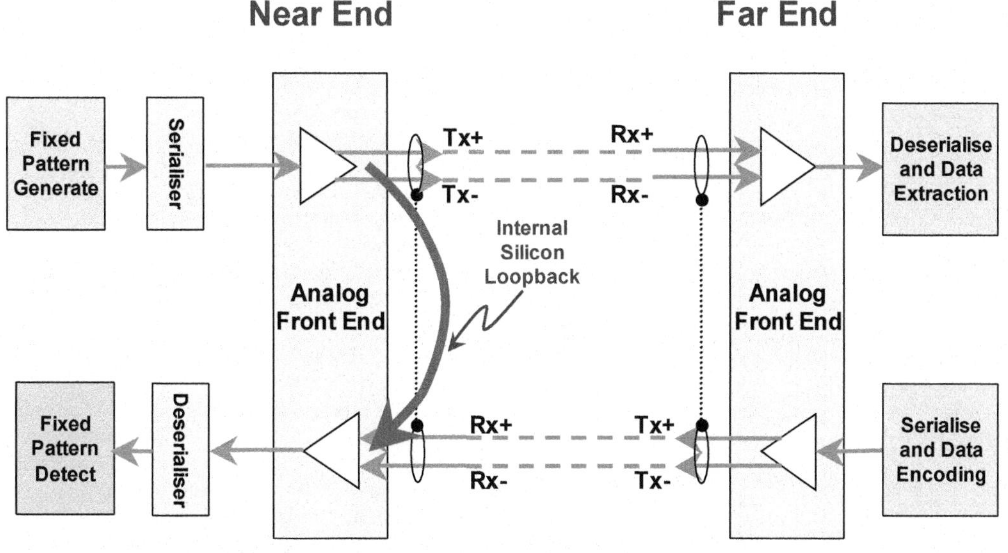

Figure 6.13 Near-End Analog.

6.3.5 *Impedance Matching*

The host adapter shall provide impedance-matching circuits to ensure termination for both its TX and RX as per the electrical parameters of Table 6.6.

Device peripherals shall provide impedance terminations, as per the specified parameters above, and may adapt their termination impedance to that of the host.

The host adapter, since it is given the first opportunity to calibrate during the power-on sequence, cannot assume that the far end of the cable is calibrated yet. For this reason, the host adapter must utilize a separate reference to perform calibration. In cabled systems, the cable provides the optimal impedance reference for calibration.

Using Time Domain Reflectometry (TDR) techniques, the host may launch a step waveform from its transmitter, so as to get a measure of the impedance of the transmitter, with respect to the cable, and adjust its impedance settings as necessary.

In a mobile system environment, where the cable is small or nonexistent, the host adapter must make use of a separate reference (such as an accurate off-chip resistor) for the calibration phase.

The device, on the other hand, may assume that the termination on the far side (host side) of the cable is fully calibrated and may make use of this as the reference. Using the host termination as the calibration reference allows the devices operating in both the desktop and the mobile system environments to use the same hardware.

Signals generated for the impedance calibration process shall not duplicate the OOB COMWAKE, COMINIT, or COMRESET sequences. Signals generated for the impedance calibration process shall not exceed the normal operating voltage levels. See the power management section for suggested times to perform calibration during power-on.

Table 6.6 Characteristic Impedance

	Nom	Min	Max	Units	Comments
Tx pair differential impedance	100	85	115	Ohm	As seen by a differential TDR with 100 ps (max) edge looking into connector (20%–80%). Measured with TDR in differential mode.
Rx pair differential impedance	100	85	115	Ohm	As seen by a differential TDR with 100 ps (max) edge looking into connector (20%–80%). Measured with TDR in differential mode.
Tx single-ended impedance		40		Ohm	As seen by TDR with 100 ps (max) edge looking into connector (20%–80%). TDR set to produce simultaneous positive pulses on both signals of the Tx pair. Single-ended impedance is the resulting (even mode) impedance of each signal. Both signals must meet the single-ended impedance requirement. Must be met during all possible power and electrical conditions of the PHY, including power off and power ramping.
Rx single-ended impedance		40		Ohm	As seen by TDR with 100 ps (max) edge looking into connector (20%-80%). TDR set to produce simultaneous positive pulses on both signals of the Rx pair. Single-ended impedance is the resulting (even mode) impedance of each signal. Both signals must meet the single-ended impedance requirement. Must be met during all possible power and electrical conditions of the PHY, including power off and power ramping.

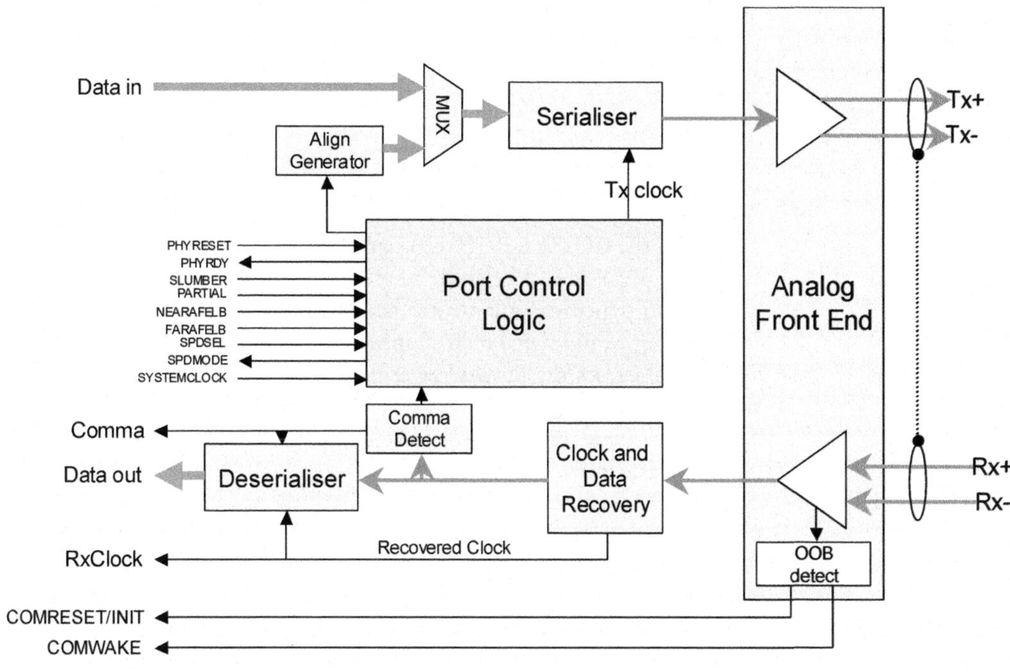

Figure 6.14 Port logic.

6.3.6 Physical Layer Electronics (PHY)

Block Diagram (Informative)

Figure 6.14 is a block diagram provided as a reference for the following sections in this chapter. Although informative in nature, the functions of the blocks described provide the basis upon which the normative specifications apply. The individual blocks provided are representative of an approach to this design and are provided as an example of one possible implementation.

Terms including "signal" in the name are used as defined terms in state diagrams and text descriptions to describe Physical layer behavior and are not intended to require a specific implementation.

Analog front end	This block is the basic interface to the transmission line. It consists of the high-speed differential drivers and receivers as well as the OOB signaling circuitry.
Control block	This block is a collection of logic circuitry that controls the overall functionality of the Physical plant circuitry.
Fixed pattern source	This block provides the support circuitry that generates the patterns as needed to implement the ALIGN primitive activity.
Fixed pattern detect	This block provides the support circuitry to allow proper processing of the ALIGN primitives.
Data extraction block	This block provides the support circuitry to separate the clock and data from the high-speed input stream.
TX clock	This signal is internal to the Physical plant and is a reference signal that regulates the frequency at which the serial stream is sent via the high-speed signal path

TX +/TX –	These signals are the outbound high-speed differential signals that are connected to the serial ATA cable.
RX +/RX –	These signals are the inbound high-speed differential signals that are connected to the serial ATA cable.
DATAIN	Data sent from the Link layer to the Physical layer for serialization and transmission.
PHYRESET	This input signal causes the PHY to initialize to a known state and start generating the COMRESET OOB signal across the interface.
PhyRdy signal	Indication PHY has successfully established communications. The PHY is maintaining synchronization with the incoming signal to its receiver and is transmitting a valid signal on its transmitter.
Slumber signal	This signal causes the Physical layer to transition to the Slumber mode for power management.
Partial signal	This signal causes the Physical layer to transition to the Partial mode for power management.
NEARAFELB	NEARAFELB causes the PHY to loop back the serial data stream from its transmitter to its receiver.
FARAFELB	FARAFELB causes the PHY to loop back the serial data stream from its receiver to its transmitter.
SPDSEL	SPDSEL causes the control logic to automatically negotiate for a usable interface speed or sets a particular interface speed. The actual functionality of this input is vendor specific and varies from manufacturer to manufacturer.
SPDMODE	SPDMODE is the output signal that reflects the current interface speed setting. The actual functionality of this signal is vendor specific and varies from manufacturer to manufacturer.
SYSTEMCLOCK	This input is the reference clock source for much of the control circuit and is the basis from which the transmitting interface speed is established.
COMMA	This signal indicates that a K28.5 character was detected in the inbound high-speed data stream.
DATAOUT	DATAOUT is data received and deserialized by the PHY and passed to the Link layer.
RX CLOCK/	Recovered clock—This signal is derived from the high-speed input data signal and determines when parallel data has been properly formed at the DATAOUT pins and is available for transfer to outside circuitry.
OOB signal detector	This block decodes the OOB signal from the high-speed input signal path.
ComInit signal	The ComInit signal is an indication from the host OOB detector that ComInit sequence is being detected.
ComReset signal	The ComReset signal is an indication from the device OOB detector that the ComReset sequence is being detected.
ComWake signal	The ComWake signal is an indication from the OOB detector that the ComWake sequence is being detected.

Analog Front End Block Diagram

Table 6.7 provides descriptions of the major functional blocks of the front end logic.

Figure 6.15 shows the internal functions of the analog front end logic, including the transmitter and receiver along with the OOB detection logic.

Figure 6.16 demonstrates how the differential signal pairs cross over in the cable so that the Tx and Rx signals always appear on the same connector pin locations.

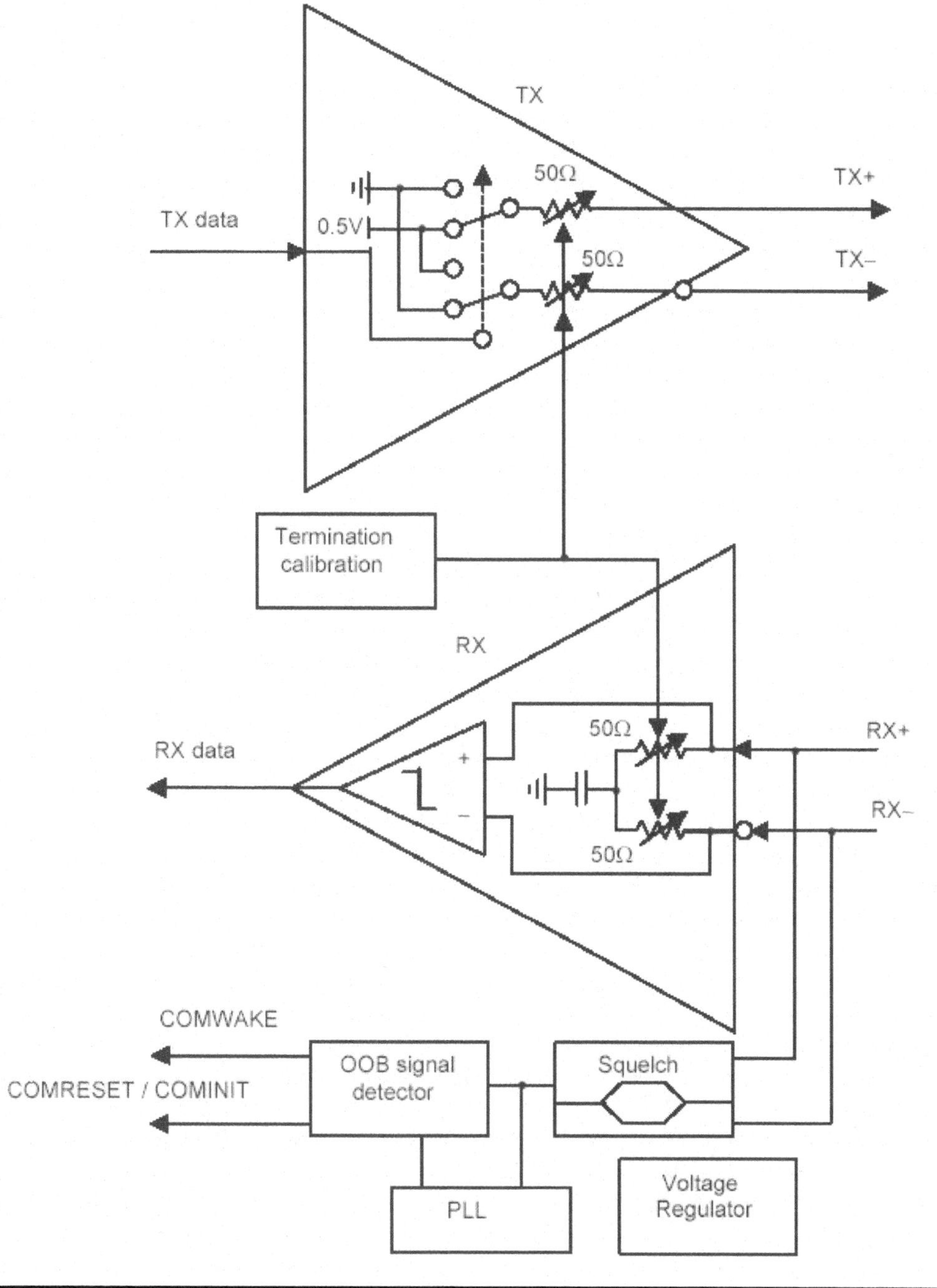

Figure 6.15 Analog Front End Logic. (*Source:* www.SATA-IO.org)

6.4 Electrical Specifications

The serial transport Physical layer electrical requirements are depicted in Table 6.8. The electrical performance is specified at the mated connector pair and includes the effects of the PCB.

Table 6.7 Analog Front End (AFE) Block Diagram Description

TX	This block contains the basic high-speed driver electronics
RX	This block contains the basic high-speed receiver electronics
Termination calibration	This block is used to establish the impedance of the RX block in order to properly terminate the high-speed serial cable.
Squelch	This block establishes a limit so that detection of a common mode signal can be properly accomplished.
OOB signal detector	This block decodes the OOB signal from the high-speed input signal path.
PLL	This block is used to synchronize an internal clocking reference so that the input high-speed data stream may be properly decoded.
Voltage Regulator	This block stabilizes the internal voltages used in the other blocks so that reliable operation may be achieved. This block may or may not be required for proper operation of the balance of the circuitry. The need for this block is implementation specific.
TX + / TX –	These signals are the outbound high-speed differential signals that are connected to the serial ATA cable.
RX + / RX –	These signals are the inbound high-speed differential signals that are connected to the serial ATA cable.
TxData	Serially encoded 10b data attached to the high-speed serial differential line driver.
RxData	Serially encoded 10b data attached to the high-speed serial differential line receiver.
COMWAKE	The ComWake signal is an indication from the OOB detector that the ComWake sequence is being detected.
COMRESET/COMINIT	The ComReset signal is an indication from the device OOB detector that the ComReset sequence is being detected.

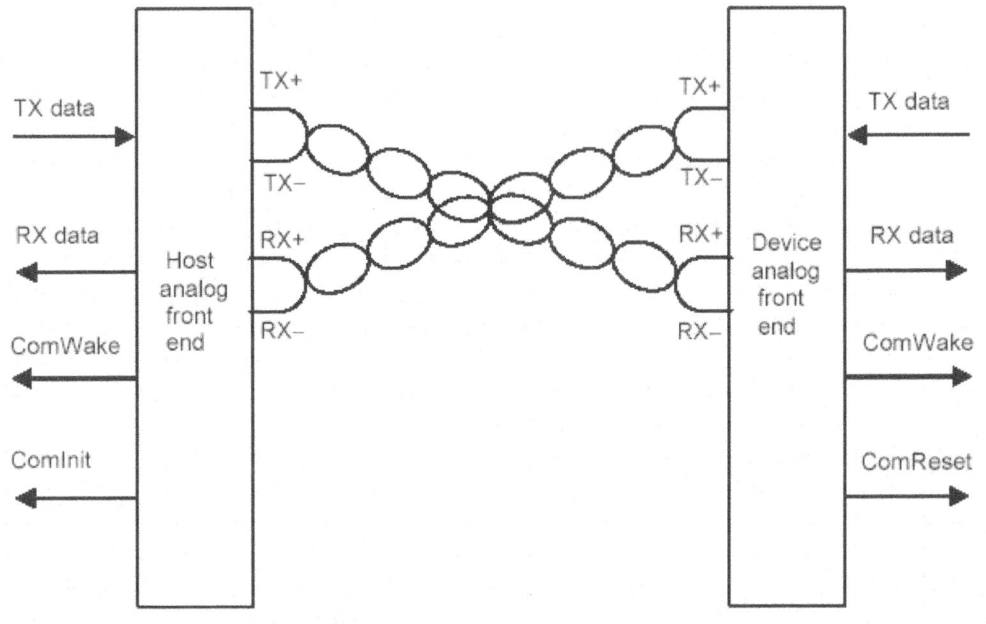

Figure 6.16 Analog front end cabling.

Table 6.8 Electrical Specifications

	Nom	Min	Max	Units	Comments
T,UI		666.43	670.12	ps	Operating data period (nominal value architecture specific)
t_{rise}	0.3	0.15	0.41	UI	20%–80% at transmitter
t_{fall}	0.3	0.15	0.41	UI	80%–20% at transmitter
$V_{cm,dc}$	250	200	450	mV	Common mode DC level measured at receiver connector. This spec only applies to direct-connect designs or designs that hold the common-mode level. AC coupled designs may allow the common mode to float. See $V_{cm,ac\ coupled}$ requirements
$V_{cm,ac\ coupled\ TX}$		0	2.0	V	Open circuit DC voltage level of each signal in the TX pair at the IC side of the coupling capacitor in an AC-coupled PHY. This requirement shall be met during all possible power and electrical conditions of the PHY, including power off and power ramping. Transmitter common mode DC levels outside this range may be used, provided that the following is met: The common mode voltage transients measured at the TX pins of the connector into an open-circuit load during all power states and transitions shall not exceed a +2.0V or –2.0V change from the CM value at the beginning of each transient. Test conditions shall include system power supply ramping at the fastest possible power ramp.
$V_{cm,ac\ coupled\ RX}$		0	2.0	V	Open circuit DC voltage level of each signal in the RX pair at the IC side of the coupling capacitor in an AC-coupled PHY. Must be met during all possible power and electrical conditions of the PHY, including power off and power ramping. Receiver common mode DC levels outside this range may be used, provided that the following is met: The common mode voltage transients measured at the RX pins of the connector into an open-circuit load during all power states and transitions shall not exceed a +2.0V or –2.0V change from the CM value at the beginning of each transient. Test conditions shall include system power supply ramping at the fastest possible power ramp.
F_{CM}		2	200	MHz	All receivers must be able to tolerate sinusoidal common-mode noise components inside this frequency range with an amplitude of $V_{cm,ac}$.
$T_{settle,CM}$			10	ns	Maximum time for common-mode transients to settle to within 10% of DC value during transitions to and from the quiescent bus condition.

(Continued on following page)

Table 6.8 Electrical Specifications (*Continued*)

	Nom	Min	Max	Units	Comments
$V_{diff,tx}$	500	400	600	mV_{p-p}	+/– 250 mV differential nominal. Measured at Serial ATA connector on transmit side.
$V_{diff,rx}$	400	325	600	mV_{p-p}	+/– 200 mV differential nominal. Measured at Serial ATA connector on receive side.
$C_{ACcoupling}$			12	nF	Coupling capacitance value for AC-coupled TX and RX pairs.
TX DC clock frequency skew		–350	+350	ppm	Specifies the allowed ppm tolerance for TX DC frequency variations around the nominal 1.500GHz. Excludes the +0/–5000ppm SSC downspread AC modulation per 0.
TX AC clock frequency skew		–5000	+0	ppm	Specifies the allowed ppm extremes for the SSC AC modulation, subject to the "Downspread SSC" triangular modulation (30–33kHz) profile per 0. *Note:* Total TX Frequency variation around nominal 1.500G includes [TXDC] + [TX AC] ppm variations.
TX differential skew			20	ps	(Nominal value architecture specific.)
Squelch detector threshold	100	50	200	mV_{p-p}	Minimum differential signal amplitude.

The electrical portion of the Physical layer includes the driver, receiver, PCB, and mated connector pair. Unless otherwise specified, all measurements shall be taken through the mated connector pair. Driver and receiver designs compensate for the effects of the path to/from the I/O connector.

6.4.1 Electrical Goals, Objectives, and Constraints

The Serial ATA Specification defines a cabled interconnect configuration and the corresponding PHY signaling parameters for that specific configuration.

- The parameters defined in the Serial ATA Specification do not directly apply to a backplane interconnect that has transmission characteristics substantially different from the cabled interconnect.
- This section defines the host PHY parameters to accommodate a backplane interconnect of up to 18 inches in length.
- The solution is constrained to isolate all compensation for the losses and reflection characteristics of the backplane interconnect to the host-side PHY, while leaving the device-side PHY wholly unchanged.

The Serial ATA Specification defines required signaling levels at the Serial ATA connector, as seen in Figure 6.17. These parameters do not directly apply to the pins of the controller electronics, and it is the responsibility of the interface component supplier to account for losses between the controller IC and the connector interface as part of the design collateral that accompanies the component.

This section provides additional host controller PHY parameter recommendations in order to promote timely and cost-effective solutions addressing the specific backplane requirements of storage

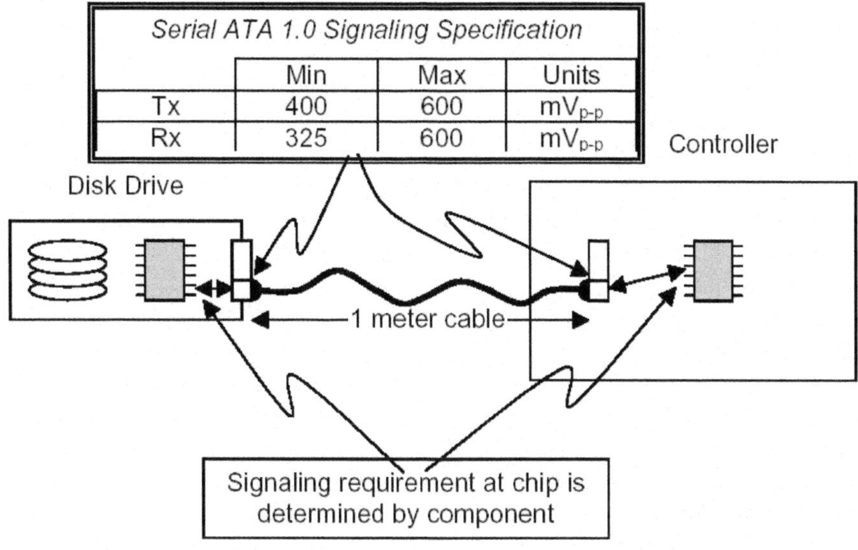

Figure 6.17 Serial ATA legacy signaling specification. (*Source:* www.SATA-IO.org)

subsystems. The overriding premise is that the device-side signal specification (at its connector) is an immutable given. All burdens for these new applications are placed upon the host-side controller.

Storage Arrays

The application of Serial ATA to storage subsystems may include integration into backplane-based designs, typically 19-inch standard racks. A storage system may consist of an array of devices mounted side by side along the front panel of a 19-inch rack in order to allow any one of the devices to be removed. A backplane routes signals to each device from a central controller. In the worst case, this controller could be positioned at one end of the device array, requiring routing from the controller silicon device to a connector on the backplane and then across the backplane. This type of application, therefore, defines a physical interconnect between the controller chip and the SATA device connector, which consists of up to 18 inches of etch and one connector (see Figure 6.18).

Recommendation

In order for a host controller to be able to reliably transmit to and receive from a SATA device within a backplane environment consisting of up to 18 inches of FR4 0.012-inch trace, greater minimum transmit levels (Table 6.9) and smaller minimum acceptable receive sensitivity levels are required. However, when that same host controller is applied in less lossy environments, the controller may run the risk of exceeding the maximum specified signal levels. This implies that it may be necessary for such a host controller to have a means of adjusting its signaling level as appropriate for the particular application. For example, the host controller may use an external set resistor that determines transmit current level.

Asynchronous Signal Recovery (Optional)

The SATA specification does not explicitly call out the PHY behavior for asynchronous signal recovery since the intended usage model for hot plugging was the insertion of a device into a receptacle where power would be applied as part of the insertion process.

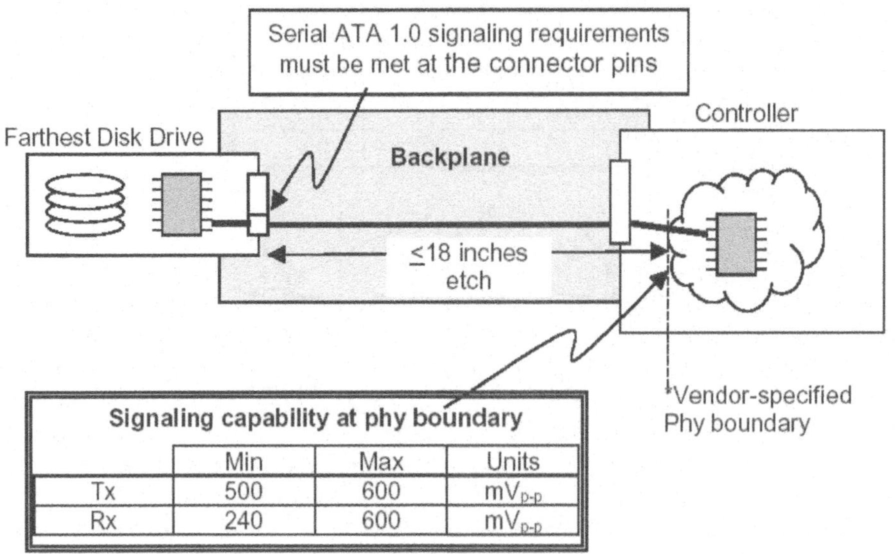

Figure 6.18 Serial ATA enhancements to signal level—second generation. (*Source:* www.SATA-IO.org)

* The vendor-specified PHY boundary encompasses design-specific support elements, including, but not limited to, coupling capacitors, compensating resistors, and tuned PCB trace geometries as part of the supplied controller component.

Table 6.9 Host-Controller 1.5 Gbps PHY Specification Recommendations for Backplane Application

	Nom	Min	Max	Units	Comments
Vdiff,tx	—	~~400~~ 500	600*	mVp-p	At host controller PHY boundary
Vdiff,rx	—	~~325~~ 240	600	mVp-p	At host controller PHY boundary

* Maximum transmit voltage may be increased above the specified maximum if the means for ensuring that the specified maximum receiver voltage is not exceeded at the far end of the interconnect is provided. All other parameters are as specified in the SATA Specification.

PHYs may optionally support Serial ATA II asynchronous signal recovery for those applications that the SATA usage model of device insertion into a receptacle (power applied at time of insertion) does not apply to.

Asynchronous signal recovery will require modifications to the following state machines:

- When the signal is lost, both the host and the device may attempt to recover the signal. If the device attempts to recover the signal before the host by issuing a COMINIT, it is unclear what state the drive is in and whether the drive will return its signature.
- If a host supports asynchronous signal recovery, when the host receives an unsolicited COMINIT, the host shall issue a COMRESET to the device.
- An unsolicited COMINIT is a COMINIT that was not in response to a preceding COMRESET, as defined by the host not being in the HP2:HR_AwaitCOMINIT state when the COMINIT signal is first received.

- As a consequence of the COMRESET, the device shall return its signature and will be in an unambiguous state.
- When a COMRESET is sent to the device in response to an unsolicited COMINIT, the host shall set the Status register to 0x7F and shall set all other task file registers to 0xFF.
- When the COMINIT is received in response to the COMRESET, the Status register value shall be updated to either 0xFF or 0x80 to reflect that a device is attached.

6.4.2 Rise/Fall Times

Output rise and fall times are measured between 20% and 80% of the signal (see Signal Rise and Fall Times in Figure 6.19). Rise and fall time requirements apply to differential transitions, for both in band and out of band signaling.

Differential Voltage/Timing (EYE) Diagram

The EYE diagram (see Figure 6.21) is more of a qualitative measurement than a specification. Any low-frequency (trackable) modulation must be tracked by the oscilloscope to prevent measurement error caused by benign EYE closure. It is also unrealistic to try to capture sufficient edges to guarantee the 14 sigma RJ requirement in order to achieve an effective 10^{-12} BER. Nonetheless, this method is useful and easy to set up in the lab.

If t3-t1 is set equal to $DJ+6*RJ_{1sigma}$, then less than one in every 750 edges cross the illegal region for a well-designed transmitter. By controlling the number of sweeps displayed, a quick health check may be performed with minimal setup. Note that this criteria is informative, and these conditions are not sufficient to guarantee compliance but are to be used as part of the design guidelines.

Jitter Output/Tolerance Mask

The spectral jitter comply to the requirements, as indicated in the Jitter Output/Tolerance Graph in Figure 6.20. A_x are the maximum peak to peak transmitter output and the minimum peak to peak receiver tolerance requirements as measured from data edge to any following data edge up to n_x*UI later (where x is 0,1, or 2).

Transmitter output data edge to data edge timing variation from t_o to t_y shall not exceed the value computed by the following:

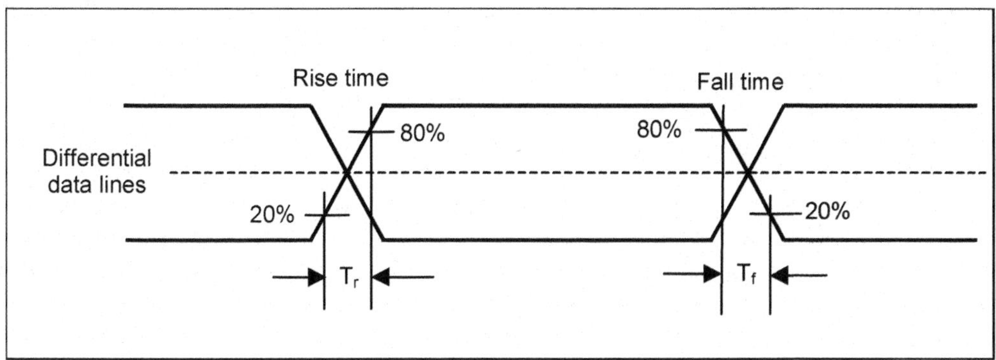

Figure 6.19 Signal rise and fall times. (*Source:* www.t10.org)

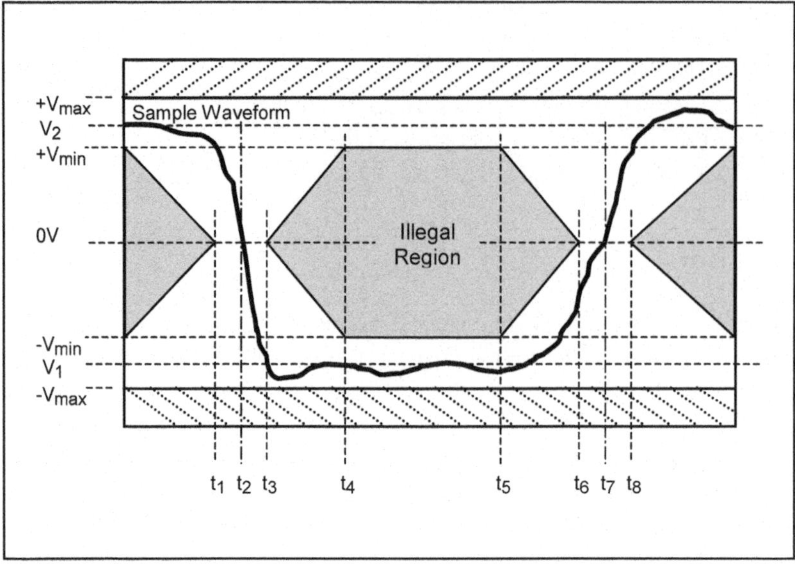

Name	Definition	Notes
t_{jitter}	t3–t1	t3–t1 = t8 - t6
T	t7–t2	t2–t1 = t3–t2 t7–t6 = t8–t7
V_{diff}	V_2–V_1	

Figure 6.20 Jitter diagram. (*Source:* www.t10.org)

- y is an integer from 1 to n_x;
- t_y is the time between the data transition at t_0 and a data transition y*UI bit periods later.

This measurement may be made with an oscilloscope having a histogram function or with a Timing Interval Analyzer (TIA). A receiver shall meet the error rate requirement under the maximum allowed jitter conditions.

Figure 6.21 shows a measured signal waveform at the end of a 1-meter electrical cable. This example has a good eye opening that does not violate the mask and provides the receiver with a wide margin for sampling the received bit stream.

Spread Spectrum Clocking (SSC)

A definition of the spread spectrum functionality follows:

a. All transmitter timings (including jitter, skew, min-max clock period, output rise/fall time) shall meet the existing non–spread spectrum specifications, when spread spectrum is on.
b. Because the minimum clock period cannot be violated, the transmitter shall adjust the spread technique to not allow for modulation above the nominal frequency. This technique is often called "down-spreading." An example of a triangular frequency modulation profile is shown in Figure 6.22. The modulation profile in a modulation period can be expressed, as follows:

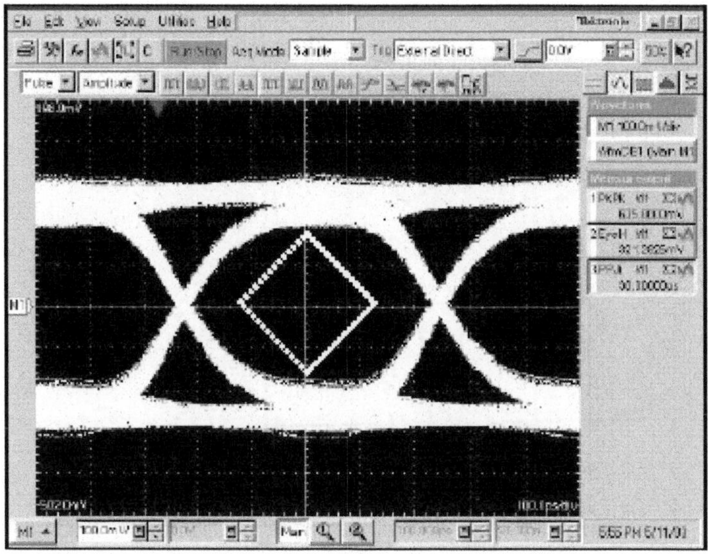

Figure 6.21 Eye diagram example.

$$f = \begin{cases} (1-\delta)\mathrm{f}_{\mathrm{nom}} + 2\mathrm{f}_{\mathrm{nom}} \cdot \delta \cdot \mathrm{f}_{\mathrm{nom}} \cdot \mathrm{t} & \text{when } 0 < \mathrm{t} < \dfrac{1}{2\mathrm{f}_{\mathrm{m}}} \\ (1-\delta)\mathrm{f}_{\mathrm{nom}} + 2\mathrm{f}_{\mathrm{nom}} \cdot \delta \cdot \mathrm{f}_{\mathrm{nom}} \cdot \mathrm{t} & \text{when } \dfrac{1}{2\mathrm{f}_{\mathrm{m}}} < \mathrm{t} < \dfrac{1}{\mathrm{f}_{\mathrm{m}}} \end{cases}$$

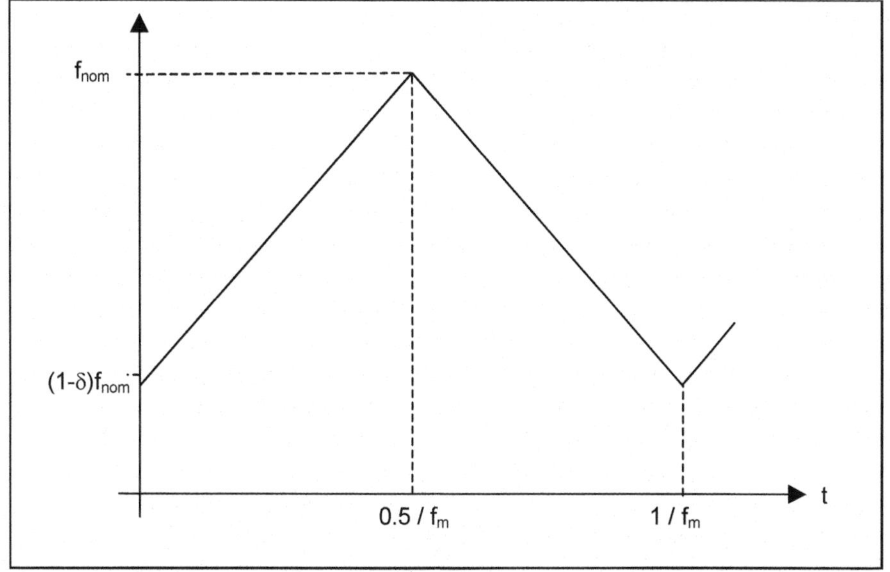

Figure 6.22 Triangular frequency modulation profile. (*Source:* www.t10.org)

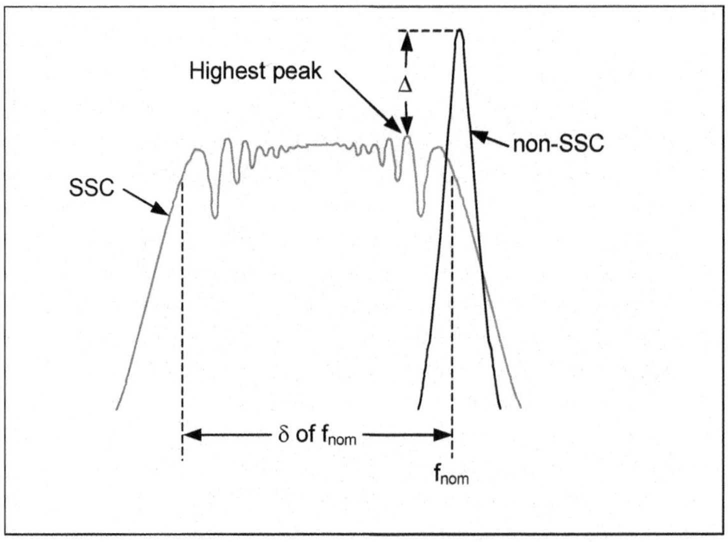

Figure 6.23 Spectral fundamental frequency comparison. (*Source:* www.t10.org)

where f_{nom} is the nominal frequency in the non-SSC mode, f_m is the modulation frequency, δ is the modulation amount, and t is time.

c. For triangular modulation, the clock frequency deviation (δ) is required to be no more than 0.5% "downspread" from the corresponding nominal frequency—that is, +0%/–0.5%. The absolute spread amount at the fundamental frequency is shown in Figure 6.23 as the width of its spectral distribution (between the –3 dB roll-off). The ratio of this width to the fundamental frequency cannot exceed 0.5%. This parameter can be measured in the frequency domain using a spectrum analyzer.

d. To achieve sufficient system-level EMI reduction, it is desired that SSC reduce the spectral peaks in the non-SSC mode by the amount specified in Table 6.10. The peak reduction Δ is defined, as shown in Figure 6.23, as the difference between the spectral peaks in SSC and non-SSC modes at the specified measurement frequency.

It is recommended that a spectrum analyzer be used for this measurement. The spectrum analyzer should have measurement capability out to 1 GHz. The measured SSC clock needs to be fed into the spectrum analyzer via a high-impedance probe that is compatible with the spectrum analyzer. The output clock should be loaded with 20 pF capacitance. The resolution bandwidth of the spectrum analyzer needs to be set at 120 kHz to comply with with FCC EMI measurement requirements. The video band

Table 6.10 Desired Peak Amplitude Reduction by SSC

Clock freq.	Peak reduction Δ	Measurement freq.
66 MHz	7 dB	466 MHz (7th harmonics)
75 MHz	7 dB	525 MHz (7th harmonics)

Note: The spectral peak reduction is not necessarily the same as the system EMI reduction. However, this relative measurement gives the component-level indication of SSC's EMI reduction capability at the system level.

needs to be set at higher than 300 kHz for appropriate display. The resolution bandwidth, in case of a measurement equipment limitation, may be set at 100 kHz. The display should be set with maximum hold. The corresponding harmonic peak readings should be recorded in both the non-SSC and the SSC modes and be compared to determine the magnitude of the spectral peak reduction.

SSC Basics

SATA is one of the first high-speed serial interfaces to use SSC to further reduce electromagnetic emissions.

- SSC is the intentional modulation of the reference clock.
- Instead of having a pure reference clock at a fixed reference frequency, a SSC varies its frequency within a specified spreading range.
- The modulation is done at a slower rate relative to the reference frequency so that the receiver can track the ever-changing clocking frequency.

Although SSC presents some design challenges, the benefits outweigh the design costs:

- SSC is a well-known technique for reducing EMI.
- SSC yields about a 6db in overall improvement, which equates to about a quarter of the original frequency.
- This reduction results from the emissions being spread out over a broader frequency range such that they are not concentrated at a single center frequency.
- Although the total energy radiated is largely unchanged, by spreading the emissions over a broader range of frequencies, the energy at any given frequency is reduced.
- Because the FCC requirements are based on maximum radiated energy for a given frequency and not the total energy radiated for all frequencies, SSC provides a substantial benefit.

Alternatives to SSC:

- Alternatives to SSC would require additional shielding and termination to improve the cables and connectors.
- Such shielding of connectors and cables makes the interconnect more costly.
- Although the SSC burdens the PHY designer, the overall SSC clocking is modest when compared to the cost of fully shielding the cables and connectors.

Downspreading:

- Downspreading refers to the fact that the modulation varies the reference frequency downward from only the nominal frequency.
- The reference clock frequency starts at the maximum clocking frequency and steadily drifts downward until it reaches the minimum frequency, at which point it does an about face and drifts steadily back up again.
  - Modulation occurs at a rate of 30 to 33 kHz.
  - The bit clock for the serial stream has a frequency of about 750 MHz, so the SSC modulation rate is about 25,000 times slower than the signaling rate.
- Downspreading, compared to center-spreading, has the disadvantage that the effective center frequency is reduced slightly when SSC is enabled, but has the benefit that it avoids overclocking the interface.
- The center frequency is reduced by about half of the downspread sweep, or about 0.25%.

6.5 Review Questions

1. What does the host controller originate to force a hard reset in the device?

   a. COMINIT
   b. COMRESET
   c. COMWAKE
   d. COMSAS

2. How does the drive request a communication initialization?

   a. COMINIT
   b. COMRESET
   c. COMWAKE
   d. COMSAS

3. What OOB signaling is used to bring the PHY out of a power-down state?

   a. COMINIT
   b. COMRESET
   c. COMSAS
   d. COMWAKE

4. How does an analyzer tell which OOB sequence is being transmitted?

   a. Duration of ALIGN characters
   b. Number of OOB sequences
   c. Duration of OBB sequence
   d. Duration of idle period

5. What immediately follows the OOB initialization sequence?

   a. Device sends a signature
   b. Device transitions to an idle condition and transmits scrambled data
   c. Speed negotiation
   d. Host sends signature to device

6. What does the signature determine about the device?

   a. The device name
   b. The supported link speeds
   c. If it supports the PACKET- or REGISTER-driven command set
   d. It supports NCQ

7. During speed negotiation what does the host send first?

   a. ALIGN (0) @ its fastest link rate
   b. D10.2 @ its fastest link rate
   c. ALIGN (0) @ its slowest link rate
   d. D10.2 @ its slowest link rate

8. During speed negotiation, what does the device send?

    a. 2048 ALIGN (0) @ its fastest link rate
    b. 2048 D10.2 @ its fastest link rate
    c. 2048 ALIGN (0) @ its slowest link rate
    d. 2048 D10.2 @ its slowest link rate

9. What does the host transmit once it detects ALIGN (0) on its receiver?

    a. ALIGN (0) @ its fastest link rate
    b. ALIGN (0) @ its slowest link rate
    c. ALIGN (0) @ detected link rate
    d. ALIGN (0) @ next link rate in speed list

10. Which item is not a valid BIST?

    a. Far-End Retimed
    b. Far-End Analog
    c. Near-End Retimed
    d. Near-End Analog

11. What is the impedance of the transmission line?

    a. 100 ohm
    b. 50 ohm
    c. 200 ohm
    d. 150 ohm

Chapter 7

Error Handling

Objectives

Once you complete this chapter you will comprehend the different types of Serial ATA errors and error handling procedures for all protocol layers.

- Protocol layers and types of errors include the following:
    - Application (Software) Errors: Bad status in Command Block Status or SError registers, command timeout
    - Transport Errors: Frame, protocol, and internal errors
    - Link Errors: Invalid state transitions and data integrity errors
    - Phy Errors: No device present, OOB protocol error, or internal error (loss of sync)
- This chapter will also demonstrate how to read and decipher state diagrams.

7.1 Error Handling

The layered architecture of the serial implementation of ATA extends to error handling as well. As indicated in Figure 7.1, each layer in the protocol stack has, as input error, indications from the next lower layer (except for the Physical layer, which has no lower layer associated with it). Each layer has its local error detection capability to identify errors specific to that layer based on the data received from the lower and higher layers. Each layer performs local recovery and control actions and may forward error information to the next higher layer in the stack.

Error responses are classified into the following four categories:

- Freeze
- Abort
- Retry
- Track/ignore

The error handling responses described in this section are not comprehensive and are included to cover specific known error scenarios, as well as to illustrate typical error control and recovery actions.

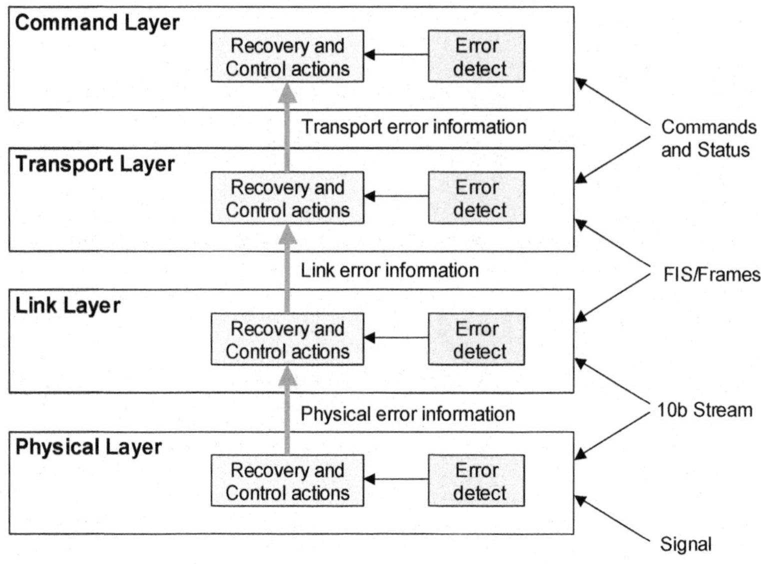

Figure 7.1 Error recovery and control action hierarchy.

7.1.1 Error Handling Responses

For the most **severe error conditions** in which state has been critically perturbed in a way that it is not recoverable, the appropriate error response is to freeze and rely on a reset or similar operation to restore all necessary states to normal operation.

For error conditions that are expected to be **persistent**, the appropriate error response is to abort and fail the attempted operation. Such failures usually imply notification up the stack in order to inform the host software of the condition.

For error conditions that are **transient** and not expected to persist, the appropriate response is to retry the failed operation. Only failed operations that have not perturbed system state are allowed to be retried. Such retries may either be handled directly by the recovery and error control actions in the relevant layer or may be handled by the host software in response to error information being conveyed to it.

Noncritical recoverable conditions may either be tracked or ignored. Such conditions include those that were recovered through a retry or other recovery operation at a lower layer in the stack. Tracking such errors may often be beneficial in identifying a marginally operating component or other imminent failure.

7.1.2 Phy Error Handling Overview

There are three primary categories of error that the Physical layer detects internally:

- No device present
- OOB signaling sequence failure
- Phy internal error (loss of synchronization of communications link)

A **no device present** condition results from a physical disconnection in the media between the host adapter and device, whether intermittent or persistent. The host adapter will detect device presence as part of the interface reset sequence.

An **OOB signaling failure** condition arises when the sequence of OOB signaling events cannot be completed, prior to declaration of the "Phy Ready" state. OOB signaling sequences are required when emerging from a power management "partial" or "slumber" state, from a "loopback/BIST" test state, or from initial power-up. OOB signaling sequences are used to achieve a specifically ordered exchange of COMRESET, COMINIT, COMWAKE, and ALIGN patterns to bring the communications link up between host adapter and device.

A **Phy Internal Error** can arise from a number of conditions. Regardless of whether it is caused by the characteristics of the input signal or an internal error unique to the implementation, it will always result in the loss of the synchronization of the communications link.

Phy Error Control Actions

No Device Present

Due to the nature of the physical interface, it is possible for the Phy to determine that a device is attached to the cable at various times. As a direct result, the Phy is responsible for detecting the presence of an attached device, and this presence will be reported in the SStatus register such that the host software can respond appropriately.

During the interface initialization sequence, an internal state bit "Device Detect" will be cleared in the host adapter when the COMRESET signal is issued. The "Device Detect" state bit will be set in the host adapter when the host adapter detects a COMINIT signal from the attached device. The "Device Detect" state bit corresponds to the device presence detect information in the SStatus register.

Note that device presence and communications established are separately reported in the SStatus register in order to encompass situations in which an attached device is detected by the Phy, but the Phy is unable to establish communications with it.

OOB Signaling Sequence Failure

The Phy does not have any time-out conditions for the interface reset signaling sequence. If a device is present, the Phy will detect device presence within 10ms of a power-on reset (i.e.,COMINIT must be returned within 10ms of an issued COMRESET). If a device is not present, the Phy is not required to time-out and may remain in the reset state indefinitely until the host software intervenes. Upon successful completion of the interface initialization sequence, the Phy is ready, active, and synchronized, and the SStatus register bits will reflect this condition.

Phy Internal Error

There are several potential sources of errors categorized as "Phy Internal Errors." In order to accommodate a range of implementations without making the software error handling approach implementation dependent, all the different potential sources of internal Phy errors are combined for the purpose of reporting the condition in the SError register. This requirement does not preclude each vendor from implementing their own level of error diagnostic bits, but those will reside in vendor-specific register locations.

Phy internal errors that result in the Phy becoming not ready (the PhyRdy signal being negated) and the corresponding SStatus register and SError registers bits will be updated.

The "Phy Ready Change" bit, as defined in the SError Register, is updated as per its definition.

Error Reporting

Phy errors are reported to the Link layer in addition to being reflected in the SStatus register and SError registers.

7.1.3 Link Error Handling Overview

Error Detection

There are two primary categories of errors that the Link layer detects internally:

- Invalid state transitions
- Data integrity errors

Invalid state transition errors can arise from a number of sources and the Link layer responses to many such error conditions. Data integrity errors generally arise from noise in the physical inter-connects.

Error Control Actions

Errors detected by the Link during a transmission are handled by accumulating the errors until the end of the transmission and reflecting the reception error condition in the final R_ERR/R_OK handshake. Specific scenarios are listed in the following sections.

Invalid State Transitions

Invalid state transitions are handled through the return to a known state, for example, where the Link state diagrams define the responses for invalid state transition attempts. Returning to a known state is achieved through two recovery paths depending on the state of the system.

If the invalid state transitions are attempted during the transmission of a frame (after the receipt of an SOF), the Link will signal negative acknowledgment (R_ERR) to the transmitting agent.

If the invalid state transition is not during a frame transmission, the Link will go directly to the idle state and await the next operation.

The following paragraphs outline the requirements during specific state transition scenarios and their respective Link error control actions:

> Following reception of one or more consecutive X_RDY at the receiver interface, if the next control character received is not SOF, the Link will notify the Transport Layer of the condition and transition to the idle state.
>
> Following the transmission of X_RDY, if there is no returned R_RDY received, no Link recovery action will be attempted. The higher-level layers will eventually time out and reset the interface.

On receipt of an unexpected SOF, when the receiving interface has not yet signaled readiness to receive data with the R_RDY signal, that receiving interface will remain in the idle state issuing SYNC primitives until the transmitting interface terminates the transmission and also returns to the idle state.

If the transmitter closes a frame with EOF and WTRM, and receives neither a R_OK nor an R_ERR within a predetermined time-out, no Link recovery action will be attempted. The higher-level layers will eventually time out, and reset the interface.

If the transmitter signals EOF, and a control character other than SYNC, R_OK, or R_ERR is received, the Link layer will persistently continue to await reception of a proper terminating primitive.

Data Integrity Errors

Data integrity errors are handled by signaling the Transport Layer in order to potentially trigger a transmission retry operation, or to convey failed status information to the host software. In order to return to a known state, data integrity errors are usually signaled via the frame acknowledgment handshake, at the end of a frame transmission, before returning to the idle state.

The following paragraphs outline the requirements during specific data integrity error scenarios and the respective Link error control actions:

On detection of a CRC error at the end of receiving a frame (at EOF), the Link Layer will notify the Transport Layer that the received frame contains a CRC error. Furthermore, the Link Layer will issue the negative acknowledgment, R_ERR, as the frame status handshake, and will return to the idle state.

On detection of a disparity error or other 8b10b coding violation during the receipt of a frame, the Link Layer will retain this error information, and at the close of the received frame the Link Layer will provide the negative acknowledgment, R_ERR, as the frame handshake, and will notify the Transport Layer of the error.

The control actions are the same for coding violations as for CRC errors.

Error Reporting

Link error conditions are reported to the Transport layer in a vendor-specific interface between the Link and Transport Layers. Additionally, Link errors are reported in the Interface Error register.

7.1.4 Transport Error Handling

Overview

The Transport layer communicates errors to the software and/or performs local error recovery and initiates control actions, such as retrying FIS transmissions.

The Transport layer informs the Link layer of detected errors so that the Link layer can reflect Transport errors in the R_ERR/R_OK handshake at the end of each frame. Devices will reflect any

R_ERR frame handshakes in the command ending status for Data FISes reflected in the transmitted register FIS that conveys the operation ending status. The Transport layer also reflects any error information encountered in the Interface Error register.

The Transport may retry any non-Data FIS transmission, provided the system state has not changed as a result of the corresponding failure, and may retry any number of times. For scenarios in which repeated retry operations persistently fail, the host software will time out the corresponding command and perform recovery operations.

Error Detection

In addition to the error information passed to it by the Link layer, the Transport layer internally detects the following categories of errors:

- Internal errors
- Frame errors
- Protocol and state errors

There are several kinds of internal errors to the Transport layer, including overflow/underflow of the various speed-matching FIFOs. Internal errors are handled by failing the corresponding transaction, and returning to a state equivalent to a failed transaction (e.g., the state that would result from a bad CRC).

The Transport layer detects several kinds of frame errors, including reception of frames with incorrect CRC, reception of frames with invalid TYPE field, and reception of ill-formed frames (such as a register frames that are not the correct length). Frame errors are handled by failing the corresponding transaction and returning to a state equivalent to a failed transaction (such as the state that would result from a bad CRC).

Protocol and state transition errors stem from devices not following the serial implementation of ATA protocol and include errors such as the PIO count value not matching the number of data characters subsequently transferred and errors in the sequence of events.

Protocol and state transition errors are handled by failing the corresponding transaction and returning to a state equivalent to a failed transaction (such as the state that would result from a frame being received with a bad CRC).

Error Control Actions

Internal Errors

Internal errors are handled by failing the corresponding transaction and either retrying the transaction or notifying the host software of the failure condition in order to ultimately generate a host software retry response. The following are specific internal error scenarios and their corresponding Transport error control actions:

> If the receive FIFO overflows, the Transport layer will signal frame reception negative acknowledgment by signaling the Link layer to return R_ERR during the frame acknowledgment handshake. Subsequent actions are equivalent to a frame reception with erroneous CRC.
>
> If the transmit FIFO underruns, the Transport layer will close the transmitting frame with an EOF and CRC value that is forced to be incorrect in order to ensure that the recipient of the

corrupted frame also executes appropriate error control actions. This scenario results only from a transmitter design problem and should not occur for properly implemented devices.

Frame Errors

Frame errors may be handled in one of two ways, depending on whether the error is expected to be transient or persistent and whether system state has been perturbed.

For error conditions expected to be transient (such as a CRC error), and for which the system state has not been perturbed, the Transport layer may retry the corresponding transaction any number of times until, ultimately, a host time-out and software reset or other error recovery attempt is made.

For error conditions that are not a result of a transient error condition (such as an invalid TYPE field in a received FIS), the error response is to fail the transaction and report the failure.

The following are specific frame error scenarios and their corresponding Transport error control actions:

If the Transport receives an FIS with an invalid CRC signaled from the Link layer, the Transport layer will signal the Link layer to negatively acknowledge frame reception by asserting R_ERR during the frame acknowledgment handshake.

The transmitter of a negatively acknowledged frame may retry the FIS transmission provided the system state has not been perturbed. Frame types that may be retransmitted are:

- Register – Host to Device
- Register – Device to Host
- DMA Activate – Device to Host
- First Party DMA – Device to Host
- PIO Setup – Device to Host
- Set Device Bits – Device to Host

Because data transmission FISes result in a change in the host bus adapter's internal state, either through the DMA controller changing its state or through a change in the remaining PIO repetition count, data transmission FISes will not be retried.

The Transport layer is not required to retry those failed FIS transmissions that do not change system state, but the Transport layer may attempt retry any number of times. For conditions that are not addressed through retries, such as persistent errors, the host software will eventually time out the transaction and reset the interface.

If the Transport Layer detects reception of an FIS with unrecognized TYPE value, the Transport Layer will signal the Link Layer to negatively acknowledge the frame reception by asserting R_ERR during the frame acknowledgment handshake.

If the Transport Layer detects reception of a malformed frame, such as a frame with incorrect length, the Transport Layer will signal the Link Layer to negatively acknowledge the frame reception by asserting R_ERR during the frame acknowledgment handshake.

Protocol and State Transition Errors

Protocol and state errors stem from devices not following defined protocol. Such errors may be handled by failing the corresponding transactions and returning to a known state. Since such errors are not caused by an environmental transient, no attempt to retry such failed operations should be made. The following are specific frame error scenarios and their corresponding Transport error control actions:

> If the PIO transfer count expires, and two symbols later is not the EOF control character (the CRC falls between the last data character and the EOF), the transfer count stipulated in the PIO Setup FIS did not match the size of the subsequent data payload. For this data-payload/transfer-count mismatch, the Transport Layer will signal the Link Layer to negatively acknowledge frame reception by asserting R_ERR during the frame acknowledgment handshake.

Error Reporting

The Transport Layer reports errors to the host software via the Shadow Status and Shadow Control registers. Devices communicate Transport error information to the host software via transmitting a register FIS to update the ATA Shadow Status and Shadow Error register values.

Transport error conditions that are not handled/recovered by the Transport layer will set the error bit in the Shadow Status register and update the value in the Error register through the transmission of an appropriate Register FIS.

Host Transport error conditions will result in the status and error values in the SStatus and SError register being updated with values corresponding to the error condition and will result in the Link layer being notified to negatively acknowledge the offending FIS during the final reception handshake.

7.1.5 Software Error Handling Overview

The software layer error handling is in part defined by the error outputs of the specific command, as defined in Chapter 3: SATA Application Layer in this book. In addition, there are superset error-reporting capabilities supported by the Transport layer through SCRs, and software may take advantage of those error-reporting capabilities to improve error handling for Serial ATA.

Error Detection

There are three error detection mechanisms by which software identifies and responds to SATA errors:

- Bad status in the Command Block Status register
- Bad status in the serial implementation of ATA SError register
- Command failed to complete (time-out)

Conditions that return bad status in the Command Block Status register, but for which no serial implementation interface error information is available, correspond to the error conditions specified in the command descriptions. Such error conditions and responses are defined in the command descriptions, and there is no unique handling of those in the serial implementation of ATA. Errors in this category include command errors, such as attempts to read from an LBA past the end of the disk, as well as device-specific failures such as data not readable from the given LBA. These failures are not related to the serial implementation of the ATA interface, and thus no the serial implementation of ATA-specific interface status information is available for these error conditions. Only the status information returned by the device is available for identifying the source of the problem, in addition to any available SMART data that might apply.

Transport layer error conditions, whether recovered or not, are reflected in the SStatus and SError registers. The host Transport layer is responsible for reflecting error information in the SStatus and

SError registers, whereas the device Transport layer is responsible for reflecting unrecovered errors in the Shadow Status and Error registers through transmission of appropriate Register FISes.

Commands that fail to complete are detected by the host driver software through a time-out mechanism. Such time-outs may not result in status or error information for the command being conveyed to the host software, and the software may not be able to determine the source or cause of such errors.

Error Control Actions

Conditions that return bad status in the Shadow Status register, but for which no interface error information in the SError register is available, will be handled as defined in the parallel implementation.

Conditions that return interface error information in the SError register are handled through four basic responses, as follows:

- Freeze
- Abort/Fail
- Retry (possible after reset)
- Track/ignore

Nontransient Errors

Error conditions that are not expected to be transient or to succeed with subsequent attempts should result in the affected command being aborted and failed. Failure of such commands should be reported to higher software layers for handling. Scenarios in which this response is appropriate include attempts to communicate with a device that is not attached and failure of the interface to successfully negotiate communications with an attached device.

Transient Errors

Error conditions that are expected to be transient should result in the affected command being retried. Such commands may either be retried directly or may be retried after an interface and/or DEVICE RESET, depending on the particular error value reported in the SError register. Scenarios in which this response is appropriate include noise events resulting in CRC errors, 8b10b code violations, or disparity errors.

Recoverable Errors

Conditions that are recoverable and for which no explicit error handling is required may be tracked or ignored. Tracking such errors allows subsequent fault isolation for marginal components and accommodates possible recovery operations. Scenarios in which this response is appropriate include tracking the number of Phy synchronization losses in order to identify a potential cable fault or to accommodate an explicit reduction in the negotiated communications rate.

7.2 State Diagram Conventions

State diagrams will be as depicted in Figure 7.2.

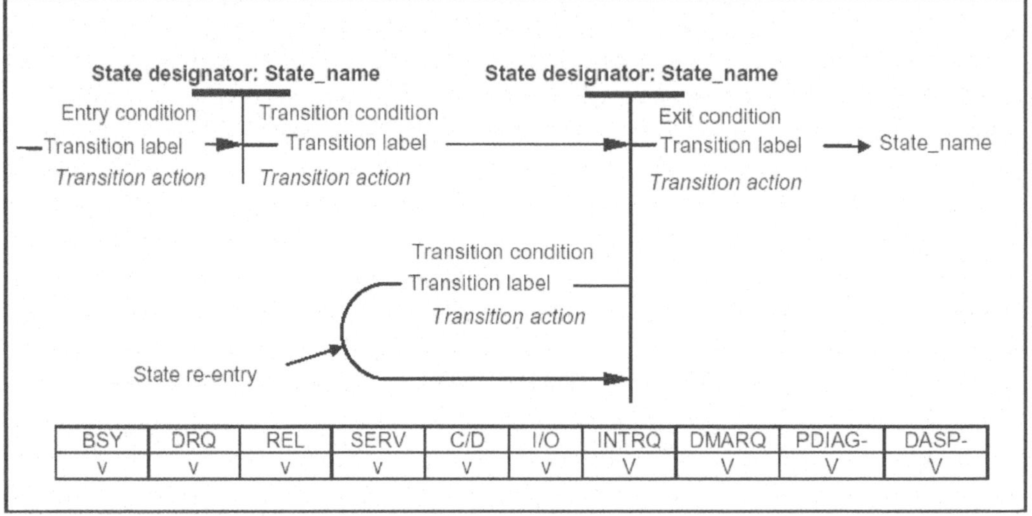

BSY	DRQ	REL	SERV	C/D	I/O	INTRQ	DMARQ	PDIAG-	DASP-
V	V	V	V	V	V	V	V	V	V

Figure 7.2 State diagram nomenclature and operation. (*Source:* www.t10.org)

Each state is identified by a state designator and a state name.

- The state designator is unique among all states in all state diagrams in this document.
- The state designator consists of a set of letters that are capitalized in the title of Figure 7.2 containing the state diagram followed by a unique number.

The state name is a brief description of the primary action taken during the state, and the same state name may appear in other state diagrams.

- If the same primary function occurs in other states in the same state diagram, they are designated with a unique letter at the end of the name.
- Additional actions may be taken while in a state, and these actions are described in the state description text.

In device command protocol state diagrams, the state of bits and signals that change state during the execution of this state diagram are shown under the state designator:state_name, and a table is included that shows the state of all bits and signals throughout the state diagram as follows:

v = bit value changes
1 = bit set to one
0 = bit cleared to zero
x = bit is don't care
V = signal changes
A = signal is asserted
N = signal is negated
R = signal is released
X = signal is don't care

Each transition is identified by a transition label and a transition condition.

1. The transition label consists of the state designator of the state from which the transition is being made, followed by the state designator of the state to which the transition is being made.
2. In some cases, the transition to enter or exit a state diagram may come from or go to a number of state diagrams, depending on the command being executed.
   a. In this case, the state designator is labeled xx.
3. The transition condition is a brief description of the event or condition that causes the transition to occur and may include a transition action, indicated in italics, that is taken when the transition occurs.
4. This action is described fully in the transition description text.

Upon entry to a state, all actions to be executed in that state are executed. If a state is re-entered from itself, all actions to be executed in the state are executed again.

Transitions from state to state will be instantaneous.

7.2.1 Power-on and COMRESET Protocol Diagram

If the host asserts Hard Reset by transmitting a COMRESET Sequence, the device will execute the hardware reset protocol regardless of the power management mode or the current device command layer state (see Figure 7.3).

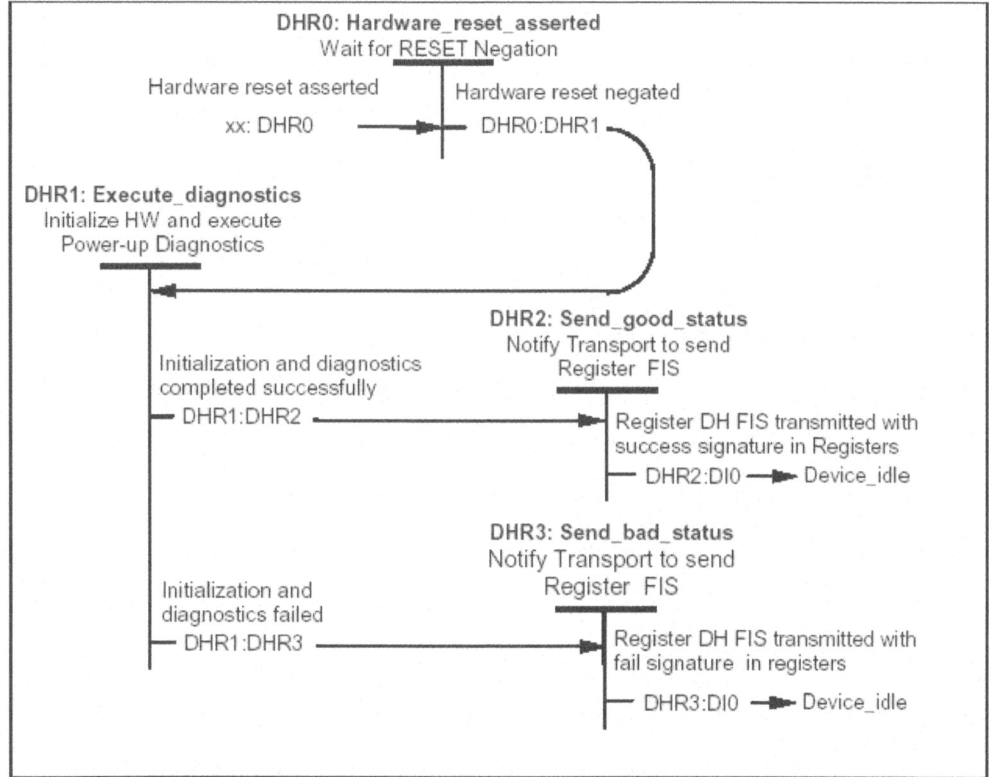

Figure 7.3 Power-on state diagram. (*Source:* www.t10.org)

DHR0: Hardware_reset_asserted: This state is entered when the Transport layer indicates that the COMRESET signal is asserted.

When in this state, the device awaits the negation of the COMRESET signal.

Transition DHR0:DHR1: When the Transport layer indicates that the COMRESET signal has been negated, the device will transition to the DHR1: Execute_diagnostics state.

DHR1: Execute_diagnostics: This state is entered when the Transport layer indicates that the COMRESET signal has been negated.

When in this state, the device initializes the device hardware and executes its power-up diagnostics.

Transition DHR1:DHR2: When the device hardware has been initialized and the power-up diagnostics successfully completed, the device will transition to the DHR2: Send_good_status state.

Transition DHR1:DHR3: When the device hardware has been initialized and the power-up diagnostics failed, the device will transition to the DHR3: Send_bad_status state.

DHR2: Send_good_status: This state is entered when the device hardware has been initialized and the power-up diagnostics successfully completed.

When in this state, the device requests that the Transport layer transmit a Register FIS to the host. The device signature will be set in the registers.

Transition DHR2:DI0: When the Transport layer indicates that the Register FIS has been transmitted, the device will transition to the DI0: Device_Idle state.

DHR3: Send_bad_status: This state is entered when the device hardware has been initialized and the power-up diagnostics failed.

When in this state, the device requests that the Transport layer transmit a Register FIS to the host. The device signature will be set in the registers.

Transition DHR3:DI0: When the Transport layer indicates that the Register FIS has been transmitted, the device will transition to the DI0: Device_Idle state.

7.2.2 Host Phy Initialization State Machine

Note: The following section demonstrates how the state machines are shown in the Serial ATA Specification.

Reception of a COMINIT signal will cause the host to reinitialize communications with the device and will unconditionally force the Host Phy state machine to transition to the HP2B:HR_AwaitNoCOMINIT state, regardless of other conditions. Reception of COMINIT is effectively an additional transition into the HP2B:HR_AwaitNoCOMINIT state that appears in every Host Phy state. For the sake of brevity, this implied transition has been omitted from all of the states.

Figure 7.4 (HP1–HP2B) only cover the first few states of the Intitialize Sequence state machine. They are simply provided to demonstrate how to read the state machines in the Serial ATA Specification.

HP1: HR_Reset	Transmit COMRESET[2, 3]		
	1. Power-on reset and explicit reset request deasserted	→	HR_AwaitCOMINIT
	2. Power-on reset or explicit reset request asserted	→	HR_Reset
	NOTE : 1. This state is entered asynchronously any time in response to power-on reset or an explicit reset request. 2. Must transmit COMRESET for a minimum of 6 bursts (and a multiple of 6) 3. As described in section 6.7.4.2, COMRESET may be transmitted for the duration of this state, or it may be transmitted starting in this state and cease transmission after departure of this state, or it may be transmitted upon departure of this state.		

HP2: HR_AwaitCOMINIT	Interface quiescent		
	1. COMINIT detected from device	→	HR_AwaitNoCOMINIT
	2. COMINIT not detected from device	→	HR_AwaitCOMINIT

HP2B: HR_AwaitNoCOMINIT	Interface quiescent		
	1. COMINIT not detected from device	→	HR_Calibrate
	2. COMINIT detected from device	→	HR_AwaitNoCOMINIT
	NOTE : 1. This state is entered asynchronously any time in response to COMINIT unless during a power-on reset or an explicit reset request (in which case HP1 is entered).		

Figure 7.4 Host Phy Reset state machine. (*Source:* www.t10.org)

7.3 Review

This section reviews the topics covered in Chapter 7.

1. SATA takes a layered approach to error detection and handling.
2. Error responses fall into one of four categories:
   a. Freeze
   b. Abort
   c. Retry
   d. Track/ignore
3. There are three primary categories of error that the Physical layer detects internally:
   a. No device present
   b. OOB signaling sequence failure
   c. Phy internal error (loss of synchronization of communications link)
4. There are two primary categories of errors that the Link layer detects internally:
   a. Invalid state transitions
   b. Data integrity errors
5. In addition to the error information passed to it by the Link layer, the Transport layer internally detects the following categories of errors:
   a. Internal errors
   b. Frame errors
   c. Protocol and state errors
6. There are three error detection mechanisms by which software identifies and responds to the SATA errors:
   a. Bad status in the Command Block Status register

   b. Bad status in the serial implementation of the ATA SError register
   c. Command failed to complete (time-out)
7. Nontransient Errors
   a. Error conditions that are not expected to be transient or to succeed with subsequent attempts should result in the affected command being aborted and failed.
8. Transient Errors
   a. Error conditions that are expected to be transient should result in the affected command being retried.
9. Review Figure 7.3 (below) and answer the following questions:
   a. When does the device notify the transport layer to send good status?
   b. When the device successfully completes Initialization and diagnostics the state machine directs the device to transition from DHR1: Execute_diagnostics to DHR2: Send_good_status.

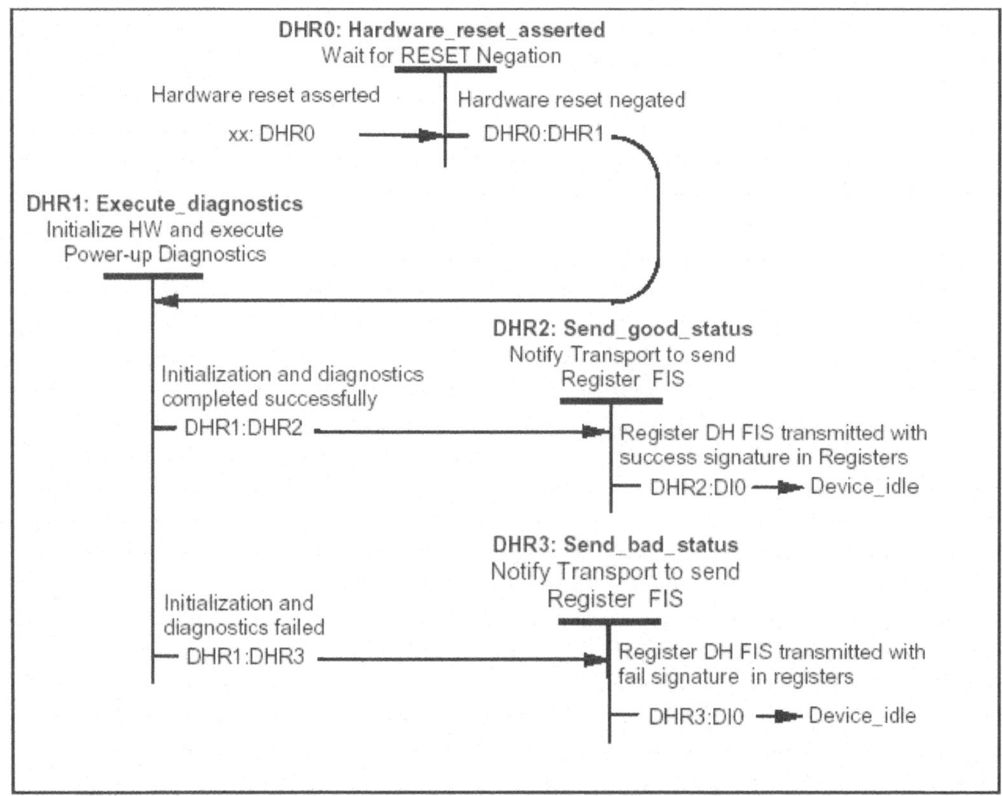

Figure 7.3 (see Section 7.2.1) Power-on state diagram. (*Source:* www.t10.org)

Chapter 8

Cables and Connectors

Objectives

This chapter covers the various connection methods available in the SATA architecture. Due to its vast adaptability, SATA has become more than just another disk technology. The SATA-IO has developed a number of different form factors, power advancements, and consumer applications solutions that include the following (see Figure 8.1):

- Micro SATA connector for 1.8-inch HDD
- Internal Slimline
- LIF-SATA (Low Insertion Force)
- mSATA
- SATA USM (Universal Storage Module)
- Micro SSD
- Internal M.2
- SATA Express

8.1 SATA Connectivity for Device Storage

Legacy SATA storage provides host systems with low-cost, high-capacity storage. From an application perspective, SATA has provided the capacity storage tier, and Serial Attached SCSI (SAS) has served the Enterprise. Today, however, SATA device manufacturers have added a middle tier of SATA devices that fills the gap between the high-cost Enterprise class and less-reliable Client class storage.

8.1.1 Client and Enterprise SATA Devices

The SATA Specifications have added numerous connectors that all serve one application requirement or another. The original legacy SATA and SAS connectors were plug compatible in a SAS backplane.

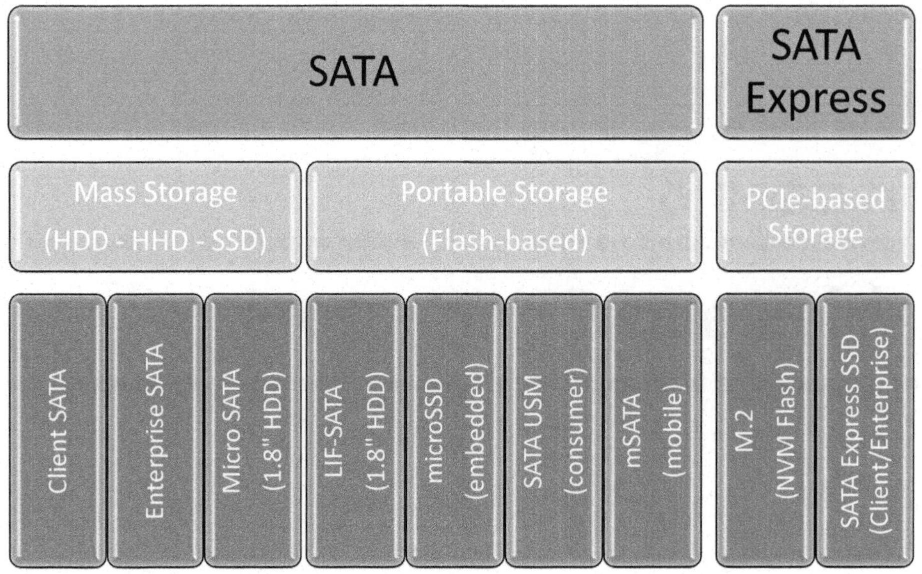

Figure 8.1 SATA connectivity and applications.

This allowed for multi-tier storage enclosures that could be populated with both high-performance SAS HDDs and low-cost, high-capacity SATA HDDs.

The easiest method to determine if the disk interface is SAS or SATA is to look for the gap (mind the gap) on the device-side connector (see Figure 8.2). If there is a gap between the power pins (right) and signal pins (left), then the device has a SATA interface and inherits numerous operational characteristics of SATA HDDs (i.e., high capacity, low cost, low performance).

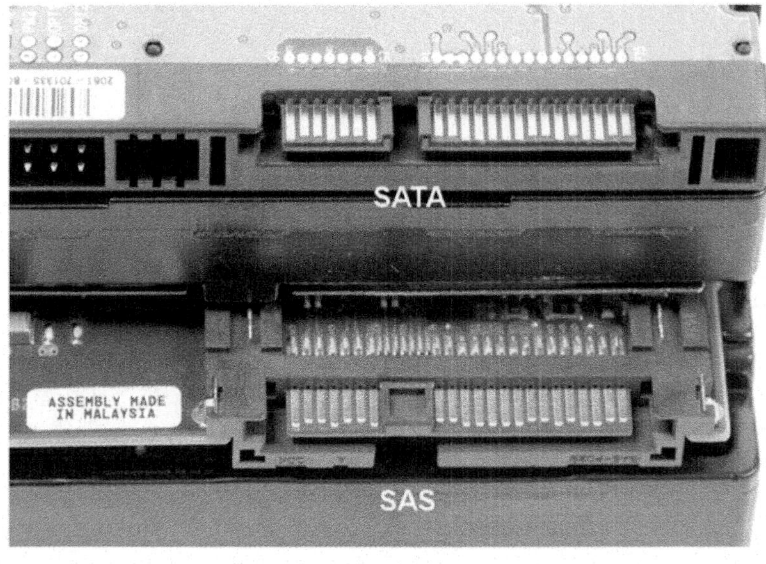

Figure 8.2 The SATA disk has a gap; the SAS disk gap is filled in with pins on the underside.

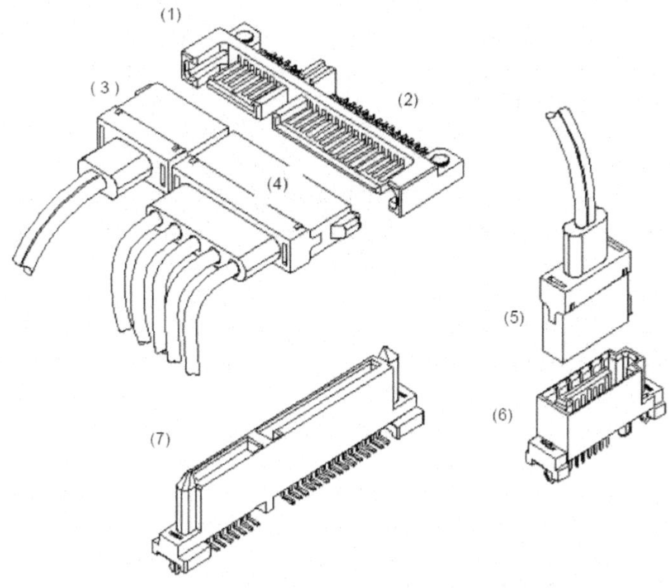

Figure 8.3 SATA device and backplane connectors. (*Source:* www.sata-io.org)

A serial implementation of an ATA device may be either directly connected to a host or connected to a host through a cable. Figure 8.3 shows the device signal and power plugs. The connector and cabling depict the original legacy SATA connectivity alternatives, as follows:

1. Device signal plug segment or connector
2. Device power plug segment or connector
3. Signal cable receptacle connector, to be mated with (1)
4. Power cable receptacle connector, to be mated with (2)
5. Signal cable receptacle connector, to be mated with (6)
6. The host signal plug connector
7. Backplane connector mating directly with device plug connector (1) and (2)

For direct connection:

- The device plug connector, shown as (1) and (2) in Figure 8.3, is inserted directly into a host receptacle (or backplane) connector, as illustrated in (7).
- The device plug connector and the host receptacle connector incorporate features that enable the direct connection to be hot pluggable and blind mateable.

For connection via cable:

- The device signal plug connector, shown as (1) in Figure 8.3 mates with the signal cable receptacle connector on one end of the cable, as illustrated in (3).
- The signal cable receptacle connector on the other end of the cable is inserted into a host signal plug connector, shown as (6) in Figure 8.3.
- The signal cable wire consists of two twinax sections in a common outer sheath.

Figure 8.4 Drive connectors.

There is also a separate power cable for the cabled connection.

- A Serial ATA power cable includes a power cable receptacle connector, shown as (4) in Figure 8.4, on one end and may be directly connected to the host power supply on the other end or may include a power cable receptacle on the other end.
- The power cable receptacle connector on one end of the power cable mates with the device power plug connector, shown as (2) in Figure 8.4.
- The host end of the power cable was not covered in the standard.

Device Connectors

In Figure 8.5, notice the length of the signal pins for both Power and Signal segments. All SATA devices have two connector mating sequences:

- All ground pins are connected first.
    - These are the longest pins on the connector, which facilitates hot plugging the device.
- This is followed by all signal and power supply pins.
    - Pin length differences ensures that grounds mate approximately 1 ms prior to all signals.

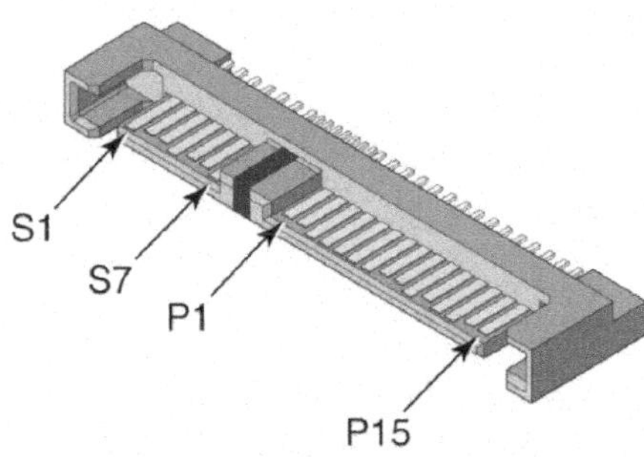

Figure 8.5 Mechanical drawing of a device-side connector. (*Source:* www.sata-io.org)

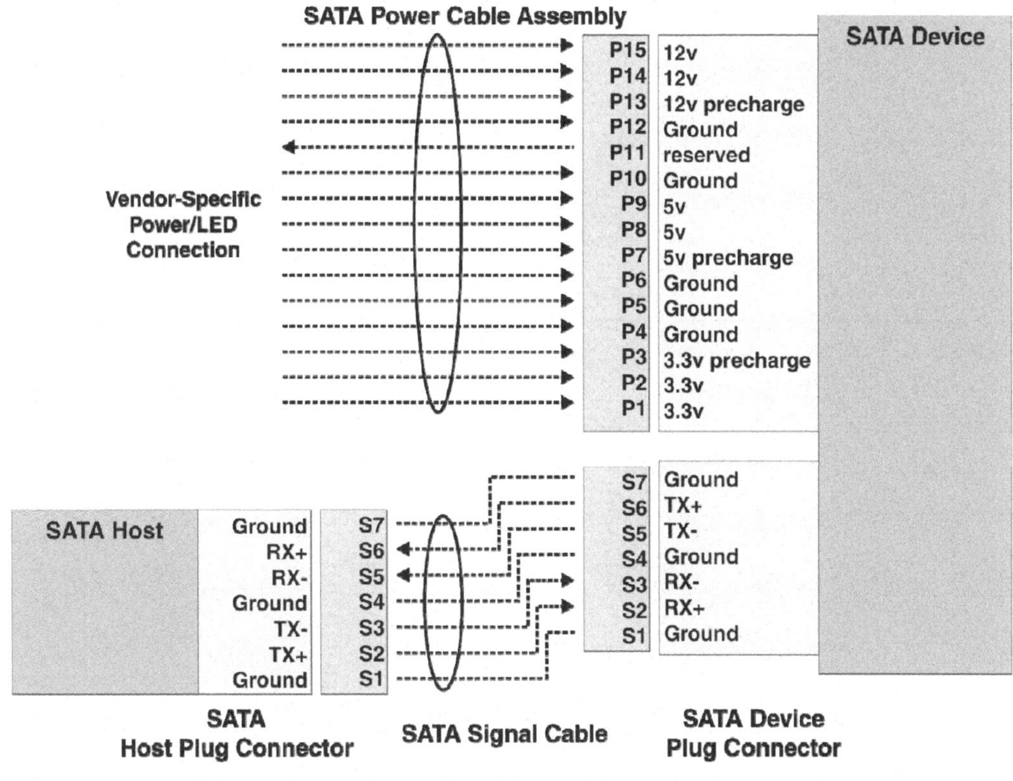

Figure 8.6 SATA connector pin outs and cable assemblies. (*Source:* www.sata-io.org)

Standard SATA Connector Pin Outs

Figure 8.6 provides the pin out configuration for a standard SATA device connector.

SATA/SAS Compatibility

This section shows the connector similarities and differences between SATA and SAS connectors.

- The SAS connector design is similar to the SATA connector.
  - This was purposely done to allow two different tiered storage devices to share the same backplane.
- The SAS connector (Figure 8.7) was slightly modified by filling the gap between the power and signal segments.
  - Connections for a second SAS port have been added on the bottom of the connector for high-availability applications.

Due to the lack of a gap in the SATA backplane connector (Figure 8.8, right connector), SAS devices will not plug into a SATA backplane because a molded plastic segment prevents insertion.

- The SATA device will, however, plug into a SAS backplane.

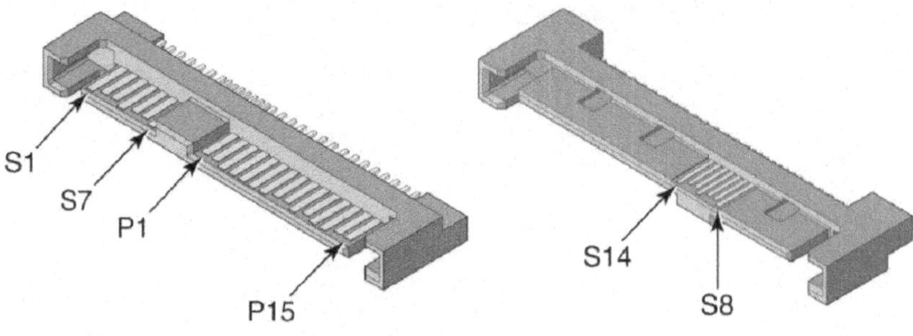

Figure 8.7 SAS device connector. (*Source:* www.t10.org)

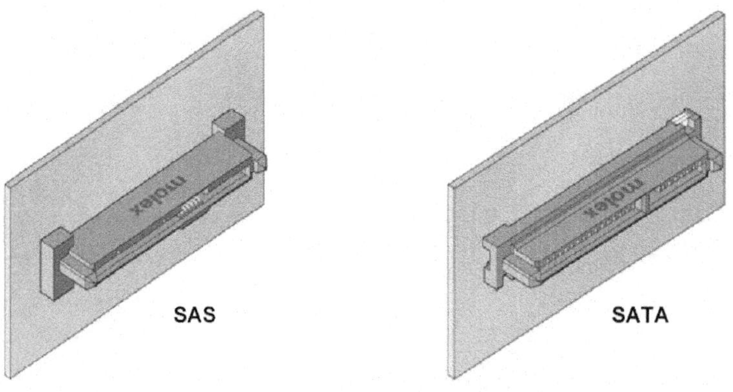

Figure 8.8 SAS and SATA backplane. (*Source:* www.molex.com)

8.1.2 Micro SATA

The internal Micro SATA connector (Figure 8.9) was introduced around 2007 and was specifically designed to enable the connection of a slim 1.8-inch form factor HDD to the Serial ATA interface.

- The internal Micro Serial ATA connector uses the 1.27-mm pitch configuration for both the signal and power segments.
- The signal segment has the same configuration as the internal standard Serial ATA connector.
- The power segment provides the present voltage requirement support of 3.3 V and includes a provision for a future voltage requirement of 5 V.
- In addition, there is a reserved pin, P7.
- Finally, there are two optional pins, P8 and P9, for vendor specific use.

The internal Micro SATA connector is designed with staggered pins for hot plug backplane (non-cabled) applications.

- A special power segment key is located between pins P7 and P8.
- This feature prevents insertion of other Serial ATA cables.

Care should be taken in the application of this device so that excessive stress is not exerted on the device or connector.

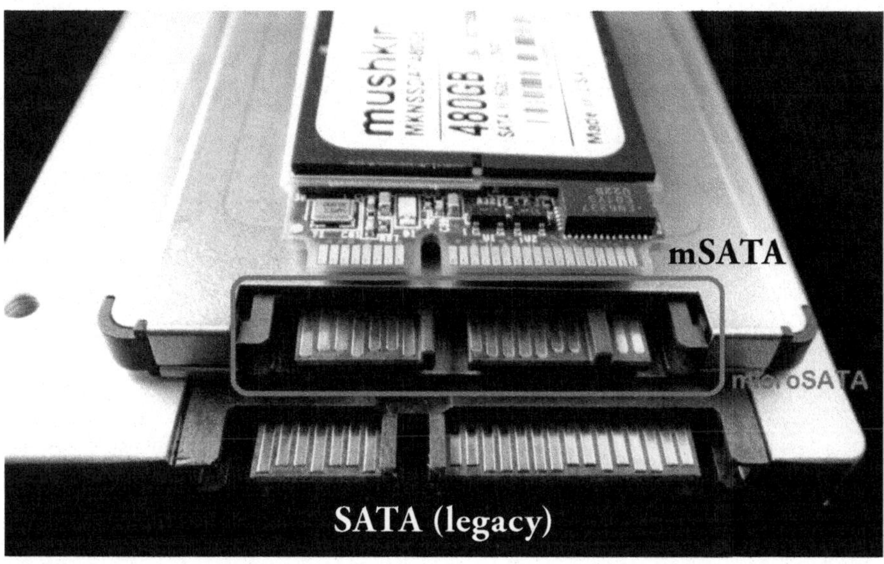

Figure 8.9 mSATA, microSATA, and SATA interfaces.

- Backplane configurations should pay particular attention so that the device and connector are not damaged due to excessive misalignment.

The Micro SATA connector shown in Figure 8.9 enables space savings by providing for smaller 1.8-inch hard disk drives (HDD or SSD). The specification supports the following capabilities:

a. Gen1 (1.5 Gbps), Gen2 (3.0 Gbps), and Gen3 (6.0 Gbps) transfer rates;
b. backplane (i.e., direct connection) and cable attachment usage models;
c. hot plugging in backplane (non-cabled) applications;
d. 8.0-mm and 5.0-mm slim 1.8-inch form factor HDDs;
e. 3.3 V, with 5 V to meet future product requirements; and
f. optional pins, P8 and P9, for vendor-specific use.

Table 8.1 provides the pin assignments for the Micro SATA connector. Micro SATA supports hot plugging therefore connector pins mate at different times to properly protect device during hot removal and insertion.

8.1.3 mSATA

This section defines the requirements of an mSATA configuration with a Serial-ATA interface. The specification supports the following capabilities:

a. Gen1 (1.5 Gbps) and Gen2 (3 Gbps) transfer rates;
b. mSATA;
c. 3.3 V;
d. 4 vendor pins; and
e. 2 vendor pins, for drive or SSD manufacturing usage.

Table 8.1 Micro SATA Pin Assignments

	Name	Type	Description	Cable Usage	Backplane Usage
Signal Segment Key					
Signal Segment	S1	GND	Ground	1st Mate	2nd Mate
	S2	A+	Differential Signal Pair A	2nd Mate	3rd Mate
	S3	A-		2nd Mate	3rd Mate
	S4	GND	Ground	1st Mate	2nd Mate
	S5	B-	Differential Signal Pair B	2nd Mate	3rd Mate
	S6	B+		2nd Mate	3rd Mate
	S7	GND	Ground	1st Mate	2nd Mate
Central Connector Gap					
Power Segment					
Power Segment	P1	V33	3.3 V Power	2nd Mate	3rd Mate
	P2	V33	3.3 V Power, Pre-charge	1st Mate	2nd Mate
	P3	GND	Ground	1st Mate	1st Mate
	P4	GND	Ground	1st Mate	1st Mate
	P5	V5	5 V Power, Pre-charge	1st Mate	2nd Mate
	P6	V5	5 V Power	2nd Mate	3rd Mate
	P7	DAS/ DHU	DAS/Direct Head Unload/ Vendor Specific	2nd Mate	3rd Mate
	Key	Key	Key	N/C	N/C
	P8	Optional	Vendor specific	2nd Mate	3rd Mate
	P9	Optional	Vendor specific	2nd Mate	3rd Mate

The mSATA connector is designed to enable connection of a new family of small form factor devices to the Serial ATA interface.

- All mSATA physical dimensions are under the control of JEDEC and provided in SATA as informative (Figure 8.10).
- See JEDEC MO-300 for all physical requirements (www.jedec.org/standards-documents/docs/mo-300).

mSATA devices save a tremendous amount of space, operate at extremely low temperatures, and draw minimal power. These devices can be used as a system cache or to serve hot data, resident applications, or as a boot device (see Figure 8.11).

Figure 8.10 mSATA SSD. (*Source:* www.anandtech.com/show/6710/intel-ssd-525-review-240gb)

mSATA Pin Signal Definition

Table 8.2 defines the signal assignment of the mSATA connection. This connection does not support hot plug capability, so there is no connection sequence specified.

There are a total of 52 pins:

a. 5 pins for 3.3 V source;
b. 3 pins for 1.5 V source;
c. 13 pins for GND;
d. 4 pins for transmitter/receiver differential pairs;
e. 1 pin for device activity/disable staggered spin-up;

Figure 8.11 mSATA application example. (*Source:* www.legitreviews.com/intel-nuc-dc3217by-review-w-windows-8_2083/3)

Table 8.2 mSATA Connector Pin Assignments

Pin	Type	Description
P1	Reserved	No Connect
P2	+3.3 V	3.3 V Source
P3	Reserved	No Connect
P4	GND	Ground
P5	Reserved	No Connect
P6	+1.5 V	1.5 V Source
P7	Reserved	No Connect
P8	Reserved	No Connect
P9	GND	Ground
P10	Reserved	No Connect
P11	Reserved	No Connect
P12	Reserved	No Connect
P13	Reserved	No Connect
P14	Reserved	No Connect
P15	GND	Ground
P16	Reserved	No Connect
P17	Reserved	No Connect
P18	GND	Ground
P19	Reserved	No Connect
P20	Reserved	No Connect
P21	GND	Ground
P22	Reserved	No Connect
P23	+B	Host Receiver Differential Signal Pair
P24	+3.3 V	3.3 V Source
P25	-B	Host Receiver Differential Signal Pair
P26	GND	Ground
P27	GND	Ground
P28	+1.5 V	1.5 V Source
P29	GND	Ground
P30	Two Wire Interface	Two Wire Interface Clock
P31	-A	Host Transmitter Differential Signal Pair
P32	Two Wire Interface	Two Wire Interface Data
P33	+A	Host Transmitter Differential Signal Pair
P34	GND	Ground
P35	GND	Ground
P36	Reserved	No Connect
P37	GND	Ground
P38	Reserved	No Connect
P39	+3.3 V	3.3 V Source

(Continued on following page)

Table 8.2 mSATA Connector Pin Assignments (*Continued*)

Pin	Type	Description
P40	GND	Ground
P41	+3.3 V	3.3 V Source
P42	Reserved	No Connect
P43	Device Type	Shall be a No Connect on mSATA Devices
P44	DEVSLP	Enter/Exit DevSleep
P45	Vendor	Vendor Specific / Manufacturing Pinb
P46	Reserved	No Connect
P47	Vendor	Vendor Specific / Manufacturing Pin
P48	+1.5 V	1.5 V Source
P49	DAS/DSS	Device Activity Signal / Disable Staggered Spinup
P50	GND	Ground
P51	Presence Detection	Connected to GND by a 0 ohm to 220 ohm Resistor on device
P52	+3.3 V	3.3 V Source

f. 1 pin for presence detection;
g. 2 pins for Vendor Specific/Manufacturing;
h. 2 pins for Two Wire Interface;
i. 19 reserved pins (no connect); and
j. 1 pin to indicate mSATA use (no connect).

SATA Connector Placement

- The SATA standard defines the connector location on a drive as shown in Figure 8.12.
    - This connector placement is also observed by SAS drives for compatibility.

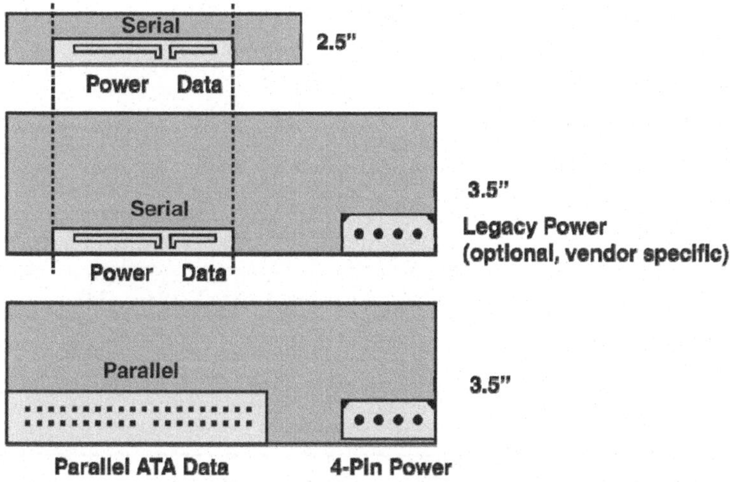

Figure 8.12 SATA connector placement. (*Source:* www.t10.org)

- The connector location is identical between 2.5-inch and 3.5-inch HDDs.
    - This enables 2.5-inch drives to be used in a 3.5-inch enclosure, with the proper mounting carrier.
- See SFF-8223, SFF-8323, and SFF-8523 for the connector locations on common form factors.
- The device connector accommodates 3.3, 5, and 12V power lines.
    - The 12V power supply is provided for the powering of 5.25-inch disk drive motors (obsolete).
    - The 5V power supply is provided for the powering of 3.5-inch disk drive motors.
    - The 3.3V power supply is provided for the powering of the drive electronic circuitry.

8.1.4 LIF-SATA

This internal LIF-SATA connector is designed to enable connection of a new family of slim 1.8-inch form factor devices to the SATA interface.

The internal low insertion force (LIF) Serial ATA connector (see Figure 8.13) uses the 0.5-mm pitch configuration for both the signal and power segments.

- The signal segment has the same configuration as the internal standard Serial ATA connector, but the power segment provides the present voltage requirement support of 3.3 V, as well as provision for a future voltage requirement of 5 V.
- In addition, there is P8 (i.e., defined as Device Activity Signal/Disable Staggered Spin-up).
- Finally, there are six vendor pins—P18, 19, 20, and 21—for HDD customer usage and P22 and P23 for HDD manufacturing usage.

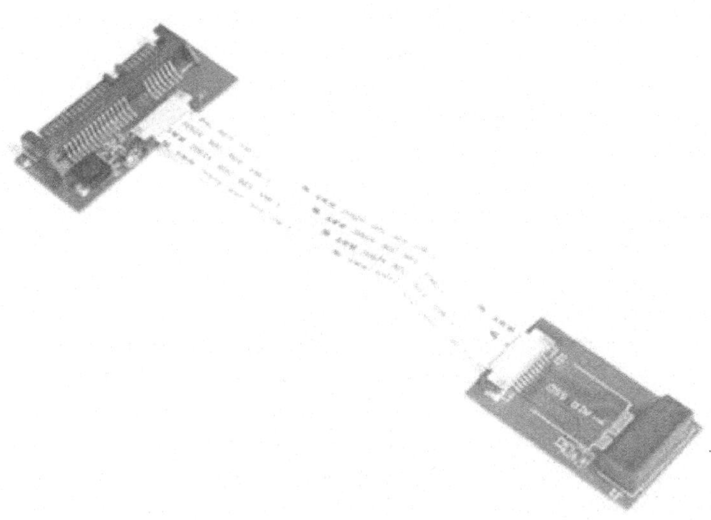

Figure 8.13 Low Insertion Force SATA connector example. (*Source:* http://lghttp.27756.nexcesscdn.net/80BBFD/magento/media/extendware/ewminify/media/inline/23/0/apl-sata-lif-st.jpg)

LIF-SATA Pin Signal Definition

The specification supports the following capabilities:

a. Gen1 (1.5 Gbps) and Gen2 (3 Gbps) transfer rates;
b. FPC usage models;
c. 8.0-mm and 5.0-mm slim 1.8-inch form factor (FF) HDDs;
d. 3.3 V, with 5 V to meet future product requirements;
e. vendor pins—P18, P19, P20, and P21—reserved for HDD customer usage; and
f. vendor pins—P22 and P23—for HDD manufacturing usage.

This LIF-SATA is only for internal 8.0-mm and 5.0-mm slim 1.8-inch form factor devices.

Specification note: It is expected that the LIF-SATA interfaces comply with Gen1i and Gen2i Specifications. The LIF-SATA connector (Table 8.3) can only be mated with a FPC cable, and the compliance test points will be measured after the mated assembly.

Table 8.3 LIF Connector Pin Assignments

Name	Type	Description
P1	GND	Ground
P2	V33	3.3 V Power
P3	V33	3.3 V Power
P4	GND	Ground
P5	V5	5 V Power
P6	V5	5 V Power
P7	GND	Ground
P8	DAS/ DSS/ Vendor Specific	Device Activity Signal Disable Staggered Spinup Vendor Specific
P9	GND	Ground
P10	GND	Ground
P11	A+	Differential Signal Pair A
P12	A-	
P13	GND	Ground
P14	B-	Differential Signal Pair B
P15	B+	
P16	GND	Ground
P17	GND	Ground
P18	Vendor	Vendor Specific
P19	Vendor	Vendor Specific
P20	Vendor	Vendor Specific
P21	DHU	Direct Head Unload
P22	Vendor	Vendor Specific - Mfg pin
P23	Vendor	Vendor Specific - Mfg pin
P24	GND	Ground

8.1.5 microSSD™

The SATA microSSD™ standard for embedded solid state drives (SSDs) offers a high-performance, low-cost, embedded storage solution for mobile computing platforms, such as ultra-thin laptops. The microSSD specification eliminates the module connector from the traditional SATA interface, enabling developers to produce a single-chip SATA implementation for embedded storage applications.

The specification defines a new electrical pin-out that allows SATA to be delivered using a single ball grid array (BGA) package (see Figure 8.14).

- The BGA package sits directly on the motherboard, supporting the SATA interface without a connecting module (see Figure 8.15).
- By eliminating the connector, the microSSD standard enables the smallest physical SATA implementation to date, making it an ideal solution for embedding storage in small form factor devices.

Figure 8.14 microSSD BGA example. (*Source:* http://americas.micross.com/products-services/standard-plastic-packages/microssd.stml)

Figure 8.15 microSSD BGA packaging and mounting.

8.1.6 M.2 SATA

M.2 (formerly known as NGFF) is a small form factor card and connector that supports applications such as WiFi, WWAN, USB, PCIe, and SATA, as defined in the PCI-SIG specification (see Figure 8.16).

- An M.2 card with a SATA or PCIe interface will typically be an SSD, suitable for very thin devices such as Ultrabooks or tablets.
- The SATA v3.2 specification standardizes the SATA M.2 connector pin layout.
- Detailed form factor specifications will be contained in the PCI-SIG M.2 Specification.

Figure 8.16 M.2 cards are available in single-sided or double-sided versions. (*Source:* www.sata-io.org)

The definition supports the following capabilities:

- SATA transfer rates:
  - Gen1 (i.e., 1.5 Gbps);
  - Gen2 (i.e., 3.0 Gbps); and
  - Gen3 (i.e., 6.0 Gbps).

and

- PCI Express:
  - V1 (i.e., 2.5 GT/s per lane);
  - V2 (i.e., 5 GT/s per lane); and
  - V3 (i.e., 8 GT/s per lane).

The purpose of the mated connector impedance requirement is to optimize signal integrity by minimizing reflections.

- The host may support the use of PCIe or SATA signaling over the same interconnect.
- The nominal characteristic differential impedance of PCIe is 85 ohm, whereas the nominal characteristic differential impedance of SATA is 100 ohm.

For SSD devices, the M.2 specification describes in detail a set of board sizes (e.g., 22 mm × 42 mm, 22 mm × 80 mm), connector heights (e.g., 2.25 mm, 2.75 mm, and 4.2 mm), and keying options for use in M.2 SSD modules.

The Serial ATA Specification 3.2 has defined five board sizes for the M.2 alternative (see Table 8.4).

Table 8.4 Module Types and Dimensions (mm)

Type	Height (A)	Width	Applications
2230	30	22	Cache – NVRAM – Flash
2242	42		Cache & SSD
2260	60		SSD & Cache
2280	80		SSD
22110	110		SSD

Other than different dimensions, all M.2 devices have the same basic layout and connectors as seen in Figure 8.17. Figure 8.18 shows an example of an 80 mm × 20 mm SSD, and Figure 8.19 shows how an M.2 device can be mounted on a motherboard.

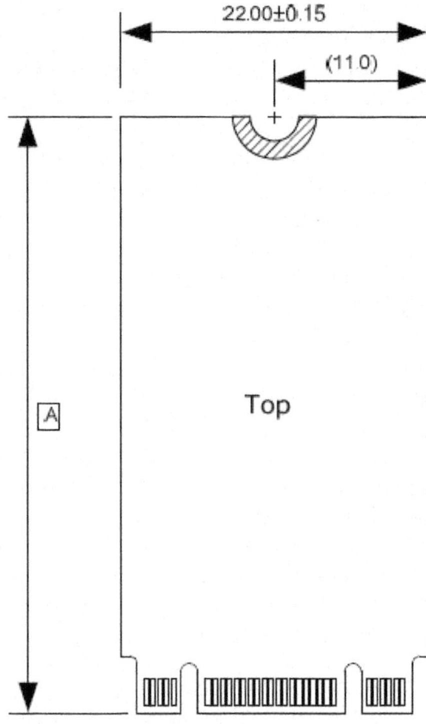

Figure 8.17 Module layout. (*Source:* www.pcisig.com/home/)

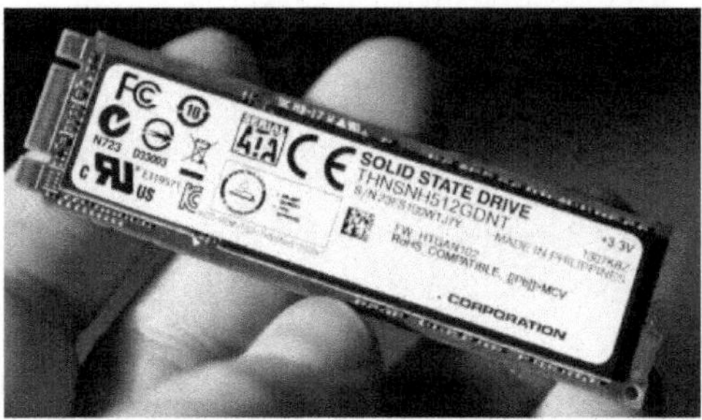

Figure 8.18 SATA M.2 SSD example. (*Source:* www.thessdreview.com/our-reviews/toshiba-hg5d-series-sata-m-2-ssd-review-512gb/)

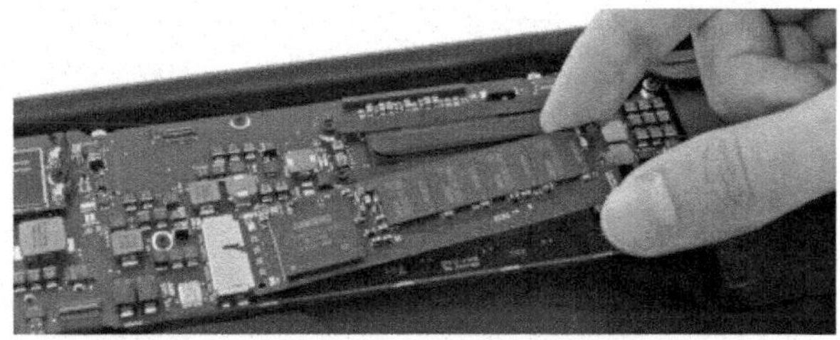

Figure 8.19 SATA M.2 application example. (*Source:* www.tested.com/tech/pcs/456464-why-ssds-are-transitioning-sata-pcie-next-gen-form-factor/)

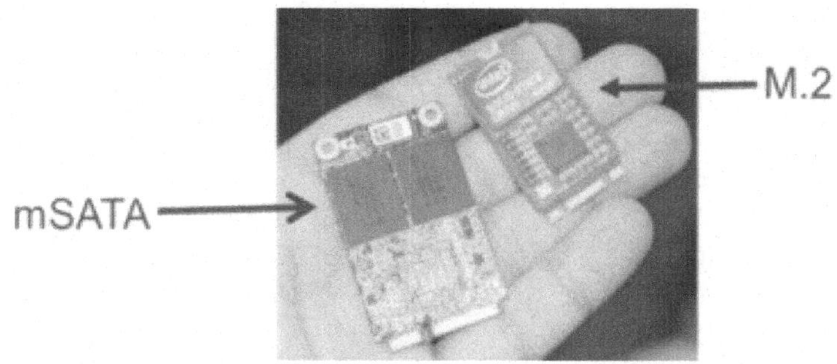

Figure 8.20 M.2 and mSATA comparison. (*Source:* www.anandtech.com/show/6293/ngff-ssds-putting-an-end-to-proprietary-ultrabook-ssd-form-factors)

M.2—mSATA Comparison

Where mSATA took advantage of an existing form factor and connector, M.2 has been designed to maximize the usage of the card space while minimizing the footprint (see Figure 8.20).

	mSATA	M.2
Width (mm)	30	22
Length (mm)	50.95	30, 42, 60, 80, 110
PCB Space—L × W (mm)	1528.5	660, 924, 1320, 1760, 2420

8.2 SATA Universal Storage Module (USM)

The SATA USM receiver uses a custom SATA receptacle connector to attach to the module.

- The connectors shall be capable of a minimum of 1,500 insertion/removal cycles.
- The SATA USM receptacle connector is available in vertical and horizontal PCB mounting configurations.

The receptacle connector mating area is compliant with the standard SATA backplane connector, with the following four exceptions:

a. Side-mounted retention springs to improve the connector retention (optional);
b. anti-wiggle bumps to reduce cable deflection (optional);
c. two alignment ribs on each of the long outside surfaces (required); and
d. extended reach/length to properly attach the SATA universal storage module (required).

Revision 3.1 of the SATA specification introduced a variety of performance improvements, reliability enhancements, and new features to expand the functionality and convenience of SATA-based storage devices. In addition to providing new power management capabilities and mechanisms to maximize device efficiency, Revision 3.1 also includes the SATA Universal Storage Module (USM) specification for implementing removal and expandable portable storage applications.

SATA is more than just a disk interface. SATA-IO also supports and fosters the efforts of the gaming and portable media industry.

Figure 8.21 SATA USM drive. (*Source:* www.pcpop.com/doc/0/685/685159.shtml)

- Introduced: Jan. 4, 2011
- USM defines slot connectors for data modules with an integrated SATA interface and defined form factor(s) in Small Form Factor (SFFs) Standards
- Modules plug in to TVs, DVRs, game consoles, and computers for portable storage applications
- First powered 6Gb/s SATA connector for consumer applications
- First industry standard consumer level cable-free external storage
- Self-contained expandable storage (see Figure 8.21) without the need for separate cabling and power

USM was designed to meet the storage needs of a variety of digital media (music, video, photos, movies, etc.), games, and applications for use in:

- PCs and Laptops
- Digital Video Recorders (DVR)/Set-Top Boxes (STB)
- TVs
- Game Consoles
- Video Surveillance

Figure 8.22 shows an example of a USM application wherein multiple interface connections are available for storage capacity expansion. This device allows a USM SATA device to copy files to/from USB devices or an SD card. The USM/USB hub connects to the host (TV, Game Console, etc.) through a Gigabit Ethernet connection, and attached devices will appear as disk devices (or storage devices) to the operating systems.

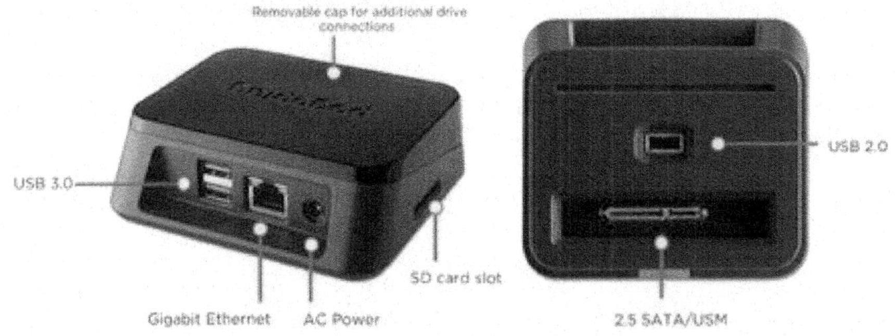

Figure 8.22 USM application. (*Source:* www.tomsguide.com/us/Pogoplug-Series-4-Cloud-Storage-preview-review,news-13531.html)

Figure 8.23 USM Slim™ profile. (*Source:* www.seagate.com/external-hard-drives/portable-hard-drives/standard/seagate-slim-mac-usm/)

USM Slim™

The USM Slim™ Specification defines a 9-mm form factor, making it an ideal storage solution for notebooks, tablets, and other portable devices (see Figure 8.23). This specification allows manufacturers to develop external storage offerings that seamlessly pair with these thin and light mobile devices so that consumers can still have instant access to their music, movies, photos, and other content at any time or place.

8.3 SATA Cabling

Figure 8.24 illustrates how signals and grounds are assigned in direct connect and cabled configurations.

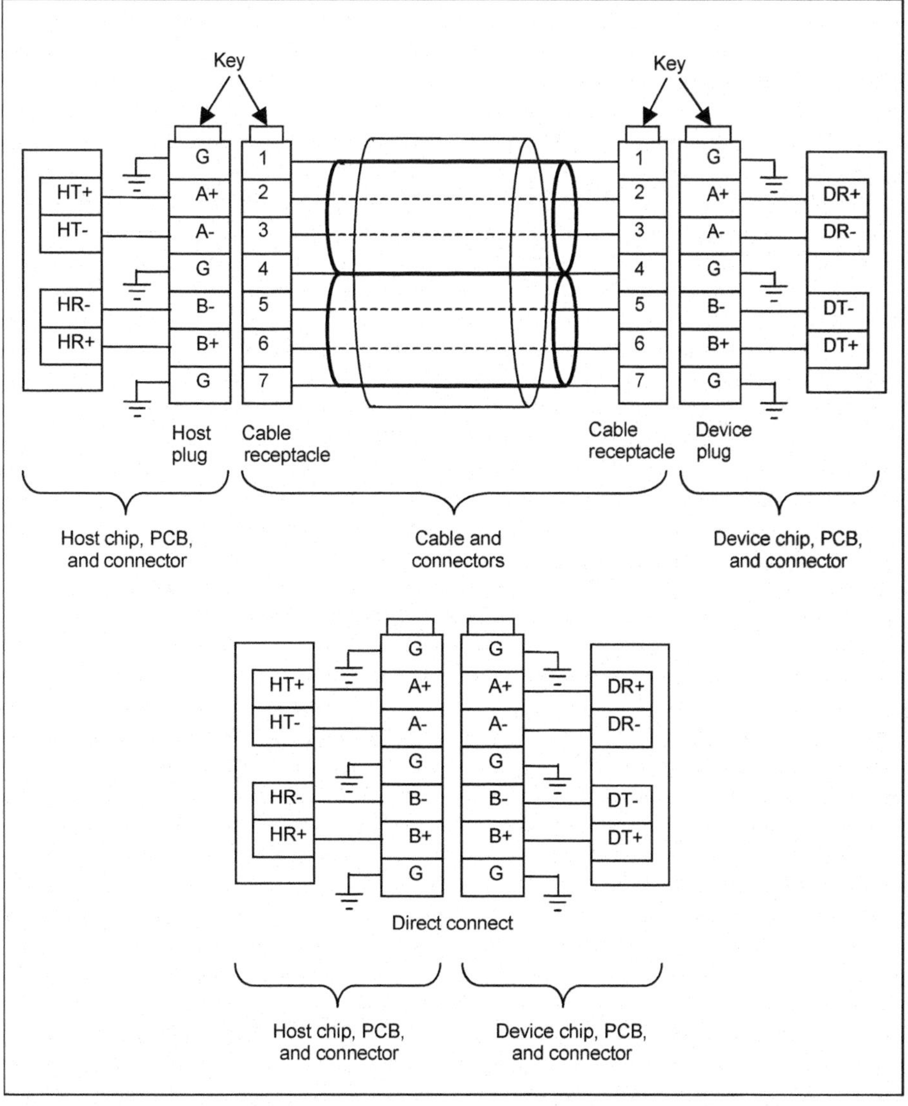

Figure 8.24 SATA cable assembly. (*Source:* www.sata-io.org)

8.3.1 *Cable Assemblies*

The Serial ATA cable consists of four conductors in two differential pairs.

- If necessary, the cable may also include drain wires to be terminated to the ground pins in the Serial ATA cable receptacle connectors.
- The cable size may be 30 to 26 AWG.
- The cable maximum length is one meter.

The standard does not specify how to construct a standard cable for a serial transport.

- Any cable that meets the electrical requirements is acceptable.
- The connector and cable vendors have the flexibility to choose cable constructions and termination methods based on performance and cost considerations.

Cable Construction Example

Figure 8.25 shows a cross section of the SATA cable with cable construction and electrical and physical specifications.

Figure 8.26 displays the different layers of conductors, drains, individual signal insulation to minimize crosstalk, shielding to contain and protect against EMI, and a protective jacket.

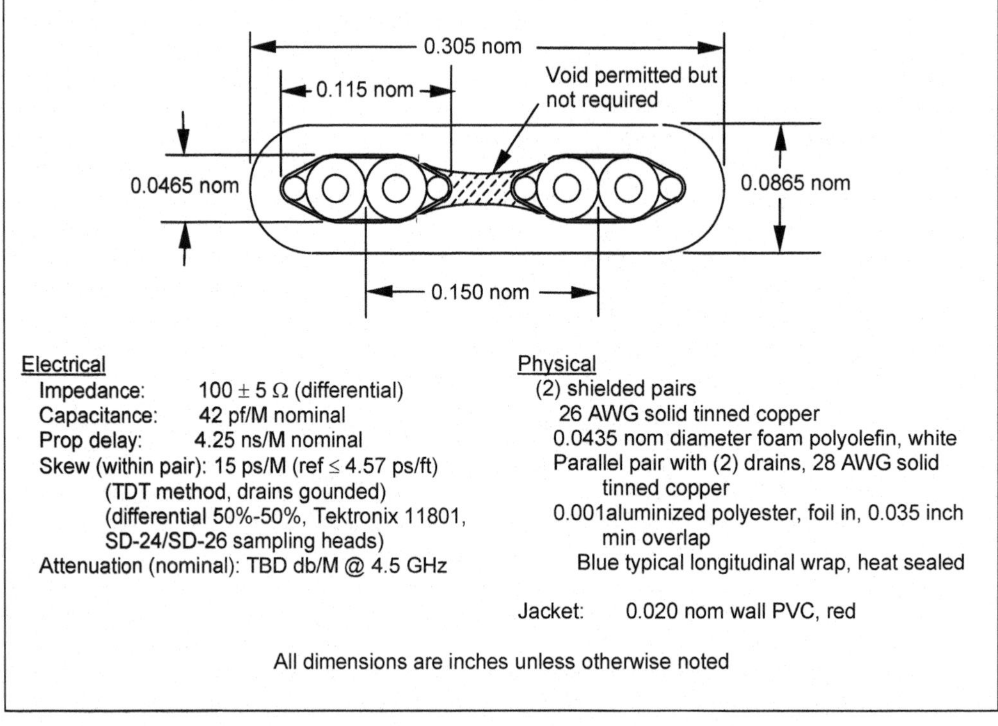

Figure 8.25 Cross section and cable specifications. (*Source:* www.sata-io.org)

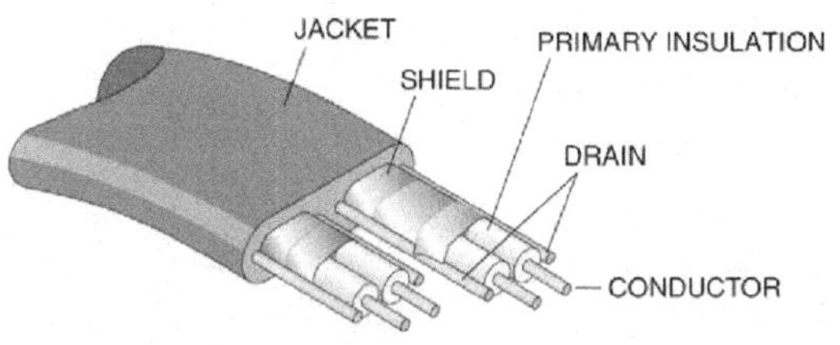

Figure 8.26 SATA cable construction example. (*Source:* www.sata-io.org)

8.3.2 Internal 1 m Cabled Host to Device

In this application, Gen1i, Gen2i, or Gen3i electrical specifications compliant points are located at the Serial ATA mated connectors on both the host controller and device (Figure 8.27). The cable requirement for this alternative operates at 1.5 Gb/s, 3.0 Gb/s, and 6.0 Gb/s.

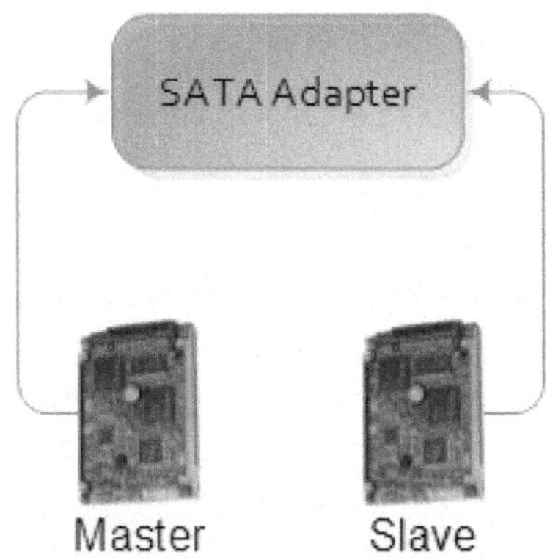

Figure 8.27 Basic SATA connectivity.

Latching Connectors

This section provides details concerning recent developments in the area of attaching external SATA devices. The connectors that can be used are optional and contain latching mechanisms (see Figure 8.28) that can prevent the accidental removal of cables.

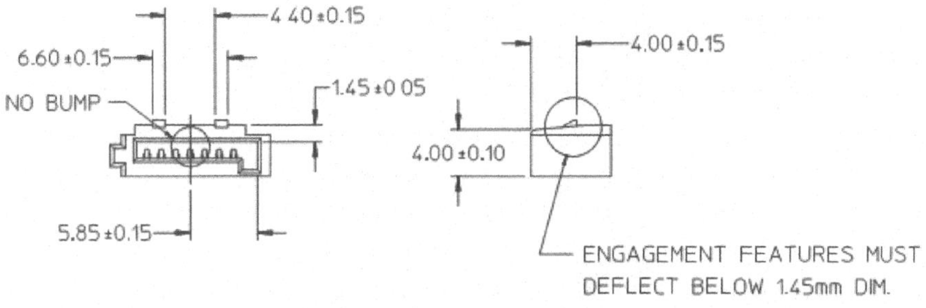

Figure 8.28 Latching Signal Cable Receptacle connector interface dimensions. (*Source:* www.sata-io.org)

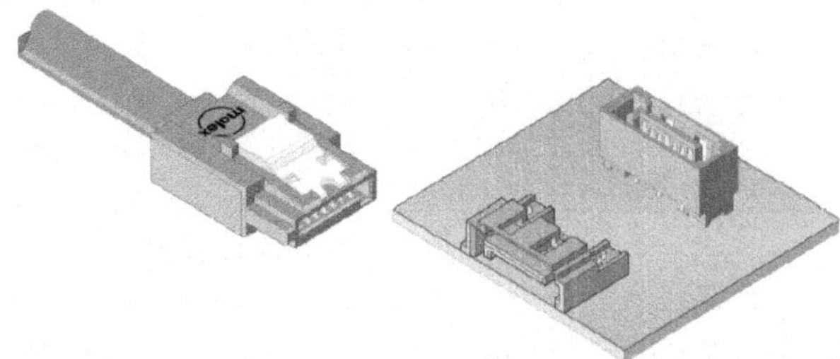

Figure 8.29 Latching Connector. (*Source:* www.sata-io.org)

Later generations of the Serial ATA Specifications added a latching connector (Figure 8.29). This connector would latch (audible snapping sound) into the host controller or motherboard and prevent the accidental cable disconnections.

When laying out multiple host connections on a controller or motherboard, designers must plan for cabling clearances and cable installation. Figure 8.30 shows the industry standard method and dimension requirements for placing four SATA connectors on the PCB. Figure 8.31 shows a computer motherboard with four SATA II connections that conform to the Serial SATA Specifications and T-13 Standards.

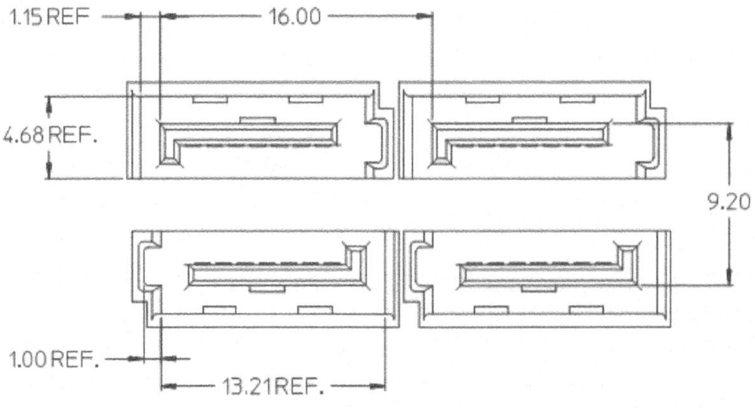

Figure 8.30 Host plug connector clearance and orientation. (*Source:* www.sata-io.org)

Figure 8.31 SATA motherboard connections with proper clearance/orientation. (*Source:* http://vermeulen.ca/computer-gigabyte-ma78gm.html)

Figure 8.32 shows the latching connector that is used to attach power to a device (disk). This connector provides pre-charges and extended ground pins to facilitate hot plugging.

Latching Power Cable Receptacle

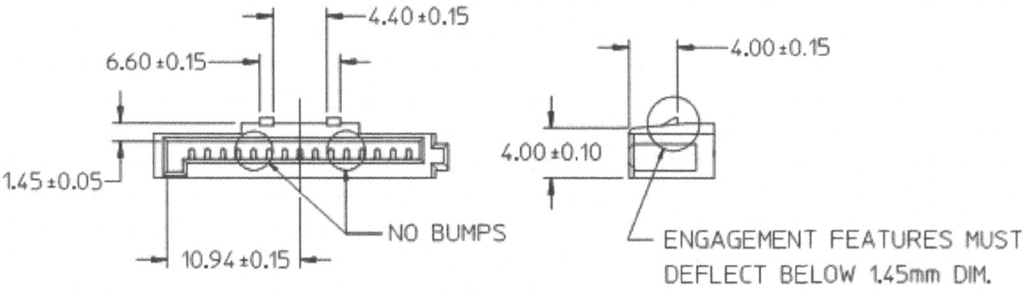

Figure 8.32 Latching Power Cable. (*Source:* www.t13.org)

8.3.3 SATA Cable

This section outlines the cables used in the SATA signal segments. The original SATA cables are non-latching and were intended for internal use only. Since its inception, SATA has now added a latching cable connector to be used for external device connections. The SATA cable shown in Figure 8.33 is a standard (non-latching) cable connector. Figure 8.34 shows standard straight and 90-degree cabling with latching connectors.

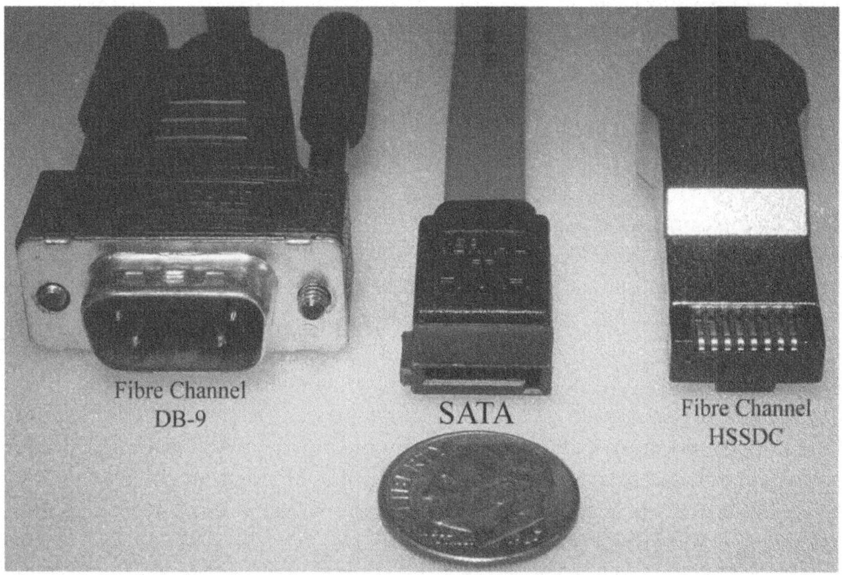

Figure 8.33 SATA (non-latching) as compared to Fibre Channel.

Figure 8.34 SATA internal latching connectors. (*Source*: www.misco.co.uk/Product/167191/LINDY-SATA-Cable-Latching-Right-Angled-90-Connector-0-7m)

8.3.4 External SATA

External SATA provides SATA with an external SATA connection up to 2 m using shielded cables and connectors. When first developed, this was known as eSATA, but that was changed as soon as SATA Express was defined (SATAe).

High-performance, cost-effective expansion storage:

1. External SATA enables usage models outside the box
2. Up to 3 times faster than existing external storage solutions: USB 3.0
3. Robust and user-friendly external connection

Better cable:

- Extra shielding layer
- Ground shield (for ESD/EMI)
- Longer (2 m vs. 1 m)
- Keyed connector end (won't plug into internal connector on host or drive)
- Hot Plug

Different connector:

- Better retention (springs like USB)
- Better shielding (to reduce EMI)
- Recessed connector (to prevent electrostatic discharge [ESD])
- Keyed (to prevent unshielded internal cables from being used)

External SATA connections, originally known as the Type B+ connector, have been seen on host cards for both PCs and Macs. The external SATA Specification uses a different connector that unfortunately looks similar to prior connectors. These connectors can be easily mistaken on first glance for the more common SATA Type A connector (see Figure 8.35). Because of that you will need to pay particular attention to the cables you select. You can select between 3 foot and 6 foot cables and between Type A and eSATA cabling, depending on the connectors on your chosen host card.

Internal-to-external cable conversion can be handled by special headers, as seen in Figure 8.36, or the system will require some type of SATA controller that has External SATA connections.

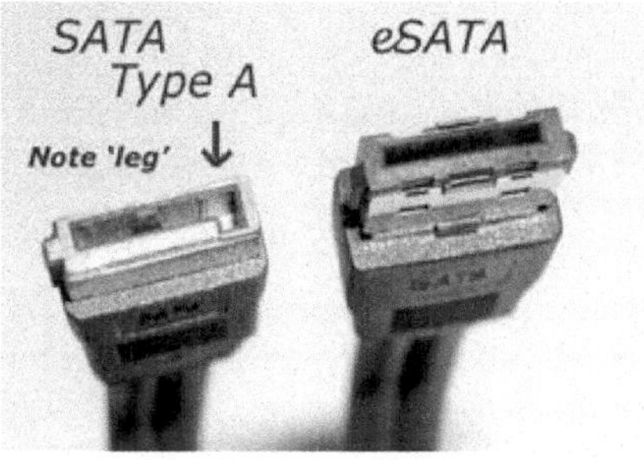

Figure 8.35 SATA and eSATA connector comparison. (*Source*: www.sata-io.org)

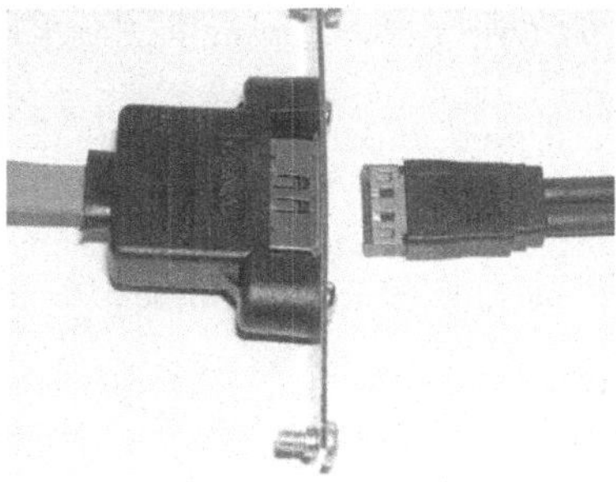

Figure 8.36 Internal SATA to External SATA. (*Source:* www.sata-io.org)

Figure 8.37 shows External SATA cabling of both male and female connectors. Notice the special construction of the cabling components, including extra grounding and shielding, higher quality connectors, and an apparent increase in jacket quality.

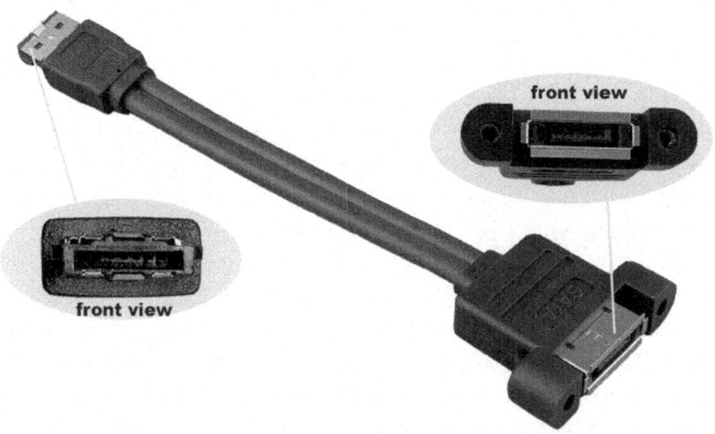

Figure 8.37 External SATA cable assembly. (*Source:* www.sata-io.org)

8.4 SATA Express

SATA Express defines electrical and mechanical requirements for SATA Express, which is a PCI Express (PCIe) connection to the existing standard 3.5-inch and 2.5-inch disk drive form factors for Client applications.

- SATAe is intended to provide a smooth transition path from SATA to PCIe storage, leveraging both PCIe Specifications and 3.5-inch and 2.5-inch drive mechanical standards.

8.4.1 SATA Express Connector Goals

SATA Express is developed with the following characteristics:

1. PCIe connection to Client PCIe storage devices;
2. standardized connectors and form factors, fitting in the existing 3.5-inch and 2.5-inch drive mechanical enclosures;
3. either PCIe lanes or an SATA port on the host so that the host connector works with either a PCIe or a SATA storage device (see Figure 8.38);
4. both cabling and direct connection solutions to support desktop and notebook product needs; and
5. no PCIe reference clock sent over the cable to allow the continued use of the low-cost SATA-like cable solutions.

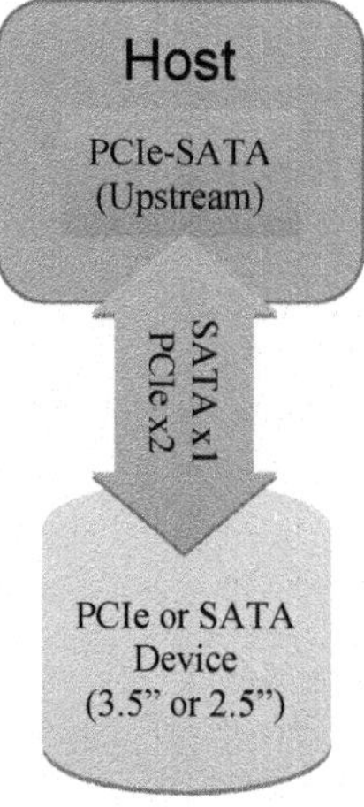

Figure 8.38 SATA Express connectivity.

8.4.2 SATA Express Is Pure PCIe

- There is no SATA link or transport layer, so there's no translation overhead—users will see the full performance of PCIe.
- SATA Express is the standardization of PCIe as an interface for a Client storage device.
  - Standards provide interoperability and interchangeability, meaning that devices from different manufacturers work together and can be interchanged with no loss of functionality.
  - To achieve these goals, SATA Express needs standard connectors and common operating system drivers.

The SATA-IO defined the SATA Express host and device connectors as follows:

- Both connectors are slightly modified standard SATA connectors and are mechanically compatible with today's SATA connector, as demonstrated in Figure 8.39.
- This plug compatibility is important, as it enables SATA and PCIe to co-exist.
- The new host connector supports
  - up to two SATA ports;
  - or up to two PCIe lanes.
- There is a separate signal, driven by the device that tells the host if the device is SATA or PCIe.
  - Allows host to configure the proper device drivers so the host knows what "language" to speak (Command Set).
  - Thus, the motherboard can have a single connector that supports a SATA drive or a PCIe drive.

Figure 8.39 SATA Drive and SATA Express (SATAe) PCIe Drive model (defined by the SATA-IO). (*Source:* www.sata-io.org)

8.4.3 SATA Express Scope

SATA Express is a form factor specification that focuses on connectors and cables for PCIe storage. The connectors on the host are backward compatible extension of SATA connectors. The overall mechanical form factors are compliant with the 3.5- and 2.5-inch form factors. SATA Express defines the following:

a. Pin list and signal assignment of the connectors
b. Mechanical definition of the connectors and cable interfaces
c. Limited electrical specification (largely, a reference to PCIe and legacy SATA Specifications)

SATA Express Mechanical

This was the first defined in the Serial ATA Specification 3.2. SATA Express connectors are as follows (see Figure 8.40):

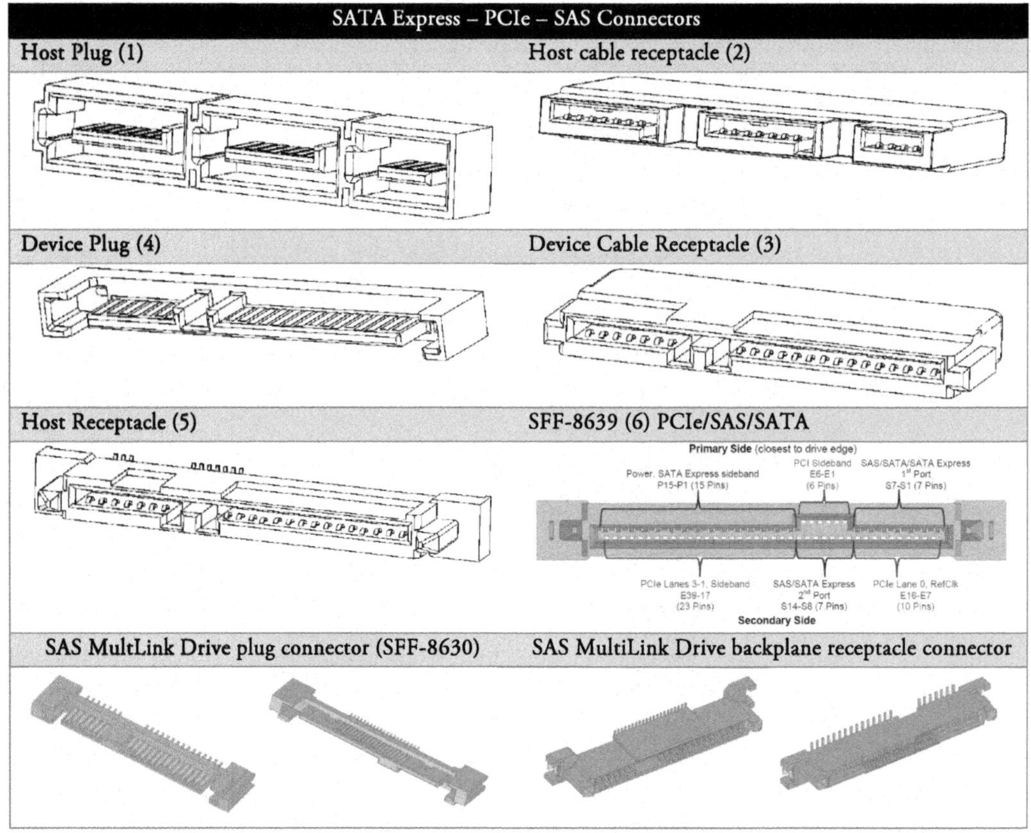

Figure 8.40 SATA Express, PCIe and SAS connectors. (*Source:* www.t10.org and www.t13.org)

1. Host plug connector.
2. Host cable receptacle connector.
3. Device cable receptacle connector.
4. Device plug connector.
5. Host receptacle connector.
6. SFF-8639 is not a Serial ATA–specified connector; however, it can be attached to SATA Express device receptacles and plugs and, therefore, be documented for completeness.
   - It was created by the Small Form Factor Committee working group T10 to support multiple interfaces through the same connector—that is, a multipurpose connector.
   - It supports up to four lanes of PCI Express (SATA Express), two SAS ports, and legacy SATA devices, as well as a new MultiLink SAS alternative (which will most likely be obsolete due to SFF-8639).
7. The SAS MultiLink plugs and receptacles can have up to four signal segments physical links
   - TX0+, TX0-, RX0+, and RX0- signal segment 0 physical link (differential signal pairs).
   - TX1+, TX1-, RX1+, and RX1- signal segment 1 physical link, if any.
   - TX2+, TX2-, RX2+, and RX2- signal segment 2 physical link, if any.
   - TX3+, TX3-, RX3+, and RX3- signal segment 3 physical link, if any.

The fantasy of a single connector scheme might actually materialize with the advent of the SFF-8639. If SATA Express survives the momentum of NVM Express, the numerous connector plugs and receptacles will only confuse matters. Table 8.5 provides a matrix of receptacle and plugs and device types that can connect to one another.

Table 8.5 Connector Mating Matrix (SATA-IO)

Receptacles / Plugs	SATAe Host Cable (2)	SATAe Device Cable (3)	SATAe Host (5)	SATA Cable (legacy)	SFF-8639 Backplane (6)	SAS MultiLink (7)
SATAe Host (1)	✓			✓		
SATAe Device (4)		✓	✓		✓ [a]	✓ [a]
SATA Device (legacy)		✓	✓	✓	✓ [b]	✓ [b]

[a] SATA Express Device Plug mates with SFF-8639 and MultiLink connectors—host must support PCIe devices.
[b] SATA Device Plug mates with SFF-8639 and MultiLink connectors—host must support SATA devices.
NOTE – SATA Express hosts support both PCIe and SATA devices.

SATA Express Device Plug Connector

The SATA Express device plug connector is physically similar to the standard SAS device connector defined in SFF-8680 (see Figure 8.41).

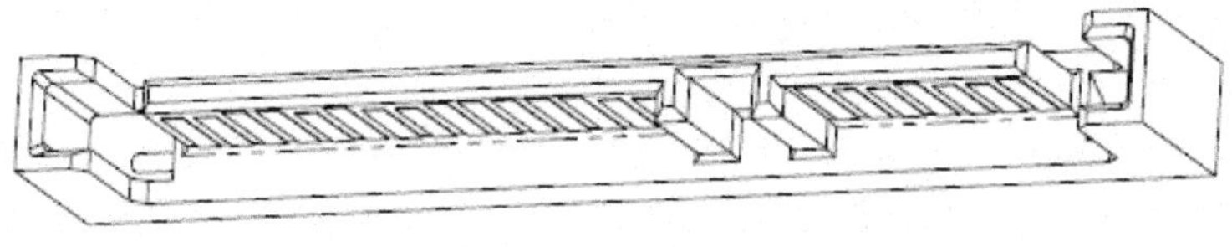

Figure 8.41 SATA Express device plug connector. (*Source:* www.sata-io.org)

The SATA Express device plug connector mates with the following:

a. SATA Express host receptacle connector
b. SATA Express device cable receptacle connector
c. SFF-8639 backplane connector
d. SFF-8482, SFF-8630, and SFF-8680 SAS receptacle connectors

SATA Express Host Receptacle Connector

The SATA Express host receptacle connector (see Figure 8.42) is identical to the SATA Express device cable receptacle connector, except for the following differences:

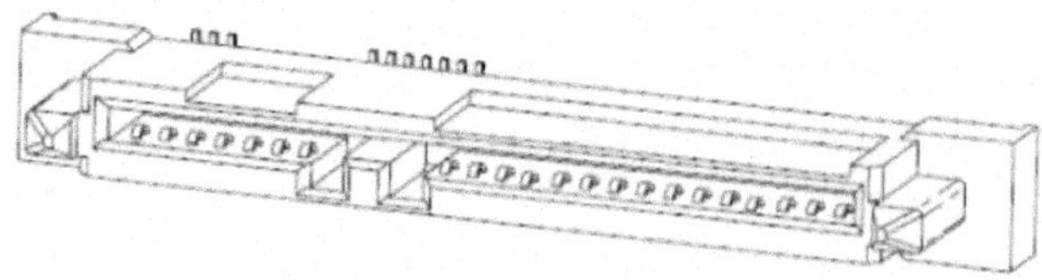

Figure 8.42 SATA Express host receptacle connector. (*Source:* www.sata-io.org)

The SATA Express host receptacle connector mates with the following connectors:

a. SATA Express device plug connector
b. SATA device plug connector

SATA Express Host Plug Connector

The SATA Express host plug connector (see Figure 8.43) may be considered an extension of the SATA host plug connector. It is essentially two SATA host plug connectors joined together, plus an additional section for sidebands (PCIe Reset, Interface Detect).

The SATA Express host plug connector mates with the following connectors:

a. SATA Express host cable receptacle connector
b. SATA single lane cable receptacle connector

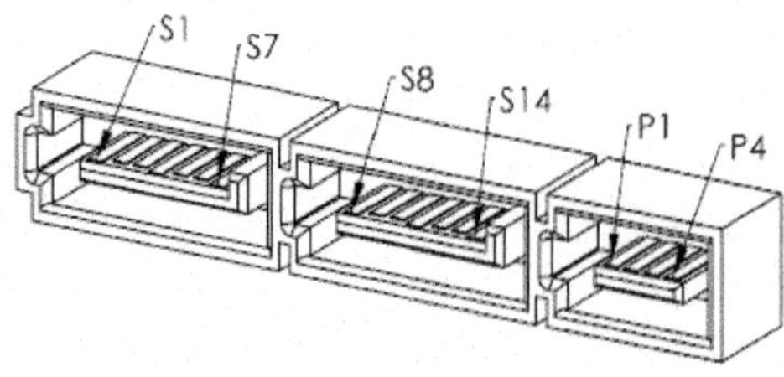

Figure 8.43 SATA Express host plug connector. (*Source:* www.sata-io.org)

SATA Express Host Cable Receptacle Connector

The SATA Express host cable receptacle connector (see Figure 8.44) mates with only the SATA Express host plug connector.

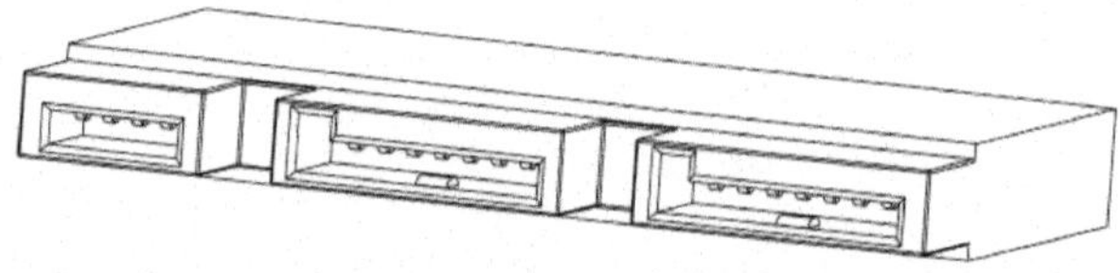

Figure 8.44 SATA Express host cable receptacle connector. (*Source:* www.sata-io.org)

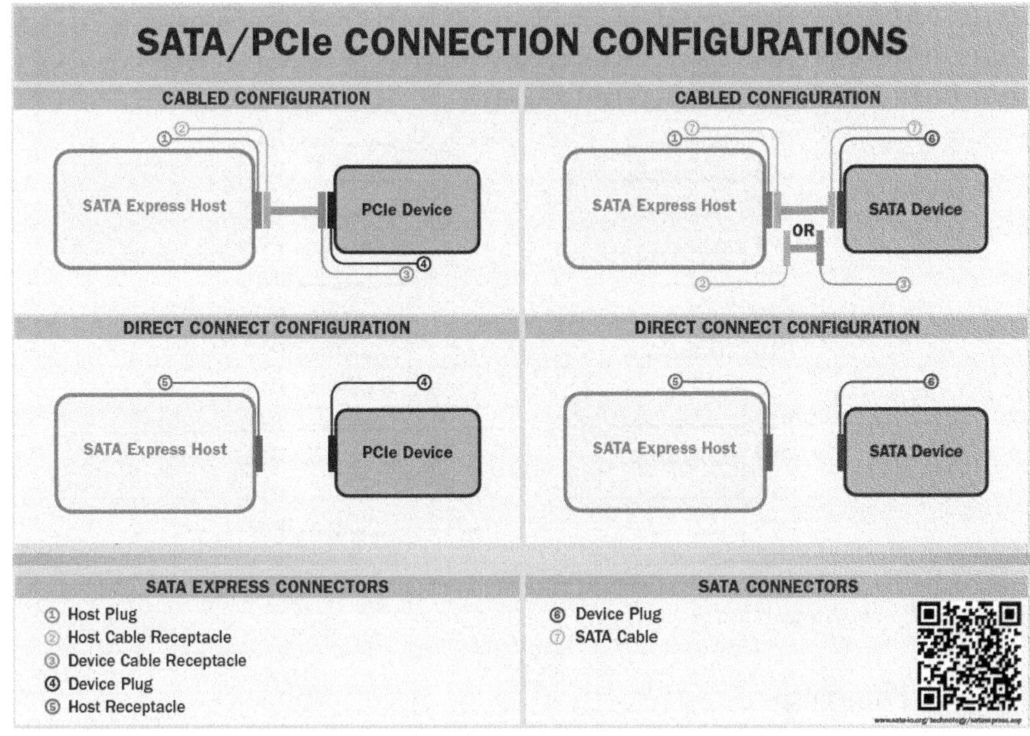

Figure 8.45 SATA/PCI Express configurations. (*Source:* www.sata-io.org)

Figure 8.45 shows numerous SATA/PCIe connections configuration. Host and device can be either cabled or directly connected.

Figure 8.46 shows a motherboard implementation of SATA Express. The SATA Express scheme provides two device connections and PCIe sidebands.

Figure 8.46 SATA Express Host Plug (1) and Cable Receptacle (2). (*Source:* www.techspot.com/news/55662-sata-express-101.html)

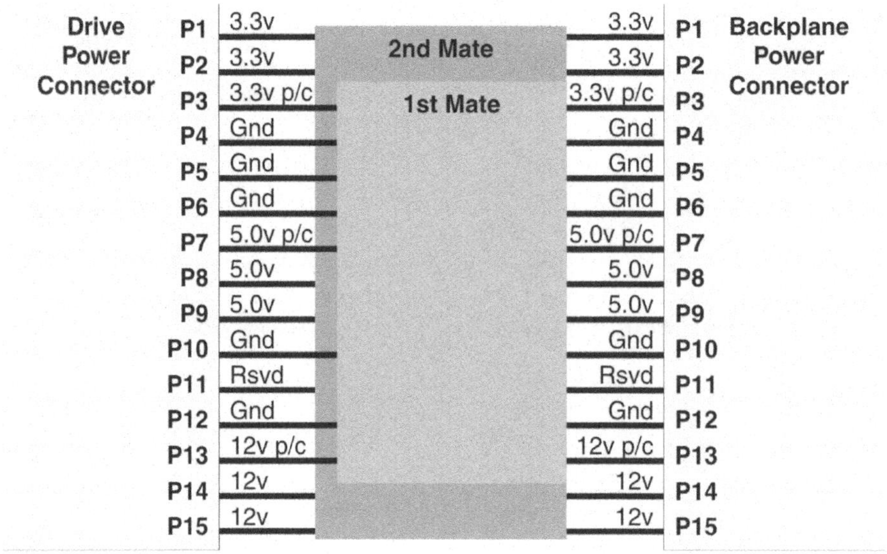

Drive Power Connector		2nd Mate / 1st Mate		Backplane Power Connector
P1	3.3v	2nd Mate	3.3v	P1
P2	3.3v		3.3v	P2
P3	3.3v p/c	1st Mate	3.3v p/c	P3
P4	Gnd		Gnd	P4
P5	Gnd		Gnd	P5
P6	Gnd		Gnd	P6
P7	5.0v p/c		5.0v p/c	P7
P8	5.0v		5.0v	P8
P9	5.0v		5.0v	P9
P10	Gnd		Gnd	P10
P11	Rsvd		Rsvd	P11
P12	Gnd		Gnd	P12
P13	12v p/c		12v p/c	P13
P14	12v		12v	P14
P15	12v		12v	P15

Figure 8.47 Two-stage mate sequence.

8.5 Hot Plugging

Cables employ a two-stage mating sequence (see Figure 8.47):

- Ground and pre-charge pins contact first, followed by the power pins.
- The power plug for cabled interfaces is not designed for hot plugging and cannot be safely hot plugged.
- The signal connector may be safely hot plugged.

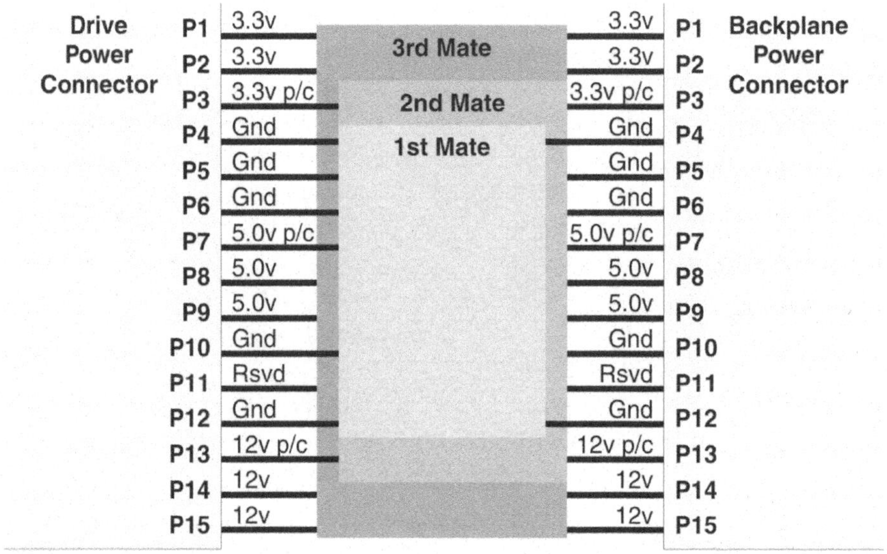

Drive Power Connector		3rd Mate / 2nd Mate / 1st Mate		Backplane Power Connector
P1	3.3v	3rd Mate	3.3v	P1
P2	3.3v		3.3v	P2
P3	3.3v p/c	2nd Mate	3.3v p/c	P3
P4	Gnd	1st Mate	Gnd	P4
P5	Gnd		Gnd	P5
P6	Gnd		Gnd	P6
P7	5.0v p/c		5.0v p/c	P7
P8	5.0v		5.0v	P8
P9	5.0v		5.0v	P9
P10	Gnd		Gnd	P10
P11	Rsvd		Rsvd	P11
P12	Gnd		Gnd	P12
P13	12v p/c		12v p/c	P13
P14	12v		12v	P14
P15	12v		12v	P15

Figure 8.48 Three-stage mate sequence for backplanes.

Backplane Power Mating Sequence

- SAS and SATA both accommodate hot plugging in a backplane environment using the direct device mating connector.
- This connector provides a three-stage mating sequence for the connector power segment (see Figure 8.48).

8.6 Review

1. SATA disks provide high-capacity mass storage to Client and Enterprise applications.
2. SAS disks provide Enterprise class mass storage to data center applications.
3. SAS disks cannot plug into a SATA backplane connector.
4. SATA disks can plug into a SAS backplane connector.
5. SATA devices have separate power and signal lines on the same connector.
6. SATA devices support hot plugging.
7. Micro SATA was specifically designed to enable connection of slim 1.8-inch form factor disks.
8. mSATA was designed to enable a new class of small form factor devices that mainly accommodated SSDs.
9. SATA microSSD™ was developed for high-performance, low-cost, embedded solid-state drives (SSDs), and embedded storage solutions for mobile computing platforms, such as ultra-thin laptops.
10. M.2 (formerly known as NGFF) is a small form factor card and connector that supports applications such as WiFi, WWAN, USB, PCIe, and SATA, as defined in the PCI-SIG specification (www.pcisig.org/home).
11. M.2 defines five board sizes for use in SSD or caching modules.
12. SATA Universal Storage Module (USM) was designed to meet the storage needs of a variety of digital media applications, including PCs and laptops, Digital Video Recorders (DVR)/Set-Top Boxes (STB), TVs, game consoles, and video surveillance.
13. SATA Express is purely PCI Express when it comes to transaction, link, and phy layers.
14. SATA Express allows for both legacy SATA and PCIe-based storage (SATAe) on the same physical interface.
15. The SATA Express device plug mates with all standard SAS connectors, as well as the new SFF-8639 multi-interface connector.

Chapter 9

Port Multipliers and Selectors

Objectives

This chapter will cover details concerning hardware that provides device connectivity and high availability to SATA system designs. SATA has limited connectivity and availability:

- Point-to-point serial links
    - Only allow for a single host to device connection
- Single-ported devices providing only one path to the device
    - If that port or path fails, then the device is no longer available to the host system

The Port multiplier and selector topic list includes the following:

- Port multipliers
    - Address connectivity
    - PM ports
    - Addressing
    - FIS modification, delivery, and Link layer protocol
    - Port and General Status and Control Registers (PSCRs and GSCRs)
    - PM commands, resets, error handling, and link power management
- Port selectors
    - Address availability
    - Active port selectors
    - BIST and resets
    - Power management, hot plug, OOB signaling, and speed negotiation

9.1 Port Multiplier

The Port Multiplier (PM) is a device that provides connectivity to the basic SATA architecture. Considering that SATA is a point-to-point interface, only one device can connect to a single host port.

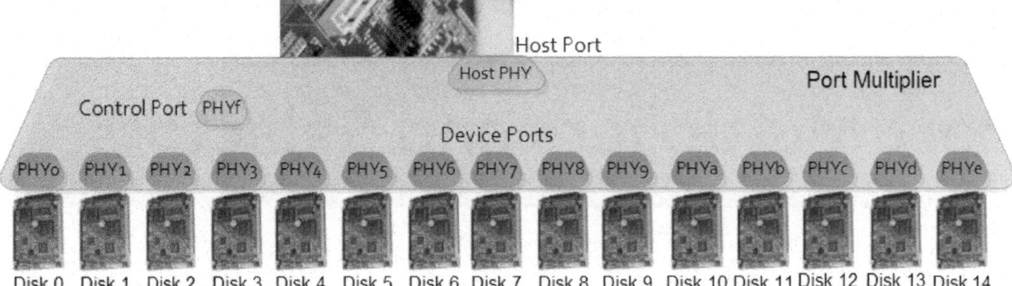

Figure 9.1 PM configuration.

PMs were made to allow for up to 15 devices (see Figure 9.1) to be accessed through a single host SATA port. SATA PMs are similar to SAS expanders and Fibre Channel loop switches in that these devices provide connectivity to point-to-point serial technologies.

9.1.1 Port Multiplier Characteristics

Hosts device drivers (System) must be Port Multiplier aware in order to access devices attached to the host via a PM. Only host systems are impacted by PMs; disks or any devices that attach to the PM are completely unaware of it.

- Port address
  - The control port and each device port present on a Port Multiplier have a port address.
  - The port address is used to route FISes between the host and a specific device or the control port.
- Device firmware and hardware are not affected by Port Multipliers.

PM Ports

There are three ports associated with Port Multipliers (PMs), as follows:

- Control port
  - The Port Multiplier has one port address reserved for control and status communication with the Port Multiplier itself. The control port has port address Fh.
- Device port(s)
  - A device port is a port that can be used to connect a device (a disk or SSD) to the PM.
    - A PM may support up to 15 device ports.
    - Device port addresses are sequentially numbered starting at zero and until all device ports have port addresses.
    - The device port address has a value less than the total number of device ports supported by the PM.

- Host port
    - The host port is the port that is used to connect the PM to the host. There is only one active host port.

9.2 PM Operational Overview

The Port Multiplier uses four bits, known as the PM Port field, in all FIS types to route FISes between the host and the appropriate device.

- Using the PM Port field, the Port Multiplier can route FISes to up to 15 Serial ATA devices from one active host.
- The PM Port field is filled in by the host on a host to device FIS with the port address of the device to route the FIS to.
- For a device to host FIS, the PM Port field is filled in by the Port Multiplier with the port address of the device that is transmitting the FIS.
- Device port addresses start at zero and are numbered sequentially higher until the last device port address has been defined.
- The control port, port address Fh, is used for control and status communication with the Port Multiplier itself.

In order to utilize all devices connected to a Port Multiplier, the host must have a mechanism to set the PM Port field in all transmitted FISes.

The Port Multiplier maintains a set of general purpose registers and also maintains the Serial ATA superset Status and Control Registers for each device port. The control port supports two commands that are used to read and write these registers:

- Read Port Multiplier
- Write Port Multiplier

Additional Port Multiplier features include the following:

- Support of booting with legacy software on device port 0h
- Support of staggered spin-up
- Support of hot plugging

9.2.1 Addressing Mechanism

The PM uses four bits, known as the PM Port field, in all FIS types to route FISes between the host and the appropriate device.

1. Using the PM Port field, the Port Multiplier can route FISes to up to 15 Serial ATA devices from one active host.
2. The PM Port field is filled in by the host on a host to device FIS with the port address of the device to route the FIS to.
3. For a device to host FIS, the PM Port field is filled in by the Port Multiplier with the port address of the device that is transmitting the FIS.
4. The PM Port field is reserved in SATA devices that don't support PMs.

FIS Modifications

The first Dword of all FIS types is shown in Figure 9.2. When SATA-IO designed the Port Multiplier, they needed a method to address up to 16 devices. This was accomplished by reclaiming four "Reserved" bits in byte 1 of the first Dword. This field became the **PM Port** field, as seen in Figure 9.3.

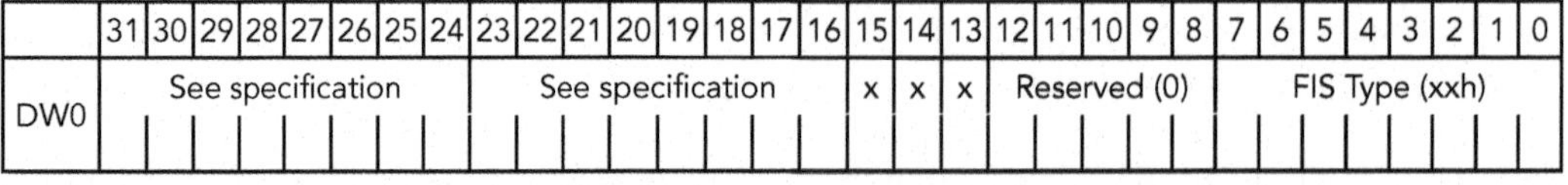

	31	30	29	28	27	26	25	24	23	22	21	20	19	18	17	16	15	14	13	12	11	10	9	8	7	6	5	4	3	2	1	0
DW0	See specification								See specification								x	x	x	Reserved (0)					FIS Type (xxh)							

Figure 9.2 Standard FIS without PM Port Field.

	31	30	29	28	27	26	25	24	23	22	21	20	19	18	17	16	15	14	13	12	11	10	9	8	7	6	5	4	3	2	1	0
DW0	See specification								See specification								x	x	x	x	**PM Port**				FIS Type (xxh)							

Figure 9.3 FIS with PM Port Field.

FIS Type

- Defines FIS type-specific fields and FIS length

PM Port

- Specifies the port address that the FIS should be delivered to or is received from
- Ports 0 to 14 are device ports

0000 = Port 0	1000 = Port 8
0001 = Port 1	1001 = Port 9
0010 = Port 2	1010 = Port A (10)
0011 = Port 3	1011 = Port B (11)
0100 = Port 4	1100 = Port C (12)
0101 = Port 5	1101 = Port D (13)
0110 = Port 6	1110 = Port E (14)
0111 = Port 7	1111 = Port F (15) Control Port

Transmission from Host to Device

To transmit a FIS to a device connected to a Port Multiplier

- The host will set the PM Port field in the FIS to the device's port address.
- The host will then start transmitting the FIS to the Port Multiplier in accordance with the Transport and Link state machines.

When a Port Multiplier receives a FIS over the host port

- The Port Multiplier will check the PM Port field in the FIS to determine the port address that the FIS should be transmitted over.
- If the FIS is destined for the control port, the Port Multiplier will receive the FIS and perform the command or operation requested.

If the FIS is destined for a device port, the Port Multiplier will perform the following procedure:

1. The Port Multiplier will determine if the device port is valid. If the device port is not valid, the Port Multiplier will issue a SYNCP primitive to the host and terminate reception of the FIS.
2. The Port Multiplier will determine if the X bit is set in the device port's PSCR[1] (SError) register. If the X bit is set, the Port Multiplier will issue a SYNCP primitive to the host and terminate reception of the FIS.
3. The Port Multiplier will determine if a collision has occurred. A collision is when a reception is already in progress from the device that the host wants to transmit to. If a collision has occurred, the Port Multiplier will finish receiving the FIS from the host and will issue an R_ERRP primitive to the host as the ending status. The Port Multiplier will follow the procedures outlined in the collision condition section to clear collision.
4. The Port Multiplier will initiate the transfer with the device by issuing an X_RDYP primitive to the device. A collision may occur as the Port Multiplier is issuing the X_RDYP to the device if the device has just decided to start a transmission to the host. If the device starts transmitting X_RDYP to the Port Multiplier, a collision has occurred. If a collision has occurred, the Port Multiplier will finish receiving the FIS from the host and will issue an R_ERRP primitive to the host as the ending status. The Port Multiplier will follow the procedures outlined in the collision condition section to clear.
5. After the device issues R_RDYP to the Port Multiplier, the Port Multiplier will transmit the FIS from the host to the device. The Port Multiplier will not send R_OKP status to the host until the device has issued an R_OKP for the FIS reception. The R_OKP status handshake will be interlocked from the device to the host.

If an error is detected during any part of the FIS transfer, the Port Multiplier will ensure that the error condition is propagated to the host and the device.

The transfer between the host and Port Multiplier is handled separately from the transfer between the Port Multiplier and device; only the end-of-frame R_OKP handshake is interlocked. The Port Multiplier will ensure that the flow control signaling latency requirement is met for all FIS transfers on a per link basis.

Specifically, the Port Multiplier will ensure that the flow control signaling latency is met between

1. The host port and the host it is connected to
2. Each device port and the device that it is connected to

If no error is detected during the FIS transfer, the Port Multiplier will not alter the FIS transmitted to the device.

- The Port Multiplier is not required to check or recalculate the CRC.
- The Port Multiplier will have Link and Phy layer state machines that comply with the Serial ATA Link and Phy layer state machines.

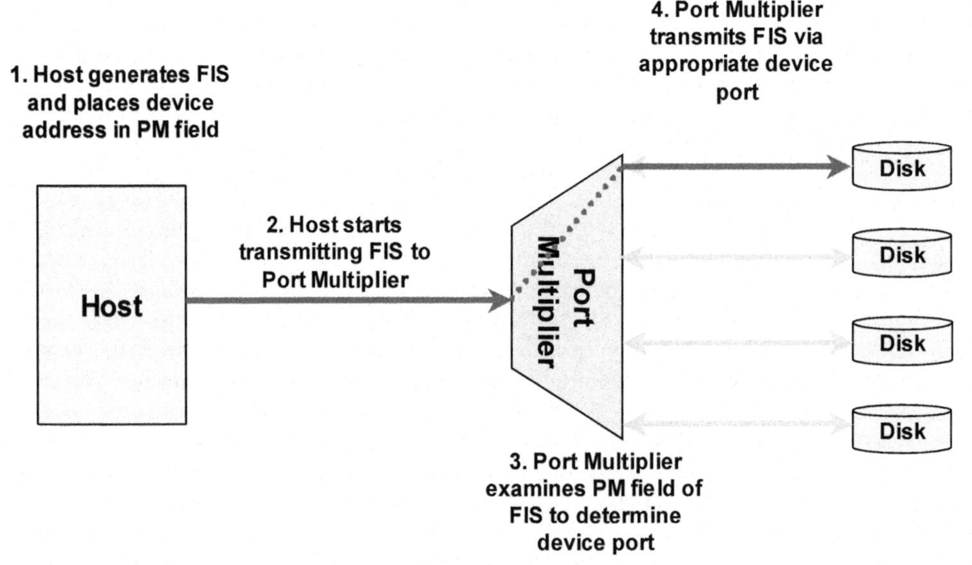

Figure 9.4 Host to Device FIS transmission.

Host to Device FIS Transmission

When transferring a FIS from the host to a device (disk) through a PM (see Figure 9.4), the

1. Host places PM port field into FIS
2. Host transmits FIS to PM
3. PM examines PM field for Port number
4. PM transmits FIS to device

Example of Primitive Protocol

When a Port Multiplier forwards a FIS, all it needs is the first Dword to determine which device port it will forward the FIS through. This forwarding process can start immediately or the PM could have buffers to capture either a portion or all of the FIS prior to routing. Figure 9.5 shows how the Link Layer protocol between the Host, PM, and device behaves during FIS transmission.

Transmission from Device to Host

The device behavior is the same, regardless of whether it is connected directly to the host or is connected to the host via a Port Multiplier.

When a device wants to transmit a FIS to the host, the Port Multiplier will perform the following procedure:

1. After receiving an X_RDYP primitive from the device, the Port Multiplier will determine if the X bit is set in the device port's PSCR[1] (SError) register. The Port Multiplier will not issue an R_RDYP primitive to the device until the X bit is cleared to zero.

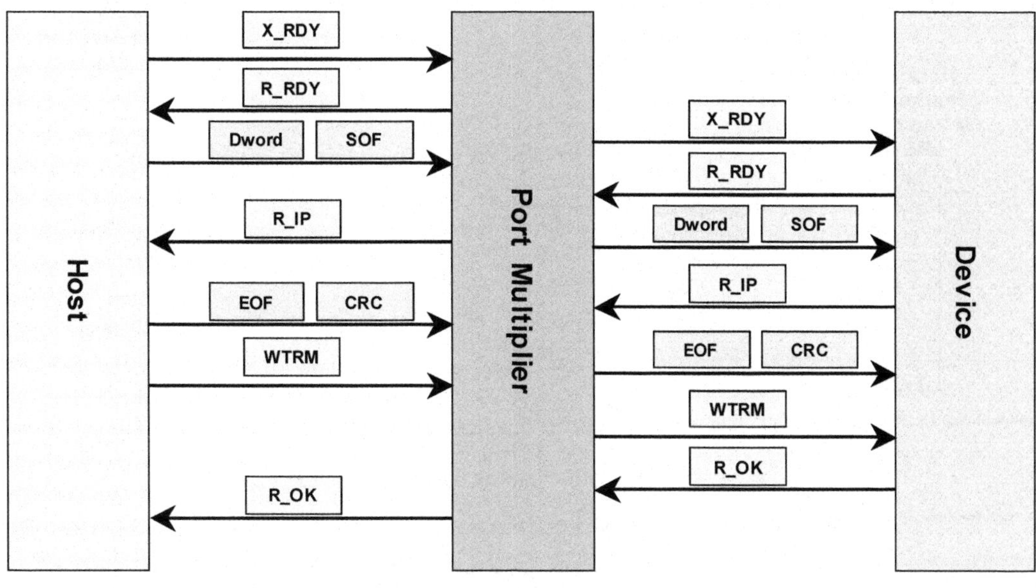

Figure 9.5 FIS primitive protocols (host and device side).

2. The Port Multiplier will receive the FIS from the device. The Port Multiplier will fill in the PM Port field with the port address of the transmitting device. The Port Multiplier will transmit the modified FIS to the host with a recalculated CRC. The Port Multiplier will check the CRC received from the device. If the CRC from the device is invalid, the Port Multiplier will corrupt the CRC sent to the host to ensure that the error condition is propagated. Specific details on how the Port Multiplier corrupts the CRC in this error case are covered in the Serial ATA Specification.
3. The Port Multiplier will issue an X_RDYP primitive to the host to start the transmission of the FIS to the host. After the host issues R_RDYP to the Port Multiplier, the Port Multiplier will transmit the FIS from the device to the host. The Port Multiplier will not send R_OKP status to the device until the host has issued an R_OKP for the FIS reception. The R_OKP status handshake will be interlocked from the device to the host.

Device to Host FIS Transmission

When transferring a FIS from the device to a host through a PM (see Figure 9.6):

1. Device sends FIS with PM port field = 0.
2. PM receives FIS and fills in PM Port field.
3. PM starts to transmit FIS to host.
4. PM recalculates and replaces CRC, prior to end of frame (EOF).
5. Host receives FIS, checks CRC, uses PM port field to associate FIS with Device.

If an error is detected during any part of the FIS transfer, the Port Multiplier will ensure that the error condition is propagated to the host and the device.

The Port Multiplier may wait for an X_RDYP/R_RDYP handshake with the host prior to issuing an R_RDYP to the device to minimize buffering.

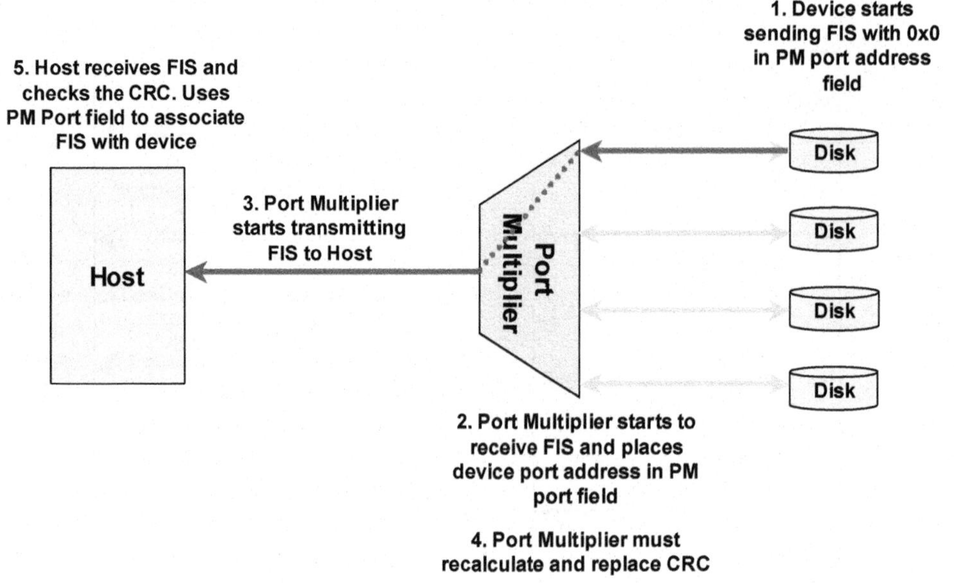

Figure 9.6 Device to host FIS protocol sequence.

1. The transfer between the device and Port Multiplier is handled separately from the transfer between the Port Multiplier and the host; only the end-of-frame R_OKP handshake is interlocked.
2. The Port Multiplier will ensure that the flow control signaling latency requirement is specified for all FIS transfers on a per link basis.

Specifically, the Port Multiplier will ensure that the flow control signaling latency is met between

- The host port and the host it is connected to
- Each device port and the device that it is connected to

The Port Multiplier will have Link and Phy layer state machines that comply with the SATA Link and Phy layer state machines.

9.2.2 Policies

FIS Delivery

The end-of-frame handshake will be interlocked between the host and the device. Specifically, the Port Multiplier will not issue an R_OKP to the initiator of a FIS before the target of a FIS has issued an R_OKP. The Port Multiplier will propagate R_OKP and R_ERRP from the target of the FIS to the initiator of a FIS.

If a transmission fails before the R_OKP handshake is delivered to the initiator, the Port Multiplier is responsible for propagating the error condition. Specifically, the Port Multiplier will propagate SYNCP primitives received during a FIS transmission end-to-end to ensure that any error condition encountered in the middle of a FIS is propagated.

If there is a FIS transfer ongoing and the link between the Port Multiplier and the active device becomes inoperable, the Port Multiplier should issue SYNCP primitives to the host until the host responds with a SYNCP primitive in order to fail the transfer. Failing the transfer upon detecting an inoperable link allows the host to proceed with recovery actions immediately, thereby eliminating latency associated with a timeout.

Port Priority

The Port Multiplier will ensure that an enabled and active device port is not starved. The specific priority algorithm used is implementation specific.

The control port will have priority over all device transfers. While a command is outstanding to the control port, no device transmissions will be started by the Port Multiplier until the command outstanding to the control port is completed.

9.3 FIS Delivery Mechanisms

This section provides a reference for one method that a Port Multiplier may use to satisfy the FIS delivery policies outlined.

9.3.1 Starting a FIS Transmission

If a device on a Port Multiplier asserts X_RDYP, the Port Multiplier has selected that device for transmission next, and the host port is not busy, the Port Multiplier will

1. Issue X_RDYP to the host;
2. Wait for the host to respond with R_RDYP
3. Issue R_RDYP to the device after the host issues R_RDYP

Then the transmission to the host may proceed.

If the host asserts X_RDYP to the Port Multiplier and the Port Multiplier does not have X_RDYP asserted to the host, the Port Multiplier responds with R_RDYP to the host. The Port Multiplier receives the first Dword of the FIS payload from the host. If the Port Multiplier Port specified is a device port that is enabled on the Port Multiplier, the Port Multiplier issues an X_RDYP over the device port specified and proceeds to transmit the entire FIS to the device. If a collision occurs during this process, the Port Multiplier will follow the procedures outlined in the section Collisions.

Status Propagation (Informative)

If there is an on-going FIS transmission between the host and a device, the Port Multiplier only issues R_OKP, R_ERRP, and SYNCP if it has first received that primitive from the host or device, unless a collision occurs or an invalid port is specified.

The Port Multiplier will not convey R_OKP or R_ERRP to the initiator of a FIS until the target of the FIS has issued R_OKP or R_ERRP, once the end-to-end transmission has commenced.

If the initiator or target of a FIS transmission issues a SYNCP primitive during a FIS transfer, this primitive will be propagated in order to ensure that the error condition is propagated to either end.

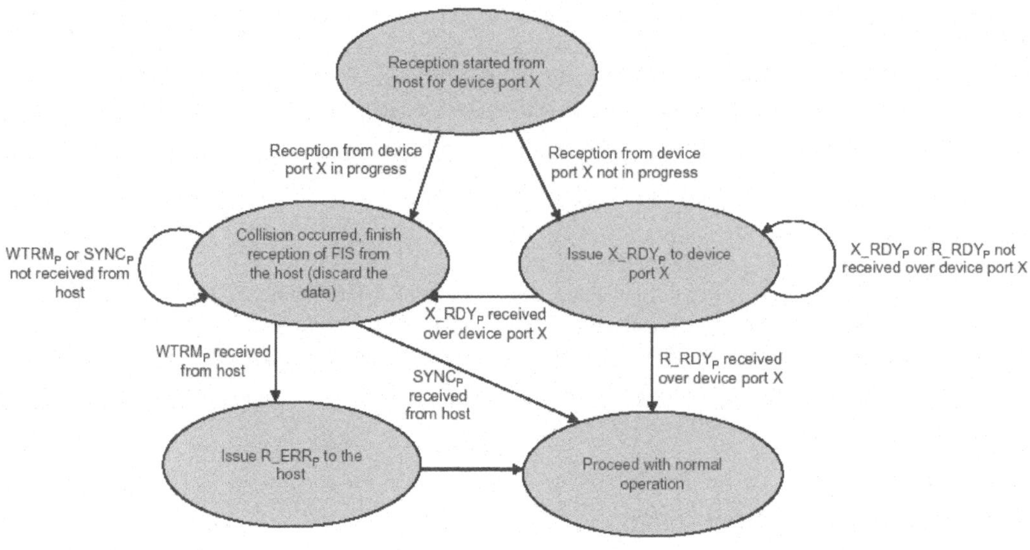

Figure 9.7 PM collision state machine. (www.SATA-IO.org)

Collisions

A collision is when the Port Multiplier has already started a reception from the device that the host wants to transmit to. A collision also occurs when the device issues an X_RDYP primitive at the same time that the Port Multiplier is issuing an X_RDYP primitive to that device. All collisions are treated as an X_RDYP/X_RDYP collision; the host will lose all such collisions and must retransmit its FIS at a later time.

A collision only occurs when the host is trying to issue another Serial ATA native queued command to a device that has native queued commands outstanding. The Serial ATA native command queuing protocol guarantees that the host will never be transmitting a Data FIS when the collision occurs. This means that the Port Multiplier can safely issue an error to the host and that the host will retry the failed FIS transmission at a later point in time.

When the Port Multiplier detects a collision, the Port Multiplier will finish reception of the FIS from the host and will issue an R_ERRP to the host for the end-of-frame handshake. The Port Multiplier will discard the FIS received from the host. The host must attempt to retry the FIS transfer that failed (Figure 9.7).

X_RDYP/X_RDYP collisions that occur on the host port will be handled in accordance with the Link layer state diagrams in the SATA specification.

9.3.2 Booting with Legacy Software

Booting is accommodated off of the first port of the Port Multiplier, device port 0h, without any special software or hardware support. An HBA that does not support attachment of a Port Multiplier will still work with the device on device port 0h.

If a system requires fast boot capability, it should ensure that both the BIOS and OS driver software are Port Multiplier aware. If legacy software is used in the presence of a Port Multiplier and there is no device present on device port 0h, legacy software will detect a device as present but will never receive a

Register FIS with the device signature. Waiting for the device signature may cause a legacy BIOS not to meet fast boot timing requirements.

Staggered Spin-up Support

The Port Multiplier will disable all device ports on power-up or upon receiving a COMRESET signal over the host port. This feature allows the host to control when each device spins up.

Refer to SATA on how to enumerate a device, including devices that may be spun down because the port is disabled.

9.3.3 Hot Plug Events

Port Multiplier handling of hot plug events is defined by the hot plug state machines for the host port and device port.

Upon receiving a COMRESET signal from the host, the Port Multiplier will perform an internal reset. As part of the internal reset, the Port Multiplier will update the Serial ATA superset Status and Control Registers for each device port.

Upon receiving a COMINIT signal from a device, the Port Multiplier will update the Serial ATA superset Status and Control Registers for that port. The Port Multiplier will set the X bit in the DIAG field of the device port's PSCR[1] (SError) register to mark that device presence has changed. If the Port Multiplier has not received a FIS for the control port after the last COMRESET and an unsolicited COMINIT signal was received over device port 0h, the Port Multiplier will propagate the COMINIT signal to the host. For all other cases, the COMINIT signal will not be propagated to the host.

If the X bit in the DIAG field of the PSCR[1] (SError) register is set for a device port, the Port Multiplier will disallow FIS transfers with that port until the X bit has been cleared. When the Port Multiplier is disallowing FIS transfers with a device port, the Port Multiplier must ensure that the FISes the device is attempting to transmit are not dropped.

It is recommended that host software frequently query GSCR[32] to determine if there has been a device presence change on a device port.

Hot Plug State Machine for Host Port

Cabled hot plug of the Port Multiplier host port will be supported since Port Multipliers may not share the same power supply as the host. Therefore, the Port Multiplier will periodically poll for host presence by sending periodic COMINIT signals to the host after transitioning to the HPHP1:NoComm state.

The state machine enables device port 0 after communication has been established over the host port and also clears the X bit in the DIAG field of the PSCR[1] (SError) register for device port 0 after it is set when operating with legacy software. In addition, the state machine will propagate unsolicited COMINIT signals received from device port 0 to the host when operating with legacy software. These accommodations allow device port 0 to work with legacy host software.

When operating with Port Multiplier aware software, the state machine treats device port 0 exactly like every other device port.

Hot Plug State Machine for Device Port

The device port hot plug behavior is exactly the same as the behavior for a host controller supporting the device hot plug directly. There is no change to the device, and there is no change to the usage model.

9.4 Link Power Management

In accordance with Serial ATA 3.2, the Port Multiplier must support reception of PMREQP primitives from the host and from attached devices. If the Port Multiplier does not support the power management state requested, the Port Multiplier will respond to the PMREQP primitive with a PMNAKP primitive.

As described in the SATA specification for Link power management, before the Port Multiplier delivers a FIS, the Port Multiplier will check the state of the link and issue a COMWAKE signal if the link is in partial or slumber. The Port Multiplier will accurately reflect the current state of each device port link in the port-specific registers (specifically PSCR[0] (SStatus) for each port).

The Port Multiplier will not propagate a PMREQP primitive received on a device port. A PMREQP primitive received from a device only affects the link between that device and the Port Multiplier. If the Port Multiplier receives a PMREQP primitive over a device port, the Port Multiplier will perform the following actions:

- The Port Multiplier will respond to the device with PMACKP or PMNAKP.
- If the Port Multiplier responds to the device with PMACKP:
    - The Port Multiplier will transition the link with that device to the power state specified.
    - The Port Multiplier will update the port-specific registers for that device port to reflect the current power management state of that link.

The Port Multiplier will propagate a PMREQP primitive received from the host to all active device ports if the Port Multiplier responds to the PMREQP primitive with PMACKP. In this case, the Port Multiplier will propagate the PMREQP primitive to all device ports that have PhyRdy set. If a device responds to a PMREQP primitive with PMNAKP, this event will only affect the link with that device. The host may interrogate the Port Multiplier port-specific registers to determine which device ports are in a power-managed state.

If the Port Multiplier receives a PMREQP primitive from the host, the specific actions the Port Multiplier will take are as follows:

- The Port Multiplier will respond to the host with PMACKP or PMNAKP.
- If the Port Multiplier responds to the host with PMACKP:
    - The Port Multiplier will transition the link with the host to the power state specified.
    - The Port Multiplier will issue PMREQP primitives to all device ports that have PhyRdy set.
        - If a device responds with PMACKP, the Port Multiplier will transition the link with that device to the power state specified and will update the Port Multiplier port-specific registers for that port.
        - If a device responds with PMNAKP, the Port Multiplier will not take any action with that device port.

The Port Multiplier will wake device links on an as-needed basis. A COMWAKE signal from the host is not propagated to the devices. The Port Multiplier will issue a COMWAKE signal to a device when a FIS needs to be delivered to that device.

The Port Multiplier may issue a PMREQP to the host if all device ports are in partial/slumber or are disabled.

9.5 Reducing Context Switching Complexity

It may complicate some host controller designs if traffic from another device is received in the middle of certain FIS sequences. For example, if a host receives a DMA Activate FIS from one device, it may be awkward for the host to receive a FIS from another device before it is able to issue the Data FIS to the first device.

The Port Multiplier will provide the host with the opportunity to transmit before initiating any pending device transmissions. The Port Multiplier will not assert an X_RDYP primitive to the host until the Port Multiplier has received at least two consecutive SYNCP primitives from the host.

If the host wants to transmit prior to receiving another FIS, the host should issue one SYNCP primitive between the end of the last FIS transmission and the start of the next FIS transmission.

One possible implementation is to transition to the host Link layer state HL_SendChkRdy defined in the SATA specification immediately, regardless of whether the host is ready to proceed with the next transmission. The host would then only transition out of the HL_SendChkRdy state when the host is ready to proceed with the transmission and the R_RDYP primitive is received.

9.6 Error Handling and Recovery

The host is responsible for handling error conditions in the same way it handles errors when connected directly to a device. The host is responsible for detecting commands that do not finish and performing error recovery procedures as needed. The exact host software error recovery procedures are implementation specific.

The Port Multiplier is not responsible for performing any error recovery procedures. The Port Multiplier will return R_ERRP for certain error conditions, as described by the SATA state diagrams. The Port Multiplier will update GSCR[32] and PSCR[1] (SError) for the device port that experiences an error.

9.6.1 Command Timeout

If a command times out, the host may check PSCR[1] (SError) to determine if there has been an interface error condition on the port that had the error. If there has been a device presence change on that port as indicated by the X bit in the DIAG field of PSCR[1] (SError) for the port, the host should re-enumerate the device on that port. The host software error recovery mechanism after a command timeout is implementation specific.

9.6.2 Disabled Device Port

If the host transmits a FIS to a device port that is disabled or that has FIS transfers disallowed due to the X bit being set in the DIAG field of PSCR[1] (SError) for that port, the Port Multiplier will not perform an R_OK or R_ERR handshake at the end of FIS reception and will instead terminate the FIS reception by issuing SYNCP primitives to the host. The host is responsible for detecting that the command did not finish and performing error recovery procedures, including clearing the X bit as necessary.

9.6.3 Invalid Device Port Address

An invalid device port address is a device port address that has a value greater than or equal to the number of device ports that the Port Multiplier supports. If the host specifies an invalid device port address as part of a FIS transmission, the Port Multiplier will issue a SYNCP primitive and terminate reception of the FIS. The host is responsible for detecting that the command did not finish and performing normal error recovery procedures.

9.6.4 *Invalid CRC for Device-Initiated Transfer*

On a device-initiated transfer, the Port Multiplier must recalculate the CRC since it modifies the first Dword of the FIS. The Port Multiplier will check the original CRC sent by the device. If the original CRC is invalid, the Port Multiplier will invert the recalculated CRC to ensure that the CRC error is propagated to the host. The inversion can be done by XORing the recalculated CRC with FFFFFFFFh. The Port Multiplier will also update the error information in PSCR[1] (SError) for the device port that experienced the error.

9.6.5 *Data Corruption*

If the Port Multiplier encounters an 8b10b decoding error or any other error that could affect the integrity of the data passed between the initiator and target of a FIS, the Port Multiplier will ensure that the error is propagated to the target. The Port Multiplier may propagate the error by corrupting the CRC for the FIS. The Port Multiplier will also update the error information in PSCR[1] (SError) for the device port that experienced the error.

Unsupported Command Received on the Control Port

If an unsupported command is received on the control port, the Port Multiplier will respond with a Register FIS that has the values shown in Table 9.1 for the Status and Error Registers.

Table 9.1 Register Values for an Unsupported Command

Register	7	6	5	4	3	2	1	0
Error	Reserved					ABRT	Reserved	
Status	BSY	DRDY	DF	na	DRQ	0	0	ERR

ABRT = 1 BSY = 0 DRDY = 1 DF = 0 DRQ = 0 ERR = 1

BIST Support

A Port Multiplier may optionally support BIST. A Port Multiplier that supports BIST will only support BIST in a point-to-point manner. A Port Multiplier that supports BIST will not propagate a BIST Activate FIS received on one port over another port. The host determines that a Port Multiplier supports BIST by checking GCSR[64].

To enter BIST mode over the host connection with a Port Multiplier, the host will issue a BIST Activate FIS to the Port Multiplier control port. The host will not issue a BIST Activate FIS to a device port.

To enter BIST mode over a device connection, the device will issue a BIST Activate FIS to the Port Multiplier. The Port Multiplier will intercept the BIST Activate FIS and enter BIST mode.

Upon entering the BIST mode with the device, the Port Multiplier will update the PSCR[0] (SStatus) register for that port to reflect that the link has entered BIST mode, as specified in the SATA specification.

Initiation of BIST by a Port Multiplier is vendor unique.

Asynchronous Notification

A Port Multiplier may optionally support asynchronous notification as defined in Serial ATA 3.2. If asynchronous notification is enabled, a Port Multiplier will send an asynchronous notification to the host when a bit transitions from zero to one in GSCR[32] of the Global Status and Control Registers. The Port Multiplier may send one notification for multiple zero to one-bit transitions. The asynchronous notification will only be sent when there is no command currently outstanding to the control port. If there is a command outstanding to the control port when an asynchronous notification needs to be sent, the Port Multiplier will first complete the command and then send the asynchronous notification. The Port Multiplier will set the PM Port field in the Set Device Bits FIS to the control port to indicate that the Port Multiplier itself needs attention.

Support for the asynchronous notification feature is indicated in GSCR[64], and it is enabled using GSCR[96].

9.6.6 Command-Based Switching

Host designs should give careful consideration to support of asynchronous notification in command-based switching designs. If a command-based switching HBA has no explicit accommodation for asynchronous notification, then the host should not enable asynchronous notification on the control port or on any attached device.

Phy Event Counters

A Port Multiplier may optionally support the Phy event counters feature that is defined in the SATA specification. If Phy event counters is enabled, a Port Multiplier will store supported counter information for all of the ports that are enabled, including the host port. It is not required that the same list of Phy event counters be implemented on every port.

Support for the Phy event counters feature is indicated in GSCR[64], and it is enabled using bit 0 in GSCR[34]. The counter values will not be retained across power cycles. The counter values will be preserved across COMRESET and software resets that occur on any port.

Counter Identifiers

Each counter has a 16-bit identifier. The format of these identifiers (see Table 9.2) is defined in the Serial ATA II: Extensions to Serial ATA 3.2. The valid identifiers include each of the existing Phy event counters, along with the additional counters specific to the Port Multiplier defined in this section.

Support for some counters defined in the Serial ATA II: Extensions to SATA specification may not be logical for all ports due to the Port Multiplier architecture—for example, the counters with identifiers 001h and 00Ah. For the Port Multiplier implementation, all counters other than 000h are

Table 9.2 Identifier Register Descriptions

Identifier (Bits 11:0)	Mandatory/ Optional	Description
000h	Mandatory	Will be implemented on both host and device ports
001h-BFFh	Optional	May be implemented on host or device ports
C00h	Optional	Number of transmitted H2D non-Data FIS R_ERR ending status due to collision
C01h	Optional	Number of Signature D2H Register FISes that were transmitted to the host from the control port
C02h	Optional	The number of corrupted CRC values that were transmitted to the host
C03h-FFFh	Reserved	Reserved for future counter definition

optional. A counter identifier value of 000h will indicate the end of the Phy event counter list implemented in the GSCR or PSCR registers. The counter with identifier 000h will have no counter value.

All counter values consume a multiple of 16 bits, with a maximum of 64 bits. Each counter will be allocated a single register location. The register location will contain both the identifier for the counter implemented along with the value of the Phy event counter.

Reading Counter Values

Initially, the host may obtain the mapping of counters implemented along with their sizes by submitting reads starting at GSCR[256] (or PSCR[256]) to obtain the identifiers for the counters on a per port basis. Once the identifier of 000h is reached, this signifies the end of the list of counters implemented. The format of this read is a Read Port Multiplier command with the RS1 bit of the Device/Head register cleared to zero. The value in the PortNum field determines which set of counters is to be accessed. Device port counter information for ports 0h–Eh can be retrieved by using port numbers 0h–Eh, respectively. Host port counter information can be retrieved by using the control port number (Fh). RegNum[15:0] will be set to the specific register location to be read. The output of the Read Port Multiplier command contains the identifier of the counter. When Phy event counters are enabled, a Port Multiplier will return identifier 000h as the first counter for a port on which no Phy event counters are implemented.

To read the counter value itself, the Read Port Multiplier command is sent with the RS1 bit of the Device/Head register set to one. The value in the PortNum field determines which set of counters are to be accessed. Device port counter information for ports 0h–Eh can be retrieved by using port numbers 0h–Eh, respectively. Host port counter information can be retrieved by using the control port number (Fh). RegNum[15:0] will be set to the specific register location to be read.

The output of the Read Port Multiplier command will contain the value of the counter, up to 64 bits in length. If the counter value read is less than 64 bits in length, the value returned by the Port Multiplier will have the upper bits padded with zeroes.

All counters are one-extended up to a 16-bit multiple (i.e., 16, 32, 48, 64 bits), once the maximum counter value has been reached. For example, when returned as a 16-bit counter, if a counter is only physically implemented as 8 bits when it reaches the maximum value of 0xFF, it will be one-extended to 0xFFFF. The counter will stop (and not wrap to zero) after reaching its maximum value.

Upon any read to Phy event counter register space that is at or beyond the identifier 000h location, the Port Multiplier will return an error status with the ERR bit set to one and the BSY and DRQ bits cleared to zero in the Status field of the FIS. The ABRT bit will also be set to one in the Error field.

Counter Reset Mechanisms

There are three mechanisms the host can use to explicitly cause the Phy event counters to be reset. The first mechanism uses the Write Port Multiplier command to clear counters on an individual counter basis. The PortNum field determines which set of counters are to be accessed.

Device port counter information for ports 0h–Eh can be written by using port numbers 0h–Eh, respectively. Host port counter information can be written by using the control port number (Fh).

The RegNum field will be set to the register location (counter) that is to be written. The Value field within the command will contain what will be written into the register, in this case all zeroes.

The second mechanism allows for a counter to be reset following a read to that counter register.

If the RS2 bit in the Device/Head register is set to one for a read to any counter, the Port Multiplier will return the current counter value for the command and then reset that value upon successful transmission of the Register FIS. If retries are required, upon unsuccessful transmission of the FIS it is possible that the counter value may be changed before the retransmission of the counter value.

The third mechanism is a global reset function by writing appropriate bits in GSCR[34] to reset all Phy event counters for specific ports. A host may reset all Phy event counters by writing FFFFh to bits 31-16 of GSCR[34].

9.7 Port Multiplier Registers

The Port Multiplier registers are accessed using Read/Write Port Multiplier commands issued to the control port.

9.7.1 General Status and Control Registers

The control port address is specified in the PortNum field of the Read/Write Port Multiplier command to read/write the General Status and Control Registers.

Static Configuration Information

The Static Configuration Information section of the General Status and Control Registers contains registers that are static throughout the operation of the Port Multiplier. These registers are read only and may not be modified by the host (see Table 9.3).

Table 9.3 Static Configuration Information Registers

Register	O/M	F/V	Description
GSCR[0]	M	F	Product Identifier
		F	31-16 Device ID allocated by the vendor
			15-0 Vendor ID allocated by the PCI-SIG
GSCR[1]	M		Revision Information
		F	31-16 Reserved
		F	15-8 Revision level of the Port Multiplier
		F	7-4 Reserved
		F	3 1 = Supports Port Multiplier specification 1.2
		F	2 1 = Supports Port Multiplier specification 1.1
		F	1 1 = Supports Port Multiplier specification 1.0
		F	0 Reserved
GSCR[2]	M		Port Information
		F	31-4 Reserved
		F	3-0 Number of exposed device fan-out ports
GSCR[3] –GSCR[31]	O	F	Reserved

Key:
O/M = Mandatory/optional requirement.
M = Support of the register is mandatory.
O = Support of the register is optional.
F/V = Fixed/variable content.
F = The content of the register is fixed and does not change.
V = The contents of the register is variable and may change.
X = The content of the register may be fixed or variable.

Register 0: Product Identifier

- Support for this register is mandatory.
- The register identifies the vendor that produced the Port Multiplier and the specific device identifier.
- Bit 15-0 will be set to the vendor identifier allocated by the PCI-SIG of the vendor that produced the Port Multiplier.
- Bit 31-16 will be set to a device identifier allocated by the vendor.

Register 1: Revision Information

- Support for this register is mandatory. The register specifies the specification revision that the Port Multiplier supports; the Port Multiplier may support multiple specification revisions.
- The register also specifies the revision level of the specific Port Multiplier product identified by Register 0.
- Bit 0 is reserved and will be cleared to zero.
- Bit 1 when set to one indicates that the Port Multiplier supports Port Multiplier specification version 1.0.
- Bit 2 when set to one indicates that the Port Multiplier supports Port Multiplier specification version 1.1.
- Bit 3 when set to one indicates that the Port Multiplier supports Port Multiplier specification version 1.2.
- Bits 7–4 are reserved and will be cleared to zero.
- Bits 15–8 identify the revision level of the Port Multiplier product identified by Register 0.
- This identifier is allocated by the vendor.
- Bits 31–16 are reserved and will be cleared to zero.

Register 2: Port Information

- Support for this register is mandatory.
- This register specifies information about the ports that the Port Multiplier contains, including the number of exposed device fan-out ports.
- The number of exposed device ports is the number of device ports that are physically connected and available for use on the product.
- Bits 3–0 specify the number of exposed device fan-out ports.
  - A value of zero is invalid.
  - The control port will not be counted.
- Bits 31–4 are reserved and will be cleared to zero.

Registers 3–31: Reserved

- Registers 3–31 are reserved for future Port Multiplier definition and will be cleared to zero.

9.7.2 Status Information and Control

The Status Information section of the General Status and Control Registers contains registers (see Table 9.4) that convey status information and control operation of the Port Multiplier.

Table 9.4 Status Information GSCR

Register	O/M	F/V	Description	
GSCR[32]	M		Error Information	
		F	31-15	Reserved
		V	14	OR of selectable bits in Port 14 PSCR[1] (SError)
		V	13	OR of selectable bits in Port 13 PSCR[1] (SError)
		V	12	OR of selectable bits in Port 12 PSCR[1] (SError)
		V	11	OR of selectable bits in Port 11 PSCR[1] (SError)
		V	10	OR of selectable bits in Port 10 PSCR[1] (SError)
		V	9	OR of selectable bits in Port 9 PSCR[1] (SError)
		V	8	OR of selectable bits in Port 8 PSCR[1] (SError)
		V	7	OR of selectable bits in Port 7 PSCR[1] (SError)
		V	6	OR of selectable bits in Port 6 PSCR[1] (SError)
		V	5	OR of selectable bits in Port 5 PSCR[1] (SError)
		V	4	OR of selectable bits in Port 4 PSCR[1] (SError)
		V	3	OR of selectable bits in Port 3 PSCR[1] (SError)
		V	2	OR of selectable bits in Port 2 PSCR[1] (SError)
		V	1	OR of selectable bits in Port 1 PSCR[1] (SError)
		V	0	OR of selectable bits in Port 0 PSCR[1] (SError)
GSCR[33]	M	V	Error Information Bit Enable	
			31-0	If set, bit is enabled for use in GSCR[32]
GSCR[34]	O	V	Phy Event Counter Control	
			31	Host port global counter reset
			30	Port 14 global counter reset
			29	Port 13 global counter reset
			28	Port 12 global counter reset
			27	Port 11 global counter reset
			26	Port 10 global counter reset
			25	Port 9 global counter reset
			24	Port 8 global counter reset
			23	Port 7 global counter reset
			22	Port 6 global counter reset
			21	Port 5 global counter reset
			20	Port 4 global counter reset
			19	Port 3 global counter reset
			18	Port 2 global counter reset
			17	Port 1 global counter reset
			16	Port 0 global counter reset
			15-1	Reserved
			0	Phy event counters enabled
GSCR[35] –GSCR[63]	O	F	Reserved	

Key:
O/M = Mandatory/optional requirement.
M = Support of the register is mandatory.
O = Support of the register is optional.
F/V = Fixed/variable content
F = The content of the register is fixed and does not change.
V = The contents of the register is variable and may change.
X = The content of the register may be fixed or variable.

9.7.3 Features Supported

The Features Supported section of the General Status and Control Registers contains registers that convey the optional features (see Table 9.5) that are supported by the Port Multiplier. The Features Supported registers have a one-to-one correspondence with the Features Enabled registers. All Features Supported registers are read-only.

Table 9.5 Features Supported—GSCR

Register	O/M	F/V	Description	
GSCR[64]	M		Port Multiplier Revision 1.X Features Support	
		F	31-5	Reserved
		F	4	1 = Supports Phy event counters
		F	3	1 = Supports asynchronous notification
		F	2	1 = Supports dynamic SSC transmit enable
		F	1	1 = Supports issuing PMREQP to host
		F	0	1 = Supports BIST
GSCR[65] –GSCR[95]	O	F	Reserved	
Key: O/M = Mandatory/optional requirement. M = Support of the register is mandatory. O = Support of the register is optional. F/V = Fixed/variable content F = The content of the register is fixed and does not change. V = The contents of the register is variable and may change. X = The content of the register may be fixed or variable.				

Features Enabled

The Features Enabled section of the General Status and Control Registers contains registers that allow optional features to be enabled. The Features Enabled registers have a one-to-one correspondence with the Features Supported registers. All Features Enabled registers are read/write.

Vendor Unique

The Vendor Unique section of the General Status and Control Registers contains registers that are vendor unique.

Phy Event Counters

The Phy event counters section of the General Status and Control Registers contains registers that store the data for each of the Phy event counters supported by the Port Multiplier for the host port. The Phy event counters registers contain both the identifier and counter values. All Phy event counter registers are read/write.

A value of "0" returned for a counter means that there have been no instances of that particular event.

9.7.4 Port Status and Control Registers

The Port Multiplier will maintain a set of Port Status and Control Registers (PSCRs) for each device port that it supports. These registers are port specific and contain the Serial ATA superset Status and Control Registers, along with the Phy event counters information for each port. The host specifies the device port to read or write registers for in the PortNum field of the Read/Write Port Multiplier commands. The registers are defined in Table 9.6.

Table 9.6 Port Status and Control Registers Descriptions

PSCR[0]	SStatus register
PSCR[1]	SError register
PSCR[2]	SControl register
PSCR[3]	SActive register (not implemented)
PSCR[4] - PSCR[255]	Reserved
PSCR[256] – PSCR[2303]	Phy event counter registers
PSCR[2304] – PSCR[65535]	Reserved

PSCR[0] – SStatus Register

The PSCR[0] (SStatus) register is the 32-bit Serial ATA superset SStatus register (Table 9.7) defined in the SATA specification and further defined in the Serial ATA II: Extensions to SATA specification.

Table 9.7 SStatus Register

Bits	31	30	29	28	27	26	25	24	23	22	21	20	19	18	17	16	15	14	13	12	11	10	09	08	07	06	05	04	03	02	01	00
PSCR0	**SStatus Register**																															

PSCR[1] – SError Register

The PSCR[1] (SError) register is the 32-bit Serial ATA superset SError register (Table 9.8) defined in the SATA specification and further defined in the Serial ATA II: Extensions to SATA specification.

Table 9.8 SError Register

Bits	31	30	29	28	27	26	25	24	23	22	21	20	19	18	17	16	15	14	13	12	11	10	09	08	07	06	05	04	03	02	01	00
PSCR1	**SError Register**																															

PSCR[2] – SControl Register

The PSCR[2] (SControl) register is the 32-bit Serial ATA superset SControl register (Table 9.9) defined in the SATA specification and further defined in the Serial ATA II: Extensions to SATA specification. The reset value for the DET field will be 4h.

Table 9.9 SControl Register

Bits	31	30	29	28	27	26	25	24	23	22	21	20	19	18	17	16	15	14	13	12	11	10	09	08	07	06	05	04	03	02	01	00
PSCR2	SControl Register																															

Phy Event Counters

The Phy event counter information for each of the device ports within a Port Multiplier will be contained in the Port Status and Control Registers starting at PSCR[256].

- The Phy event counters registers contain both the identifier and counter values.
- All Phy event counter registers are read/write.
- A value of "0" returned for a counter means that there have been no instances of that particular event.

9.8 Port Multiplier Command

9.8.1 Read Port Multiplier

The Read Port Multiplier command (Table 9.10) is used to read a register on a Port Multiplier. The Read Port Multiplier command must be issued to the control port.

Table 9.10 Read Port Multiplier Inputs

Register	7	6	5	4	3	2	1	0
Features	RegNum[7:0]							
Features (exp)	RegNum[15:8]							
Sector Count	Reserved							
Sector Count (exp)	Reserved							
Sector Number	Reserved							
Sector Number (exp)	Reserved							
Cylinder Low	Reserved							
Cylinder Low (exp)	Reserved							
Cylinder High	Reserved							
Cylinder High (exp)	Reserved							
Device/Head	na	RS1	RS2	na	PortNum			
Command	E4h							

PortNum Set to the port address that has register to be read.
RegNum Set to number of register to read.
RS1 Register Specific 1: This bit is register specific.
Phy event counter usage: Used to determine access to Phy event counter identifier or value. If cleared to zero upon a read to a Phy event counter register, the Port Multiplier will

return the Phy event counter identifier for that register. If set to one, the Port Multiplier will return the Phy event counter value for that register, up to 64-bits in length.
All other usage: This bit will be treated as "na" on all Read Port Multiplier commands to registers other than Phy event counter registers.

RS2 Register Specific 2: This bit is register specific.
Phy event counter usage: Used in reads to Phy event counter values. If set to one upon a read to a Phy event counter register, the Port Multiplier will return the value of the counter, followed by a reset of the counter value for the register supplied in RegNum.
All other usage: This bit will be treated as "na" on all Read Port Multiplier commands to registers other than Phy event counter registers.

na Field or register is not used.

Success Outputs

Upon successful completion, the ERR bit in the Status register will be cleared, the value in the Error register will be zero, and the value of the specified register will be returned.

Register	**7**	**6**	**5**	**4**	**3**	**2**	**1**	**0**
Error	0							
Sector Count	Value (7:0)							
Sector Count (exp)	Value (39:32)							
Sector Number	Value (15:8)							
Sector Number (exp)	Value (47:40)							
Cylinder Low	Value (23:16)							
Cylinder Low (exp)	Value (55:48)							
Cylinder High	Value (31:24)							
Cylinder High (exp)	Value (63:56)							
Device/Head	Reserved							
Status	BSY	DRDY	DF	na	DRQ	0	0	ERR

Value Set to the 32-bit value read from the register.
BSY = 0 DRDY = 1 DF = 0 DRQ = 0 ERR = 0

Error Outputs

Upon encountering an error, the Port Multiplier will set the ERR bit in the Status register and identify the error code in the Error register.

Register	7	6	5	4	3	2	1	0
Error	Reserved					ABRT	REG	PORT
Sector Count	Reserved							
Sector Count (exp)	Reserved							
Sector Number	Reserved							
Sector Number (exp)	Reserved							
Cylinder Low	Reserved							
Cylinder Low (exp)	Reserved							
Cylinder High	Reserved							
Cylinder High (exp)	Reserved							
Device/Head	Reserved							
Status	BSY	DRDY	DF	na DRQ		0	0	ERR

ABRT = 0
REG Set to one if the register specified is invalid.
PORT Set to one if the port specified is invalid.
na Field or register is not used.
BSY = 0 DRDY = 1 DF = 0 DRQ = 0 ERR = 1

9.8.2 Write Port Multiplier

The Write Port Multiplier command (Table 9.11) is used to write a register on a Port Multiplier. The Write Port Multiplier command must be issued to the control port.

Table 9.11 Write Port Multiplier Inputs

Register	7	6	5	4	3	2	1	0
Features	RegNum[7:0]							
Features (exp)	RegNum[15:8]							
Sector Count	Value (7:0)							
Sector Count (exp)	Value (39:32)							
Sector Number	Value (15:8)							
Sector Number (exp)	Value (47:40)							
Cylinder Low	Value (23:16)							
Cylinder Low (exp)	Value (55:48)							
Cylinder High	Value (31:24)							
Cylinder High (exp)	Value (63:56)							
Device/Head	na	RS1	na	na	PortNum			
Command	E8h							

Value	Set to the 32-bit value to write to the register.
RegNum	Set to number of register to write.
PortNum	Set to the port address that has register to be written.
RS1	Register Specific 1: This bit is register specific. Phy event counter usage: The Write Port Multiplier command may not be used to write to a Phy event counters' identifier value. If this bit is set to one during a write to a Phy event counter register, the counter addressed by RegNum will be written with Value. If cleared to zero during a write to a Phy event counter register, the Port Multiplier will return error status with the ERR bit set to one, and the BSY and DRQ bits cleared to zero in the Status field of the FIS. The ABRT bit will also be set to one in the Error field. All other usage: This bit will be treated as "na" on all Read Port Multiplier commands to registers other than Phy event counter registers.
na	Field or register is not used.

Success Outputs

Upon successful completion, the Port Multiplier will write the specified value to the specified register. The ERR bit in the Status register will be cleared, and the value in the Error register will be zero.

Register	7	6	5	4	3	2	1	0
Error	0							
Sector Count	Reserved							
Sector Count (exp)	Reserved							
Sector Number	Reserved							
Sector Number (exp)	Reserved							
Cylinder Low	Reserved							
Cylinder Low (exp)	Reserved							
Cylinder High	Reserved							
Cylinder High (exp)	Reserved							
Device/Head	Reserved							
Status	BSY	DRDY	DF	na	DRQ	0	0	ERR

na Field or register is not used.
BSY = 0 DRDY = 1 DF = 0 DRQ = 0 ERR = 0

Error Outputs

Upon encountering an error, the Port Multiplier will set the ERR bit in the Status register and identify the error code in the Error register.

Register	7	6	5	4	3	2	1	0
Error	Reserved					ABRT	REG	PORT
Sector Count	Reserved							
Sector Count (exp)	Reserved							
Sector Number	Reserved							
Sector Number (exp)	Reserved							
Cylinder Low	Reserved							
Cylinder Low (exp)	Reserved							
Cylinder High	Reserved							
Cylinder High (exp)	Reserved							
Device/Head	Reserved							
Status	BSY	DRDY	DF	na	DRQ	0	0	ERR

ABRT = 0
REG Set to one if the register specified is invalid.
PORT Set to one if the port specified is invalid.
na Field or register is not used.
BSY = 0 DRDY = 1 DF = 0 DRQ = 0 ERR = 1

Interrupts

The Port Multiplier will generate an interrupt in the status response to a command for the control port by setting the I bit in the Register FIS. The Port Multiplier will not generate an interrupt in response to a software reset for the control port.

9.9 Serial ATA Superset Registers Enhancements

The host controller must provide a means by which software can set the PM Port field in all transmitted FISes. This capability may be exposed to software by minor enhancements to the SControl register. In addition, a means by which software can cause a specific device port to transition to a low power management state must be provided. This capability is exposed to software by adding an additional field called SPM to the SControl.

SControl Register Enhancements

The Serial ATA interface control register (SControl – Figure 9.8) provides a means by which software controls Serial ATA interface capabilities. In order to set the PM Port field, an additional field called PMP is added to the SControl register. A field called SPM has also been added to allow the host to initiate power management states.

Bits	31	30	29	28	27	26	25	24	23	22	21	20	19	18	17	16	15	14	13	12	11	10	09	08	07	06	05	04	03	02	01	00
SCR2	Reserved (0)												PMP				SPM				IPM				SPD				DET			

Figure 9.8 Enhancement to SControl register.

DET As defined in Serial ATA 3.2

SPD As defined in Serial ATA 3.2

IPM As defined in Serial ATA 3.2

SPM The Select Power Management (SPM) field is used to select a power management state. A non-zero value written to this field will cause the power management state specified to be initiated. A value written to this field is treated as a one-shot. This field will be read as "0000."

0000 No power management state transition requested.

0001 Transition to the PARTIAL power management state initiated.

0010 Transition to the SLUMBER power management state initiated.

0100 Transition to the active power management state initiated.

All other values reserved.

PMP The Port Multiplier Port (PMP) field represents the 4-bit value to be placed in the PM Port field of all transmitted FISes. This field is "0000" upon power-up. This field is optional and an HBA implementation may choose to ignore this field if the FIS to be transmitted is constructed via an alternative method.

Reserved All reserved fields will be cleared to zero.

9.9.1 Resets and Software Initialization Sequences

Power-up

On power-up, the Port Multiplier will enter state HPHP1:NoComm in the hot plug state machine for the host port. Upon entering this state, the Port Multiplier will:

1. Clear any internal state and reset all parts of the Port Multiplier hardware
2. Place the reset values in all Port Multiplier registers, including port-specific registers

The reset values will disable all device ports.

After performing this sequence, the Port Multiplier will proceed with the actions described in the hot plug state machine for the host port.

If the Port Multiplier receives a COMRESET from the host before issuing the initial COMINIT signal to the host, the Port Multiplier will immediately perform the actions detailed in COMRESET and cease performing the power-up sequence listed above.

Resets

In Serial ATA 3.2, there are three mechanisms to reset a device: COMRESET, software reset, and the DEVICE RESET command for PACKET devices.

To reset a Port Multiplier, the host must issue a COMRESET. A Port Multiplier does not reset in response to a software reset or a DEVICE RESET command. The specific actions that a Port Multiplier takes in response to each reset type is detailed in the following sections.

COMRESET

When the Port Multiplier receives a COMRESET over the host port, the Port Multiplier will enter state HPHP1:NoComm in the hot plug state machine for the host port.

Upon entering this state, the Port Multiplier will

1. Clear any internal state and reset all parts of the Port Multiplier hardware
2. Place the reset values in all Port Multiplier registers, including port-specific registers

The reset values will disable all device ports.

After performing this sequence, the Port Multiplier will proceed with the actions described in the hot plug state machine for the host port.

There is no functional difference between COMRESET and Power-On Reset.

Software Reset

When the host issues a software reset to the control port, two Register FISes will be sent to the control port as a result. In the first Register FIS, the SRST bit in the Device Control register will be asserted. In the second Register FIS, the SRST bit in the Device Control register will be cleared.

Upon receiving the Register FIS with the SRST bit asserted, the Port Multiplier will wait for the Register FIS that has the SRST bit cleared before issuing a Register FIS with the Port Multiplier signature to the host. The Port Multiplier's behavior will be consistent with the software reset protocol described in the SATA specification. The values to be placed in the Register FIS are listed in Table 9.12.

The Port Multiplier will take no reset actions based on the reception of a software reset. The only action that a Port Multiplier will take is to respond with a Register FIS that includes the Port Multiplier signature. To cause a general Port Multiplier reset, the COMRESET mechanism is used.

Table 9.12 Port Multiplier Signature

Register	7	6	5	4	3	2	1	0
Error	00h							
Sector Count	01h							
Sector Count (exp)	00h							
Sector Number	01h							
Sector Number (exp)	00h							
Cylinder Low	69h							
Cylinder Low (exp)	00h							
Cylinder High	96h							
Cylinder High (exp)	00h							
Device/Head	00h							
Status	BSY	DRDY	DF	na	DRQ	0	0	ERR

BSY = 0   DRDY = 1   DF = 0   DRQ = 0   ERR = 0

Device Reset

A device reset command issued to the control port will be treated as an unsupported command by the Port Multiplier.

9.10 Examples of Software Initialization Sequences

This section details the sequences that host software should take to initialize a Port Multiplier device.

9.10.1 Port Multiplier Aware Software

Port Multiplier aware software will check the host's SStatus register to determine if a device is connected to the port. If a device is connected to the port, the host will issue a software reset to the control port. If the Port Multiplier signature is returned, then a Port Multiplier is attached to the port. Then the host will proceed with the enumeration of the devices on the Port Multiplier ports since Port Multiplier aware software will not require a device to be present on device port 0h in order to determine that a Port Multiplier is present.

Legacy Software

Legacy software will wait for the signature of the attached device to be returned to the host. If a device is present on device port 0h, the device connected to device port 0h will return a Register FIS to the host that contains its signature. If a device is not present on device port 0h, the legacy software will time-out waiting for the signature to be returned and will assume that a device failure has occurred. If fast boot is a requirement, the system should have a Port Multiplier aware BIOS and a Port Multiplier aware OS driver.

When legacy software is loaded, all device ports other than device port 0h are disabled. The host will only receive FISes from the device attached to device port 0h.

Boot Devices Connected to the Port Multiplier

System designers should only connect multiple boot devices to a Port Multiplier if the BIOS is Port Multiplier aware. For example, in a system that contains three bootable devices (hard drive, CD-ROM, and DVD) these devices should only be attached to the Port Multiplier if the BIOS is Port Multiplier aware or the user will not be able to boot off of the devices that are not connected to device port 0.

9.10.2 Port Multiplier Discovery and Device Enumeration

Port Multiplier Discovery

To determine if a Port Multiplier is present, the host performs the following procedure. The host will determine if communication is established on the host's Serial ATA port by checking the host's SStatus register. If a device is present, the host will issue a software reset with the PM Port field set to the control port. The host will check the signature value returned, and if it corresponds to the Port Multiplier Signature, the host knows that a Port Multiplier is present. If the signature value does not correspond to a Port Multiplier, the host may proceed with the normal initialization sequence for that device type.

The host will not rely on a device being attached to device port 0h to determine that a Port Multiplier is present.

When a Port Multiplier receives a software reset to the control port, the Port Multiplier will issue a Register FIS to the host according to the procedure. The signature value contained in the Register FIS is shown in Table 9.13.

Table 9.13 Port Multiplier Signature

	Sector Count	Sector Number	Cylinder Low	Cylinder High	Device
Port Multiplier Signature	01h	01h	69h	96h	00h

Device Enumeration

After discovering a Port Multiplier, the host will enumerate all devices connected to the Port Multiplier. The host will read GSCR[2] to determine the number of device ports on the Port Multiplier.

For each device port on the Port Multiplier, the host performs the following procedure to enumerate a device connected to that port:

1. The host will enable the device port. The host can enable a device port by setting the DET field appropriately in the device port's PSCR[2] (SControl) register. The host uses the Write Port Multiplier command to write PSCR[2] (SControl) for the device port to be enabled.
2. The host should allow for communication to be established and device presence to be detected after enabling a device port.
3. The host reads PSCR[0] (SStatus) for the device port using the Read Port Multiplier command. If PSCR[0] (SStatus) indicates that a device is present, the host queries PSCR[1] (SError) for the device port and clears the X bit indicating device presence has changed.
4. The signature generated by the device as a consequence of the initial COMRESET to the device port may be discarded if the host does not support context switching because the BSY bit may be clear when the Register FIS is received by the host. Therefore, to determine the signature of the attached device, the host should issue a software reset to the device port. If a valid signature is returned for a recognized device, the host may then proceed with normal initialization for that device type.

Cascading Port Multipliers will not be supported.

9.11 Switching Type Examples

The host may use two different switching types depending on the capabilities of the host controller.

1. If the host controller supports hardware context switching based on the value of the PM Port field in a received FIS, then the host may have commands outstanding to multiple devices at the same time.
   - This switching type is called *FIS-based switching* and results in FISes being delivered to the host from any of the devices that have commands outstanding to them.
2. If the host controller does not support hardware context switching based on the value of the PM Port field in a received FIS, then the host may only have commands outstanding to one device at any point in time.
   - This switching type is called *command-based switching*, and it results in FISes being delivered to the host only from the one device that has commands outstanding to it.

The Port Multiplier's operation is the same, regardless of the switching type used by the host.

9.11.1 Command-Based Switching

Host controllers that do not support hardware context switching utilize a switching type called command-based switching.

- To use command-based switching, the host controller has commands outstanding to only one device at any point in time.
- By only issuing commands to one device at a time, the result is that the Port Multiplier only delivers FISes from that device.

A host controller may support command-based switching by implementing the Port Multiplier Port (PMP) field in the SControl register.

- In order to use this mechanism, host software will set the PMP field appropriately before issuing a command to a device connected to the Port Multiplier.
- When host software has completed the commands with a particular device port, it will modify the PMP field before issuing commands to any other device port.
- Note that the PMP field must be set to the control port when host software would like to issue commands to the Port Multiplier itself (such as Read Port Multiplier or Write Port Multiplier).

9.11.2 FIS-Based Switching

Host controllers that support hardware context switching may utilize a switching type called FIS-based switching.

- FIS-based switching allows the host controller to have commands outstanding to multiple devices at any point in time.
- When commands are outstanding to multiple devices, the Port Multiplier may deliver FISes from any device with commands outstanding.

Host Controller Requirements

To support FIS-based switching:

1. The host controller must have context switching support.
2. The host controller must provide a means for exposing a programming interface for up to 16 devices on a single port.
3. The context switching support must comprehend not only Control Block register context but also DMA engine context and the SActive register.
4. The host controller must be able to update the context for a particular device even if the programming interface for that device is not currently selected by the host software.
5. The host controller must fill in the PM Port field in hardware-assisted FIS transmissions.
    - For example, if the host controller receives a DMA Activate from a device, it must be able to construct a Data FIS with the PM Port field set to the value in the received DMA Activate FIS.

9.12 Port Selectors

A Port Selector is a mechanism that allows two different host ports to connect to the same device in order to create a redundant path to that device (see Figure 9.9). In combination with configuring disks into RAID volumes, the Port Selector allows system providers to build fully redundant solutions. The upstream ports of a Port Selector can also be attached to a Port Multiplier to provide redundancy in a more complex topology. A Port Selector can be thought of as a simple multiplexer that provides two paths to a single SATA device.

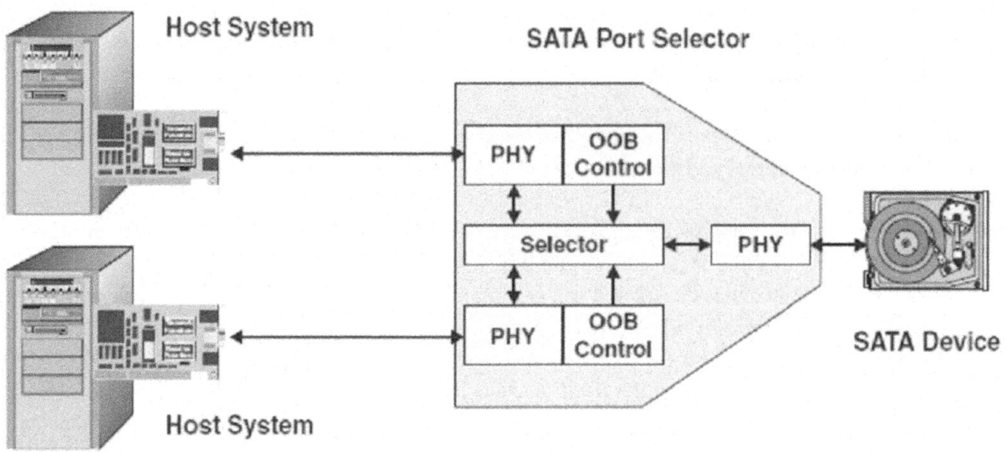

Figure 9.9 Port selector basic diagram.

9.12.1 Port Selector Definitions, Abbreviations, and Conventions

Definitions and Abbreviations

The terminology used in this specification is consistent with the terminology used in the SATA and Serial ATA II specifications, and all definitions and abbreviations defined in those specifications are used consistently in this document. Additional terms and abbreviations introduced in this specification are defined in the following sections.

Active port	The active port is the currently selected host port on the Port Selector.
Device port	The device port is the port on the Port Selector that is connected to the device.
Host port	The host port is a port on the Port Selector that is connected to a host. There are two host ports on a Port Selector.
Inactive port	The inactive port is the host port that is not currently selected on the Port Selector.
Protocol-based port selection	Protocol-based port selection is a method that may be used by a host to select the host port that is active. Protocol-based port selection uses a sequence of Serial ATA out-of-band Phy signals to select the active host port.
Quiescent power condition	Entering a quiescent power condition for a particular Phy is defined as the Phy entering the idle bus condition.
Side-band port selection	Side-band port selection is a method that may be used by a host to select the host port that is active. Side-band port selection uses a mechanism that is outside of the Serial ATA protocol for determining which host port is active. The port selection mechanism used in implementations that support side-band port selection is outside the scope of this specification.

9.12.2 Port Selector Overview

A Port Selector is a mechanism that allows two different host ports to connect to the same device in order to create a redundant paths to the device. This is done to remove single points of failures for applications that cannot afford excessive downtime.

- Only one host connection to the device is active at a time.
- The effective use of a Port Selector requires coordinated access to the device between the two host ports.
- The host(s) must coordinate to determine which host port should be in control of the device at any given point in time.
- Definition of the coordination mechanism or protocol is beyond the scope of this specification.

Once the host(s) determines the host port that should be in control of the device, the host that contains the host port to be made active will take control of the device by selecting that host port to be active.

- The active host selects a port to be active by using either a protocol-based or side-band port selection mechanism.
- A side-band port selection mechanism can be as simple as a hardware select line that is pulled high to activate one host port and low to activate the other.
- The side-band port selection mechanism is outside the scope of this specification. A protocol-based port selection mechanism uses the Serial ATA protocol to cause a switch of the active host port.

This specification defines a protocol-based port selection mechanism that uses a particular Morse coding of COMRESET signals to cause a switch of active host port.

- A Port Selector will only support one selection mechanism at any point in time.
- The externally visible behavior of a Port Selector is the same regardless of whether a protocol-based or side-band port selection mechanism is used.

A Port Selector that supports protocol-based port selection can be detected in the signal path if the optional presence detection feature is supported by the Port Selector and the host has an enhanced SError register that can latch this event. The detection mechanism for a Port Selector that supports side-band port selection is outside the scope of this specification.

9.12.3 Active Port Selection

The Port Selector has a single host port active at a time. The Port Selector will support one of two mechanisms for determining which of the two host ports is active.

1. The first mechanism is called side-band port selection. Side-band port selection uses a mechanism outside of the Serial ATA protocol for determining which host port is active.
2. The second mechanism is called protocol-based port selection. Protocol-based port selection uses a sequence of Serial ATA out-of-band Phy signals to select the active host port.

Whether a protocol-based or side-band port selection mechanism is used, the Port Selector will exhibit the behavior defined within this specification.

After selection of a new active port, the device is in an unknown state.

- The device may have active commands outstanding from the previous active host that need to be flushed.
- After an active port switch has been performed, it is strongly suggested that the active host issue a COMRESET to the device to ensure that the device is in a known state.

Protocol-Based Port Selection

Protocol-based port selection is an active port selection mechanism that uses a sequence of Serial ATA out-of-band Phy signals to select the active host port.

1. A Port Selector that supports protocol-based port selection will have no active host port selected upon power-up.
2. The first COMRESET or COMWAKE received over a host port will select that host port as active.
3. The host may then issue explicit switch signals to change the active host port.
4. Reception of the protocol-based port selection signal on the inactive host port causes the Port Selector to deselect the currently active host port and select the host port over which the selection signal is received.
5. The protocol-based port selection signal is defined such that it can be generated using the SATA superset Status and Control Registers and such that it can be received and decoded without the need for the Port Selector to include a full Link or Transport layer (i.e., direct Phy detection of the signal).

Port Selection Signal Definition

The port selection signal is based on a pattern of COMRESET out-of-band signals transmitted from the host to the Port Selector. As illustrated in Figure 9.10, the Port Selector will qualify only the timing from the assertion of a COMRESET signal to the assertion of the next COMRESET signal in detecting the port selection signal.

The port selection signal is defined as a series of COMRESET signals with the timing from the assertion of one COMRESET signal to the assertion of the next as defined in Table 9.14 and illustrated in Figure 9.11.

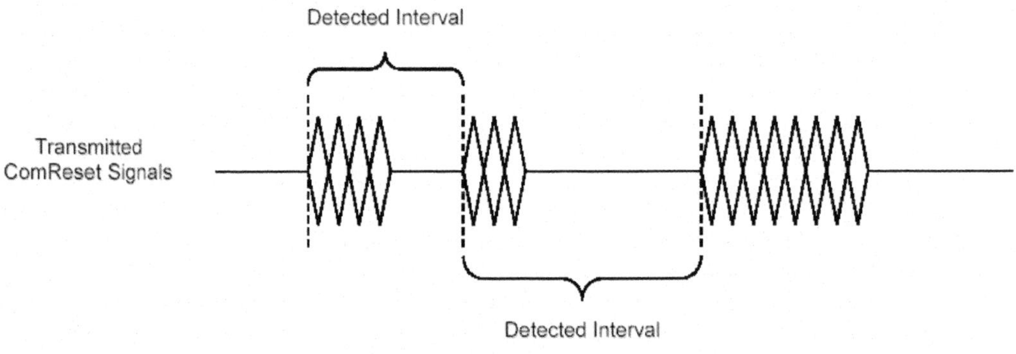

Figure 9.10 Port selection signal detection. (*Source*: www.sata-io.org)

Table 9.14 Port Selection Signal Inter-Reset Timing Requirements

	Nom	Min	Max	Units	Comments
T1	2.0	1.6	2.4	ms	Inter-reset assertion delay for first event of the selection sequence
T2	8.0	7.6	8.4	ms	Inter-reset assertion delay for the second event of the selection sequence

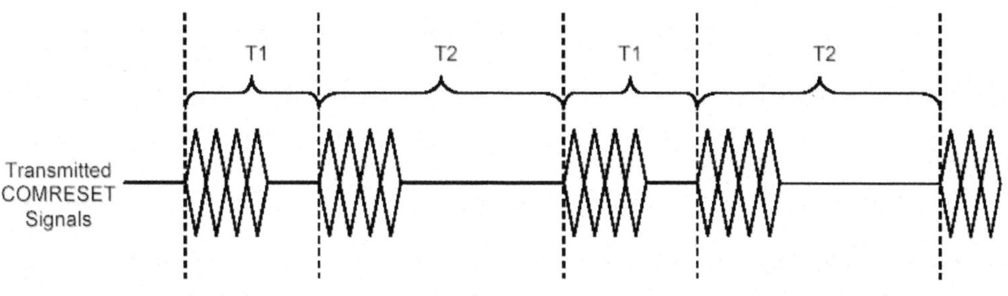

Figure 9.11 Complete port selection signal consisting of two sequences. (*Source*: www.sata-io.org)

1. The Port Selector will select the port, if inactive, on the de-assertion of COMRESET after receiving two complete back-to-back sequences with specified inter-burst spacing over that port (i.e., two sequences of two COMRESET intervals comprising a total of five COMRESET bursts with four inter-burst delays).
2. Specifically, after receiving a valid port selection signal, the Port Selector will not select that port to be active until the entire fifth COMRESET burst has been de-asserted.
3. The Port Selector is only required to recognize the port selection signal over an inactive port.
4. Reception of COMRESET signals over an active port is propagated to the device without any action taken by the Port Selector, even if the COMRESET signals constitute a port selection signal.
5. This may result in multiple device resets.

The timings detailed in Table 9.14 will be independent of the signaling speed used on the link. For example, the inter-reset timings are the same for links using Gen1 or Gen2 speeds.

The interpretation and detection of the COMRESET signal by the Port Selector is in accordance with the SATA definition, that is, the COMRESET signal is detected upon receipt of the fourth burst that complies with the COMRESET signal timing definition. The inter-reset timings referred to here for the port selection signal are from the detection of a valid COMRESET signal to the next detection of such a signal and are not related to the bursts that comprise the COMRESET signal itself.

Presence Detection

Presence detection is the ability for a host to detect that a Port Selector is present on a port.

- If a Port Selector supports presence detection capabilities, a host will be able to determine whether a Port Selector is connected to a host port.
- The host may determine this regardless of whether the Port Selector port to which it is connected is the active or inactive link.

Presence detection capabilities are defined for Port Selectors utilizing protocol-based port selection only. Systems utilizing side-band port selection must be preconfigured for side-band port selection, and, therefore, those systems can also be preconfigured to support presence detection. Presence detection is an optional feature of protocol-based Port Selectors.

SError Register Enhancement for Presence Detection

The Serial ATA interface error register (SError) as defined in the SATA specification and Serial ATA II: Extensions to Serial ATA 1.0 specification includes indications for various events that may have occurred on the interface such as a change in the PhyRdy state, detection of disparity, errors, and so on.

In order to facilitate a means for notifying host software that a Port Selector presence detection signal was received, an additional bit in the DIAG field of the SError superset register is defined, as indicated in Table 9.15.

Table 9.15 Location of A Bit in DIAG Field

Bits	31	30	29	28	27 26	25	24	23	22	21	20	19	18	17	16	15	14	13	12	11	10	09	08	07	06	05	04	03	02	01	00
SCR1	DIAG															ERR															
		R		A	X	F	T	S	H	C	D	B	W	I	N	R	R	R	R	E	P	C	T	R	R	R	R	R	R	M	I

A Port Selector presence detected.

This bit is set to one when COMWAKE is received while the host is in state HP2: HR_AwaitCOMINIT. Upon power-up reset, this bit is cleared to 0. The bit is cleared to 0 when the host writes a "1" to this bit location.

9.12.4 Side-Band Port Selection

The active host port may be selected by a side-band mechanism.

- Side-band port selection uses a mechanism outside of the Serial ATA protocol for selecting which host port is active.
- One example of a side-band port selection mechanism is a hardware select line.
- The side-band selection mechanism used is outside the scope of this specification.
- A Port Selector that supports side-band port selection will exhibit the behavior defined within this specification.

9.12.5 Behavior During a Change of Active Port

During a change of active port, the previous host connection is broken, and all internal states other than the active host port are initialized before the connection with the new active host is made.

When a new active host port is selected, the Port Selector will perform the following procedure:

1. The Port Selector will stop transmitting and enter the quiescent power condition on the previously active host port Phy (now the inactive host port).
2. The Port Selector will initialize all internal states other than the state of the selection bit for the active host port.

3. The Port Selector will enter the active power condition on the new active host port.
4. The Port Selector will allow out-of-band and in-band traffic to proceed between the new active host port and the device.

Device State after a Change of Active Port

A Port Selector may support an orderly switch to a new active host port.

- A Port Selector that supports an orderly switch ensures that primitive alignment with the device Phy is maintained during the switch to the new active host port.
- Maintaining primitive alignment ensures that PhyRdy remains present between the Port Selector and the device throughout the switch to the new active host port.

After selection of a new active port, the device may be in an unknown state.

- The device may have active commands outstanding from the previously active host port that need to be flushed.
- The new active host can issue a COMRESET to the device in order to return the device to a known state.

9.13 Behavior and Policies

9.13.1 Control State Machine

The Port Selector Control state machine is based on a model of a Port Selector consisting of three Serial ATA Phys interconnected and controlled by an overall control logic block, as depicted in Figure 9.12. For convenience, the three ports of a Port Selector are abbreviated "A," "B," and "D," corresponding to Host Port A, Host Port B, and the Device Port, respectively.

The Phy states depicted in Figure 9.12 are presumed to have the basic capabilities and controls indicated in Figure 9.13.

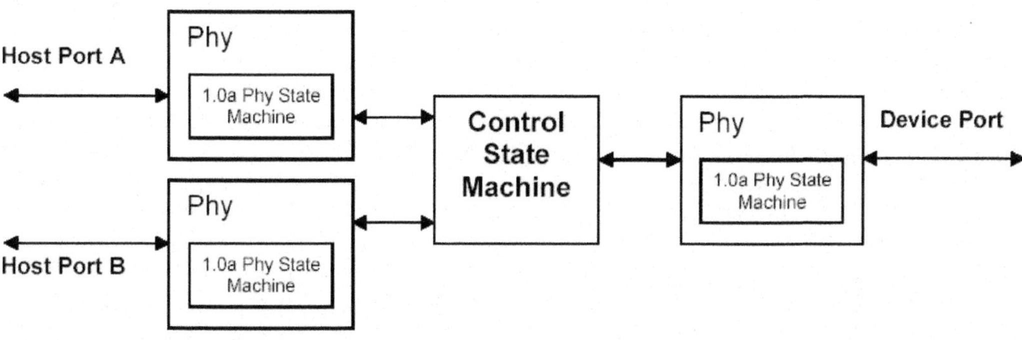

Figure 9.12 Control State Machine. (*Source:* www.sata-io.org)

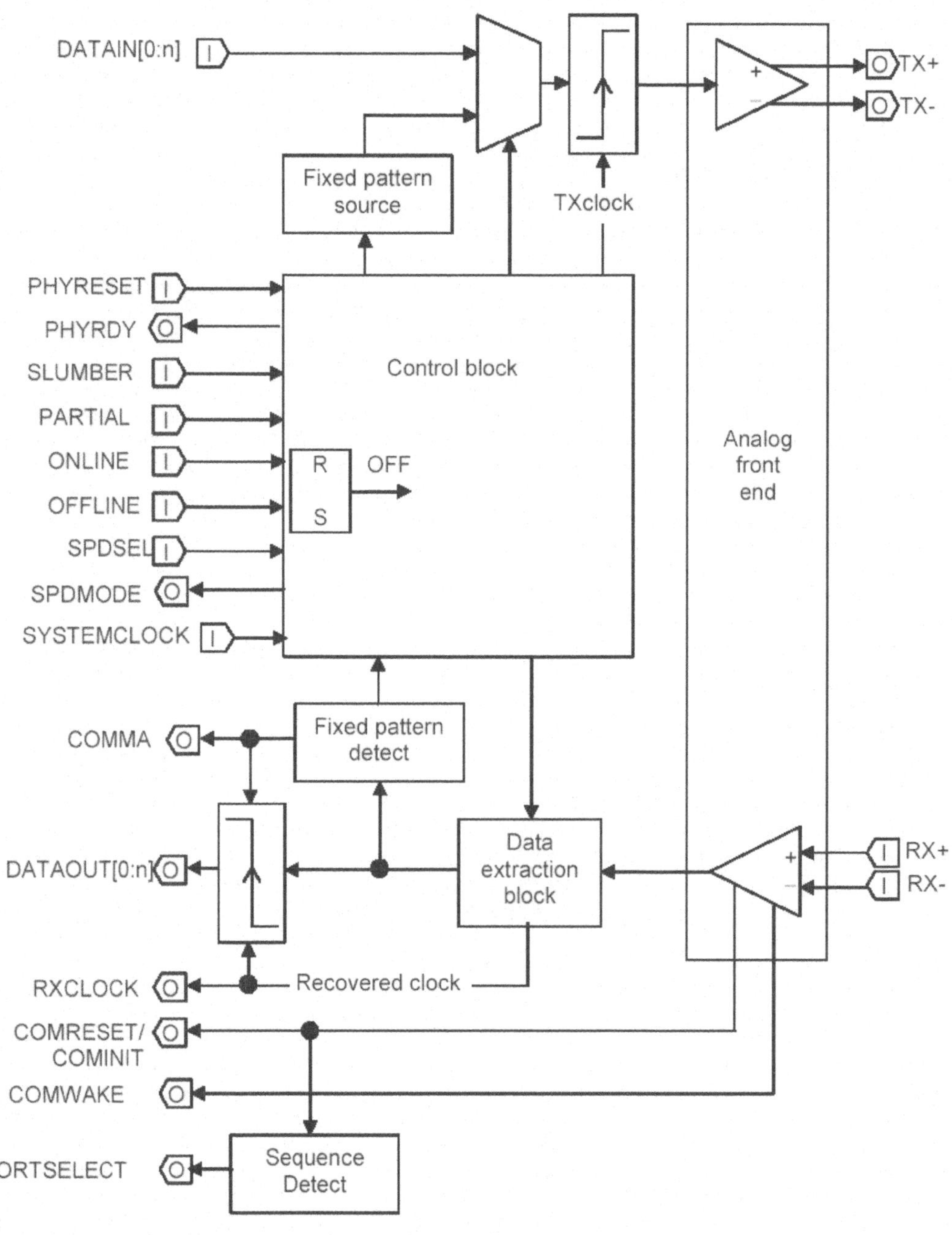

Figure 9.13 Phy Block Diagram. (*Source:* www.sata-io.org)

Changes to the original SATA Phy are as follows:

- Loopback controls were removed since those controls were not relevant in this state machine.
- Explicit ONLINE and OFFLINE signals were added, including a register latch for these signals so that the signals do not need to be asserted in every state. The SATA specification specifies a

mechanism for the host to put the Phy in offline mode using the SControl register. Therefore, it is reasonable to expect that the Phy has signals ONLINE and OFFLINE that can be utilized.

- A PORTSELECT signal was added. A Port Selector using protocol-based port selection will set the PORTSELECT signal to the output of the Sequence Detect block. The Sequence Detect block is asserted when the protocol-based selection signal is received; otherwise, the signal is de-asserted. A Port Selector using side-band port selection will set the PORTSELECT signal when a change in active port is requested.

9.14 BIST Support

A Port Selector is not required to support the BIST Activate FIS. The resultant behavior of sending a BIST Activate FIS through a Port Selector is undefined.

9.14.1 Flow Control Signaling Latency

The Port Selector must satisfy the flow control signaling latency specified in the SATA specification. The Port Selector will ensure that the flow control signaling latency is met on a per link basis. Specifically, the Port Selector will ensure that the flow control signaling latency is met between the following:

1. The Port Selector active host port and the host to which it is connected.
2. The Port Selector device port and the device to which it is connected.

The Port Selector will not reduce the flow control signaling latency budget of the active host it is connected to or the device it is connected to.

9.14.2 Power Management

The Port Selector must maintain the active host port across power management events and only allow an active host port change after receiving a valid port selection signal.

- The Phy on the inactive host port will be in the quiescent power condition. Upon detecting that the PhyRdy signal is not present for the active host port or device port, the Port Selector will place that Phy in a quiescent power condition.
- If the PhyRdy signal is not present between the device and the Port Selector, the Phy connected to the active host port will enter the quiescent power condition and squelch the Phy transmitter.
- If the PhyRdy signal is not present between the active host and the Port Selector, the Phy connected to the device will enter the quiescent power condition and squelch the Phy transmitter. During these periods while PhyRdy is not present, OOB signals will still be propagated between the active host and the device to ensure that communication can be established.

If the Port Selector is able to determine that the active host and device negotiated a SLUMBER power management transition, the Port Selector may recover from the quiescent power condition in the time defined by the SLUMBER power state. If the Port Selector is not able to determine the power state entered by the host and device, the Port Selector will recover from the quiescent power condition in the time defined by the PARTIAL power state.

Wakeup Budget

The wake-up budget out of PARTIAL or SLUMBER may increase when a Port Selector is connected to a device. When the active host Phy comes out of a low power condition, the Port Selector active host Phy may wake up before causing the Port Selector device Phy to wake up, which, in turn, will wake up the device. The host will allow the device at least 20 microseconds to wake up from the PARTIAL power management state.

Out of Band (OOB) Phy signals

The Port Selector will propagate the COMRESET received from the active host to the device. The Port Selector will propagate COMINIT received from the device to the active host port. If no active host port is selected, the Port Selector will propagate COMINIT received from the device to the active host port. The Port Selector is allowed to delay the delivery of propagated OOB signals.

The Port Selector will not respond to COMRESET signals received over the inactive host port. The inactive host port Phy will remain in the quiescent power condition when COMRESET is received over the inactive host port.

Hot Plug

The Port Selector will only generate a COMINIT over a host port when a COMINIT signal is received from the device or as part of an active speed negotiation. If a drive connected to a Port Selector is hot plugged, the drive will issue a COMINIT sequence as part of its normal power-up sequence in accordance with Serial ATA 3.2. The Port Selector will propagate the COMINIT over the active host port (or both host ports, if both host ports are inactive). The host will see the COMINIT signal and then interrogate the port to determine whether a drive is attached.

If a drive connected to a Port Selector is hot unplugged, the Port Selector will squelch the transmitter for the active host port. The active host will determine that the PhyRdy signal is no longer present and that there is no longer a drive present.

Speed Negotiation

Speed is negotiated on a per link basis. Specifically, the Port Selector will negotiate speed between the following:

1. The Port Selector active host port and the host to which it is connected.
2. The Port Selector device port and the device to which it is connected.

The Port Selector starts speed negotiation at the highest speed rate that it supports. The Port Selector then negotiates speed on each link to the appropriate supported speed. After negotiating speed on the active host link and on the device link, the Port Selector will check whether the two speeds match. If the speeds do not match, the Port Selector limits the maximum speed rate it supports to the lower of the two speeds negotiated. Then the Port Selector forces speed to be renegotiated to reach a common speed rate.

Spread Spectrum Clocking

The Port Selector will support spread spectrum clocking receive on all of its ports. The Port Selector may support spread spectrum transmit. It is recommended that a configuration jumper be used to enable/disable spread spectrum clocking if it is settable. There is no means within the Serial ATA protocol provided to enable/disable spread spectrum clocking, if it is statically configurable.

If spread spectrum clocking is used, the spreading domain between the host and the Port Selector is not required to be the same as the spreading domain between the device and the Port Selector. The signals passing through a Port Selector may be re-spread.

9.15 Power-up and Resets

9.15.1 Power-up

Upon power-up, the Port Selector will reset all internal state, including the active host port. This will cause no active host port to be selected when protocol-based port selection is used.

Presence Detection of Port Selector

For protocol-based port selection, presence detection can be performed using the optional mechanism. Presence detection for Port Selectors implementing sideband port selection is outside the scope of this specification.

9.15.2 Resets

COMRESET

When COMRESET is received over the active host port, the Port Selector will reset all internal states, the active host port will remain unchanged, the maximum speed will remain unchanged, and the COMRESET signal will be propagated to the device. The Port Selector will take no reset action upon receiving a COMRESET signal over the inactive host port, when there is no active host port selected after power-up.

Software Reset and DEVICE RESET

The Port Selector will not reset in response to receiving a Software reset or the DEVICE RESET command.

9.16 Host Implementation

9.16.1 Software Method for Protocol-Based Selection (Informative)

The preferred software method for producing a protocol-based port selection signal is detailed in this section. Software for HBAs that implement the SControl and SStatus registers may use this method to create the protocol-based port selection signal.

If the Phy is left on for long periods during the generation of the sequence, a hardware-based COMRESET polling algorithm may interfere and corrupt the sequence. This method tries to minimize any impact of host-based COMRESET polling algorithms by leaving the Phy online for very short sequences during the creation of the signal. The procedure outlined is appropriate for any HBA design that has a COMRESET polling interval greater than 25 microseconds.

1. Set Phy to offline by writing SControl.DET to 4h.
2. Set Phy to reset state by writing SControl.DET to 1h.
3. Wait 5 microseconds to allow charging time for DC-coupled Phy designs.
4. Set Phy to online state by writing SControl.DET to 0h.
5. Wait 20 microseconds to allow COMRESET burst to be transmitted to the device.
6. Set Phy to offline by writing SControl DET to 4h.
7. Wait 1.975 milliseconds to satisfy T1 timing.
8. Repeat steps 2–6.
9. Wait 7.975 milliseconds to satisfy T2 timing.
10. Repeat steps 2–6.
11. Wait 1.975 milliseconds to satisfy T1 timing.
12. Repeat steps 2–6
13. Wait 7.975 milliseconds to satisfy T2 timing.
14. Set Phy to reset state by writing SControl.DET to 1h.
15. Wait 5 microseconds to allow charging time for DC-coupled Phy designs.
16. Set Phy to online state by writing SControl.DET to 0h.
17. Wait up to 10 milliseconds for SStatus.DET = 3h.
18. If SStatus.DET ! = 3h, go to step 1 to restart the process.

9.17 Summary

The Port Multiplier (PM – Figure 9.14) is a device that provides connectivity to the basic SATA architecture.

- PMs were made to allow for up to 15 devices to be accessed through a single host SATA port.
- SATA PMs are similar to SAS expanders and Fibre Channel loop switches in that these devices provide connectivity to point-to-point serial technologies.

Figure 9.14 Port Multiplier (PM) diagram.

There are three ports associated with PMs.

1. Control port
   - The Port Multiplier has one port address reserved for control and status communication with the Port Multiplier itself. The control port has port address Fh.
2. Device port(s)
   - A device port is a port that can be used to connect a device (a disk or SSD) to the PM.
   - A PM may support up to 15 device ports.
   - Device port addresses are sequentially numbered, starting at zero and until all device ports have port addresses.
   - The device port address has a value less than the total number of device ports supported by the PM.
3. Host port
   - The host port is the port that is used to connect the PM to the host. There is only one active host port.

The Port Multiplier uses four bits, known as the PM Port field (see Table 9.16), in all FIS types to route FISes between the host and the appropriate device.

1. Using the PM Port field, the Port Multiplier can route FISes to up to 15 Serial ATA devices from one active host.
2. The PM Port field is filled in by the host on a host to device FIS with the port address of the device to route the FIS to.
3. For a device to host FIS, the PM Port field is filled in by the Port Multiplier with the port address of the device that is transmitting the FIS.

Table 9.16 PM Port Field

	31	30	29	28	27	26	25	24	23	22	21	20	19	18	17	16	15	14	13	12	11	10	9	8	7	6	5	4	3	2	1	0
DW0	See specification								See specification								x	x	x	x	**PM Port**				FIS Type (xxh)							

The Port Multiplier registers are accessed using READ PORT MULTIPLIER/WRITE PORT MULTIPLIER commands issued to the control port.

The Static Configuration Information section of the General Status and Control Registers (GSCR) contains registers that are static throughout the operation of the Port Multiplier (see Table 9.17).

- These registers are read only and may not be modified by the host.
- The registers contain Device ID, Vendor ID, PM revision level, and number of fan-out ports.

Table 9.17 PM PSCRs

PSCR[0]	SStatus register
PSCR[1]	SError register
PSCR[2]	SControl register
PSCR[3]	SActive register (not implemented)
PSCR[256] – PSCR[2303]	Phy event counter registers

The Port Multiplier will maintain a set of Port Status and Control Registers (PSCRs) for each device port that it supports.

Port Multiplier aware software and drivers are required to check the host's SStatus register to determine if a device is connected to the port.

- If a device is connected to the port, the host will issue a software reset to the control port.
- If the Port Multiplier signature is returned, then a Port Multiplier is attached to the port.

	Sector Count	Sector Number	Cylinder Low	Cylinder High	Device
Port Multiplier Signature	01h	01h	69h	96h	00h

- Then the host will proceed with the enumeration of the devices on the Port Multiplier ports.
- Port Multiplier aware software will not require a device to be present on device port 0h in order to determine that a Port Multiplier is present.

A Port Selector is a mechanism that allows two different host ports to connect to the same device in order to create a redundant path to that device.

- In combination with configuring disks into RAID volumes, the Port Selector allows system providers to build fully redundant solutions.
- The upstream ports of a Port Selector can also be attached to a Port Multiplier to provide redundancy in a more complex topology.
- A Port Selector (Figure 9.15) can be thought of as a simple multiplexer that provides two paths to a single SATA device.

Port Selector Ports and Terms

Active port	The active port is the currently selected host port on the Port Selector.
Device port	The device port is the port on the Port Selector that is connected to the device.
Host port	The host port is a port on the Port Selector that is connected to a host. There are two host ports on a Port Selector.
Inactive port	The inactive port is the host port that is not currently selected on the Port Selector.
Protocol-based port selection	Protocol-based port selection is a method that may be used by a host to select the host port that is active. Protocol-based port selection uses a sequence of Serial ATA out-of-band Phy signals to select the active host port.
Quiescent power condition	Entering a quiescent power condition for a particular Phy is defined as the Phy entering the idle bus condition.
Side-band port selection	Side-band port selection is a method that may be used by a host to select the host port that is active. Side-band port selection uses a mechanism that is outside of the Serial ATA protocol for determining which host port is active. The port selection mechanism used in implementations that support side-band port selection is outside the scope of this specification.

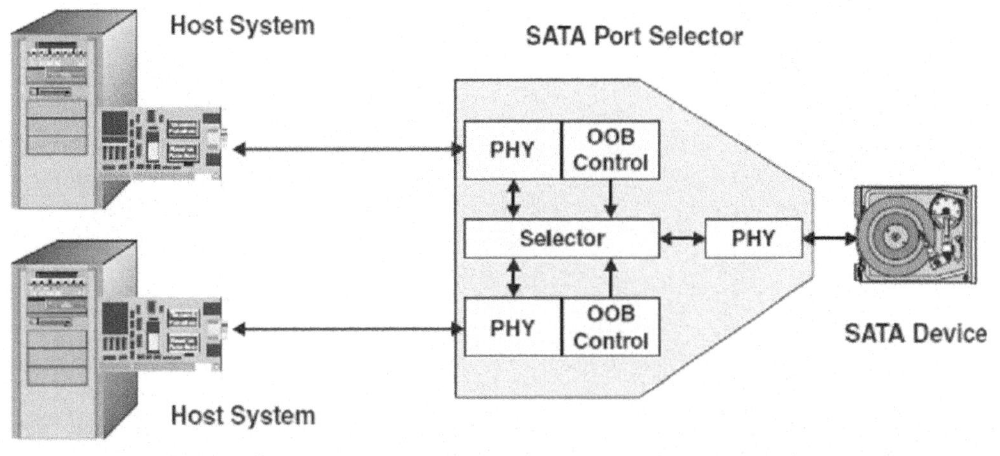

Figure 9.15 PS diagram.

Chapter 10

Software and Drivers

Objectives

This chapter covers the basic operational characteristics of the storage software stack. Whereas previous sections have concentrated on how to physically connect SATA devices and their architecture-associated protocols, this chapter addresses the most common method to create a SATA-based host implementation. Topics include the following:

- Storage I/O Stack
  - I/O manager
  - File system, volume manager, and volumes
  - Class and miniport drivers
  - Hardware: PCI bus, HBAs, and Devices
- SATA and SATA Express architectures
  - SATA protocol layers
  - PCIe protocol layers
- SATA configurations
- The Advanced Host Channel Interface (AHCI) Specification, which has been implemented in all SATA applications since 2005
- System memory structure:
  - General status and control
  - Table of Command Lists
- HBA configuration registers
- PCI Header and Configuration Space
- Memory registers
  - Generic
  - Port control
- Port register map
- HBA Memory Space Usage
- Port memory usage
  - Received FIS structure
  - Command List

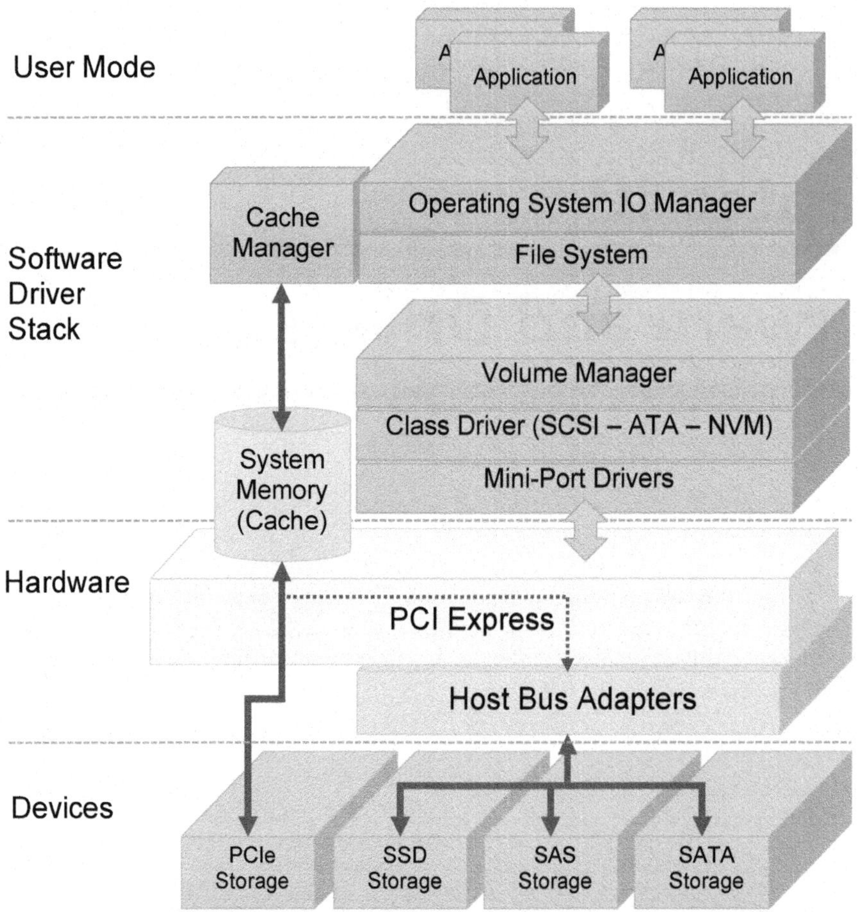

Figure 10.1 Storage software and hardware stack. (*Source:* Adapted from SNIA SSSI - PCIe Round Table, SNIA Symposium, January 2012)

10.1 Storage I/O Stack

Storage I/O must traverse many layers of software and hardware stacks as seen in Figure 10.1.

- I/Os are subject to caching mechanisms, OS task switching and timing, and driver fragmentation and coalescing
- I/O can be different at the Device and System level
  - Open file = any number of READ commands
- I/O can lose 1:1 correspondence from the original I/O and physical device I/O
- Performance is heavily influenced by the efficiency of the I/O stack and streamlining drivers for direct system memory transfers

When an application user needs to open or close a file, many software and hardware mechanisms get involved in the request. Starting with user mouse clicks or key strokes, the Application sends a request to the operating system (OS) to either open (send READ commands) or save (send WRITE commands) a file. From that point the I/O Manager takes control and performs the operations necessary to complete the user request. The sections below provide descriptions of the key components in an I/O storage stack.

10.1.1 I/O Manager

The I/O management module in the OS provides a means by which a process can communicate with the outside world, that is, the mechanism for input and output information.

- Traditionally, the I/O system is regarded as the most difficult to implement because of the different number of peripherals that can be used with a different configuration.
- In storage applications, matters are complicated by the following:
    - The numerous methods available to access storage devices (file, block, raw, object, etc.)
    - The command sets used for host-device communication (ATA, SCSI, NVMe)
    - The different types of hardware interfaces (PCIe, SATA, SAS, etc.)

10.1.2 File System

The file system is a software component that maps the logical or physical storage space (blocks, records, sectors) to named objects called files.

- File systems are normally supplied as operating system components.
- Each operating system has one or more native file systems.
    - These include the 16- or 32-bit File Allocation Table (FAT-16, FAT-32), the New Technology File System (NTFS), the Unix File System (UFS), the Hierarchical File System (HFS).
- File systems can also be implemented and marketed as independent software components.

10.1.3 Volume Manager

Volumes are implemented by a device driver called a volume manager.

- Examples include the FtDisk Manager, the Logical Disk Manager (LDM), and the VERITAS Logical Volume Manager (LVM).
- Volume managers provide a layer of physical abstraction, data protection (using some form of RAID), and performance.

Volumes

The highest level of organization in the file system is the volume.

- A file system resides on a volume.
- A volume contains at least one partition, which is a logical division of a physical disk(s).
- A volume that contains data that exists on one partition is called a simple volume, and a volume that contains data that exists on more than one partition is called a multi-partition volume.

Class Driver

A class driver is a type of hardware device driver that can operate a number of different devices of similar type.

- For example, a SCSI class driver can be used to communicate with SCSI devices across SAS or FC fabrics.
- A SATA class driver (AHCI) can use either SATA or PCIe (SATA Express) physical interfaces.
- Same command set—different hardware interfaces.

Miniport Drivers

Miniport drivers directly manage storage host bus adapters that are installed on a device. Miniport drivers perform numerous interface-specific functions:

- Discovery
  - Scan SCSI bus now becomes a query of the Name server in FC or reading Routing tables in SAS.
- Configuration
  - In FC and iSCSI applications, HBA/NIC must perform a login process and configuration details.
- Command delivery
  - Mapping of SCSI/ATA/NVMe commands into interface-dependent delivery model.
- Error handling and recovery

10.1.4 Hardware

PCIe bus

All server and host applications use PCI Express as the internal computer bus.

- PCIe uses a switch-based fabric that provides multiple endpoint connections within the host.
- It is based on PCI-SIG specifications with bandwidths up to 64 GBps.
- It utilizes standard PCI registers for communications between HBA, memory, and CPU.

HBA

The HBA provides the interface between the PCIe host bus interface and the storage device interface (PCIe ⇔ FC or PCIe ⇔ SAS).

- Some converged network adapters (CNAs) can combine TCP/IP, iSCSI, NAS, and FCoE through the same hardware.
- The SATA simple design allows the entire host function within a single integrated circuit.

Devices

Different devices serve different application requirements. For example, if your application requires 100TB of storage, using SSDs may prove to be cost prohibitive. Likewise, if your application requires 10,000 IOps then any device with rotational media would be out of the question. The following list provides a basic view of storage device types and their current availability status.

- PCIe-based storage.
  - This is still in its infancy (slope of inflated expectations) as of 2014.
  - Options include SATA Express, NVM Express, and SCSI Express.
- SATAe is the current leader in next generation PCIe-based storage; however, in the long-run Client class storage market share will be influenced or completely dominated by NVMe devices.
  - NVMe is quickly gaining market share in server market.
- SCSIe is made up of SCSI over PCIe (SOP) and PCIe Queuing Interface (PQI) standards.
  - SCSIe devices are currently in the development phase.

- SAS will remain mainstay in enterprise class HDD for the next 10 years.
    - SAS can accommodate SATA devices.
    - SAS hardware enables multiple storage tiers in same array (high-performance versus high capacity).
- SATA will provide storage for most Client applications until NVMe gains traction.
- Keep in mind that NVMe was built from the ground up for SSD.
    - SATA has always provided mass storage device tier.

10.2 SATA and SATA Express

Standard software drivers can be used regardless of which physical connection is made to the storage device. The majority of SATA-based storage applications utilize the Advanced Host Channel Interface (AHCI) specification to write device drivers. AHCI describes the following:

- HBA PCI Configuration register mapping, Power Management capabilities, Message Interrupt capabilities, Global and Port Control registers, and their use for device communication.
- How standard ATA/ATAPI command sets are mapped into memory use models, Command List structures, received FIS structures, and physical region descriptors.

SATA and SATA Express systems use the ATA or ATAPI command sets regardless of the physical interface that carries information to the device (see Figure 10.2).

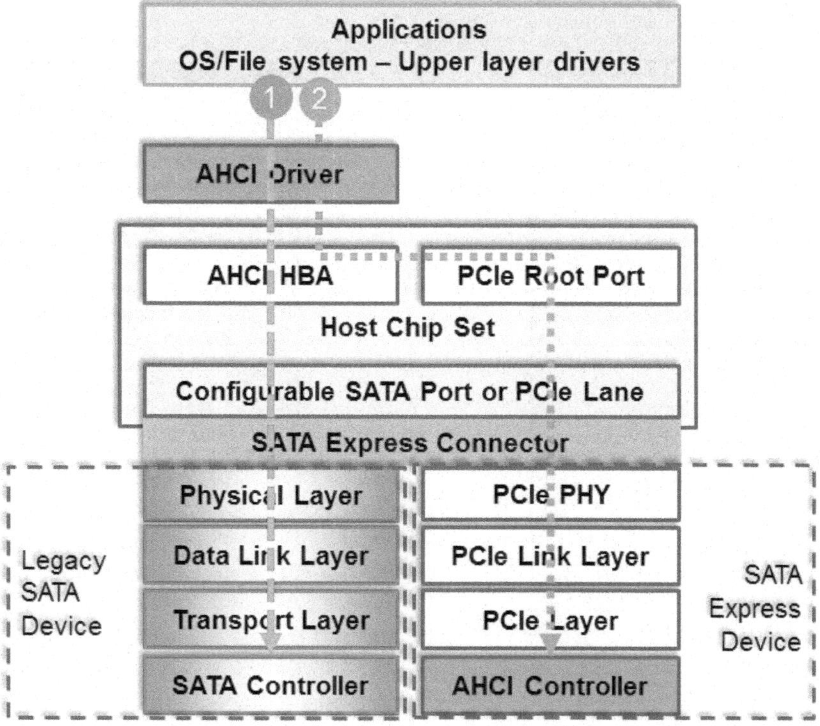

Figure 10.2 SATA versus SATA Express. (*Source:* www.sata-io.org)

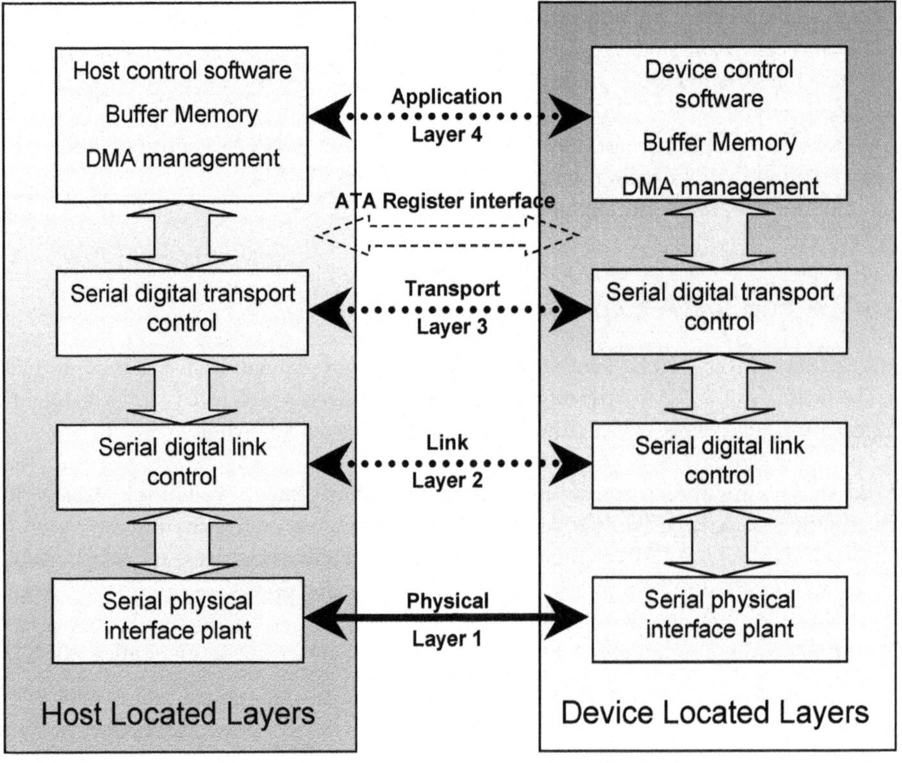

Figure 10.3 SATA protocol layers.

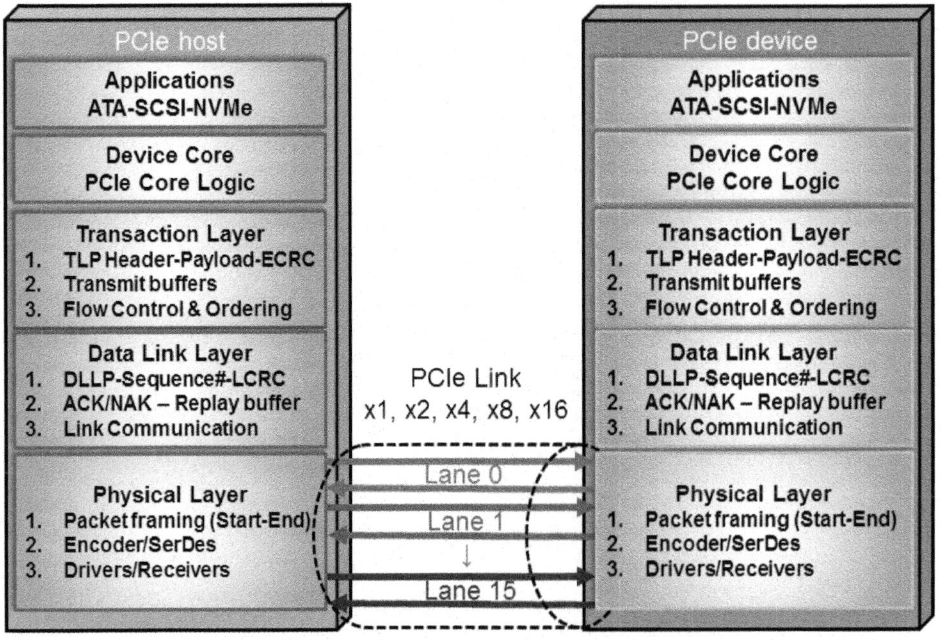

Figure 10.4 PCI Express protocol layers.

- Hardware has been designed to accommodate both types of devices to be plugged into the same physical connector.
- A special signal from the device will inform the host what type of device is present and, therefore, use the correct protocols.

Figure 10.3 shows a detailed view of SATA protocol layers and their respective functions:

- Application: Host software, command sets, memory management, DMA . . .
- Transport Layer: Frame Information Structures, queuing, data transfer protocols . . .
- Link Layer: Primitives, flow control, encoding/decoding . . .
- Physical Layer: Cables, connectors, signal levels . . .

Figure 10.4 shows a detailed view of the PCIe protocol layers and their respective functions:

- Application: Host software, command sets (ATA, NVMe, and SCSI), memory management . . .
- Transaction Layer: TLPs, header, ECRC, buffering, flow control & packet ordering . . .
- Data Link Layer: DLLP, packet numbering, ACK/NAK . . .
- Physical Layer: Packet framing, encoding/decoding, drivers/receivers . . .

10.3 SATA Configurations

Figure 10.5 shows several AHCI HBAs attached in a typical computer system.

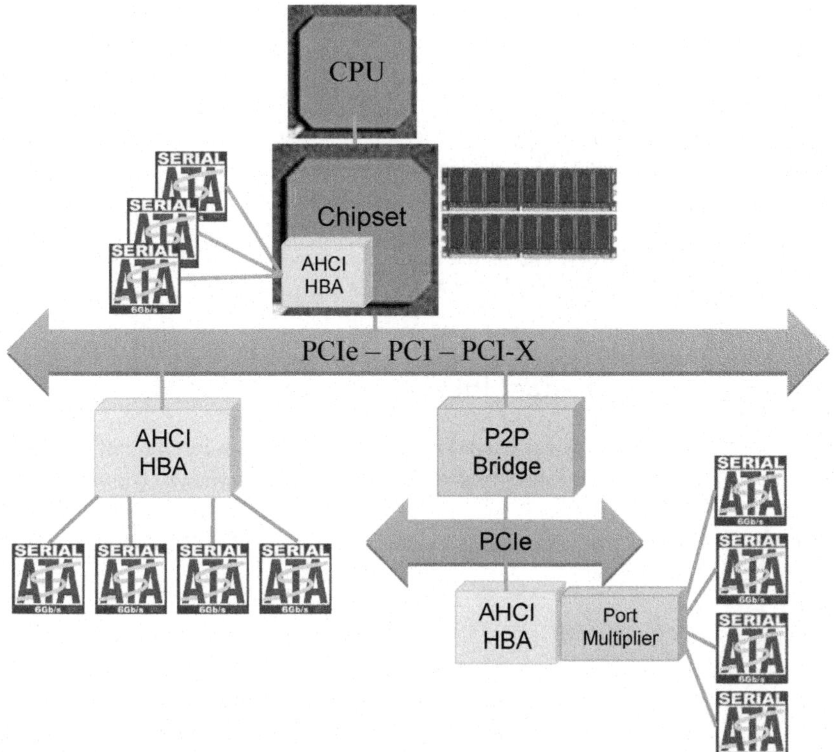

Figure 10.5 IA-Based System Diagram. (*Source:* Adapted from AHCI Specification)

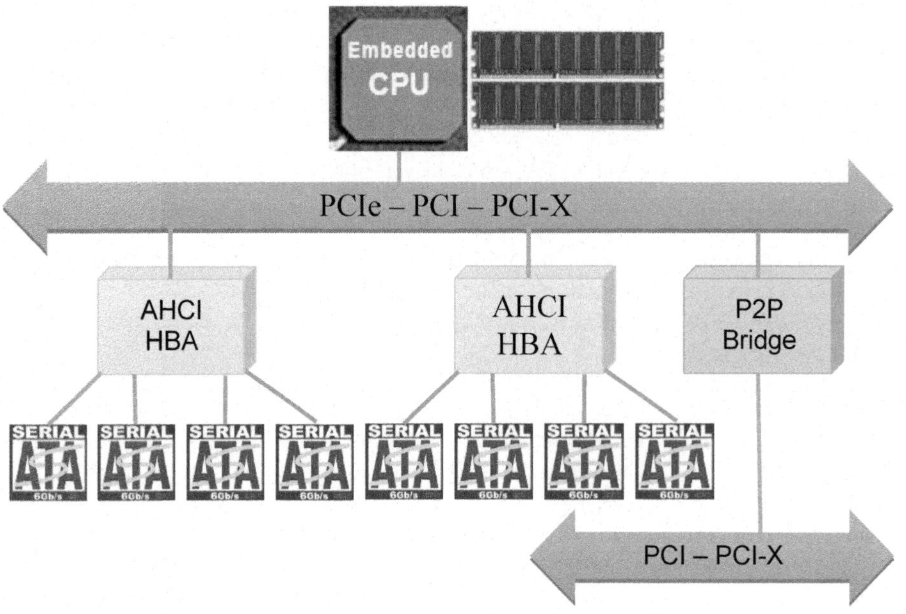

Figure 10.6 Embedded System Diagram. (*Source:* Adapted from AHCI Specification)

- One HBA is integrated in the core chipset.
- Another sits off the first available PCI/PCI-X bus.
- A final HBA sits off a second PCI bus that exists behind a PCI-PCI bridge. This last HBA has one port attached to a Port Multiplier.

Figure 10.6 shows how two HBA are connected to an embedded CPU with its own local memory. This configuration would most likely be used in a RAID-type environment.

10.4 Advanced Host Channel Interface

The AHCI specification defines the functional behavior and software interface of the Advanced Host Controller Interface, which is a hardware mechanism that allows software to communicate with Serial ATA devices.

AHCI is a PCI class device that acts as a data movement engine between system memory and Serial ATA devices.

- AHCI host devices support from 1 to 32 ports.
- An HBA must support ATA and ATAPI devices, and must support both the PIO and DMA protocols.
- An HBA may optionally support a Command List on each port for overhead reduction and support Serial ATA Native Command Queuing via the FPDMA Queued Command protocol for each device with up to 32 entries.
- An HBA may optionally support 64-bit addressing.

10.4.1 System Memory Structure

AHCI describes a system memory structure, which contains the following:

1. A generic area for control and status.
2. A table of entries describing a Command List (an HBA that does not support a Command List shall have a depth of one for this table).

Each Command List entry contains the following:

- Information necessary to program an SATA device.
- A pointer to a descriptor table for transferring data between system memory and the device.

10.4.2 AHCI Encompasses a PCI Device

AHCI contains a PCI BAR (Base Address Register) to implement native SATA features. AHCI specifies the following features:

- Support for 32 ports
- 64-bit addressing
- Elimination of master/slave handling
- Large LBA support
- Hot plug
- Power management
- HW-assisted Native Command Queuing
- Staggered spin-up
- Cold device presence detection
- Serial ATA superset registers
- Activity LED generation
- Port Multiplier

Theory of Operation

AHCI takes the basics of the scatter/gather list concept of Bus Master IDE and expands it to reduce CPU/software overhead.

- Communication between a device and software moves from the task file via byte-wide accesses to a command FIS located in system memory that is fetched by the HBA.
  - This reduces command set-up time significantly, allowing for many more devices to be added to a single host controller.
  - Software no longer communicates directly to a device via the task file.
- AHCI is defined to keep the HBA relatively simple so that it can act as a data mover.
  - An HBA implementing AHCI is not required to parse any of the ATA or ATAPI commands as they are transferred to the device, although it is not prohibited from doing so.
- All data transfers between the device and system memory occur through the HBA acting as a bus master to system memory.
  - Whether the transaction is of a DMA type or a PIO type, the HBA fetches and stores data to memory, offloading the CPU.
  - There is no accessible data port.

- All transfers are performed using DMA.
  - The use of the PIO command type is strongly discouraged.
  - PIO has limited support for errors—for example, the ending status field of a PIO transfer is given to the HBA during the PIO Setup FIS, before the data is transferred.
  - However, some commands may only be performed via PIO commands (such as IDENTIFY DEVICE).
  - Some HBA implementations may limit PIO support to one DRQ block of data per command.
- Standard mechanism for implementing a SATA command queue using the DMA Setup FIS.
  - An HBA that supports queuing has individual slots in the Command List allocated in system memory for all the commands.
  - Software can place a command into any empty slot, and upon notifying the HBA via a register access, the HBA shall fetch the command and transfer it.
  - The tag that is returned in the DMA Setup FIS is used as an index into the Command List to get the scatter/gather list used in the transfer.
- The Command List can be used by the system software and the HBA even when non-queued commands need to be transferred.
  - System software can still place multiple commands in the list, whether DMA, PIO, or ATAPI, and the HBA will walk the list transferring them.

Table 10.1 PCI Header and Capabilities Registers

Start (hex)	End (hex)	Name
00	3F	PCI Header
PMCAP	PMCAP+7	PCI Power Management Capability
MSICAP	MSICAP+9	Message Signaled Interrupt Capability

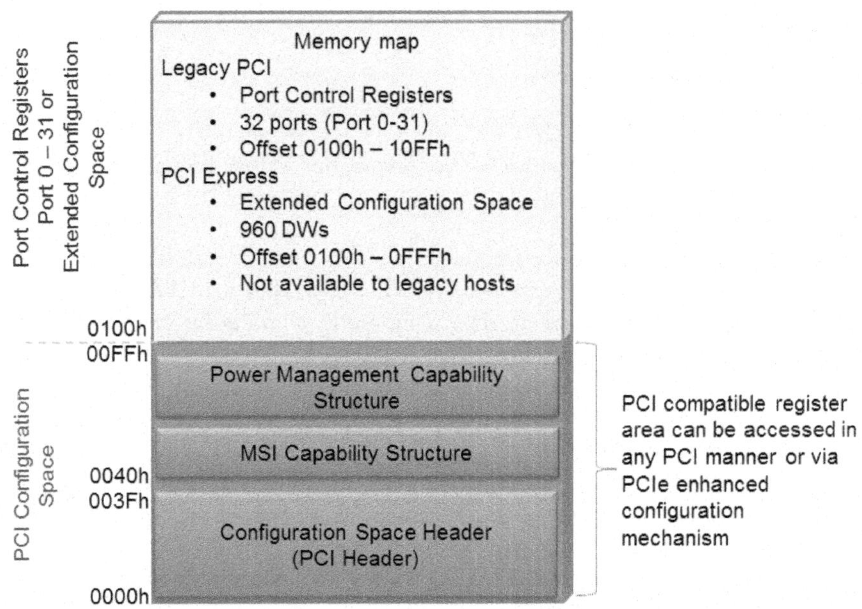

Figure 10.7 PCI Configuration space.

10.5 HBA Configuration Registers

This section details how the PCI Header and Capabilities registers are constructed for an AHCI HBA. The AHCI specification details the additional requirements of standard PCI registers and structures (Table 10.1).

AHCI implementations rely on the PCI Configuration space and Headers (Figure 10.7). Each configuration space is made up of a header and any number of capability structures (MSI, Power, etc.). The PCI header is a global register set that is applied to the entire HBA. The address range for the PCI header is 00h to 3Fh and be seen in Table 10.2. Starting at the offset 40h, you will find interrupt and power management capability registers (if any—optional).

In legacy PCI implementations, the space beyond 100h will contain up to 32 port control registers. Only Port 0 PCR is required by AHCI, and all other ports are optional. The PCRs are local registers, and their register set is only applied to the port. The PCRs provide the communication structure between the host driver and the SATA device.

In PCIe implementations, the space beyond 100h contains "Extended Configuration Space." This space will not be available to legacy implementations and will not be found in the AHCI specification. These registers can be found the PCI Express specifications.

Table 10.2 PCI Header

DWs	Offset	byte n+3	byte n+2	byte n+1	byte n
0	0000h	Identifiers (ID)			
		Device ID		Vendor ID	
1	0004h	Status Register (STS)		Command Register (CMD)	
2	0008h	Class Codes (CC)			Revision ID
		Base Class Code (BCC)	Sub Class Code (SCC)	Programming Interface (PI)	
3	000Ch	Built-in Self Test (BIST)	Header Type (HTYPE = 0)	Master Latency Timer (MTL)	Cache Line Sequence (CLS)
4	0010h	BAR0			
5	0014h	BAR1			
6	0018h	BAR2			
7	001Ch	BAR3			
8	0020h	BAR4			
9	0024h	ABAR (BAR5) AHCI Base Address			
10	0028h	CardBus CIS Pointer (CCPTR)			
11	002Ch	Subsystem Identifiers (SS)			
12	0030h	Expansion ROM Base Address (Optional) (EROM)			
13	0034h	Reserved (R)			Capabilities Pointer (CAP)
14	0038h	Reserved (R)			
15	003Ch	Maximum Latency = 00h (LAT)	Minimum Grant = 00h (GNT)	Interrupt pin (IPIN)	Interrupt line (ILINE)

The header defines the PCI base registers and their respective functions.

The identifiers **device ID** (assigned by the device vendor) and **vendor ID** (assigned by PCI-SIG) provide a method to identify a device.

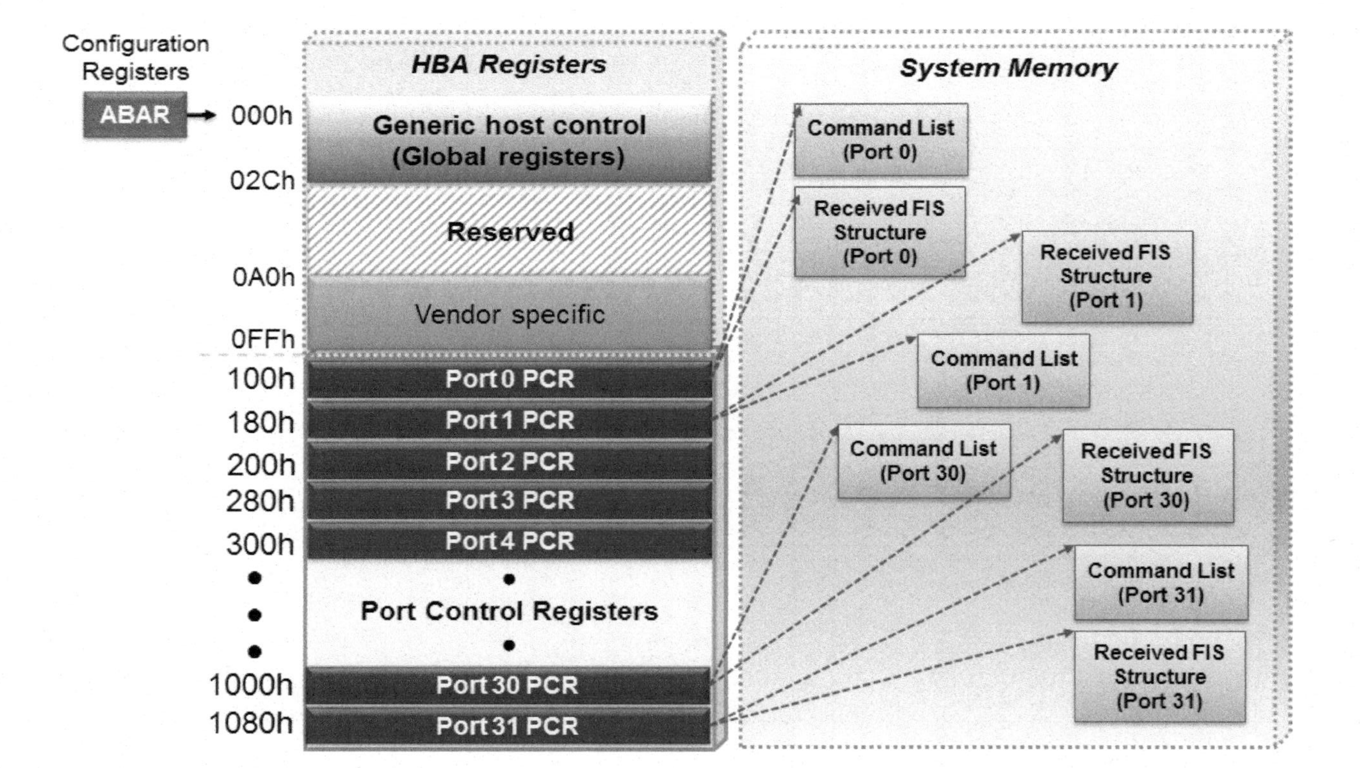

Figure 10.8 HBA memory space usage.

A **command register** provides the ability to enable memory capabilities, including bus master enable and I/O space enable.

A **status register** provides parity or system error reporting, abort interrupt status, indication that a Command List is present, and/or that an interrupt request from the device is pending.

Class code includes the following:

- Base Class Code (BCC): Indicates that this is a mass storage device.
- Sub Class Code (SCC): When set to 06h, indicates that this is a Serial ATA device.
- Programming Interface (PI): When set to 01h and the Sub Class Code is set to 06h, indicates that this is an AHCI HBA.

BARS – Other Base Addresses (Optional)

- These registers allocate memory or I/O spaces for other BARs.
- An example application of these BARs is to implement the native IDE and bus master IDE ranges for an HBA that wishes to support legacy software.

ABAR – AHCI Base Address

- This register allocates space for the HBA memory registers.
- The ABAR must be allocated to contain enough space for the global AHCI registers, the port-specific registers for each port, and any vendor-specific space (if needed).
- It is permissible to have vendor-specific space after the port-specific registers for the last HBA port.
- ABAR appears in Figure 10.8 and demonstrates how it points to the HBA system memory registers.

10.5.1 HBA Memory Registers

The memory mapped registers within the HBA exist in non-cacheable memory space.

The registers are broken into two sections:

- Global registers
- Port control

Global Registers:

- Start below address 100h
- Apply to the entire HBA

Port Control Registers:

- Are the same for all ports
- As many register banks as there are ports

Figure 10.9 demonstrates a top-down offset scheme wherein offset 00h is at the top of the register set definition. Although this is not vitally important, you should be aware that PCI-SIG specifications take a bottom-up approach, as seen in Figure 10.7 (offset 00h at the bottom of the figure). The AHCI speci-

Start	End	Symbol	Description	
00h	03h	CAP	Host Capabilities	Global registers Generic host control
04h	07h	GHC	Global Host Control	
08h	0Bh	IS	Interrupt Status	
0Ch	0Fh	PI	Ports Implemented	
10h	13h	VS	Version	
14h	17h	CCC_CTL	Command Completion Coalescing Control	
18h	1Bh	CCC_PORTS	Command Completion Coalescing Ports	
1Ch	1Fh	EM_LOC	Enclosure Management Location	
20h	23h	EM_CTL	Enclosure Management Control	
24h	27h	CAP2	Host Capabilities Extended	
28h	2Bh	BOHC	BIOS/OS Handoff Control and Status	
2Ch	*5Fh*	*Reserved*		
60h	9Fh	Reserved	Non Volatile Memory Host Channel Interface	
100h	17Fh	Port 0	Port 0 control registers	Port control registers
180h	1FFh	Port 1	Port 1 control registers	
200h	FFFh	Port 2-29	Ports 2-29 control registers	
1000h	107Fh	Port 30	Ports 30 control registers	
1080h	10FFh	Port 31	Ports 31 control registers	

Figure 10.9 HBA register map.

fication has defined the first 2Bh bytes that make up a globally defined "Generic host control" register set that applies to the entire HBA. Starting at offset 100h, you will find the port registers.

Port Control Registers

Port control registers describe the registers necessary to implement a port; all ports shall have the same register mapping (see Table 10.3).

- Port 0 starts at 100h
- Port 1 starts at 180h
- Port 2 starts at 200h
- Port 3 at 280h, etc.

The algorithm for software to determine the offset is as follows:

- Port offset = 100h + (PI Asserted Bit Position * 80h)

The Port registers contain

- Base address pointer to Command List data structure
- Base address pointer to Received FIS data structure
- Interrupt status and control

Table 10.3 Port Registers (one set per port)

Start	End	Symbol	Description
00h	03h	PxCLB	Port x Command List Base Address
04h	07h	PxCLBU	Port x Command List Base Address Upper 32-Bits
08h	0Bh	PxFB	Port x FIS Base Address
0Ch	0Fh	PxFBU	Port x FIS Base Address Upper 32-Bits
10h	13h	PxIS	Port x Interrupt Status
14h	17h	PxIE	Port x Interrupt Enable
18h	1Bh	PxCMD	Port x Command and Status
1Ch	1Fh	Reserved	Reserved
20h	23h	PxTFD	Port x Task File Data
24h	27h	PxSIG	Port x Signature
28h	2Bh	PxSSTS	Port x Serial ATA Status (SCR0: SStatus)
2Ch	2Fh	PxSCTL	Port x Serial ATA Control (SCR2: SControl)
30h	33h	PxSERR	Port x Serial ATA Error (SCR1: SError)
34h	37h	PxSACT	Port x Serial ATA Active (SCR3: SActive)
38h	3Bh	PxCI	Port x Command Issue
3Ch	3Fh	PxSNTF	Port x Serial ATA Notification (SCR4: SNotification)
40h	43h	PxFBS	Port x FIS-based Switching Control
44h	6Fh	Reserved	Reserved
70h	7Fh	PxVS	Port x Vendor Specific

- Command and Status
  - Power management, Drive LED
  - Device is ATAPI, Port Multiplier Attached
  - Hot plug capable, Power-On Device, Spin-Up Device, Start
- Task Data file maintains a copy of
  - D2H Register FIS
  - PIO Setup FIS
  - Set Device Bits FIS (BSY and DRQ are not updated with this FIS)
- Attached device signature
- SATA Control Registers
  - SCR0: SStatus – Interface Power Management, Current Speed, Device Detection
  - SCR1: SControl – set IPM, SPD, DET, and Diagnostics
  - SCR2: SError – hardware errors, protocol errors, Phy states
  - SCR3: SActive – used for Native Command Queuing
  - SCR3: SNotification – used for asynchronous event notifications
- Command Issue
  - 32-bit significant field that corresponds to a command slot.
  - Bit 0 is command slot 0, bit 1 is command slot 1, etc.
  - Set by software to indicate to the HBA that a command has been built in system memory for a command slot and may be sent to the device.
  - When the HBA receives a FIS that clears the BSY, DRQ, and ERR bits for the command, it clears the corresponding bit in this register for that command slot.

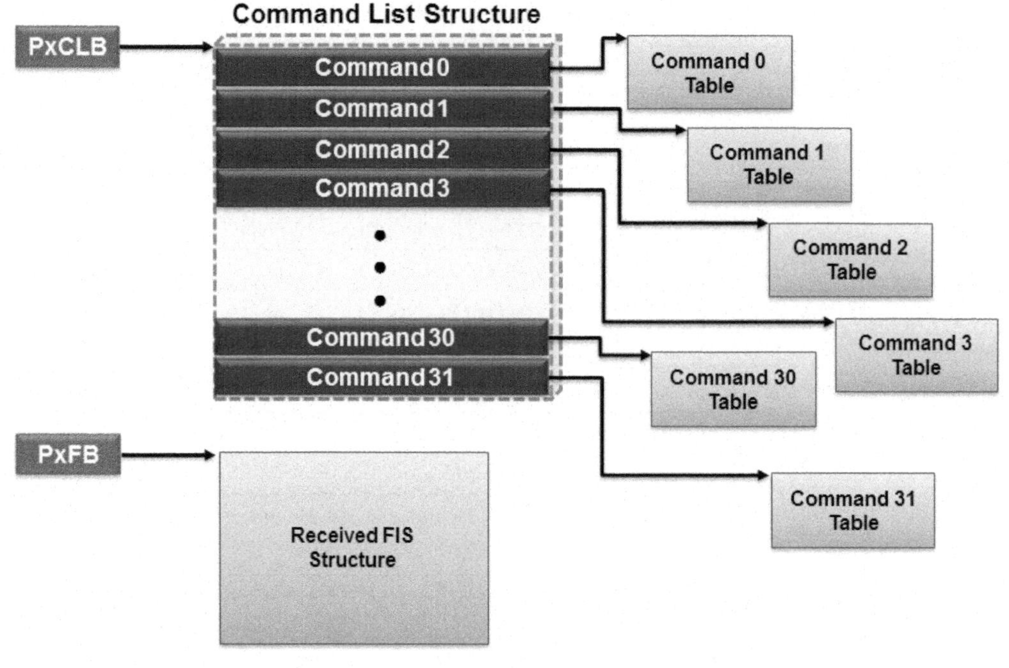

Figure 10.10 Port memory usage.

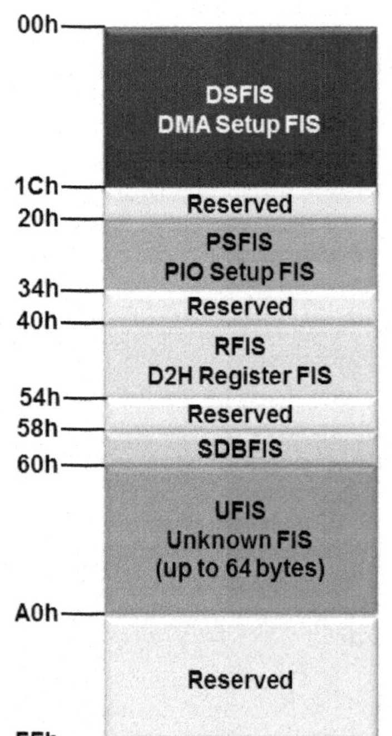

When a DMA setup FIS arrives from the device,
- HBA copies it to the DSFIS area

When a PIO setup FIS arrives from the device,
- HBA copies it to the PSFIS area

When a D2H Register FIS arrives from the device,
- HBA copies it to the RFIS area

When a Set Device Bits FIS arrives from the device,
- HBA copies it to the SDBFIS area

When an unknown FIS arrives from the device,
- HBA copies it to the UFIS area

Figure 10.11 Received FIS structure.

10.6 System Memory Model

Most communication between software and an SATA device is through the HBA via system memory descriptors.

- These descriptors indicate the status of all received and sent FISes, as well as pointers for data transfers.

Additional communication is done via registers contained within the HBA:

- Registers for each supported port
- Global control

Generic host control can be seen in Figure 10.8 and includes global host control that is applied to the entire HBA.

The format of the Port Control Registers (PCR) can be seen in Table 10.3 and includes the base address pointers to a Command List structure (Figure 10.10) and a Received FIS structure (Figure 10.11).

Port Memory Usage

There are two descriptors per port that are used to convey information.

- One is the FIS descriptor, which contains FISes received from a device.
- The other is the Command List, which contains a list of 1 to 32 commands available for a port to execute.

10.6.1 Received FIS Structure

The HBA uses an area of system memory to communicate information on received FISes (Figure 10.11).

- This structure is pointed to by PxFBU and PxFB.

FIS Arrival

- When a DMA setup FIS arrives from the device, the HBA copies it to the DSFIS area of this structure.
- When a PIO setup FIS arrives from the device, the HBA copies it to the PSFIS area of this structure.
- When a D2H Register FIS arrives from the device, the HBA copies it to the RFIS area of this structure.
- When a Set Device Bits FIS arrives from the device, the HBA copies it to the SDBFIS area of this structure.
- When an unknown FIS arrives from the device, the HBA copies it to the UFIS area in this structure.

10.6.2 Command List Structure

Each entry contains a command header (a 32-byte structure) that details the direction, type, and scatter/gather pointer of the command. The command entries provide specific memory information used during DMA transfers, including (see Figure 10.12) the following:

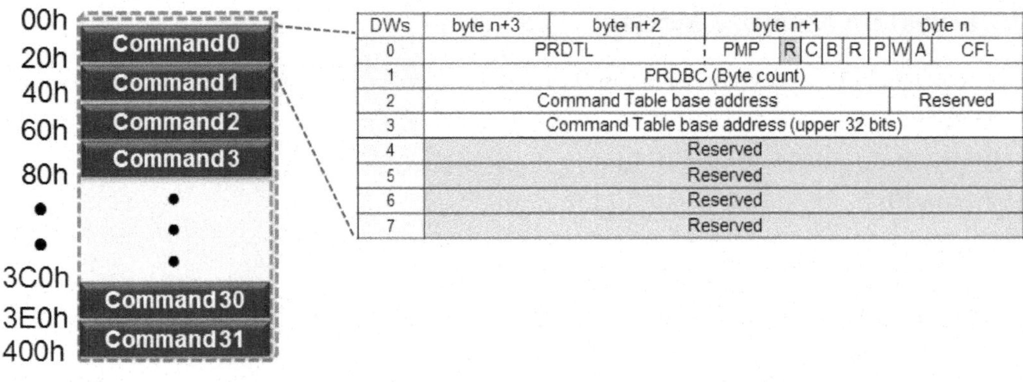

Figure 10.12 Command list structure.

- The Physical Region Descriptors Table Length (PRDTL), which tells the host when to stop fetching additional PRDs (see Table 10.4 for full descriptions)
- How many bytes are to be written or read (PRDBC) for the operation
- Where the base address of the data segment is located within memory (CTBA)
- If a Port Multiplier is being addressed (PMP)
- The length of the command FIS (CFL)

Table 10.4 Command Structure Descriptions

Definition	Name	Description
Physical Region Descriptor Table Length	PRDTL	Length of the scatter/gather descriptor table in entries, called the Physical Region Descriptor Table. Each entry is 4 DW.
Port Multiplier Port	PMP	Port number used when constructing Data FISes on transmit and to check against all FISes received for this command.
Command FIS Length	CFL	Length of the Command FIS. A "0" represents 0 DW, "4" represents 4 DW. A length of "0" or "1" is illegal. The maximum value allowed is 10h, or 16 DW.
Physical Region Descriptor Byte Count	PRDBC	Current byte count that has transferred on device writes (memory to device) or device reads (device to memory).
Command Table Descriptor Base Address	CTBA	The 32-bit physical address of the command table, which contains the command FIS, ATAPI Command, and PRD table.

Command Table

Each entry in the Command List points to a Command Table. The Command Table (Figure 10.13) provides the interface to send H2D register FISes or ATAPI commands (SCSI) to the device. The Command Table also provides PRD pointers and byte counts for data transfers.

Command FIS (CFIS)

- This is a software-constructed FIS.
- For data transfer operations, this is the H2D Register FIS format, as specified in the Serial ATA specification. Valid CFIS lengths are 2 to 16 Dwords and must be in Dword granularity.

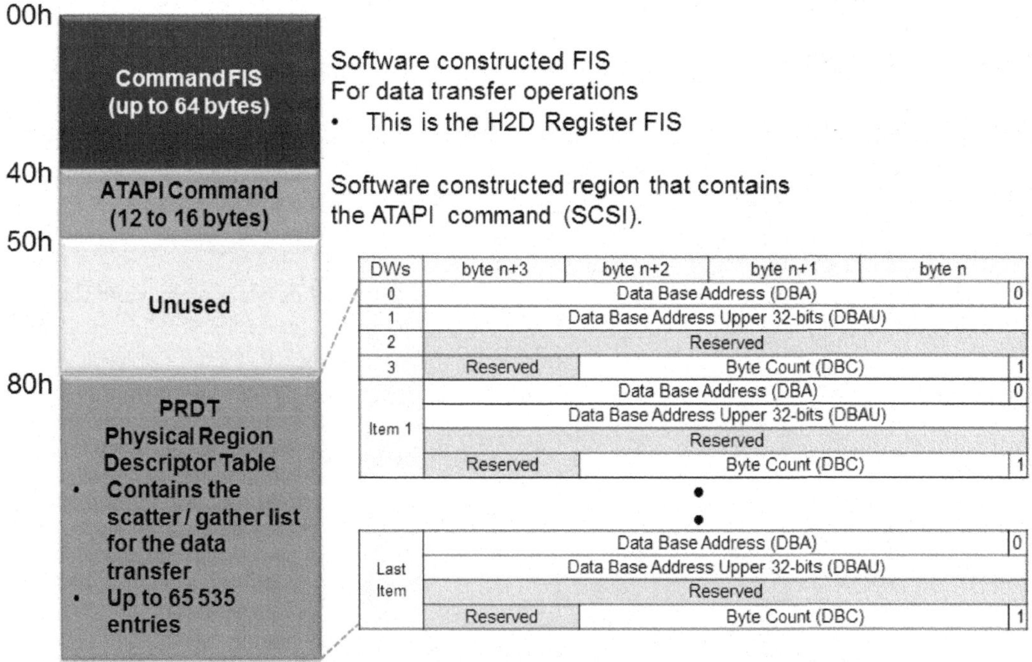

Figure 10.13 Command table.

ATAPI Command (ACMD)

- This is a software constructed region of 12 or 16 bytes in length that contains the ATAPI command to transmit if the "A" bit is set in the command header.
- The ATAPI command must be either 12 or 16 bytes in length.
- The length transmitted by the HBA is determined by the PIO setup FIS that is sent by the device requesting the ATAPI command.

Physical Region Descriptor Table (PRDT)

- This table contains the scatter/gather list for the data transfer.
- It contains a list of 0 (no data to transfer) to up to 65,535 entries.
- A breakdown of each field in a PRD table is shown below.
- Item 0 refers to the first entry in the PRD table.

Data Base Address (DBA):

- Indicates the 32-bit physical address of the data block.
- The block must be word aligned, indicated by bit 0 being reserved.

Data Byte Count (DBC):

- A "0" based value that indicates the length, in bytes, of the data block.
- A maximum length of 4MB may exist for any entry.
- Bit "0" of this field must always be "1" to indicate an even byte count.
- A value of "1" indicates 2 bytes, "3" indicates 4 bytes, etc.

10.7 Summary

10.7.1 SATA and SATA Express Architectures Review

SATA and SATA Express systems will use the ATA command sets regardless of the physical interface that carries information to the device (see Figure 10.14).

- Hardware has been designed to accommodate both types of devices to be plugged into the same physical connector.
- A special signal from the device will inform the host what type of device is present and therefore use the correct protocols.

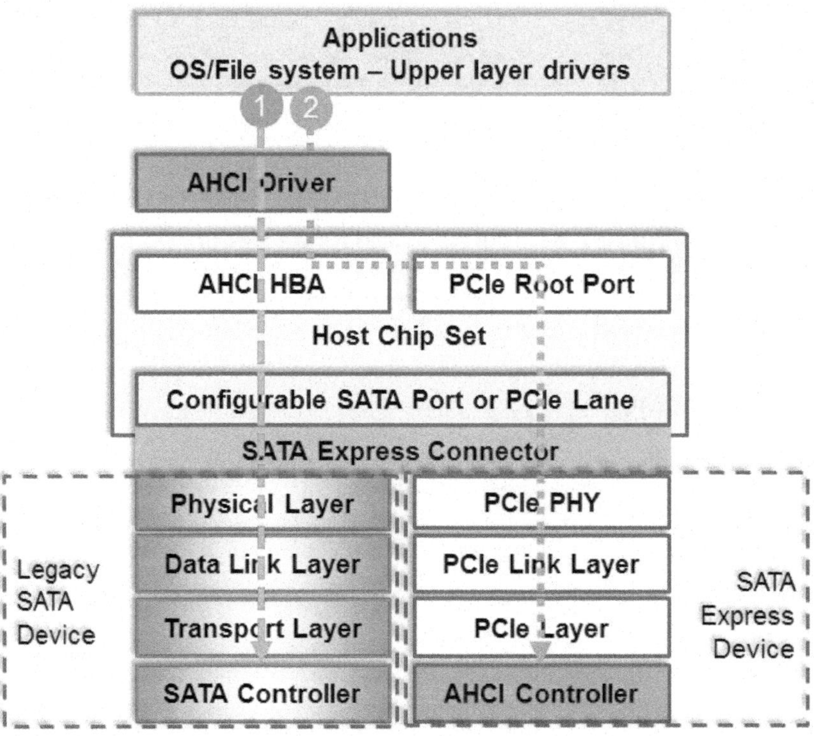

Figure 10.14 [as it appears in Section 10.2 (Figure 10.2)] SATA versus SATA Express. (*Source:* www.sata-io.org)

10.7.2 Advanced Host Channel Interface (AHCI) Review

AHCI defines the functional behavior and software interface of the Advanced Host Controller Interface that allows software to communicate with Serial ATA devices.

AHCI is a PCI class device that acts as a data movement engine between system memory and Serial ATA devices.

- AHCI host devices support from 1 to 32 ports.
- An HBA must support ATA and ATAPI devices, and must support both the PIO and DMA protocols.

- An HBA may optionally support a Command List on each port for overhead reduction, and support Serial ATA Native Command Queuing via the FPDMA Queued Command protocol for each device of up to 32 entries.

An HBA may optionally support 64-bit addressing.

AHCI Encompasses a PCI Device

It contains a PCI BAR (Base Address Register) to implement native SATA features. AHCI specifies the following features:

- Support for 32 ports
- 64-bit addressing
- Elimination of master/slave handling
- Large LBA support
- Hot plug
- Power management
- HW-assisted Native Command Queuing
- Staggered Spin-up
- Cold device presence detection
- Serial ATA superset registers
- Activity LED generation
- Port Multiplier

10.7.3 PCI Header and Configuration Space Review

AHCI implementations rely on the PCI Configuration Space and Headers (see Figure 10.15).

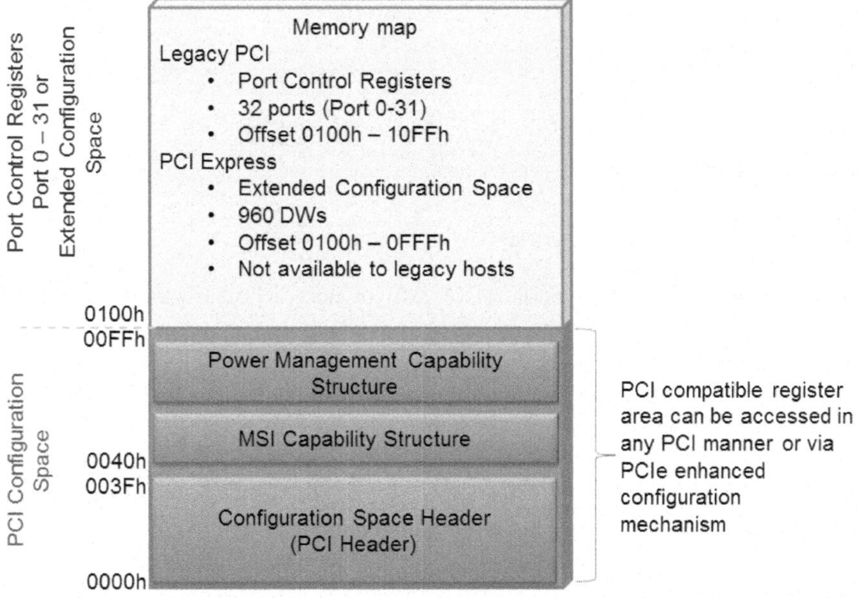

Figure 10.15 [as it appears in Section 10.4 (Figure 10.7)] PCI Configuration space.

- Each configuration space is made up of a header and any number of capability structures (MSI, Power, etc.).
- The PCI header is a global register set that is applied to the entire HBA.
- The address range for the PCI header is 00h to 3Fh, as shown in Table 10.5.
- Starting at the offset 40h, you will find interrupt and power management capability registers (if any—optional).

Table 10.5 [as it appears in Section 10.5 (Table 10.2)] PCI Header

DWs	Offset	byte n+3	byte n+2	byte n+1	byte n
0	0000h	Identifiers (ID)			
		Device ID		Vendor ID	
1	0004h	Status Register (STS)		Command Register (CMD)	
2	0008h	Class Codes (CC)			Revision ID
3	000Ch	Built-in Self Test (BIST)	Header Type (HTYPE = 0)	Master Latency Timer (MTL)	Cache Line Sequence (CLS)
4	0010h	BAR0			
5	0014h	BAR1			
6	0018h	BAR2			
7	001Ch	BAR3			
8	0020h	BAR4			
9	0024h	ABAR (BAR5) AHCI Base Address			
10	0028h	CardBus CIS Pointer (CCPTR)			
11	002Ch	Subsystem Identifiers (SS)			
12	0030h	Expansion ROM Base Address (Optional) (EROM)			
13	0034h	Reserved (R)			Capabilities Pointer (CAP)
14	0038h	Reserved (R)			
15	003Ch	Maximum Latency = 00h	Minimum Grant = 00h	Interrupt pin	Interrupt pin

10.7.4 Memory Registers Review

The memory mapped registers within the HBA exist in non-cacheable memory space, as seen in Figure 10.16.

The registers are broken into two sections:

- Global registers
- Port control

Global Registers

- Start below address 100h
- Apply to the entire HBA

Port Control Registers

- Are the same for all ports
- As many register banks as there are ports

Start	End	Symbol	Description	
00h	03h	CAP	Host Capabilities	Global registers Generic host control
04h	07h	GHC	Global Host Control	
08h	0Bh	IS	Interrupt Status	
0Ch	0Fh	PI	Ports Implemented	
10h	13h	VS	Version	
14h	17h	CCC_CTL	Command Completion Coalescing Control	
18h	1Bh	CCC_PORTS	Command Completion Coalescing Ports	
1Ch	1Fh	EM_LOC	Enclosure Management Location	
20h	23h	EM_CTL	Enclosure Management Control	
24h	27h	CAP2	Host Capabilities Extended	
28h	2Bh	BOHC	BIOS/OS Handoff Control and Status	
2Ch	*5Fh*	*Reserved*		
60h	9Fh	Reserved	Non Volatile Memory Host Channel Interface	
100h	17Fh	Port 0	Port 0 control registers	Port control registers
180h	1FFh	Port 1	Port 1 control registers	
200h	FFFh	Port 2-29	Ports 2-29 control registers	
1000h	107Fh	Port 30	Ports 30 control registers	
1080h	10FFh	Port 31	Ports 31 control registers	

Figure 10.16 [as it appears in Section 10.5.1 (Figure 10.9)] HBA register map.

Port Register Map Review

Port registers describe the registers necessary to implement a port; all ports shall have the same register mapping, as seen in Figure 10.17.

- Port 0 starts at 100h
- Port 1 starts at 180h
- Port 2 starts at 200h
- Port 3 at 280h, etc.

Port Control Registers

Port control registers describe the registers necessary to implement a port; all ports shall have the same register mapping (see Table 10.6 for a list of port control registers).

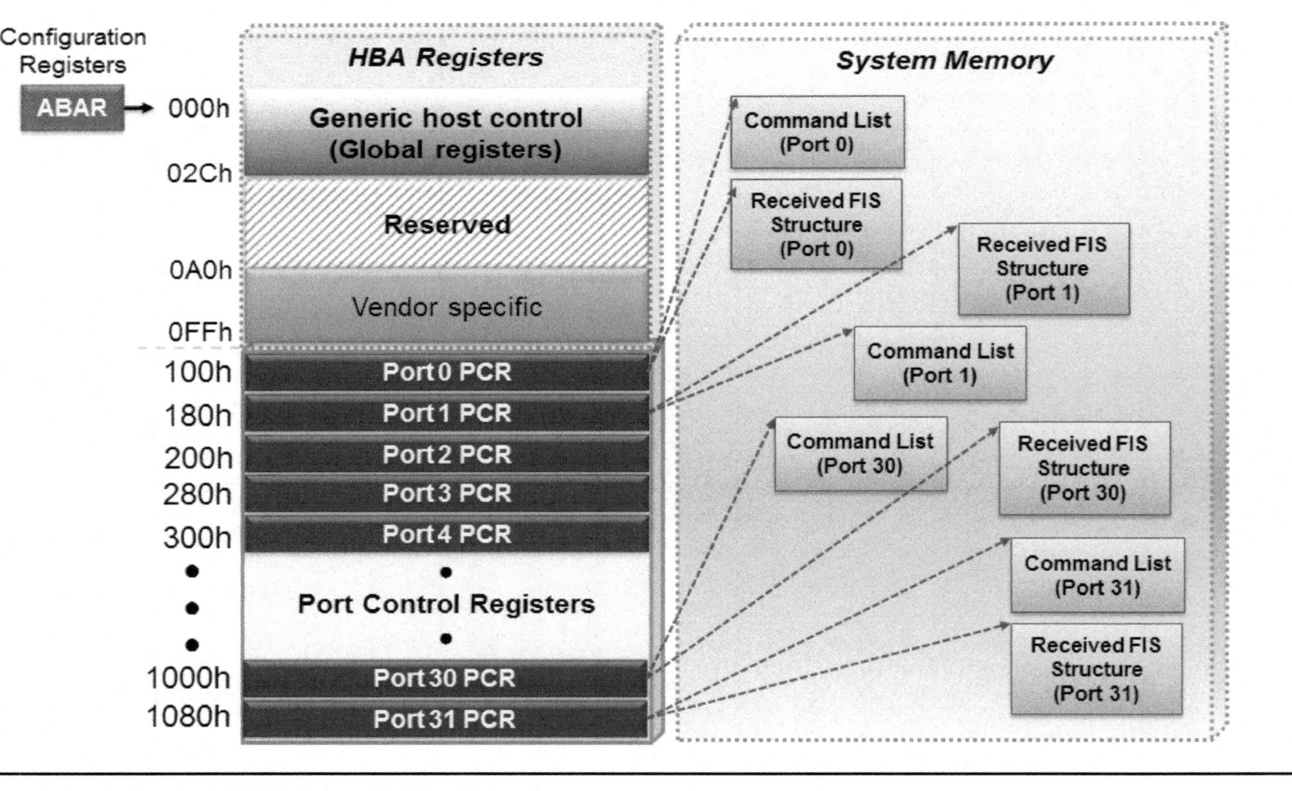

Figure 10.17 [as it appears in Section 10.5 (Figure 10.8)] HBA memory space usage.

Table 10.6 [as it appears in Section 10.5.1 (Table 10.3)]
Port Registers (one set per port)

Start	End	Symbol	Description
00h	03h	PxCLB	Port x Command List Base Address
04h	07h	PxCLBU	Port x Command List Base Address Upper 32-Bits
08h	0Bh	PxFB	Port x FIS Base Address
0Ch	0Fh	PxFBU	Port x FIS Base Address Upper 32-Bits
10h	13h	PxIS	Port x Interrupt Status
14h	17h	PxIE	Port x Interrupt Enable
18h	1Bh	PxCMD	Port x Command and Status
1Ch	1Fh	Reserved	Reserved
20h	23h	PxTFD	Port x Task File Data
24h	27h	PxSIG	Port x Signature
28h	2Bh	PxSSTS	Port x Serial ATA Status (SCR0: SStatus)
2Ch	2Fh	PxSCTL	Port x Serial ATA Control (SCR2: SControl)
30h	33h	PxSERR	Port x Serial ATA Error (SCR1: SError)
34h	37h	PxSACT	Port x Serial ATA Active (SCR3: SActive)
38h	3Bh	PxCI	Port x Command Issue
3Ch	3Fh	PxSNTF	Port x Serial ATA Notification (SCR4: SNotification)
40h	43h	PxFBS	Port x FIS-based Switching Control
44h	6Fh	Reserved	Reserved
70h	7Fh	PxVS	Port x Vendor Specific

10.7.5 HBA Memory Space Usage Review

Most communication between software and an SATA device is through the HBA via system memory descriptors (see Figure 10.17), which

- indicates the status of all received and sent FISes,
- as well as pointers for data transfers.

Additional communication is done via registers contained within the HBA:

- Registers for each supported port
- Global control for all ports

10.7.6 Port Memory Usage Review

There are two descriptors per port that are used to convey information.

- The Command List structure (see Figure 10.18), which contains a list of 1 to 32 commands that are available for a port to execute.

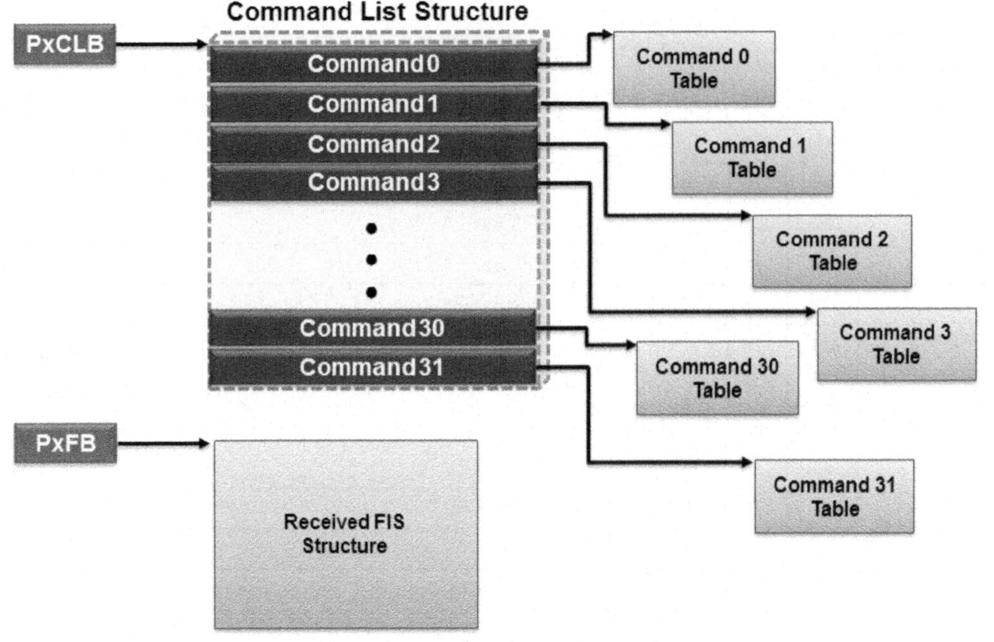

Figure 10.18 [as it appears in Section 10.5.1 (Figure 10.10)] Port memory usage.

- The Received FIS structure (see Figure 10.19), which contains FISes received from a device.

Receive FIS Structure

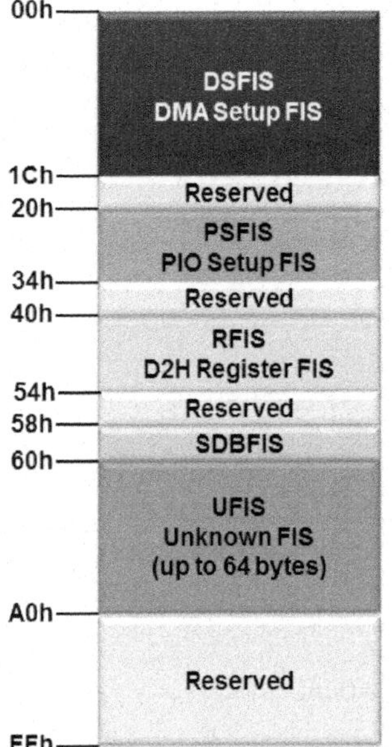

When a DMA setup FIS arrives from the device,
- HBA copies it to the DSFIS area

When a PIO setup FIS arrives from the device,
- HBA copies it to the PSFIS area

When a D2H Register FIS arrives from the device,
- HBA copies it to the RFIS area

When a Set Device Bits FIS arrives from the device,
- HBA copies it to the SDBFIS area

When an unknown FIS arrives from the device,
- HBA copies it to the UFIS area

Figure 10.19 [as it appears in Section 10.5.1 (Figure 10.11)] Received FIS structure.

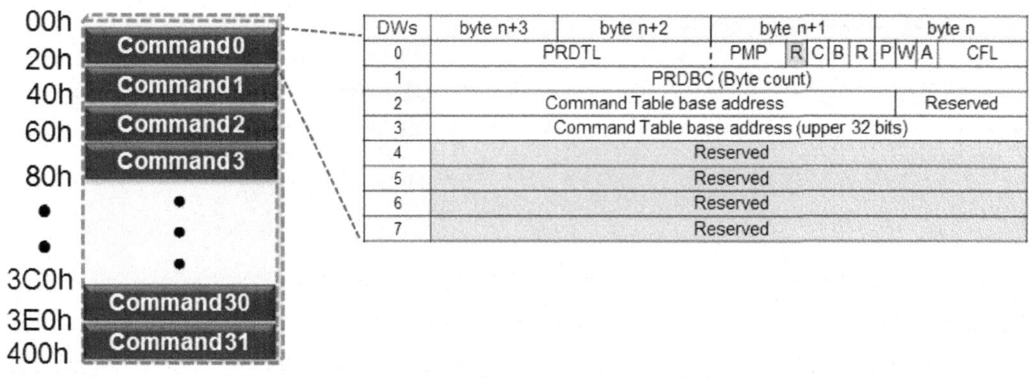

Figure 10.20 [as it appears in Section 10.6.2 (Figure 10.12)] Command list structure.

10.7.7 Command List Structure Review

Each command List structure has defined a 32-entry command structure (see Figure 10.20) that contains

- a command header, which is a 32-byte structure;
- details about the direction, type, and scatter/gather pointer of the command.

The Command List Review

Figure 10.21 shows the Command List structure, which contains the following:

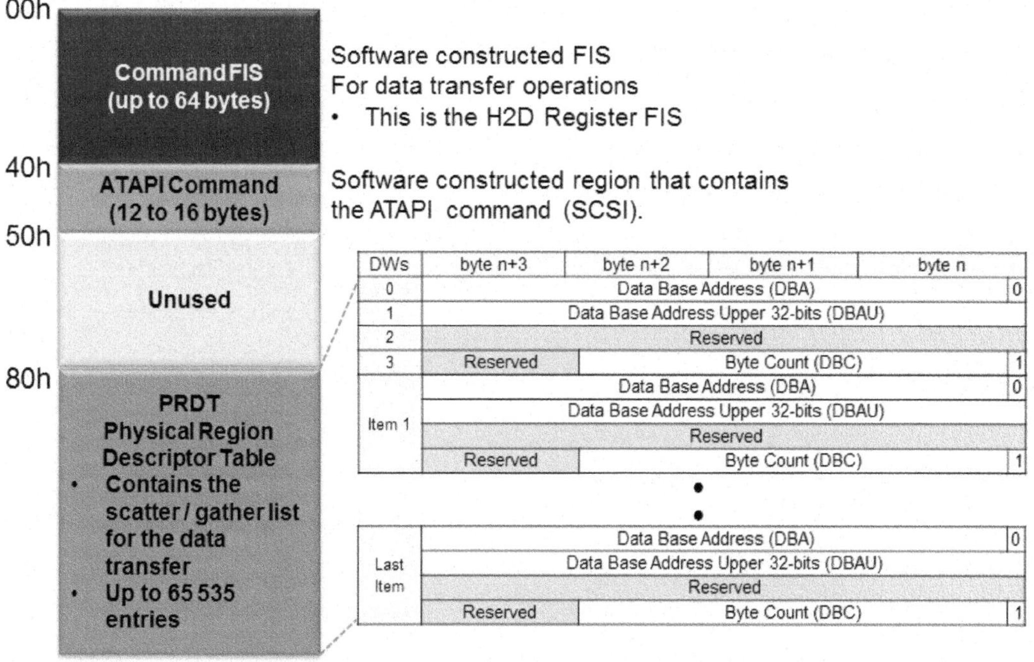

Figure 10.21 [as it appears in Section 10.6.2 (Figure 10.13)] Command list.

1. Command FIS (64 bytes)
2. ATAPI Command (12 or 16 bytes)
3. Physical Region Descriptor Table (up to 64k entries)
    - PRDT contains the scatter/gather list for the data transfer

Appendix A: Chapter Review Answers

This Appendix includes the answers to the Review Questions that appear at the end of Chapters 2 through 6.

Chapter 2

1. (b) ATA-7 Volume 3
2. (c) INCITS T-13
3. (b) www.sata-io.org
4. (c) Transport
5. (a) PHY
6. (d) Application
7. (b) Link
8. (h) COMMAND
9. (b) SECTOR COUNT
10. (g) STATUS
11. (a) ERROR
12. (g) All of the above
13. (b) To pass SCSI commands to devices
14. (c) Differential signal pairs
15. (c) 8b/10b
16. (c) 32-bits
17. (b) Begins with K28.3 followed by three data dwords
18. (d) Scrambled dwords
19. (i) X_RDY
20. (c) R_IP
21. (h) WTRM

22. (e) R_RDY
23. (d) R_OK
24. (d) Frame Information Structure
25. (a) Register transfer host to device
26. (b) Register transfer device to host

Chapter 3

1. (c) SATA-IO
2. (d) INCITS T-13
3. (b) ACS-8
4. (a) AST-8
5. (c) C8h
6. (c) EXT Commands use 48-bit addressing
7. (c) Inputs
8. (a) IDENTIFY DEVICE

Chapter 4

1. (c) Frame Information Structure
2. (b) 34h
3. (b) Command
4. (c) Status

Chapter 5

1. (c) 32-bits
2. (b) Data and (d) Primitive
3. (b) K28.3
4. (b) All Data characters
5. (c) HOLD
6. (e) WTRM
7. (a) X_RDY
8. (d) R_OK
9. (e) R_IP
10. (d) R_RDY
11. (d) 8192
12. (b) HOLDA and (c) HOLD
13. (b) To minimize the EMI impact of long sequences of repeated primitives
14. (a) ALIGN
15. (c) Scrambling

16. (c) Power Management
17. (b) 8b10b
18. (c) D15.6

Chapter 6

1. (b) COMRESET
2. (a) COMINIT
3. (d) COMWAKE
4. (d) Duration of idle period
5. (c) Speed negotiation
6. (c) If it supports the PACKET- or REGISTER-driven command set
7. (d) D10.2 @ its slowest link rate
8. (a) 2048 ALIGN (0) @ its fastest link rate
9. (a) ALIGN (0) @ its fastest link rate
10. (c) Near-End Retimed
11. (a) 100 ohm

Index

B

C

D

O

P

Q

R

S